Computer and Information Science Applications in Bioprocess Engineering

NATO ASI Series

Advanced Science Institutes Series

A Series presenting the results of activities sponsored by the NATO Science Committee, which aims at the dissemination of advanced scientific and technological knowledge, with a view to strengthening links between scientific communities.

The Series is published by an international board of publishers in conjunction with the NATO Scientific Affairs Division

A **Life Sciences** B **Physics**	Plenum Publishing Corporation London and New York
C **Mathematical and Physical Sciences** D **Behavioural and Social Sciences** E **Applied Sciences**	Kluwer Academic Publishers Dordrecht, Boston and London
F **Computer and Systems Sciences** G **Ecological Sciences** H **Cell Biology** I **Global Environmental Change**	Springer-Verlag Berlin, Heidelberg, New York, London, Paris and Tokyo

PARTNERSHIP SUB-SERIES

1. **Disarmament Technologies**	Kluwer Academic Publishers
2. **Environment**	Springer-Verlag / Kluwer Academic Publishers
3. **High Technology**	Kluwer Academic Publishers
4. **Science and Technology Policy**	Kluwer Academic Publishers
5. **Computer Networking**	Kluwer Academic Publishers

The Partnership Sub-Series incorporates activities undertaken in collaboration with NATO's Cooperation Partners, the countries of the CIS and Central and Eastern Europe, in Priority Areas of concern to those countries.

NATO-PCO-DATA BASE

The electronic index to the NATO ASI Series provides full bibliographical references (with keywords and/or abstracts) to more than 50000 contributions from international scientists published in all sections of the NATO ASI Series.
Access to the NATO-PCO-DATA BASE is possible in two ways:

– via online FILE 128 (NATO-PCO-DATA BASE) hosted by ESRIN, Via Galileo Galilei, I-00044 Frascati, Italy.

– via CD-ROM "NATO-PCO-DATA BASE" with user-friendly retrieval software in English, French and German (© WTV GmbH and DATAWARE Technologies Inc. 1989).

The CD-ROM can be ordered through any member of the Board of Publishers or through NATO-PCO, Overijse, Belgium.

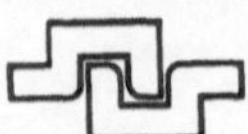

Series E: Applied Sciences - Vol. 305

Computer and Information Science Applications in Bioprocess Engineering

edited by

Antonio R. Moreira

and

Kimberlee K. Wallace

University of Maryland Baltimore County (UMBC),
Baltimore, Maryland, U.S.A.

Kluwer Academic Publishers

Dordrecht / Boston / London

Published in cooperation with NATO Scientific Affairs Division

Proceedings of the NATO Advanced Study Institute on
Use of Computer and Informatic Systems in Bioprocess Engineering
Ofir, Portugal
May 18–29, 1992

A C.I.P. Catalogue record for this book is available from the Library of Congress.

ISBN-13: 978-94-010-6564-1 e-ISBN-13: 978-94-009-0177-3
DOI: 10.1007/978-94-009-0177-3

Published by Kluwer Academic Publishers,
P.O. Box 17, 3300 AA Dordrecht, The Netherlands.

Kluwer Academic Publishers incorporates the publishing programmes of
D. Reidel, Martinus Nijhoff, Dr W. Junk and MTP Press.

Sold and distributed in the U.S.A. and Canada
by Kluwer Academic Publishers,
101 Philip Drive, Norwell, MA 02061, U.S.A.

In all other countries, sold and distributed
by Kluwer Academic Publishers Group,
P.O. Box 322, 3300 AH Dordrecht, The Netherlands.

Softcover reprint of the hardcover 1st edition 1996

NATO ADVANCED STUDY INSTITUTE
USE OF COMPUTER AND INFORMATIC SYSTEMS IN BIOPROCESS ENGINEERING
OFIR, PORTUGAL
MAY 18-29, 1992

MAIN SPONSORS:

NATO	North Atlantic Treaty Organization
NSF	National Science Foundation

OTHER SPONSORS:

ESB	Escola Superior de Biotecnologia
UMBC	University of Maryland Baltimore County

DIRECTORS:
Augusto Medina, Escola Superior de Biotecnologia, Portugal
Antonio R. Moreira, UMBC, University of Maryland Baltimore County, USA

SCIENTIFIC ORGANIZING COMMITTEE:
Augusto Medina, Escola Superior de Biotecnologia, Portugal
Antonio R. Moreira, UMBC, University of Maryland Baltimore County, USA
Carlos Rolz, Center for Scientific and Technological Studies, Guatemala

CONTENTS

LIST OF PARTICIPANTS

Lecturers:

Henry R. Bungay
Department of Chemical Engineering,
Rensselaer Polytechnic Institute
Troy, New York 12180
United States

Joaquim M.S. Cabral
Departamento de Engenharia Quimica
Instituto Superior Tecnico
1096 Lisbon Codex
Portugal

Sebastiao Feyo De Azevedo
Departamento de Engenharia Quimica
Faculdade de Engenharia da Universidade do Porto
Rua dos Bragas
4099 Porto Codex
Portugal

Cornelis D. De Gooijer
Wageningen Agricultural University
Food and Bioprocess Engineering Group
P.O. Box 8129
6700 EV Wageningen
The Netherlands

Gerard Goma
Institut National des Sciences Appliquees
Departement de Genie Biochimique et Alimentaire
URA-CNRS 544
Complex Scientifique de Rangueil
31077 Toulouse Cedex
France

Augustin Lopez-Munguia
Instituto de Biotecnologia
Universidad Nacional Autonoma de Mexico
Apartado Postal 510-3
Cuernavaca, Morelos
Mexico

Juan Mata-Alvarez
Department d'Enginyeria Quimica
Universitat de Barcelona
Marti i Franques 1,6p. 08028
Barcelona Spain

Ferda Mavituna
Department of Chemical Engineering
University of Manchester Institute of Science and Technology
Manchester, M60 1 QD
United Kingdom

Antonio Moreira
Department of Chemical and Biochemical Engineering
University of Maryland Baltimore County
Baltimore, Maryland 21228
United States

Anton Moser
Institut fur Biotechnologie, TU Graz
A-8010 Graz, Petersgasse 12
Austria

Carlos Rolz
Central American Research Institute for Industry
P.O. Box 1552
Guatemala 01901
Central America

Karl Schugerl
Institut fur Technische Chemie
Univeritat Hannover
Callinst. 3
D-3000 Hannover
Germany

Raymond E. Spier
School of Biological Sciences
University of Surrey
Guildford, Surrey
United Kingdom

Eric J. VanDamme
Laboratory of General and Industrial Microbiology
Faculty of Agricultural Sciences
University of Gent
Gent, Belgium

Participants:

Rafael H. Aramendis
National Science Council
Biotechnology Program
Transversal 9A
133-28 Bogota
Colombia

Victor M.C.F. Balcao
Escola Superior de Biotecnologia
Universidade Catolica Portuguesa
Rua Antonio Bernardino de Almeida
4200 Porto
Portugal

Florinel Balteanu
University of Pitesti
Department of Electronics
Vasile Milea Sq.
Pitesti RO-0300
Romania

Frank Bechmann
Technical University Munich
Department of Brewery Plants and Food
Packaging
D8050 Freising
Weinstephan 12
Germany

Isabel Maria Pires Belo
Universidade do Minho
Engenharia Biologica
Largo Do Paco
4700 Braga Codex
Portugal

Scott Bennyhoff
Escola Superior de Biotecnologia
Centro de Calculo e Tecnologias e Informacao
Rua Antonio Bernardino de Almeida
4200 Porto
Portugal

Gulnar Birol
Bogazici University
Department of Chemical Engineering
PK2 80815
Bebek Istanbul
Turkey

Paula M.L.E. Castro
University of Kent at Canterbury
Biological Laboratory
Canterbury CT2 7NJ
United Kingdom

Lenore S. Clesceri
Department of Biology
Rensselaer Polytechnic Institute
Troy, New York 12180
United States

Teresa M.R. Da Silva
Escola Superior de Biotecnologia
Universidade Catolica Portuguesa
Rua Antonio Bernardino de Almeida
4200 Porto
Portugal

Christian Destruhaut
Institut National des Sciences Appliquees
Complex Scientifique de Rangueil
31077 Toulouse Cedex
France

M.S. Doulah
Polytechnic of Wales
Science and Chemical Engineering
Pontypridd Mid Glamorgan, CF37 1 DL
United Kingdom

Murat Elibol
University of Manchester (UMIST)
Department of Chemical Engineering
P.O. Box 88
Manchester M60 1QD
United Kingdom

German Espinosa
Instituto de Biotechnologia
Universidad Nacional Autonoma de Mexico
APDO Postal 510-3
Cuernavaca, Morelos
Mexico

Maria C. Ferreira
Departamento de Engenharia Biologica
Universidade do Minho
Largo do Paco
4719 Braga Codex
Portugal

Alexandre M. Figueiredo
University of Porto
Faculty of Engineering
Department of Chemical Engineering
Rua dos Bragas
4099 Porto Codex
Portugal

Ana Freitas
Escola Superior de Biotecnologia
Universidade Catolica Portuguesa
Rua Antonio Bernardino de Almeida
4200 Porto
Portugal

Peter A. Gostomski
Department of Chemical Engineering
Rensselaer Polytechnic Institute
Troy, New York 12180
United States

Dimitrios G. Hatzinikolaou
National Technical University of Athens
Biosystems Technology Laboratory
Department of Chemical Engineering
Div. IV
5 Iroon Polytechniou Str.
15700 Zografou, Athens
Greece

Nese Ilgin
University of Gazi
TIP Fakultesi
Nukleer TIP ABD
Besevler Ankara
Turkey

David A. Janes
University of the West Indies
Department of Chemical Engineering
Saint Augustine, Trinidad and Tobago
West Indies

Erhan Kiziltan
Ankara University
TIP Fakultesi
Biyofizik ABD
06100 Sihhiye Ankara
Turkey

Andrej Lenhart
Warsaw Agricultural University
Department of Food Engineering
02-766 Warsaw
UL. Nowoursynowska 166
Poland

Teresa Lopes
Escola Superior de Biotecnologia
Universidade Catolica Portuguesa
Rua Antonio Bernardino de Almeida
4200 Porto
Portugal

F. Xavier Malcata
Escola Superior de Biotecnologia
Universidade Catolica Portuguesa
Rua Dr. Antonio Bernadino de Almeida
4200 Porto
Portugal

Alvaro Luiz Mathias
Faculdade De Engenharia Da Universadade Do Porto
Portugal

Goncalo L. Medeiros
Universidade do Porto
Departamento de Engenharia Quimica
Rua dos Bragas
4099 Porto Codex
Portugal

Ana M.A. Nabais
Instituto Superior Tecnico
Departamento de Engenharia Quimica
Av. Rovisco Pais
1906 Lisboa Codex
Portugal

Filomena Oliveira
Departamento de Engenharia Quimica
Faculdade de Engenharia da Universidade do Porto
Rua dos Bragas
4099 Porto Codex
Portugal

Kutlu Ozergin-Ulgen
Bogazici University
Department of Chemical Engineering
P.K. 280815
Bebek Istanbul
Turkey

Milan Polakovic
University of Linkoping
Department of Physics and Measurement Technology
S-581 83 Linkoping
Sweden

John Prior
Biotechnology Process Engineering Center
Department of Chemical Engineering
Massachusetts Institute of Technology
Cambridge, Massachusetts
United States

Thomas Radler
Technical University Munich
Department of Brewery Plants and Food Packaging
D8050 Freising
Weinstephan 12
Germany

Giles Roux
Universate Paul Sabatier of Toulouse
LAAS, CNRS-7
Avenue du Colonel Roche
31077 Toulouse Cedex
France

Luisa M.D.L.F. Santos
Imperial College
Department of Chemical Engineering and Chemical Technology
London SW7 2BY
United Kingdom

Kay Saunders
Bristol Polytechnic
Department of Biotechnology
Coldharbor Lane
Frenchay, Bristol BS16 1QY
United Kingdom

Aldo Schepers
Wageningen Agricultural University
Department of Food and Bioprocess Engineering
Postbus 8129
6700 EV Wageningen
The Netherlands

Marc Siegel
University of Waterloo
Department of Chemical Engineering
Waterloo Ontario N2L 3G1
Canada

Elvan Solay
Bogazici University
Department of Chemical Engineering
PK2 80815
Bebek Istanbul
Turkey

Gary Stanlake
Biology Department
Hardin-Simmons University
Drawer N, HSU Station
Abilene, Texas 79698
United States

Maria Teresa Tavares
Universidade do Minho
Engenharia Biologica
Largo Do Paco
4700 Braga Codex
Portugal

Erik M. Torringa
Agrotechnological Research Institute (ATO-DLO)
Haagsteeg 6
PO Box 17
NL 6700 AA Wageningen
The Netherlands

Harold Van Ryswyk
Olympialaan 12
4625 CT Bergen op Zoom
The Netherlands

Kimberlee K. Wallace
Department of Chemical and Biochemical Engineering
University of Maryland Baltimore County
Baltimore, Maryland 21228
United States

Albrecht Weber
Institut for Food Technology
Universita for Bodenkultur-Vienna
Peter Jordan Str, 82
A-1190 Wien
Austria

Ray Y.K. Yang
Bioreaction Engineering Laboratory
Department of Chemical Engineering
West Virginia University
Morgantown, West Virginia
United States

Preface

This book includes the edited version of the majority of the lectures presented at the NATO ADVANCED STUDY INSTITUTE on "USE OF COMPUTER AND INFORMATIC SYSTEMS IN BIOPROCESS ENGINEERING" held in Ofir, Portugal, May 18-29, 1992. This ASI focused on the engineering concepts and principles associated with biological processing as well as the utilization of personal computers and informatic systems for bioprocess analysis.

Biotechnology has been labeled in recent years as one of the key technologies of the last two decades of the 20th century. It will continue to be a prominent area of science as we enter the 21st century and will offer boundless solutions to problems which range from food and agricultural production to pharmaceutical and medical applications as well as environmental and bioremediation problems. However, biological processes are very complex in nature as, in many instances, the prevailing mechanisms of action and control in biological systems are either unknown or yet poorly understood. This means that adequate techniques for data acquisition and analysis, leading to appropriate modeling and simulation packages which can be superimposed on the engineering principles of biological systems, need to be routine tools for future biotechnologists.

This ASI was attended by sixty-two participants representing sixteen countries. The format of the institute included presentations by invited lecturers, hands-on computer practice tutorials and four workshop sessions. Two sessions with short communications by a number of student participants were also organized during the ASI. These activities all contributed to the excellent success of the Institute and promoted a strong exchange of information and flow of ideas between the leading scientists and the young budding scholars in attendance.

This volume is divided into five sections, each focusing on a different key field of biotechnology; however, considerable overlap exists between the various divisions:

Part 1: Instrumentation Systems
Part 2: Enzyme Technology
Part 3: Environmental Biotechnology
Part 4: Food Applications
Part 5: Metabolic Engineering

Part 1 focuses on the instrumentation systems required for the successful operation of biological processes. Examples of current methods of data acquisition on bench scale illustrate the vast assortment of instrumentation strategies available, while addressing the scale-up problems associated with pilot and production scale operations. This data, combined with other aspects of engineering design, can be incorporated into modeling and simulation packages to solve a multitude of biotechnology problems. Refinement of artificial intelligence systems, in future years, will enhance the capabilities of process monitoring and control, leading to the solution of problems of even greater complexity.

Modeling of enzyme reactors requires a knowledge of reactor design coupled with an in-depth understanding of enzyme kinetics and stability, as described in Part 2. Papers included in this section address the operation of various reactor types such as immobilized systems, membrane bioreactors and stirred tank reactors.

One of the major campaigns of today's world is environmental awareness. In the past, waste materials were often buried and forgotten, only to become major ecological problems years later. As described in Part 3, biotechnology plays a pivotal role in the clean-up of environmental hazards, as well as the prevention of future disasters. This section contains papers which describe the application of biological systems to various situations impacting the quality of our environment. These include the breakdown of organic wastes, the utilization of biomass resources and wastewater treatment, among others.

Part 4 includes papers dealing with modeling and simulation of heat and mass transfer processes associated with widely used food technology operations, namely freezing and drying. These systems are extremely complex to describe mathematically in light of the complexity of food product matrices and the lack of adequate transport data for such systems.

The papers in Part 5, relating to metabolic engineering, further illustrate the complexity of biological systems. Successful modeling of bioprocesses requires coupling of the knowledge of the intracellular and extracellular environments with the basic engineering principles of reactor design. Information on such relationships provides insight into possible biosynthetic bottlenecks. Incorporation of this information into simulation routines, provides a mechanism for assessing the effect of various process variables, including oxygen availability, substrate concentrations and pH, on culture productivity.

This ASI and book would not have been possible without the financial support of the Scientific Affairs Division of NATO and the National Science Foundation. The Institute Directors wish to thank all the lecturers and participants in the ASI for their valuable contributions to the discussions and presentations made during the various sessions. We are also indebted to the staff of the Escola Superior de Biotecnologia for the outstanding support and attention to all minute details of the organization of the ASI which provided for a smooth running conference and two very enjoyable weeks in Portugal. We also wish to thank the secretarial staff of the Department of Chemical and Biochemical Engineering at the University of Maryland Baltimore County for their assistance with the many tasks associated with developing the ASI and accomplishing the publication of the proceedings.

Baltimore, August 1995

Antonio R. Moreira

Kimberlee K. Wallace

PART 1:

INSTRUMENTATION SYSTEMS

COMPUTER-BASED STUDIES ON BIOPROCESS ENGINEERING

I - TOOLS FOR PROCESS ANALYSIS

S. FEYO DE AZEVEDO, P. PIMENTA, F. OLIVEIRA
Departamento de Engenharia Química
Faculdade de Engenharia da Universidade do Porto
Rua dos Bragas 4099 Porto Codex Portugal

ABSTRACT

In this paper we review aspects of the application of computers in the analysis of bioprocesses. Concepts and methods concerning model development are examined. Basic numerical tools and details on least squares estimation, particularly statistical analysis of estimates are discussed. A process simulator of the dynamics of non-linear systems is presented and shown through the application to case studies to be an efficient tool for teaching and research in process dynamics and control.

1. Introduction

Nowadays computers are simply essential tools for all sorts of research. Even if only as word processors employed by Arts students to write their essays on the prose and verse of Camões, Cervantes or Shakespeare. Or as databases for Law students.

In Applied Sciences we like to use mathematical relationships to represent (to model) basic mechanisms or global processes. Research and development in chemical and biochemical engineering both at process and at elementary mechanism level are for many years now based in 'modelling and experiment'.

The applications of computers in those domains can be grouped in different (complementary) ways, depending on the criteria adopted. It makes sense nowadays to speak of 'on-line' and 'off-line' applications or of 'real-time' and 'simulation-time'.

We propose to divide computer applications between those which represent essentially tools for process analysis and those which aim at or are the basis for implementing desired forms of process operation. The former concern mainly model development and aim at the understanding of the mechanisms and of the input-output internal relationships. They are mostly conducted 'off-line' and in 'simulation time' but laboratory and process data acquisition represent examples of such applications which are carried out on-line and in real-time. The

A.R. Moreira and K.K. Wallace (eds.), Computer and Information Science Applications in Bioprocess Engineering, 3-26.

latter, concerning process operation, are usually on-line, real-time applications but many exceptions such as process off-line optimization can be singled out.

There is an obvious 'fuzziness' in the classifications suggested. Life is to a large extent ruled by 'fuzzy' concepts where heuristic knowledge plays a central role. Such understanding is actually the basis for the latest emerging methodologies and computer applications which aim at improving simultaneously process understanding and process operation. Knowledge-based systems and 'fuzzy control' strategies are today on that sense the tip of the iceberg which constitutes the uncertain future of new developments.

In this two-part paper we shall address a number of computer applications concerning process analysis and operation. In part I concepts and methods related to model development and process simulation will be presented. Questions concerning parameter estimation will be examined. Special attention will be given to simulation of the dynamics of multiple-input, multiple-output non-linear systems. The use of computers in process operation will be addressed in part II. Aspects of process monitoring, software sensors and adaptive control will be reviewed. A case-study concerning baker's yeast fermentation will serve to illustrate the theoretical analysis.

2. Model Development

Aspects of model development have been recently reviewed, with application to catalytic reactors (Feyo de Azevedo et al., 1990). The methodology described finds as well application to bioreactors.

Shinnar (1978) has suggested a simple classification which distinguishes between 'predictive' and 'learning' models. The former have more of an industrial use and are employed to determine:

1. The scale-up from pilot to full-scale plant.
2. The study of the behaviour of new feedstock and new catalysts (enzymes).
3. The prediction of the effects of different operating regimes on the reactor yield.
4. The analysis of dynamic trends for the purpose of control.
5. The optimization of operating conditions.

Predictive structures may have a stochastic or a deterministic character. Examples of the former are the time-series ARIMA (Auto-Regressive-Integrated-Moving-Average) models of Box and Jenkins (1971) or the more recent neural networks models which are finding application to biochemical processes (Kosko, 1992; Cooney et al., 1991; Rivera and Karim, 1992). In the modelling of bioreactors use is generally made however of structures with a deterministic character where application is made of principles of thermodynamics, chemical and biochemical kinetics and transport phenomena, in an attempt to reflect the true physical process. The predictive model is characterized as being as simple as possible, containing the minimum of necessary information, involving the estimation of as few parameters as possible and, very importantly, requiring an acceptable computational time for its solution.

Computational time is not a prime factor in a learning model whose features can be summarized as:

i. Providing a structural relation between parameters which are characteristic of more elementary mechanisms.
ii. Helping to understand the simpler models by indicating their relationships with the more complex models.
iii. Providing a diagnostic procedure of elementary causes of unexpected system behaviour.
iv. Providing a guide for experimentation by suggesting which parameters dominate in the predictions of the behaviour of the reactor.
v. Providing a better understanding of the system which may lead to new development and/or improvements in the process and its design.

There may be more than one model which satisfy criteria specified at the beginning of an analysis. The set of models which generally satisfy the quantitative and qualitative criteria adopted is termed the 'permissible model space'. System identification and parameter estimation emerge naturally as the problems of determining model structures belonging to the model space and of estimating the particular values of the characteristic parameter involved. For most deterministic approaches the structure is assumed to be known and the task is to estimate model parameters. For complex models with heavier computational loads, a pragmatic simple approach involves model adaptation with steady state data and an off-line least squares minimization procedure.

A large number of control strategies are based in stochastic models such as those proposed by Box and Jenkins (1971) and by Åström and Wittenmark (1984, 1989). On-line determination of process parameters is a key step in adaptive control strategies. Here parameter estimation is employed in the larger context of system identification. The main elements for such identification are selection of model structure, experiment design, parameter estimation and model validation.

Having accepted a model structure the question remains (more for deterministic than for stochastic modelling) of the confidence related to the model and to particular estimated parameters. Because of the complex nature of the physical and chemical (deterministic) models concerned, an integral element of the study should be an assessment of the effect that variation of parameters has upon the model predictions. This includes the effects of uncertainties in the estimation of parameters (e.g. kinetic and transport parameters in reactor modelling) (Atherton et al., 1975; Priestley and Agnew, 1975; Feyo de Azevedo and Wardle, 1989). Traditionally, the analysis of the sensitivity of models to small perturbations is termed 'local sensitivity analysis'. When a measure of the sensitivity of the solution to variations in a parameter is combined in an appropriate manner with an estimate of the actual degree of uncertainty in the value of that parameter, it can be determined which parameters (through both their sensitivity and uncertainty) have the most influence on the solution. The same applies to assert which operating variables have the most influence in process output (Feyo de Azevedo and Wardle, 1989).

Hence, the development of a model is an iterative process where the objective of the application for which modelling is carried out should always be present. The application of the different phases of the analysis is not necessarily sequential and is characterized by a continuous feedback of information among the different steps. Moreover, modelling ends up as essentially an exercise of statistics where there is a further clear link, this one between modelling and experimentation. This stems not only from the aims attributed to predictive models but also from the strategy adopted in the different phases of the development.

Experiments are carried out at laboratory, pilot or industrial scale. A clear distinction should be made as to the objectives of experiments at each of these levels.

Laboratory experiments generally aim at studying basic mechanisms as much isolated as possible from interacting phenomena.

The main objectives for pilot plant experiments are usually :

a) Definition of process parameters.
b) Development of methods for on-line data interpretation (identification, parameter estimation).
c) Process model discrimination.
d) Test of optimization and control strategies.

Finally, at industrial scale the guidelines are that analysis should lead to :

α) Quality control and product uniformity.
β) Process monitoring - data interpretation and alarming.
χ) Reliability and simplicity of operation.
δ) Control.

Computers play an obvious major role in the exercise. Many questions relate to their use - which computer and which languages to employ ?, which basic numerical methods and numerical routines are more frequently necessary ?, which optimization strategies, how to select them ?, which integration routines ?, how to deal with distributed models ?, what about process simulators ?, how realistic and how difficult to use are simulation packages ?, what about applications of computers on-line and in real time ?, is experiment and process monitoring easy to implement ?, and system identification ? Finally, computer control, which alternatives ?, shall we be able to adapt automatically the strategies to the obvious (and the non-obvious) process changes ?, shall we be able to gather heuristic everyday knowledge in simple structures and with simple rules ?, can we use such knowledge (so often fuzzy) to improve process performance ?

Throughout this paper we shall provide some information concerning these questions.

3. Basic tools - Computers, Languages and Numerical Routines

3.1. Computers and Languages

Can we be affirmative about which computer, which environment and which language ?

If a tool or technique is to be widely disseminated both in university and in industry and not only in few countries of this world, then it should be developed to run in a cheap computer under a friendly environment. On the lower side of the scale we have today the IBM PC compatible, 80486-based machines and MS-DOS 5.0. They serve for larger and larger applications, but there are clear limitations related to their single-user, single-task characteristics, together with the lack of real-time built-in capabilities of the operating systems. Above these there are the UNIX-based PCs and workstations. The latter can be essentially viewed as a more powerful single-user, multi-task system with advanced graphics capabilities.

The potential and power of the new architectures that enable parallel computing and of UNIX and other multi-tasking environments is thus not to be questioned. The development of new applications may require such computational capacities before possibly being scaled down to PC-based versions. The use of shared and distributed memory supercomputers allow large-scale engineering problems to be solved. The availability of networks where PCs are employed as terminals and where workstations and mainframes are connected means that the researcher will have the best of the two worlds without leaving his desk. The most recent version of the X11/AT (an X server from Integrated Inference Machines Co.) allows the User to have simultaneous MS-DOS Windows and X-Windows sessions with on-screen information interchange. It is useful to point out some numbers. In average our applications run 10 to 15 times faster in an 80486-based PC at 33 MHz as compared to its performance in an 80386 machine with coprocessor at 25 MHz. In an Apollo workstation (70 MIPS, 25 MFlops) they run 10 to 15 times faster than in the 80486. The latter costs less than one tenth of the former and less than twice the price of an 80386.

Transportability, costs of machines to serve as platforms and cost of development (including here the cost to pass information from one student to another) are the main aspects to consider on the decision of which language to employ. FORTRAN and C are the main languages to adopt. If carefully written (avoiding some specificities of different compilers) a source programme is directly compilable by DOS and UNIX compilers. A third language should not be neglected - that is BASIC, more specifically Quick Basic 4.5 or any compatible version above. It is a friendly environment, allows structured programming and contains good graphical and I/O capacities. The yield ratio efficiency/development time is probably the highest. The limitation is that for the moment it is not transportable to UNIX, hence size and computational times are limited by PC standards.

3.2. Numerical Methods and Routines

There are libraries in the market with all the necessary numerical support for process analysis. NAG and IMSL are well known packages. However they are not always available, particularly if software is to be developed and supplied to a third partie. Also, there are tools which constitute powerful programming languages for scientific and technical computations. MATLAB (Math Works Inc.) is an example of one such simulation language. It solves many numeric problems in a fraction of the time required to write a program in BASIC, FORTRAN or C. But again there may be problems with transportability and in general these languages lack the flexibility of the traditional software in aspects such as communications and interfacing and interacting with the User. Hence, these all are tools which complement our own personal libraries of basic routines.

Excellent references are available which describe in detail methods and algorithms for the many numerical tasks required in process analysis. Three out of many are - Gerald (1977), Chapra and Canale (1988), Press et al. (1989).

In our practice of process analysis we had at some point to develop routines for solving :

i. Systems of linear algebraic equations.
ii. Matrix inversion.
iii. Tridiagonal systems of equations.
iv. Interpolation problems, particularly employing cubic splines.
v. Non-linear algebraic equations.
vi. Sets of non-linear algebraic equations.
vii. Polynomial equations.
viii. Linear least-squares parameter estimation.
ix. Forward integration of sets of ordinary differential equations (SEDO).
x. Non-linear least-squares parameter estimation on static and dynamic models.
xi. Model discretization with orthogonal collocation as intermediate steps in the solution of boundary-value problems and partial differential equations.

We employ for years Gaussian elimination with partial pivoting (Gerald, 1977). Double precision (in QB and FORTRAN) is clearly necessary for ill conditioned problems. Matrix inversion routines are based on the 'economical' solution of n sets of equations for the same coefficient matrix. Sets of tridiagonal equations are solved with Gaussian elimination without pivoting (the Thomas method). Numerical linear algebra is required in all stages of process analysis, in optimization and in control.

Non-linear equations are easily solved usually with Steffensen´s method (fixed-point iteration with Aitken´s acceleration). In these and other iterative algorithms much help to convergence is provided by heuristic support, particularly imposing certain bounds on the domain.

We do not have general routines for the solution of sets of non-linear equations. Large systems usually require a first step of system decomposition. The solution for that employs heuristic rules and it is problem-dependent. An efficient but time-consuming approach is to consider the general system $\underline{f}(\underline{x})=0$, take $\underline{x}$ as parameters and solve the non-linear least-squares parameter estimation problem with the objective function $F = \sum f_i^2(x)$. The Gauss-Seidel type approach is intuitive and simple to implement for small systems, but much attention must be given to convergence. Modified Newton´s methods such as the so-called matrix-updating methods (e.g. Broyden´s method) are the other alternatives where essentially the difficulty is to guarantee convergence.

Cubic spline interpolation (de Boor, 1978) generally gives the smoothest interpolation. It is widely employed in constructing graphical solutions from a limited number of points. Furthermore cubic splines are the basis for orthogonal collocation on finite elements.

Bairstow's method (Press et al., 1989) together with routines for polynomial deflation provide an efficient iteration solution to extract all real and complex zeros of a nth degree polynomial.

Linear least-squares parameter estimation on linear and non-linear models require essentially numerical linear algebra. The static problem is obviously a particular case of the general dynamic problem. For the latter, simple well established recursive algorithms are available in the literature (Najim and Muratet, 1987; Åström and Wittenmark, 1989). Least squares parameter estimation is possibly the most employed technique in experimental research and as such it will be discussed in more detail in the following sections.

For the forward integration of sets of ordinary differential equations we have been obtaining for several years robust performances with our routines based on embedded 4th/5th order Runge-Kutta type formulae. The first is Sarafayan's method which employs Butcher's formulae (Lapidus and Seinfeld, 1971). The other is Felbhergs method (Chapra and Canale, 1988). Both methods provide an estimate of the truncation errors which serve to control the integration step. More powerful methods (hence more demanding in computation time) are proposed by Gear (1971).

Detailed presentation of orthogonal collocation is out of scope of this paper. The method is however worth to be singled out since it is widely employed in the analysis of non-linear processes. It has been applied to problems involving discretization in one and two space directions (Villadsen and Sorensen, 1969; Ferguson and Finlayson, 1970; Michelsen and Villadsen, 1972; Young and Finlayson, 1973) and is thoroughly documented in textbooks and reviews (Finlayson, 1972, 1974, 1980; Villadsen and Michelsen, 1978; Morbidelli et al, 1983). A package on orthogonal collocation should include routines for discretization, interpolation and quadrature, for linear, cylindrical and spherical geometries, with and without simmetry.

The method is widely applied in the analysis of tubular reactors (Georgakis et al., 1977; Jorgensen and Jensen, 1989; Feyo de Azevedo et al., 1990) which presents many similarities with the same kind of processes used in biotechnology. Luttman et al (1985), Isaacs et al (1986), and Isaacs and Thoma (1992) employed orthogonal collocation in the simulation of air-lift bioreactors. The interest of the method in the analysis of bioreactors has also been emphasized recently by Dochain et al (1992).

End routines for partial differential equations tend to be problem dependent and as such they are left out of this analysis. It is however worth to mention that more and more there are packages available for 'classes of problems', employing not only finite-difference methods but also orthogonal collocation methods and finite-element methods (Costa et al., 1986; Brebbia, 1985).

For studies on identification and control employing classical languages it is important to make sure that a number of support routines are available, otherwise the programming burden is too high. These are very specialised routines as compared with those required for basic process analysis.

The NAG group provides SLICOT, a library of FORTRAN 77 subroutines for control systems analysis and design. It contain 68 user-callable routines (information dated March 1992), which can be as well called by interactive packages such as MATLAB by using appropriate shells. Routines available cover topics such as: state-space model reduction, convolution and deconvolution signals, Kalman filtering at different situations, Ricatti equation, pole placement methods, matrix transformation, eigenvalues and eigenvectors and least squares methods.

One relevant final comment. For a given problem there will be more than one possible method. There is no such thing as a 'universal method' on the sense that accurate solution will always be obtained. And for a class of problems, there is not a 'best method'. By the way, what is a 'best method'? A method that is robust? A method that is fast in convergence to the solution within a given accuracy?

4. Least-squares parameter Estimation

4.1. INTRODUCTION

The validity of a model depends on its capacity to predict observed behaviour.

We assume we have a general model of the form :

$$\underline{y}^{(j)} = \underline{f}(\underline{x}^{(j)}, \underline{\theta}^{(j)}) + \underline{\varepsilon}^{(j)} \tag{1}$$

where $\underline{y}^{(j)}$ is the m×1 vector of observations under conditions given by $\underline{x}^{(j)}$.
$\underline{x}^{(j)}$ is the q×1 vector of independent variables for the set of conditions specified by j
$\underline{\theta}^{(j)}$ is the p×1 vector of unknown parameters.
and $\underline{f}$ is the m×1 vector of non-linear functions.
$\underline{\varepsilon}$ is the m×1 vector on errors including the full discrepancies between y and f (measurement errors, errors in the independent variables and model inadequacies)

The set of parameters $\underline{\theta}$ contain information on the system. The values for such parameters depend on the system itself but also on the criteria accepted to judge the quality of the model.

Gauss formulated the principle of least-squares at the end of the eighteenth century and used it to determined the orbit of planets. According to this principle the unknown parameters of a mathematical model should be chosen in such a way that 'the sum of the squares of the differences between the observed and the computed values multiplied by numbers that measure the degree of precision, is a minimum'.

The Gauss formulation corresponds to determine the estimated vector $\hat{\theta}$ that minimizes -

$$F(\hat{\theta}) = \underline{\varepsilon}^{(j)T}\, \underline{W}\, \underline{\varepsilon}^{(j)} \tag{2}$$

where $\underline{W}$ is a diagonal matrix of weights.

Least squares parameter estimation is based on the assumption that the error structure is a normal distribution and that gross errors are not present. It is also generally accepted that errors in the independent variable can be minimized and thus neglected. Gross error detection is important because their presence results in biased estimates (Hlavacek, 1977; Mah and Tamhane, 1992; Tamhane and Mah, 1985, Tjoa and Biegler, 1991a, b). The analysis for implicit models and the estimation on the case of non-linear models with errors in all variables are treated by Schwetlick and Tiller (1985) and by Dovi and Paladino (1989).

The statistical study of parameter estimation includes the analysis of variances, residues, and joint and/or individual confidence intervals of the estimates. Only then a decision on model validation can be taken (Draper, 1966; Nash and Walker-Smith, 1989).

Real-time parameter estimation is extensively described by Åström and Wittenmark (1989), including the estimation of parameters in dynamical systems. The theory on linear least squares and on recursive least squares to be described next follows closely their approach.

4.2. Linear Least Squares

The method is particularly simple if the model has the property of being linear in the parameters. In such case equation (1) is written for one set of observations $(y, x_1, ..., x_q)$ as:

$$y = \varphi_1(\underline{x})\,\theta_1 + ... + \varphi_p(\underline{x})\,\theta_p + \varepsilon \tag{3a}$$

or

$$y = \varphi^T \underline{\theta} + \varepsilon \tag{3b}$$

For m observations we can further define :

$$\underline{Y}^T = [y_1, y_2, ... , y_m] \tag{4a}$$

$$\underline{\varepsilon}^T = [\varepsilon_1, \varepsilon_2, ... , \varepsilon_m] \tag{4b}$$

$$\text{also } \Phi = \begin{bmatrix} \varphi_1(1) & \varphi_2(1) & ... & \varphi_p(1) \\ ... & ... & ... & ... \\ \varphi_1(m) & \varphi_2(m) & ... & \varphi_p(m) \end{bmatrix} \tag{4c}$$

If weights are identically taken to be one and the vector of observations is $\underline{Y}$, then the objective function $F(\underline{\theta})$ takes the form :

$$F(\underline{\theta}) = \underline{\varepsilon}^T \underline{\varepsilon} \tag{5a}$$

$$\text{where } \quad \underline{\varepsilon} = \underline{Y} - \hat{\underline{Y}} = \underline{Y} - \underline{\Phi}\,\underline{\theta} \tag{5b}$$

The function will be minimal for

$$\underline{\Phi}^T \underline{\Phi}\,\underline{\theta} = \underline{\Phi}^T Y \tag{6}$$

from where

$$\theta = \hat{\theta} = [\underline{\Phi}^T \underline{\Phi}]^{-1} \underline{\Phi}^T \underline{Y} \tag{7}$$

Equations (6) are called the 'normal equations'. A detailed statistical and geometric interpretation of the least squares problem is provided by Åström and Wittenmark (1989).

4.3. Recursive Least Squares

Frequently, we have situations where observations are obtained sequentially in real-time. Parameters can be updated through the use of recursive computations, in order to save computational time.

In the following we assume that we have the estimated vector of parameters $\hat{\theta}(m)$ after m observations, and that we have available a new observation y(m+1). It is possible to estimate $\hat{\theta}$ (m+1) from the previously obtained $\hat{\theta}(m)$. Assuming that the matrix $\underline{\Phi}(t)$ has full rank for all $t \geq m$, then $\hat{\theta}(m+1)$ satisfies the following recursive equations -

$$\hat{\theta}(m+1) = \hat{\theta}(m) + K(m+1)\,(y(m+1)-\varphi^T(m+1)\,\hat{\theta}(m)) \tag{8}$$

$$K(m+1) = P(m+1)\varphi(m+1) = P(m)\varphi(m+1)\,(I + \varphi^T(m+1)P(m)\varphi(m+1))^{-1} \tag{9}$$

$$P(m+1)=P(m)-P(m)\varphi(m+1)(I+\varphi^T(m+1)P(m)\varphi(m+1))^{-1}\,\varphi^T(m+1)P(m) =$$

$$= (I - K(m+1)\,\varphi^T(m+1))\,P(m) \tag{10}$$

The initial (pxp) matrix P(m) is obtained from -

$$P(m) = [\underline{\Phi}^T(m)\;\;\underline{\Phi}(m)\,]^{-1} \tag{11}$$

4.3.1. Slowly Time-varying Parameters: Forgetting factor. A simple approach to weight differently

Often parameters are time-varying. For the cases where slow variations are assumed the problem can be solved by simply replacing the least squares objective function (Eq. 5a) by

$$F(\theta) = \frac{1}{2}\sum_{i=1}^{m} \lambda^{m-i}\,(y(i) - \varphi^T(i)\theta)^2 \tag{12}$$

λ is a parameter (bounded $0<\lambda<1$) which allows different weighing of observations. The most recent data is given unit weight, with the weight of the old data tending exponentially to zero. λ is known as 'exponential forgetting factor'.

Assuming the same conditions as in the previous case, recursive least squares with exponential forgetting is given by:

$$\hat{\theta}(m+1) = \hat{\theta}(m) + K(m+1)\,(y(m+1)-\varphi^T(m+1)\,\hat{\theta}(m)) \tag{13}$$

$$K(m+1) = P(m+1)\varphi(m+1) = P(m)\varphi(m+1)\,(\lambda I + \varphi^T(m+1)P(m)\varphi(m+1))^{-1} \tag{14}$$

$$P(m+1)= (I - K(m+1)\,\varphi^T(m+1))\,P(m)/\lambda \tag{15}$$

4.4. Non-linear Least Squares

The non-linear problem is solved employing non-linear optimizing routines.

For objective functions with a least squares structure it is shown (Gill et al, 1981) that the Hessian matrix H ($H_{ij} = \partial^2 F/\partial\theta_i\partial\theta_j$, i, j = 1, p) is given by

$$\underline{H} = 2\,[\underline{J}^T\,\underline{J} + \sum_{i=1}^{m} \varepsilon_i\,\underline{G}(i)\,]_\theta \tag{16}$$

where $\underline{G}^{(i)}$ is the (p×p) Hessian matrix of $\varepsilon_i(\theta)$
and $J(\theta)$ is the (m×p) Jacobian matrix of $\underline{\varepsilon}(\underline{\theta})$

In the neighbourhood of the solution the norm $\|\underline{\varepsilon}(\underline{\theta})\|$ is often small compared with $\|\underline{J}^T\underline{J}\|$, hence

$$\underline{H} \cong 2\,\underline{J}^T\,\underline{J} \tag{17}$$

If $F(\underline{\theta})$ has continuous second derivatives, then $\underline{H}(\underline{\theta})$ must be positive semi-definite at any unconstrained minimum of F. Hence, some variant of the Gauss-Newton method can be applied, requiring only estimates of first derivatives.

We have examined a number of alternative methods. Our experience favours the variant proposed by Meyer and Roth (1971) (Feyo de Azevedo et al, 1988; Salcedo et al., 1990).

4.5. STATISTICAL ANALYSIS - CONFIDENCE INTERVALS

4.5.1. *Individual Confidence Intervals (ICI).* The additional advantage of Gauss-Newton methods is that it provides immediately the necessary data for analysis of confidence intervals.

It is known (Gill et all, 1981) that an unbiased estimate of the (p×p) covariance matrix $\underline{C}(\underline{\theta})$ is given by-

$$\underline{C}(\underline{\theta}) = \left[\frac{2\,F}{m-p}\,H^{-1}\right]_{\theta=\hat{\theta}} \tag{18}$$

An unbiased estimate of the (p×1) vector υ of the parameter variances is, for the i-th element

$$\upsilon_i(\hat{\theta}) = C_{ii}(\hat{\theta}) \tag{19}$$

and the standard deviation is estimated by-

$$s_i(\hat{\theta}) = \upsilon_i(\hat{\theta})^{1/2} \tag{20}$$

Finally, if $\underline{\theta}^*$ is the true solution then the 100(1-β) individual confidence interval (ICI) on $\underline{\theta}$ is

$$\hat{\theta}_i - s_i(\hat{\theta})\,t_{(\beta/2,\,m-p)} < \underline{\theta}_i^* < \hat{\theta}_i + s_i(\hat{\theta})\,t_{(\beta/2,\,m-p)} \tag{21}$$

or in terms of the percent relative confidence interval %ICI -

$$\%ICI_i = \pm\frac{s_i(\hat{\theta})\,t_{(\beta/2,\,m-p)}}{\hat{\theta}_i} \times 100 \tag{22}$$

4.5.2. *Joint Confidence Region (JCR).* The joint confidence region (JCR) where k out of the p elements of $\underline{\theta}^*$ are of interest may be described by,

$$(\underline{\theta}_k - \hat{\underline{\theta}}_k)^T\,\underline{C}_k^{-1}\,(\underline{\theta}_k - \hat{\theta}_k) \le K_a \tag{23}$$

where $\underline{\theta}_k$, $\underline{\hat{\theta}}_k$ are the (k×1) vectors of the parameters of interest.
$\underline{C}_k$ is the (k×k) matrix obtained by omitting from the covariance matrix $C(\hat{\theta})$ the rows and columns corresponding to the parameters which are of no interest.

and $$K_a = k\,F(k, m\text{-}p, 1\text{-}a) \qquad (24)$$

where F(k, m-p, 1-a) is the 'upper a-point' of the F(k, m-p) distribution.

For large values of (m-p), say larger than 120, K_a may also be obtained from tables for the chi-square distribution and is defined by,

$$\Pr[\,\chi_k \leq K_a\,] = a \qquad (25)$$

where χ^2 is a random variable distributed with k degrees of freedom and 'a' is the desired confidence level for the confidence region.

The meaning of the confidence region at a confidence level 'a' defined by eq. 23 is that on repeating the experiment the probability is 'a' that the confidence region (which varies from experiment to experiment) will include the constant true value.

4.5.3. *Construction of the JCR ellipse.* For the case of two parameters the region defined by eq. 23 is an ellipse.

If that ellipse is described by the equation -

$$(\theta_1 - s, \theta_2 - t)\begin{pmatrix} p & q \\ q & r \end{pmatrix}\begin{pmatrix} \theta_1 - s \\ \theta_2 - t \end{pmatrix} = K \qquad (26)$$

Then (i) The center of the ellipse is at (s,t)
(ii) The major axis has a slope

$$m_1 = \frac{r - p - \sqrt{(r-p)^2 + 4q^2}}{2q} \qquad (27)$$

and a semi-length

$$a = \left\{ \frac{K}{2(rp-q^2)} \left[r + p + \sqrt{(r-p)^2 + 4q^2} \right] \right\}^{1/2} \qquad (28)$$

(iii) the minor axis has a slope

$$m_2 = \frac{r - p + \sqrt{(r-p)^2 + 4q^2}}{2q} \qquad (29)$$

and a semi-length

$$b = \left\{ \frac{K}{2(rp-q^2)} \left[r + p - \sqrt{(r-p)^2 + 4q^2} \right] \right\}^{1/2} \qquad (30)$$

the foci are at distances ±C from the center along the major axis where $C^2 = a^2 - b^2$.

The ellipse is the locus of all points such that the sum of their distances from the two foci is 2a. For computer plotting the following parametric equations may be used

$$\theta_1 = s + \frac{1}{z}(u - m_1 v) \tag{31a}$$

$$\theta_2 = t + \frac{1}{z}(m_1 u + v) \tag{31b}$$

where $z = \sqrt{1+m_1^2}$, $u = a \cdot \cos(\alpha)$, $v = b \cdot \sin(\alpha)$ and $0 \le \alpha \le 2\pi$

4.6. Parameter Estimation in Differential Equations

Developing an efficient method for solving parameter estimation problems of dynamic models or of steady-state distributed model is an important part of the development and improvement of process modeling (this implies that here we are dealing with deterministic models).

Optimization strategies for differential-algebraic equations have been discussed by Cuthrell and Biegler (1987, 1989) and by Tjoa and Biegler (1989a). The problem formulation is simple. Gauss-Newton type methods should be employed and each function evaluation implies the solution of the SODE. It is clear that the success of this scheme depends heavily on this step of solving the SODE, particularly if these are stiff. Model discretization and model reduction may lead to simplified optimization but for that it is necessary first to prove the validity of such transformations.

4.6.1. *An Application - Identification of third order models from input-output experimental data.* The existing and expected availability of computational power have definitely encouraged time-domain analysis.

The type of off-line process identification suggested by the well known process reaction curve technique (Seborg et al., 1989.; Stephanopoulos, 1984.) can be much improved not only to extend to 3rd order processes but mainly to derive results from any type on experimental non-stationary input data.

We consider the general 3rd order linear model with phase lead and dead time represented in the Laplace domain by -

$$Y(s) = \frac{K\,(\tau_1 s+1)\, e^{-\tau_0 s}}{(\tau_p s+1)\,(\tau^2 s^2+2\zeta\tau s+1)}\; X(s) \tag{32}$$

We introduce a working variable $Y_a(s)$ such that

$$Y_a(s) = \frac{K\, e^{-\tau_0 s}}{(\tau_p s+1)\,(\tau^2 s^2+2\zeta\tau s+1)}\, X(s) \tag{33a}$$

and $$Y(s) = (\tau s+1)\; Y_a(s) \tag{33b}$$

The corresponding equations in the time domain are -

$$a_3 \frac{d^3y_a(t)}{dt^3} + a_2 \frac{d^2y_a(t)}{dt^2} + a_1 \frac{dy_a(t)}{dt} + a_0y_a(t) = b_0x(t-\tau_0) \tag{34}$$

and $$y(t) = b_1 \frac{dy_a(t)}{dt} + y_a(t) \tag{35}$$

where
$$a_3 = \tau_p \tau^2$$
$$a_2 = \tau^2 + 2\zeta\tau\tau_p \qquad b0 = k$$
$$a_1 = \tau_p + 2\zeta\tau \qquad b1 = \tau 1$$
$$a_0 = 1$$

By doing $y_1 = y_a$, the state space representation is -

$$\frac{dy_1}{dt} = y_2 \tag{36a}$$

$$\frac{dy_2}{dt} = y_3 \tag{36b}$$

$$\frac{dy_3}{dt} = [\, b_0\, x(\tau-\tau_0) - a_2y_3 - a_1y_2 - a_0y_1] / a_3 \tag{36c}$$

and $$y(t) = b_1y_2 + y \tag{37}$$

With initial conditions $y_1 = y_2 = y_3 = y = 0$.

Having available m observations (x_j, y_j) from input-output data a least-squares parameter estimation scheme is conceptually simple to describe - a Gauss-Newton type scheme is employed. The objective function is of the form of eq. (2) -

$$F = \sum_{i=1}^{m} (\hat{y}_i - y_i)^2$$

For each set of parameters F is obtained from the numerical integration of the state-space model. In such integration x(t) is approximated by linear interpolation by cubic spline interpolation or by curve fitting (polynomial, exponential or Fourier type), the option depending on a previous study of the data.

Some aspects should be further emphasized. The number of parameters is very large. Care should be taken when estimating more than three parameters at a time since experience shows that above this number a multitude of local optima tend to appear. Also, the dead time parameter is particularly difficult to estimate, being a source of problem ill-conditioning.

The overall approach is readily adapted to simpler versions of the linear model (i. e., a simplified version of eq. 32), particularly the lead-lag problem with dead time.

5. Process Simulators and Simulation Languages

5.1. AN OVERVIEW

Process simulation is nowadays a well established primary tool for design and for basic decision making in all areas of process industries. The demand for simulators is high and the supply is responding to that demand. We can find in the market and in the literature general-purpose and dedicated simulators developed by companies and research groups. They aim at process analysis, at process operation and at operator training. More and more they run in microcomputers. For those requiring workstations and UNIX there are versions for PCs.

In this overview we shall not distinguish between process analysis and operation because tools are adaptable, hence it generally makes no sense to have a dividing line.

ASPEN/SP (JSD Simulation Services, MIT) and CHEMCAD (Chemstation) are examples of 'Plant Simulators' with versions available for PCs. They include modules for a very large number of unit operations. They are particularly applicable to design and to steady-state optimization.

Simulators for specific operator training are described by Schewchuk et al (1992), by Spuhler (1992) and by Machietto and Kessianides (1992).

MATLAB (Moller 1987; The Mathworks Inc. 1992) and SIMNON (Elmqvist et al, 1986) are languages for process simulation. They represent well today's availability of tools -

The former is an interactive system whose mathematical base tool is a matrix. It has a library of macros that can implement several control system analysis, as well as modelling techniques. MATLAB does not allow the on-line change of the system parameters. With the latest version it is possible to make on-line data acquisition. The latter is essentially a simulation language to solve ordinary differential equations and difference equations and to simulate system dynamics. SIMNON does not allow the on-line change of system parameters. The latest version allows real-time data acquisition through RS232. Clarke (1991) provides a good example of the usefulness of MATLAB in the simulation of model based control systems. SIMNON can as well be employed in the analysis of control schemes.

Other packages more directed to specific domains are reported. BIOSIM (Cadman and Davison, 1989) deals with dynamic simulation of bioreactors. MERLINO is directed to model base predictive control (Clarke and Scattoloni, 1990); ASCET (Eppinger et al, 1990) is an integrated control system design tool which allows on-line interaction, changing model structure and parameters. PC-TACT (de Keyser, 1990) is a software for training instrumentation and control engineers, providing 'real-time' user interface. It is interesting to note that earlier in 1975 Ord-Smith and Stephenson divided simulators between block oriented (such as PC-TACT) and expression oriented (e. g. BEDSOCS - Bradford Educational Simulation Language for Continuous Systems). EASY5, Matrix-X, TUTSIM, ACSL, CC, CODAS, CYPROS, SFPACK, ProSimulator are further examples of packages for simulation of dynamic systems and control.

It is clear that researchers have today a variety of software tools for their studies on process dynamics and control. A possible advice to give is that the package or language chosen should be adequately documented, otherwise one risks not to take full advantage of it.

5.2. MIMOSA-MULTIPLE INPUT, MULTIPLE OUTPUT SYSTEMS ANALYSIS

Key characteristics for the practical interest of a process simulator are its capacities for (i) simulation in real-time, (ii) interaction with the User (operator), (iii) communications via A/D, D/A cards and serial interfacing; (iv) incorporation of features observed in real systems (noises and dead times); (v) simulation of time-varying process parameters.

We have developed a package for real-time simulation which includes the features just singled out.

'MIMOSA - Multiple Input, Multiple Output System Analysis', is a PC-based simulator of the dynamic behaviour of non-linear MIMO systems. The strategy is developed in the time domain and simulation can be optionally performed in 'real-time', in 'scaled-time' or in 'simulation time'. 'Scaled time' is an optional form of operation where one second of real time corresponds to 'scale' seconds of simulation, 'scale' being an user-defined parameter. The option 'simulation time' corresponds to the form of operation where no time control is produced.

Virtually all processes represented mathematically by sets of non-linear algebraic-differential equations can be described by the simulator. Exceptions, not yet found in our practice, may be cases of large number of ill-conditioned equations where the numerical strategies employed, albeit tested and robust, may experience difficulties.

The User defines in very simple terms his own process (algebraic-differential) equations, control inputs, loads, measured variables and process parameters. The latter can be changed on-line during the simulation. Protocols for input/output are included for A/D, D/A and RS232 communications.

5.2.1. *Process Representation.*The process as simulated by MIMOSA is represented in Fig. 1. The relevant variables for the purpose of process implementation are grouped in six main vectors, viz-

c[ ] - control inputs
u[ ] - process loads (disturbances)
ud[ ] - delayed process loads
p[ ] - process parameters
y[ ] - 'system state variables'
f[ ] - derivatives of 'system state variables'

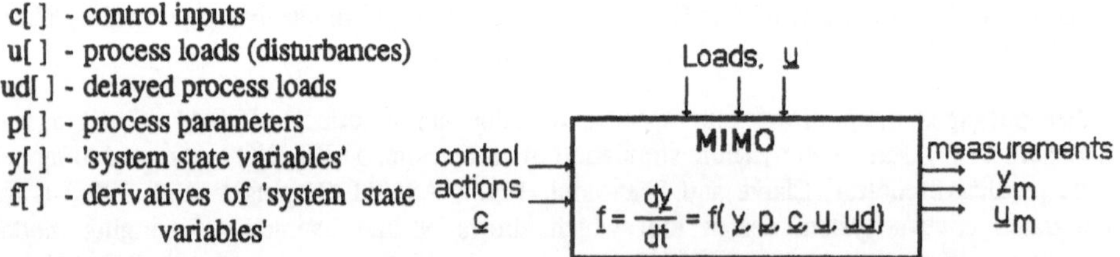

Fig. 1. Representation of a MIMO process.

The 'system state vector' - y[ ] is a concept specific of MIMOSA. It is composed of all variables which have their first derivative explicitly represented in the set of model equations. This includes the formal state variables and all the other variables which are the result of adding to the formal state the dynamic equations for the sensors and final control elements. All system state variables are eligible for output via a D/A interface or an RS232 port, depending on the version of the package being employed. The output is thus constituted by the measured state variables -y_m- and the measured loads -u_m. Both are subsets of y.

The following general characteristics should be noted, viz-

(i) All control inputs (c[ ]) can be monitored. Such variables may be changed manually through the keyboard (with Communications set to 'off') or from an 'external controller' (via A/D cards or RS232).

(ii) The User will be able to monitor all variables declared as loads to the process (u[]). Such variables may be changed manually through the keyboard. This includes step, sine or ramp functions.

(iii) All parameters named as p[ ] in the model equations can be monitored, and their values changed on-line (this will include the noises and dead-times defined in a data file, which MIMOSA automatically will add to the list of parameters and as such will make available for on-line change).

(iv) All 'system state variables' (y[ ]) can be monitored and all are eligible as output variables, with a superimposed 'user defined' random noise.

(v) MIMOSA allows for the implementation of time delays associated to the 'measurements' ($\underline{u}_m$ and $\underline{y}_m$), to the control inputs and to the load inputs. This independent characterization allows for the simulation of the simultaneous effects of 'transportation lags' (usually associated to measurements) and of process 'intrinsic' dead times (the latter associated to distributed process). Such feature enlarges significantly the range of applications with MIMOSA.

5.2.2. *Hardware Setup*. The concept that presided to its development was that of 'communications' via analog means. Such solution is represented in Fig. 2a. It has the limitation that it is expensive if a large number of I/O signals has to be employed and also that each different A/D or D/A card requires a specific 'driver' to be written. It has the advantage that real industrial controller can be employed both for training and for development. With such controllers only applications in real-time are of interest. An alternative package was developed, employing bidirectional RS232 for communications. This is represented in Fig. 2b. It is most adequate for research and development since it allows for the same cost and in standard form any number of I/O signals and allows studies in 'scaled' time and in 'simulation' time without losing the important features.

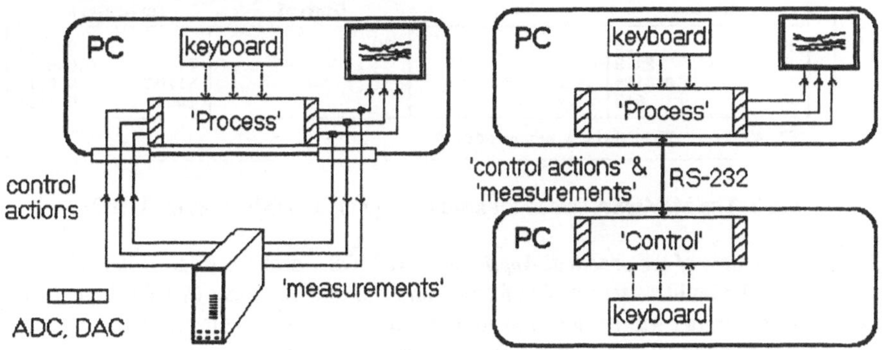

Fig. 2. MIMOSA with a) industrial controller or with b) control computer.

5.2.3. *Computer Implementation-Algorithmic Aspects and Interaction with the Operator*. The general solution of MIMOSA (coded in MS C Vs. 5.1) is based on the integration in the time-domain of the set of differential-algebraic equations which constitute the process model. This is achieved in the current version by employing a variable-step algorithm based on Felhberg's embedded 4th/5th order, Runge Kutta type formulae (Chapra and Canale, 1989).

A basic cycle of work consists on the following tasks - (i) Reads control action; (ii) Proceeds with integration; (iii) Displays and files results; (iv) Outputs measurements; (v) Checks keyboard for interaction while controlling time. Currently, with a 80486 based machine, a well-conditioned set of 20 differential equations can be processed in 0.25 seconds, i. e an observer can 'see' the process every 0.25 seconds.

The basic concept to create an operator interface was that of 'all-in-one' graphics interface where all interactive information is displayed in one single page (screen). This is a major feature for simplicity of operation.

Fig. 3 shows the 'all-in-one' approach. The screen is essentially divided in two zones. In the upper half a graphics facility emulates a 4-pen paper register. In the lower half the main variables are monitored and made available for on-line change. Further to these two zones the operator has all times the indication of functions available, such as putting communications on/off, changing variable to be plotted, starting/stopping saving results on disk, starting/stopping printing, producing a printscreen...

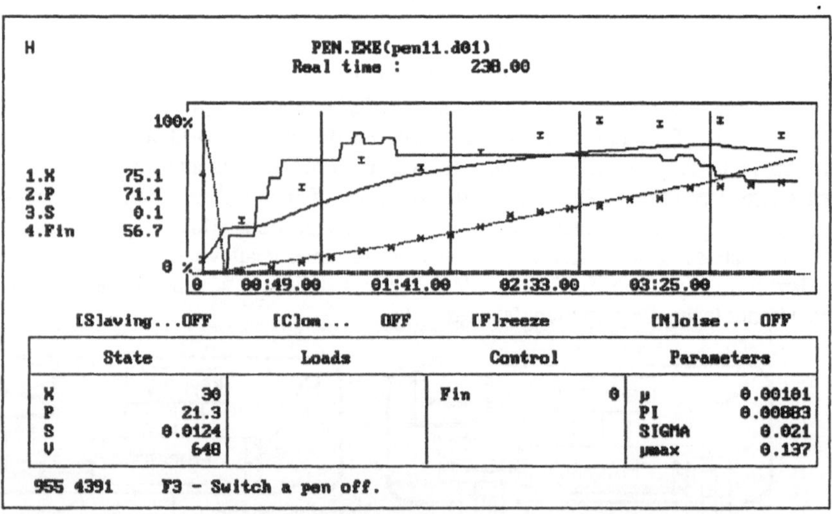

Fig. 3. The MIMOSA screen for a study on penicillin (Menezes et al, 1991).

5.2.4. *The Structure of the practical Application.* MIMOSA package (Fig. 4) is made of three files, viz - (*i*) The builder - responsible for building and running cases; (*ii*) The kernel - which by action of the builder will be linked to an object file corresponding to a specific case, giving the executable file; (*iii*) The data file with the integration control parameters.

Two simple case-studies will illustrate the working characteristics of the simulator. They represent a linear and a non-linear problem which will be treated in an unified form. As mentioned before, this is possible because the approach is developed in the time domain.

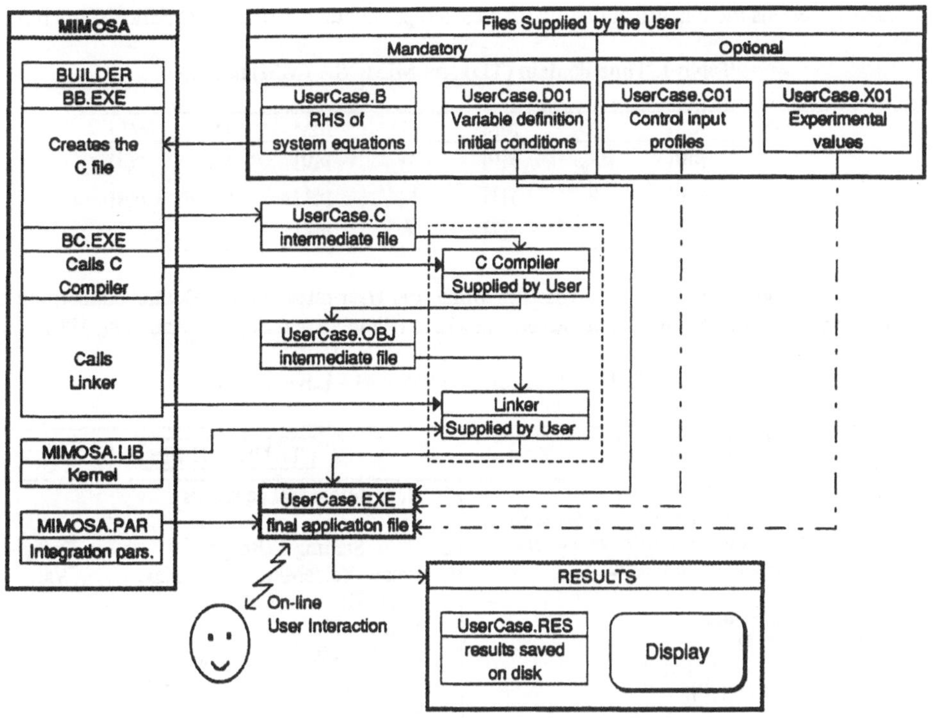

Fig. 4. The files and applications structure of MIMOSA.

5.2.5. *Case-Study 1 - Level Control.* We consider the level control of the tank represented in Fig. 5. The tank area A is constant.

The level is measured by a sensor which may exhibit a first order behaviour and a possible time delay. The final control element is also first order. The control signal to this element is C and the inflow is subject to fluctuations (hence, it is a perturbation). The set of model equations is-

$$\frac{dF_{out}}{dt} = (K_v * c - F_{out})/\tau_v \qquad (38)$$

$$\frac{dh}{dt} = \frac{1}{A} * (F_{in} - F_{out}) \qquad (39)$$

$$\frac{dh_m}{dt} = (K_m * h - h_m)/\tau_m \qquad (40)$$

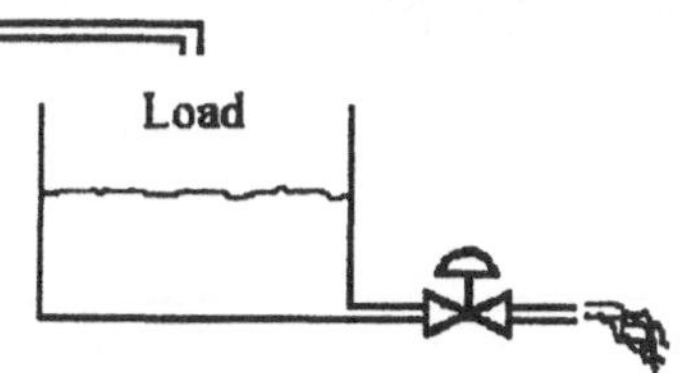

Fig. 5. Level control.

The initial conditions, which are not required to be zero, are defined in the .D01 file (Table 2)

Table 1 presents the classification of variables and parameters, according to MIMOSA's rules.

Table 1. Time domain (TD) and MIMOSA (MM) notation.

TD	MM	TD	MM	TD	MM	TD	MM
c	c[0]	F_{out}	y[0]	τv	p[0]	τ_m	p[3]
F_{in}	u[0]	h	y[1]	k_v	p[1]	k_m	p[4]
		h_m	y[2]	A	p[2]		

Table 2 presents the two plain ASCII files which the User must supply. Details concerning all rules and features of MIMOSA are given elsewhere (Pimenta and Feyo de Azevedo, 1992).

Table 2. User files for case study #1 - Level control.

TANK.B file

```
// Tank - DEMO - PP, SFA 92.04
// Comment lines must start with double dash
// All assignments must end with semi-colon
// If ... else cases can be employed

    // y0 - Valve position.
    if ( p[0] > 1.e-3)
          f[0] = (p[1]*c[0]-y[0])/p[0];
    else  y[0] = p[1]*c[0];

    // y1 - Holdup.
    f[1] = (u[0] - y[0]) / p[2];

    // y2-Measured level 1st order sensor
    if ( p[3] > 1.e-3)
          f[2] = (p[4] * y[1] - y[2])/ p[3];
    else  y[2] = p[4]*y[1];

    // Scaled Time:
    f[0] = f[0] * p[5];
    f[1] = f[1] * p[5];
    f[2] = f[2] * p[5];
```

TANK.D01 file

```
C  N. of differential equations
   3
C          Starting   Range                    Delay
C  Name  ValueMin.          Max.          Var.
   (sec/100)
   Valve    10.
   Level   500.
   Levelm 500. 300.         700.          2.
   0.
C  N. of loads/disturbances
   1
C  Name    Starting Value              Delay (sec/100 )
   In        10.                        20
C  N. of control actions
   1
C  Name   Starting Value               Delay (sec/100)
   Out      10.                         10
C  N. of parameters
   5
C  Name   Value      Comments
   TauV    1.        p0 Valve Tau
   GainV   1.        p1 Valve Gain
   Area    1.        p2 Area of section
   TauM    .2        p3 Level sensor Tau
   GainM   1.        p4 Level sensor Gain
   Scale   1.        Time factor
```

By giving the command BB TANK the builder will take care of the job of building the final executable file TANK.EXE. The intermediate TANK.C and TANK.OBJ (Fig. 4) are not relevant for the User.

The command TANK -DTANK.D01 will run the case-study with the data file TANK.D01. The User can create different data files TANK.Dxx to run with the .EXE file.

If 'communications' are switched off MIMOSA will run as a 'stand-alone' simulator. The initial control inputs will be those defined in the .Dxx file. Another useful feature included in MIMOSA consists of the plotting of experimental values supplied by the User as simulation progresses, allowing thus immediate visual comparison with the theoretical predictions.

5.2.6. *Case Study 2 - Dynamics of two tanks in series.* The scheme is represented in Fig. 6.

The following set of non-linear equations describe the system-

$$F2 = K1 * \sqrt{h1} \quad \text{(Discharge by orifice)} \quad (41)$$

$$F3 = k2 * h2 \quad \text{(Resistance)} \quad (42)$$

and, for the levels of the two tanks,

$$A1 * \frac{dh1}{dt} = F1 - F2 \quad (43)$$

$$A2 * \frac{dh2}{dt} = F2 - F3 \quad (44)$$

where h1 and h2 represent the levels and A1 and A2 the areas. Table 3 presents the .B and .D01 files.

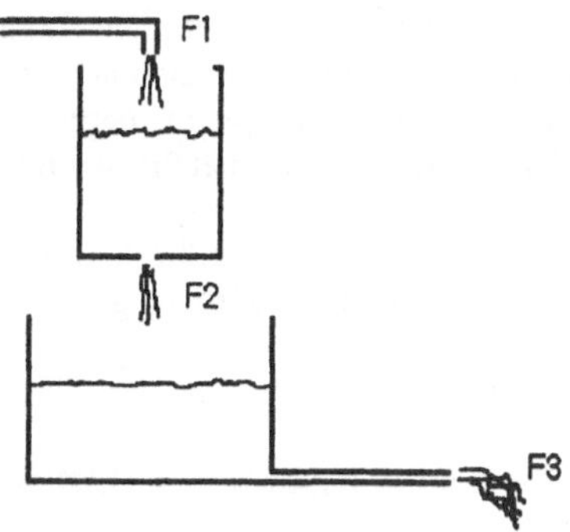

Fig. 6 - Coupled tank problem.

Table 3. ASCII files for Case Study #2 - Two tanks in series.

TWOTANK.B file

```
// Coupled Tank problem
// PP, SFA, 92.04
// Computer Simulation of Continuous Systems
// R. J. Ord-Smith & J. Stephenson
// Cambridge Computer Science Texts
// Cambridge University Press, 1975


p[4] = p[0] * sqrt(y[0]);          // F2
p[5] = p[1] * y[1];                // F3

f[0] = (u[0] - p[4])/p[2];         // h1
f[1] = (p[4] - p[5])/p[3];         // h2
```

TWOTANK.D01 file

```
C  Initial conditions - 2 tanks (serie)
C  N. of differential equations
   2
C  Name   Starting Value   Min    Max
   h1     0.               0.     1.
   h2     0.               0.     1.
C  N. of loads/disturbances
   1
C  Name   Starting Value
   F1     .05
C N. of control actions
   0
C N. of parameters
   6
C        Starting
C  Name  Value     Comments
   K1    .1        Discharge coefficient
   K2    .2        Viscous flow
   A1    1.        Tank 1 Area
   A2    2.        Tank 2 Area
   F2    0.        Tank 1 -> Tank 2 flow
   F3    0.        Tank 2 -> out flow
```

6. Final Comments

In our discussion we have favoured the use of deterministic, phenomenological models for process analysis. We have stressed the relation between model development and experiment, making clear that the validity of the model depends on its capacity to predict observed behaviour. The question of which type of model to adopt is open to alternatives when the objective is process control.

We also favoured analysis in time domain. This usually allows the unified treatment of linear and nonlinear systems. The simulator which we presented is a good example of the advantages of such approach. The kind of computational power which it requires is nowadays available in all Centers for Research and Production.

Finally, the paper stresses the role of interactive programming for the use of computers in process analysis and staff training.

References

Åström, K.J. and Wittenmark, B. (1984) 'Computer Controlled Systems-Theory and Design', Prentice Hall Inc., N.Y.

Åström, K.J. and Wittenmark, B. (1989) 'Adaptive Control', Addison-Wesley Pub. Comp., Massachusets.

Atherton, R.W., Schainker, R.B. and Ducot, E.R. (1975) 'On the Statistical Sensitivity Analysis of Models for Chemical Kinetics', AIChE J., 21, 441

Brebbia, C. A. (1985) 'Finite Element Systems - Handbook', 3rd Ed. Springher-Verlag.

Box, G.E.P. and Jenkins, G. (1971), 'Time Series Analysis, Forecasting and Control', Halden Day Inc., London.

Cadman, T. W. and Davison , S. W. (1989) 'BIOSIM-A General Purpose Tool for Simulation of Biological Fermentors'. ASEE CoED J. 9 (3), 32-37.

Carey, G.F. and Finlayson, B.A. (1975) 'Orthogonal Collocation on Finite Elements', Chem. Eng. Sci., 30, 587-596.

Chapra, S. and Canale, R. (1988) 'Numerical Methods for Engineers', 2nd Ed., Mc Graw Hill Inc., Appl. Math. Series, N.Y.

Clarke, D. W. (1991) 'Implementation of Adaptive Generalized Predictive Control', Intensive Training Course on Model Based Predictive Control, Automatic Control Lab., University of Ghent, Belgium, 7-8 Oct.

Clarke, D.W. and Scattolini, R. 'MERLINO: A Software Environment for Digital and Model Based Predictive Control', Proc. of the CIM-Europe Workshop on Computer Integrated Design of Controlled Industrial Systems, Ed. J. Richalet, Sp. Tzafestas, Paris 26-27 April 1990.

Cooney, C.L., Raju, G.K., O'Connor, G. (1991) 'Expert Systems and Neural Nets for Bioprocess Operation', Int. Symp on Bioproc. Modelling and Control, Ed.J.Morris, Newcastle, U.K., Jan 21-22.

Costa, C., Rodrigues, A. and Loureiro, J. (1986) 'Numerical Methods' in 'Ion Exchange: Science and Technology' Ed. A. Rodrigues, NATO ASI Series, Martinus Nijhoff Pub., USA.

Cuthrell, J.L. and Biegler L.T. (1987) 'On the Optimization of Differential-Algebraic Process Systems', AICh. J., 33, 1257.

de Boor, C. (1978) 'A Practical Guide to Splines', Appl. Math. Sci. Series, 27, Springer Verlag, N.Y.

de Keyser, R.M.C. 'The PC-TACT Software in Control Engineering', Proc. of the CIM-Europe Workshop on Computer Integrated Design of Controlled Industrial Systems, Ed. J. Richalet, Sp. Tzafestas, Paris 26-27 April 1990.

Dochain, D., Babary, J.P., Tali-Maamar, N. (1992) 'Modelling and Adaptative Control of Non-linear Distributed Parameter Bioreactors via Orthogonal Collocation', to be published in Automatica.

Dovi, V.G. and Paladino, O. (1989) 'Fitting of Experimental Data to Implicit Models using a Constrained Variation Algorithm', Computers and Chem. Eng., 13, 731.

Draper, N. and Smith, H. (1966) 'Applied Regression Analysis', J.Wiley, N.Y.

Elmqvist, H.; Åström, K. and Schonthal (1986) ' User's guide for MS-DOS computers - V1.0', Dep. of Automatic Control, Lun Institute of Technology.

Eppinger, A.; Kasper, R. and Heinkel, H.M. 'Hardware-in-the-Loop Design Techniques with ASCET (Advanced Simulation and Control Engineering Tool)', Proc. of the CIM-Europe Workshop on Computer Integrated Design of Controlled Industrial Systems, Ed. J. Richalet, Sp. Tzafestas, Paris 26-27 April 1990.

Ferguson, N.B. and Finlayson, B.A. (1970) 'Transient Chemical Reactor Analysis by Orthogonal Collocation', Chem Eng. J., 1, 327.

Feyo de Azevedo, S. and Wardle, A.P. (1989) 'Sensitivity Analysis Concerning the Design and Operation of a Tubular Fixed Bed Catalytic Reactor', Chem. Eng. Sci., 44, 2311-2322.

Feyo de Azevedo, S., Salcedo, R. and Gonçalves, M.J. (1988) 'Optimization Strategies in Chemical Engineering', In SIMO 88 - 'Simulation et optimization en Génie des Procédés', Ed. Domenech, S., Joulia, X., Koehret, B., Vol 2, n.6, p. 199, Lavoisier, Paris.

Feyo de Azevedo, S., Romero-Ogawa,M.A. and Wardle, A.P. (1990) 'Modelling of Tubular Fixed-bed Catalytic Reactors:A Brief Review', Trans IChemE, Vol.68, Part A, 483-502.

Finlayson, B.A. (1972) 'The Method of Weighted Residuals and Variational Principles', Academic Press, N.Y.

Finlayson, B.A. (1974) 'Orthogonal Collocation in Chemical Reaction Engineering', Cat. Rev.-Sci. Eng., 10, 69-139.

Finlayson, B.A. (1980) 'Non-linear Analysis in Chemical Engineering', McGraw-Hill Inc. N.Y.

Gear, C.W. (1971) 'Numerical Initial Value Problems in Ordinary Differential Equations', Prentice Hall, Englewood Cliffs, N.Y.

Georgakis, C., Aris, R., Amundson, R. (1977) 'Studies in the Control of Tubular Reactors I. General Considerations', Chem. Eng. Sci., 32, 1359-1369.

Gerald, C. (1977) 'Applied Numerical Analysis', Addison Wesley, Massachusetts.

Gill, P.E., Murray, W., Wright, M.H. (1981) 'Pratical Optimization', Academic Press, London.

Harmon, P. and Sawyer, B. (1990) 'Creating Expert Systems', J. Wiley, N.Y.

Hlavacek, V. (1979) 'Analysis of a Complet Plant Steady-state and Transient Behaviour', Computers and Chem. Eng., 1, 75.

Isaacs, S., Munack, A. and Thoma, M. (1986) 'Use of Orthogonal Collocation Aproximation for Parameters Identification and Control Optimization of a Distributed Parameter Biological System' , AIChe Meeting Miami Beach, USA.

Isaacs, S. and Thoma, M. (1992) 'The Adaptive Control of an Air-Lift Tower Loop Fermenter: An Application of Model-Based Control Employing a Sophisticated Process Model', Chem. Eng. Sci., 47, 943-958

Jorgensen, S.B. and Jensen, N. 'Dynamics and Control of Chemical Reactors - Selectively surveyed' (1989), Proc. of IFAC Symp. DYCORD+89, p.359, August 21-23, Maastricht, Netherlands.

Kosko, G. (1992) 'Neural Networks and Fuzzy Systems', Prentice-Hall Inc., New Jersey.

Lapidus, L. and Seinfeld, J.H. (1971) 'Numerical Solution of ODEs', Academic Press, N.Y.

Luttman, R., Munack, A. and Thoma, M. (1985) 'Mathematical Modelling, Parameter Identification and Adaptative Control of Single Cell Protein Processes in Tower Loop Bioreactors', Advances in Biochem. Eng., 32, 95-205, Spriger-Verlog.

Machietto, S. and Kessianides, S. (1992) 'An Integrated System for Computer Based Training of Process Operators', AIChE 1992 Spring Annual Meet , March 29 - April 2, paper 60f, LA, USA.

Mah, R.S.H. and Tamhane, A.C. (1982) 'Detection of Gross Errors in Process Data'.

MATLAB - Math Works Inc. (1992), Prentice Hall, Engl. Cliffs, N.Y.

Menezes, J. C., Feyo de Azevedo, S., Alves, S., Lemos, J. M. (1991) 'Modelling Penicillin-G Fermentation' (in Portuguese) Internal report, Departamento de Engenharia Química, Faculdade de Engenharia da Universidade do Porto.

Meyer, R.R and Roth, P.M. (1972) 'Modified Damped Least Squares, An Algorithm for Non-linear Estimation', J. Inst. Math. Applic., 94, 218

Michelsen, M.L. and Villadsen, J (1972) 'A Convenient Collocation Procedure for Collocation Constants', Chem. Eng. J., 4, 64-68.

Moller, C., Little, J. and Bangert, S. (1987) PC-MATLAB for MS-DOS personal computers - User's guide, The Math. Works Inc., Sherborn, USA.

Morbidelli, M., Slorti, G., Paludetto, R. and Carra, S. (1983) ' Application of the Orthogonal Collocation Method to some Chemical Engineering Problems', Ing. Chim. Ital., 19, 46-60.

Najim, K. and Muratet, G. (1987) 'Optimisation et Commande en Génie des Procédés', Masson, Paris.

Nash, J.C. and Walker-Smith (1987) 'Nonlinear Parameter Estimation, Statistics textbooks and Monographs Series V. 82', Marcel Dekker Inc., N.Y.

Pimenta, P and Feyo de Azevedo, S. (1992), "MIMOSA - User's Manual and Applications Reference Book", V 1.0 Internal Publication (in English), Departamento de Engenharia Química, Faculdade de Engenharia, Universidade do Porto.

Press, W.H., Flannery, B.P., Tentrolsky, S.A. and Vetterlung, W.T. (1989) 'Numerical Recipes', Cambridge Univ Press , Cambridge, U.K.

Priestley, A.J. and Agnew, J.B. (1975) 'Sensitivity Analysis in the Design of a Packed Bed Reactor', Ind. Eng. Chem. Proc. Des. Dev., 14, 1.

Rivera, S. and Karim, M.N. (1992) 'On-line State Estimation in Bioreactors using Recurrent Neural Networks', Proc. of the 2nd IFAC Symp. on Modelling and Control of Biotechnical Processes, Keystone, USA.

Salcedo, R., Gonçalves, M.J., Feyo de Azevedo, S. (1990) 'An Improved Random Search Algorihm for Non-linear Optimization', Computers Chem. Engng., 14, 1111-1126.

Schwetlick, H and Tiller, V. (1985) 'Numerical Methods for Estimating Parameters on Non-linear Models with Errors in the Variables', Technometries, 27, 17.

Seborg, D., Edgar, T., Melichamp, D. (1989) 'Process Dynamics and Control', J. Wiley, N.Y.

Shewchuk, C.F.; Stephenson, G.; Belletti, D. (1992) 'Innovation and Advances: The Evolution of TRAINER 2.0', AIChE 1992 Spring Annual Meet , March 29 - April 2, paper 60c, LA, USA.

Shinnar, R. (1978) 'Chemical Reactor Modelling - the Desirable and Achievable', Chem. React. Eng. Reviews - ACS Symp. Series, 72.

Spuhler, M.W. (1992) 'Use of Process specific Dynamic Simulators for Operator Training', AIChE 1992 Spring Annual Meet , March 29 - April 2, paper 60d, LA, USA.

Stephanopoulos, G. (1984) 'Chemical Process Control - An Introduction to Theory and Practice', Prentice Hall Inc., N.Y.

Tamhane, A.C. and Mah R.S.H. (1985) 'Data Reconciliation and Gross Error Detection in Chemical Process Networks', Technometrics, 27, 409-422.

Tjoa, I.B. and Biegler L.T. (1991a) 'Simultaneous Solution and Optimization Strategies for Parameter Estimation of Differential-Algebraic Equation Systems', Ind. Eng. Chem., Res 30, 376.

Tjoa, I.B. and Biegler L.T. (1991b) 'Simultaneous Strategies for Data Reconciliation and Gross Error Detection of Nonlinear Systems', Computers and Chem.Engng., 13

Villadsen, J. and Michelsen, M.L. (1978) 'Solution of Differential Equation Models by Polynomial Aproximation', Prentice Hall, N.J., USA.

Villadsen, J. and Sorensen, J.P. (1969) 'Solution of Parabolic Partial Differential Equations by a Double Collocation Method', Chem. Eng. Sci., 24, 1337-1349.

Young, L.C. and Finlayson, B.A. (1973) 'Axial Dispersion in Nonisothermal Packed Bed Chemical Reactors', Ind. Chem. Fund., 12, 421-422.

Acknowledgements - This work was partially supported by JNICT - *Junta Nacional de Investigação Científica e Tecnológica*, under contract numbers BD/224/90-IF and BD/1476/91-RM and INIC - *Instituto Nacional de Investigação Científica.*

COMPUTER-BASED STUDIES ON BIOPROCESS ENGINEERING

II - TOOLS FOR PROCESS OPERATION

S. FEYO DE AZEVEDO, P. PIMENTA, F. OLIVEIRA
Departamento de Engenharia Química
Faculdade de Engenharia da Universidade do Porto
Rua dos Bragas 4099 Porto Codex Portugal

E. C. FERREIRA
Departamento de Engenharia Biológica
Universidade do Minho
Largo do Paço 4719 Braga Codex Portugal

ABSTRACT

In this paper we review recent advances on the practice and theory of process control with particular emphasis to the operation of bioreactors. We present in detail a case-study on the modelling, model-based identification and adaptive control of fed-batch baker's yeast fermentation.

1. Practice and Theory of Process Control - An Overview

1.1. INTRODUCTION

After about thirty years of research and development digital systems and related methodologies have achieved the definite recognition of being vital tools for process operation, both in chemical and biochemical plants.

A kind of landmark in the history of process control can be recognised as having been reached at the close of the eighties. Nowadays, a large number of suppliers offer the capabilities of the digital technology, employing open architecture and standard operating systems and allowing external programming with high level languages.

For years researchers developed ideas and methods which as yet found little application in the industry. Much because these methods require computer and programming power which were not available in the industrial control equipment of the recent past.

A.R. Moreira and K.K. Wallace (eds.), Computer and Information Science Applications in Bioprocess Engineering, 27-49.

From the Smith predictor, through the state estimators to the more recent model based predictive control techniques, including also more and more improved PID controllers, methods can now be implemented at industrial scale.

The job for the future, and we shall assist to that, will be to incorporate new methodologies, some already existing and some still emerging, into those industrial equipment and thus bring such methodologies into operation.

The pace of progress depends on factors which go beyond research. It is worth to examine how the present stage has been reached in order to foresee the difficulties of the job ahead.

1.2. PROGRESS IN DIGITAL CONTROL TECHNOLOGY

The advancements in digital technology already had a most significant impact in the instrumentation and control industry.

Sensors and analytical instruments are the primary elements for process monitoring. Twork and Yacynych (1990) edited a review book on such equipment for bioprocess control. Sensors more and more employ digital technology, particularly for digital detection (e.g. - optical detection for refractometry-based sensors). Most analytical instruments have their performances improved by incorporating a microcomputer as a basic part of the equipment. Both sensors and instruments readily communicate with the control equipment via standard analogue signals or digital protocols. Yet, much research is necessary, particularly in the search for non-invasive sensors to operate on-line and in real-time. Just an example, reliable measurement of biomass has not yet been achieved.

In what concerns control systems, we have now reached the point where commercial, robust, reasonably simple to operate, computer-based control systems are available with the necessary characteristics to implement applications, on-line and in real-time. Most systems run a distributed philosophy with a central supervisory control computer linked via a digital communications data network to local modules. These can be analogue and digital input/output cards or local controllers (including programmable logic controllers) with their own I/O interfaces. A major improvement over older systems is on the large capacity to perform process monitoring (with reporting and alarm action) and to perform complex sequential control. Regulatory control is in most cases still limited to single input, single output (SISO) discrete PID, to discrete cascade control, to ratio control and to some limited feedforward control with lead-lag actions. This is a clear 'under utilisation' of the technology.

The 'step forward' which we observe in the existing commercial control systems is mainly related to the 'process control computer'. For some years the systems were practically closed to external programming in high level languages. Now, with the advances in hardware and operating systems, manufacturers are turning to open systems, employing standard hardware and software.

We find large scale applications of data acquisition and control typically based in VAX, SUN, HP or IBM stations (e. g., in oil refining, Hydrocarbon Processing, 1991). Many industrial systems are already based in microcomputers. This includes not only industrial data acquisition and monitoring systems, (e. g., IMPs from Schlumberger) but also process control systems (e.g. SCAP from SCAPE EUROPE, IA from Foxboro). SCAP runs under QNX and offers adaptive predictive control (Martin Sanchez, 1976; Martin-Sanchez *et al.*, 1984). The IA system allows programming in the C language at the 'supervisory' level.

The trend observed above applies as well to expert systems. Computer Scientists developed artificial intelligence in large part around the computer language LISP (List Processing) and more recently around PROLOG. Both process symbols and list symbols. Special purpose computer platforms ('symbolic' machines) dedicated to symbolic processing were developed assuming that this would bring superior performances. But reviewing the availability of expert systems shells (Stock, 1989; Harmon, 1990) one concludes that traditional computers and more common languages are being employed. Recent work on 'continuous process improvement through induction and analogical learning' (Saraiva and Stephanopoulos, 1992) is being developed in a traditional mainframe with FORTRAN.

1.3. PROGRESS IN PROCESS CONTROL THEORY

Years ago Foss (1973), Lee and Weekman (1976) and Kestenbaum *et al.* (1976) analysed critically the theoretical developments and trends in process control. There was at the time a recognised gap between theory and practice. Rightly, the authors generally pointed that it was up to the theory to jump and cover the gap.

The algorithms employed were based on models which hardly represented real chemical and biochemical processes. They lacked the required robustness for industrial application. And they required a computer power which was not available in control systems. Very important and pragmatic developments both in computer systems (as already seen) and in process control theory occurred since those days. The criticism of the past was largely taken into account. Methods and algorithms were developed which specifically had in mind the characteristics of real processes, viz - non-linear, often multivariable without the possibility of decoupling, ...

Bequette (1991) extensively reviews contributions for the non-linear control of chemical processes. Cadman (1991) similarly addresses the problem of bioreactor control and Lübbert (1991) produces a comprehensive description of automation in biotechnology.

A number of topics emerge on the front-line of concern -

(i) the requisites for the use of on-line techniques and for the use of packages for process monitoring
(ii) the role of mathematical models in bioengineering, for -
    - process analysis (discussed in part I)
    - monitoring the behaviour of internal variables that are impossible to measure on-line
    model based control
(iii) the problem of system identification both for stochastic and deterministic models
(iv) adaptive control, once again based on stochastic or deterministic modelling.

The problems concerning sensors and instrumentation in biotechnology are dealt with by Schügerl in his many papers (1988), including his lessons in this book. The required characteristics of packages for on-line monitoring will be discussed later in section 2.

Self tuning and adaptive control have been now for several years a reference for research and development. The current trend, which very much seems successful, is to couple a robust parameter and state estimation to a robust control algorithm.

Bastin, Dochain, Pomerleau and Perrier, published a string of papers describing model based software sensors, identification and adaptive control strategies for bioreactors (Dochain and Bastin, 1985; Bastin and Dochain, 1990; Dochain, 1991; Dochain *et al.* 1991; Dochain *et al.*, 1992; Pomerleau *et al.*, 1989; Pomerleau and Perrier, 1990; Pomerleau and Viel, 1992).

The general methodology on state and parameter estimation makes full use of the non-linear structures accepted for fermentation processes, with the particular feature that it avoids the need to know, a priori, kinetic parameters.

The concept of exact linearizing control is extensively detailed by Bastin and Dochain (1990) and will be presented in section 3 with reference to a case study with baker's yeast fermentation.

Model identification and adaptive control based on stochastic modelling of dynamical systems have as well received much attention of researchers. Beck and Young (1989) provide a very concise review on system identification and state and parameter estimation. The textbooks of Åström and Wittenmark (1984, 1989) and of Goodwin and Sin (1984) and the texts of Graupe (1987) and of Landau (1987) provide a background on concepts and theory.

In the last few years important progress has been achieved when, in independent developments, control strategies were proposed which couple methods of model based identification with the concept of controlling an a priori known process by making use of predictions of its future behaviour. Such methodology is known as Model Based Predictive Control (MBPC).

The first known version of MBPC is the adaptive predictive method (Martin Sanchez, 1976, 1986; Martin Sanchez *et al.*, 1984). Other versions are the extended-prediction self adaptive control (de Keyser and Van Canwenbergh, 1979), the generalised predictive control (Clarke *et al.*, 1987) and the MUSMAR (Mosca *et al.*, 1984). The theoretical basis, the practical advantages and the basic tools for building a MBPC strategy are clearly described by Richalet (1990), de Keyser (1990) and Clarke (1990, 1991).

Many successful industrial applications of MBPC are reported (de Keyser, 1990) and industrial control systems are now available which implement the methodology as a standard.

There are known draw-backs in the available theories and as such research goes on. An whole issue of Chemical Engineering Science (1992, Vol. 47, n. 4, 705-958) is dedicated to applications of model based control to chemical and biochemical engineering problems. Possibly the more relevant question is that MBPC requires a suitable process model. Such model must be able to represent adequately the process dynamics behaviour and yet must be simple enough not to preclude on-line identification. It often so happens that this is a difficult compromise to achieve.

Artificial neural networks represent alternative forms of representation of system dynamics which have been receiving much attention for the modelling of bioprocesses. The subject is discussed by Bungay (1992) in this book and is out of the scope of this paper. Yet a couple of comments are appropriate. Morris *et al.* (1991) apply neural nets to fed-batch penicillin fermentation and to continuous mycelial fermentation; Cooney *et al.* (1991) discuss the general application to bioprocess operation; Rivera and Karim (1992) tested neural network predictions in a bioreactor for the production of ethanol; Simutis *et al.* (1992) have recently studied the use of this method in predicting the substrate degradation during a production scale beer brewery fermentation. The conclusion of the latter application was that neural nets provide results which are similar to those obtained with the use of extended Kalman filters based on dynamic mathematical process models, but requiring significantly less development time.

Fuzzy reasoning linked to knowledge bases is another emerging methodology. Lübbert *et al.* (1992) have applied with excellent results a fuzzy supported extended Kalman filter in the description of the beer fermentation process. The textbook by Kosko (1992) presents in detailed neural networks and fuzzy theory from an unified engineering prespective.

There is today a strong interest on the application of real time knowledge based systems (RTKBS) to the process industries. The interest is as 'obvious' as that on 'adaptive' strategies. Essentially RTKBS are a form of memorising experience and using it in all possible ways - fault detection, emergency action, programmed operation, model switching, etc. In daily plant routine operators run their processes by experience. They adapt operation (reference points) manually when facing situations which are not programmed but which are not strange to their 'know-how'. Or they may well have written recipes of multiple set point profiles to implement in batch and fed-batch operation. Application of RTKBS at supervisory level in a computer based control system seems to be the future 'computerized' response for operation.

Stephanopoulos and Stephanopoulos (1986) discussed requirements of RTKBS for application to bioprocesses. Morris *et al.* (1991) and Aynsley *et al.* (1990) report results of a study on the production of penicillin. More recently, at the 5th International Conference on Computer Applications in Fermentation Technology a paper was given of the activity towards integrating the 'G2' (Gensym Corporation) real-time expert system into a process control strategy for the control of industrial fermentation (Fowler *et al.*, 1992). This paper is singled out for what it represents of a link between research and practice but it is accepted that artificial intelligence, expert systems and neural networks are somehow still futuristic methods for practical industrial control.

A closing reference on AI and neural networks. As for other subjects, there is today a news group dedicated to neural nets in the USENET worldwide communications network (comp.ai.neural-nets). It aims to be an open forum for all aspects of neural networks. Since it is user driven, both general and specific topics can be discussed.

1.4. WHAT ABOUT PID CONTROL ?

At an international process control conference in 1991 it was stated that 90% of industrial control requirements are still met by proportional-integral-derivative (PID) type controllers (Deshpande, 1992). Richalet (1990) wrote recently that 'an impressive majority of problems can be solved by simple and factual PI controllers'.

It is clear that the justification for the so-called modern algorithms has to be found on the relatively few but economically significant processes which are not so appropriately controlled by PID algorithms. This includes inverse-response, non-linear, integrating and multivariable processes.

Model based predictive control methods seem to have all the conditions to become future standards because they combine superior performance with reasonably accessible concepts to non-experts. But the simple to understand, simple to operate and reasonably efficient PID will stay in operation for many years. Looking back to its birth, fifty years ago, PID is probably the most successful concept of this century as far as process operation is of concern.

One should note that PID can well be coupled to identification stages and to model based software sensors.

For a number of years attempts have been made to improve the basic PID algorithms. Åström and Hagglund (1988) dedicate a book to summarise the state of the art of PI control and methods for the automatic tuning of such controllers. Autotuning is a concept related to adaptive control which was developed and put into commercial equipment. Examples of these are the Exact^{TM} system of Foxboro, the Statt Control Instruments Automation^{TM}, the Electromax V^{TM} from Leeds of Northamp and Turnball Controls Systems 6355 Autotuning Controller^{TM}.

Recently Omron launched a digital PID controller with internal feedforward for set point changes. Also they launched the first fuzzy feedback controller. Both types come with fixed sampling periods of 0.5 seconds which is reasonable for the vast majority of chemical and biochemical processes.

The interest of the search for improved PID action is confirmed by the work of Deshpande (1992) who proposes the Q-PID controller. It includes a lead-lag network and dead-time compensation. Digital controllers open the way to major improvements due to the flexibility of building-in new algorithms. In general, feedforward action is not taken as a standard but it normally improves significantly the behaviour of the feedback action. We believe that digital feedback-feedforward controllers (with two inputs and one output) based on the feedback PID concept would be a simple and effective control equipment.

1.5. STEPS IN SYSTEM DEVELOPMENT.

It should have emerged from this analysis that there are clear steps on the study and implementation of a control system, viz. -

i. Process monitoring - which includes data acquisition, alarms, safety on-off action and some basic data analysis.

ii. Data interpretation - which may be off-line or on-line parameter estimation and system identification.

iii. Optimisation and process control - where optimisation is usually carried out off-line, based on a deterministic representation of the process or the plant.

In the reminder of this paper we shall discuss some of our experience in industrial process monitoring and we shall present an application on model-based identification and control of a fed batch baker's yeast fermentation.

2. Process Monitoring

Process monitoring is the single application where computers can have a fast and significant impact in process operation. Production managers are usually more keen in knowing instantaneously what is happening in the plant than in having a sophisticated control.

Any medium and large scale commercial control system includes the tools for process monitoring but usually these systems are expensive and the decision on acquiring them is difficult to take.

There are available nowadays low cost hardware-software packages which aim primarily at process monitoring but also incorporate low level control capabilities. Such systems are PC-based, run under MS-DOS and have a distributed network of interface modules, with a protocol assuring communication through a single cable over distances above one kilometre.

Basic process monitoring should include scanning input, alarms and alarm action, historian report, graphical and spreadsheet-format display and finally storing data on disk. In practical terms the following general characteristics are expected from a process monitoring system:

i. Analog inputs - capability for scanning of current, voltage, resistance, temperature (thermocouple), temperature (RTD) and strain.
ii. Digital inputs - frequency, period, events and counts.
iii. Digital outputs - voltage free for alarm action and manual on-off control of external processes.
iv. Analog outputs - for limited regulatory control and for external communication.
v. System configuration - should display automatically all choices for each channel; should allow the set-up and saving of several alternative configurations for quick set-up of system operation.
vi. Headers - should allow reasonable identification of each channel.
vii. Conversions - should include linear and non-linear conversion. The latter should include polynomial, exponential (several types), hyperbolic and sinusoidal functions.
viii. Interchannel and individual calculations - for functions between input channels or for internal individual channel calculations such as averages, totalizers, maximums and minimums.
ix. Programmed timer - there would be the possibility of assigning different sampling and recording rates to individual or groups of channels.
x. Alarms - different types of alarm decisions based on level limits, rate of change limits or deviation limits. Alarms must be linked to digital output channels for on-off action.
xi. Display - a wide variety should be possible from spreadsheet-format to graphical display. The latter should include bargraph, trend plots and localized display in synoptic. Displays should be available in ´pages´.
xii. Synoptic - process drawing capabilities are required with definition of localized variable outputs.
xiii. Data recording and data printing - should be available by modules - process data, alarm and control action reports. It is particularly relevant that such data is recorded in a format directly compatible with spreadsheet packages.
xiv. Data communications - with other computers, using serial communication.

xv. Link with external subprograms - it is most convenient that software written by the user can be added to the package. With microcomputers this currently means C language and MS-Quick Basic. In the near future alternatives may be available.

Together with looking into the characteristics mentioned the potential buyer would do well in inquiring about a few more details such as - Is the package standard ? Is it available off-shelf ? For how many years is it in the market ? What are the limitationsyyh of hardware ? Is it fully PC-compatible ? Which requirements for graphical display ? What is the need and/or the flexibility of programming ? What is the available applications consultancy provided by the vendor and which technical support can we expect ? These two final questions are particularly relevant where the equipment comes from outside the buyer's home country.

3. A Case Study - Modelling, Identification and Control of Baker's Yeast Fermentation

We shall discuss in this section concepts on identification and control based in structured phenomenological models, with direct application to baker´s yeast fermentation. The basic theory in described in detail by Bastin and Dochain (1990) and by Pomerleau and Perrier (1990) and Pomerleau and Viel (1992). The simulator described in Part I (Feyo de Azevedo *et al.*, 1992b) is employed for tests concerning the identification algorithms and their robustness to noisy measurements (Feyo de Azevedo *et al.*, 1992a).

3.1. PROCESS MODELLING

The dynamical model for the fed-batch fermentor is obtained from a mass balance on the components, considering that the reactor is well mixed, the yield coefficients are constant and the dynamics of the gas phase can be neglected. The kinetic model proposed by Sonnleitner and Käppeli (1986) is employed.

3.1.1. Kinetic model. Yeast growth is characterized by three metabolic pathways, viz-

$$S + C \xrightarrow{\mu_s^o} X + G \qquad \text{(respiratory growth on glucose)} \qquad (1)$$

$$S \xrightarrow{\mu_s^r} X + E + G \qquad \text{(fermentative growth on glucose)} \qquad (2)$$

$$E + C \xrightarrow{\mu_e^o} X + G \qquad \text{(respiratory growth on ethanol)} \qquad (3)$$

with S: glucose; C: oxygen; X: biomass; E: ethanol; G: CO_2 and μ_s^o, μ_s^r, μ_e^o: specific growth rates for the three pathways. In the sequel X, S, E, C, G mean concentrations.

The metabolic pathways of oxidative growth on glucose and ethanol are governed by the respiratory capacity of the cells. If glucose is present in low concentrations and there is enough oxygen in the medium, only the oxidative metabolism takes place, being the glucose a preferable substrate rather than ethanol. On the other hand, if the glucose flow exceeds the respiratory bottleneck, part of it is catabolized oxidatively and the rest follows the fermentative catabolism, producing ethanol. In this case no oxidative growth on ethanol takes place.

The kinetics equations for baker's yeast growth (equations (1)-(3)), considered as Monod equations, are determined as follows -

i. the total specific growth rate, μ_t, is the sum of the growth rates for the three pathways -

$$\mu_t = \mu_s^o + \mu_s^r + \mu_e^o \tag{4}$$

where μ_i can be related to the corresponding substrate fluxes, q, and yield coefficients, Y, viz -

$$\mu_t = - Y_{x/s}^o q_s^o - Y_{x/s}^r q_s^r - Y_{x/e}^{oe} q_e^o \tag{5}$$

ii. as ethanol uptake is influenced by the priority of glucose uptake, which functions as an inhibitor, the specific growth rate on ethanol can be described as -

$$\mu_e^o = \mu_{e,max} \frac{E}{E+K_e} \frac{K_i}{S+K_i} \tag{6}$$

where $\mu_{e,max}$ is the maximal specific growth rate, K_i is the inhibition parameter, K_e is the saturation parameter.

However, this equations holds true only if there is an available respiratory capacity of the cells.

iii. the glucose uptake, q_s,is sligthly different because it follows two metabolic pathways: oxidative and fermentative -

$$q_s = q_s^o + q_s^r \tag{7}$$

iv. the total flux may be described by the Monod kinetics equation -

$$q_s = q_{s,max} \frac{S}{S+K_S} \tag{8}$$

where $q_{s,max}$ is the maximal specific glucose uptake rate and K_s is the saturation parameter.

The oxidative glucose uptake depends on the availability of dissolved oxygen, and may be defined as -

$$q_s^o = \frac{q_{cs}}{a} \tag{9}$$

where a is the stoichiometric coefficient of the oxygen in the respiratory pathway of glucose and $q_{c,s}$ is the oxygen uptake on glucose. This last parameter also follows a Monod kinetics -

$$q_{c,s} = q_{c,s,max} \frac{C}{C+K_C} \tag{10}$$

with -

$$q_{c,s,max} = \min (q_{c,max}, aq_S) \tag{11}$$

being $q_{c,max}$ the maximal specific oxygen uptake rate.

3.1.2. Fermentor dynamic model. The mass balances, in terms of concentration, are written as -

$$\frac{dX}{dt} = (\mu_s^o + \mu_s^r + \mu_e^o - D)\,X \tag{12}$$

$$\frac{dS}{dt} = D\,(S_{in} - S) + \left(-\frac{\mu_s^o}{Y_{x/s}^o} - \frac{\mu_s^r}{Y_{x/s}^r}\right) X \tag{13}$$

$$\frac{dE}{dt} = -D\,E + \left(\frac{\mu_s^r}{Y_{x/e}^r} - \frac{\mu_e^o}{Y_{x/e}^{oe}}\right) X \tag{14}$$

$$\frac{dC}{dt} = -D\,C + OTR + \left(-\frac{\mu_s^o}{Y_{x/c}^o} - \frac{\mu_e^o}{Y_{x/c}^{oe}}\right) X \tag{15}$$

$$\frac{dG}{dt} = -D\,G - CTR + \left(\frac{\mu_s^o}{Y_{x/g}^o} + \frac{\mu_s^r}{Y_{x/g}^r} + \frac{\mu_e^o}{Y_{x/g}^{oe}}\right) X \tag{16}$$

and the additional equation -

$$\frac{dV}{dt} = F = D\,V \tag{17}$$

In the former - D is the dilution rate and the k_i (i=1 to 9) are the yield coefficients; S_{in} is the substrate concentration in the feed; OTR is the oxygen transfer rate (defined as OTR = K_La (C^*-C) where K_La is the mass transfer coefficient and C^* is the equilibrium concentration of dissolved oxygen) and CTR is the carbon dioxide transfer rate. The dynamics of CO_2 may be assumed as being very fast relatively to the dynamics of the other components. Also, the concentration of carbon dioxide at the pH of operation is known to be low. Hence the balance equation of CO_2 can be simplified by assuming dG/dt = 0 and G = 0. Defining Q_{CO2} as the gas outflow of CO_2, this leads to -

$$Q_{CO2} = CTR = (K_7\mu_s^o + K_8\mu_s^r + K_9\mu_e^o)X \tag{18}$$

The set of equations above has a structure which can be represented by the general dynamic model proposed by Bastin and Dochain (1990) in the following matrix form:

$$\frac{d\xi}{dt} = K\varphi(\xi) - D\xi + F - Q(\xi) \tag{19}$$

where the notation emphasize that φ and Q may be time-varying and may depend on the process state ξ.

For the present case equations (12)-(16) take the matrix form:

$$\frac{d}{dt}\begin{pmatrix} X \\ S \\ E \\ C \\ G \end{pmatrix} = \begin{pmatrix} 1 & 1 & 1 \\ -k_1 & -k_2 & 0 \\ 0 & k_3 & -k_4 \\ -k_5 & 0 & -k_6 \\ k_7 & k_8 & k_9 \end{pmatrix} \begin{pmatrix} \mu_S^o \\ \mu_S^r \\ \mu_e^o \end{pmatrix} X - D\begin{pmatrix} X \\ S \\ E \\ C \\ G \end{pmatrix} + \begin{pmatrix} 0 \\ DS_{in} \\ 0 \\ OTR \\ 0 \end{pmatrix} - \begin{pmatrix} 0 \\ 0 \\ 0 \\ 0 \\ CTR \end{pmatrix} \tag{20}$$

The relevant kinetic data were taken from Sonnleitner and Käppeli (1986) (see table 1). The yield coefficients proposed by Pomerleau and Perrier (1990) were employed (see table 2).

Table 1. Kinetic parameters

Parameter	Value
$q_{s,max}$	3.5 g gluc^{-1} g biom^{-1}h^{-1}
$q_{c,max}$	0.256 g O_2^{-1} g biom^{-1}h^{-1}
$\mu_{e,max}$	0.17 h^{-1}
K_e	0.1 gl^{-1}
K_i	0.1 gl^{-1}
K_s	0.2 gl^{-1}
K_c	0.1 mgl^{-1}

Table 2. Yield coefficient values

Coefficient	Value
$Y^{o}_{x/s}$	0.49 g biom g gluc^{-1}
$Y^{r}_{x/s}$	0.05 g biom g gluc^{-1}
$Y^{r}_{x/e}$	0.10 g biom g eth^{-1}
$Y^{oe}_{x/e}$	0.72 g biom g eth^{-1}
$Y^{o}_{x/c}$	1.20 g biom g O_2^{-1}
$Y^{oe}_{x/c}$	0.64 g biom g O_2^{-1}
$Y^{o}_{x/g}$	0.81 g biom g CO_2^{-1}
$Y^{r}_{x/g}$	0.11 g biom g CO_2^{-1}
$Y^{oe}_{x/g}$	1.11 g biom g CO_2^{-1}

3.1.3. Simulation with MIMOSA. Process simulation with MIMOSA (Feyo de Azevedo *et al.*, 1992b, Part I) is implemented by writing two files, *viz* - (i) BAKER1.B file where the right hand side of the model equations are defined (Table 3); (ii) BAKER1.D01 file - data file where variable and initial values are defined (Table 4).

Simulation is available in 'real-time', in 'scaled-time' and in 'simulation-time'.

During operation, loads to the process can be generated through the keyboard of the process computer. Also, the declared time-varying parameters (including delays and measurement noises) can be changed on-line and in real time. Finally, the appropriate control actions (if it is the case) generated by an external controller (computer or standard industrial system) can be sent to the process.

Figure 1 shows a run after 21.10 hours. In the upper half of the screen a graphics facility emulating a 4-pen register is available. In the lower half all main variables are monitored and made available for on-line change. In particular it should be noted that:

i. All system state variables y[ ] can be monitored and all are eligible as output variables with a superimposed 'user-defined' random noise;
ii. All parameters named as p[ ] in the model equations can be monitored and their values changed on-line. This includes the noises and dead-times defined in a data file, which MIMOSA automatically will add to the list of parameters and as such will make available for on-line change.

Table 3. BAKER.B file.

```
/*  Modelling baker's yeast.
    Universidade do Minho, Engenharia Biológica
    Gabinete de Sistemas (CEQ) do DEQ/FEUP
        PORTO   PORTUGAL, 92.01.07
    Eugénio Campos Ferreira
    Filomena Rocha Oliveira
    Pedro Pimenta                    */

//  Checking ranges
if (y[1] <.0) y[1] =    .0;      //[Substrate]
if (y[2] <.0) y[2] =    .0;      //[Ethanol]
if (y[3] > p[23])y[3]=p[23];     //[O2diss]<C*

//  Kinetic model.
p[25] = p[6] * y[1]/(p[9]+y[1]);
if ( (p[21] * p[25]) <= p[7])
        p[28] = p[25]*y[3]/(y[3]+p[26]);
else    p[28] = p[7]/p[21]*y[3]/(y[3]+p[26]);

p[1] = p[15] * p[28];
p[2]=p[8]*y[2]/(p[10]+y[2])*p[20]/(p[20]+y[1]);
p[3]=p[18]*( p[25]-p[28] );
p[5]=(p[1]/p[16]+p[2]/p[17])*y[0];
p[4]=(p[1]/p[13]+p[3]/p[14]+p[2]/p[12])* y[0];
p[0] = p[4]/p[5];                        //RQ

//Model of transient state
f[0]=(p[1]+p[2]+p[3]-c[0])*y[0];
f[1]=c[0]*(u[0]-y[1])-
    p[1]/p[15]+p[3]/p[18])*y[0];

f[2]=-c[0]*y[2]+(-p[2]/p[11]+p[3]/p[19])*y[0];
f[3]=-c[0]*y[3]+(p[22]*(p[23]-y[3]))-p[5];
f[4] = c[0] * y[4];

if( y[1] <= .0)           //[substrate]> 0.
{y[1] = .0; if( f[1] < .0) f[1] = .0;}

if( y[2] <= .0)           //[ethanol] > 0.
{y[2] = .0; if( f[2] < .0) f[2] = .0;}

if( y[3] <= .0)           //[oxygen] > 0.
{y[3] = .0; if( f[3] < .0) f[3] = .0;}

if( y[3]>=p[23])          //[O2] < C*
{y[3] = p[23]; if (f[3]>.0) f[3]=.0;}

f[0] = f[0] * p[30];      // Scaling in time...
f[1] = f[1] * p[30];
f[2] *= p[30];
f[3] *= p[30];
f[4] *= p[30];

// Stop conditions
if (y[4] >= p[27]){c[0]=f[4]=0.;p[29]=1.;}

if (p[29]==1.)       // Stationary state
{   p[24] = fabs(f[0])+
      fabs(f[1]) + fabs(f[2]) + fabs(f[3]);
    if (p[24] < 5e-5) fim(0, "STAT");
}
```

Table 4. BAKER.D01 file.

```
C Initial Conditions for BAKER1
C No. State variables
  5
C              Initial           Variancies  Delays
C Name         Value  Min  Max   (noise)    (sec/100)
  Biomass      0.1    0.   3.     .1         0.
  Glucose      .8     0.   5.     .1
  Ethanol      .0     0.   2.     .1
  O2diss       .0066  0.   .007   5e-4       0.
  Volume       3.5
C No. Loads
  1
C Name         Initial Value
  Sin             10.
C No. control actions
  1
C Name         Starting Value
  D               .03
C No. of parameters
  31
C              Initial
C Name         Value    Comments
  RQ           1.0      p0
  μo           0.3      p1
  μe           0.       p2
  μr           .08      p3
  CER          0.0      p4
  OUR          0.0      p5
  qsmax        3.5      p6
  qO2max       .256     p7
  μemax        .17      p8
  Ks           .2       p9
  Ke           .1       p10
  Ye           .72      p11
  Yge          1.11     p12
  Ygo          .81      p13
  Ygr          .11      p14
  Yo           .49      p15
  Yo2          1.2      p16
  Yo2e         .64      p17
  Yr           .05      p18
  Yre          .1       p19
  Ki           .1       p20
  a            0.4142   p21
  Kla          100.     p22
  C*           0.007    p23
  SumDer       0.00     p24
  qs           0.0   m  p25
  Ko           .0001    p26
  Vmax         10.      p27
  qso          0.0      p28
  VMax?        .0       p29 - Flag for max. Volume
  TFact        .1       p30
```

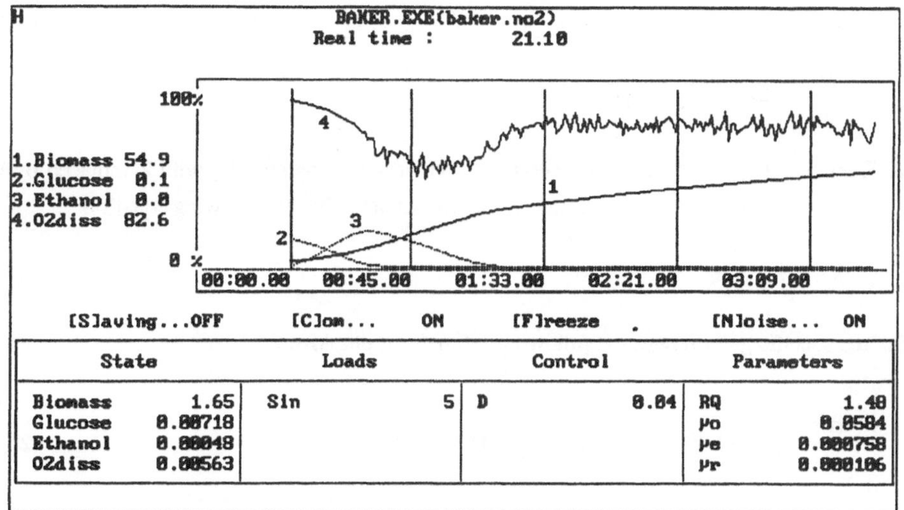

Fig. 1. Process simulation with MIMOSA.

3.2. THE OBSERVER ALGORITHM

We propose to develop an algorithm which will allow state and kinetic parameter estimation from a limited number of measurements, assuming that the yield coefficient are known.

Eq. (19) can be written as:

$$\frac{d\xi}{dt} = K\mu X - D\xi + U \tag{21}$$

in which: ξ is the vector of the bioprocess components ($\dim(\xi) = N$); μ is the specific reaction rate vector ($\dim(\mu) = M$); K is the yield coefficients matrix ($\dim(K) = N \times M$); U is the feed and the gaseous outflow rates vector ($\dim(U) = N$).

This state space representation can be divided in two partitions: the first one includes the equations related to the measured state variables (ξ_1); the second partition, the equations related to the nonmeasured state variables (ξ_2). The dynamic model is rewritten as:

$$\frac{d\xi_1}{dt} = K_1\mu X - D\xi_1 + U_1 \tag{22}$$

$$\frac{d\xi_2}{dt} = K_2\mu X - D\xi_2 + U_2 \tag{23}$$

where K_1 (a full rank matrix), K_2, U_1, U_2 correspond to the division of K and U according to each partition. A transformation Z is applied to the nonmeasured state variables partition, giving:

$$Z = \xi_2 - K_2 K_1^{-1} \xi_1 \tag{24}$$

$$\frac{dZ}{dt} = -DZ + U_2 - K_2 K_1^{-1} U_1 \tag{25}$$

A "Luenberger-type" asymptotic observer can be written using eq. (25) with Z replaced by its estimate $\hat{Z}$ and the non-measured state variables estimated by the following equation:

$$\hat{\xi}_2 = \hat{Z} + K_2 K_1^{-1} \xi_1 \tag{26}$$

For the specific growth rates the following estimator is employed:

$$\frac{d\hat{\psi}}{dt} = \hat{\mu} X - D\psi + K_1^{-1} U_1 + \omega(\psi - \hat{\psi}) X \tag{27}$$

$$\frac{d\hat{\mu}}{dt} = \gamma(\psi - \hat{\psi}) X \tag{28}$$

with $\psi = K_1^{-1}\xi_1$, a transformation to decouple the equations with respect to the specific growth rate; ω, γ are diagonal matrices containing time varying tuning parameters updated by a pole placement procedure (Pomerleau and Perrier, 1990). A discrete version of the estimator algorithms was implemented by first order Euler approximation with a sampling period of 6 minutes.

Two case studies illustrate the application. In both, the initial conditions were -

X(0) = 0.1 g/l, S(0) = 0.8 g/l, E(0) = 0 g/l, C(0) = 0.0066 g/l, V(0) = 3.5 l, $S_{in}(0)$ = 5 g/l.

The value for K_La was assumed as 60 hr^{-1}.

In the first a constant dilution rate of 0.04 hr^{-1} is assumed. The final volume of 10 L is taken as criterion to stop the feeding. Figure 1 represents the process simulation with MIMOSA.

Glucose, dissolved oxygen and ethanol were the measured variables available to the 'observer'. After about 3 hours a white noise of approximately 5% was imposed on the measurement of oxygen. The dotted lines in Fig. 2 represent the biomass and the flux of CO_2 estimated. The continuous lines represent the 'true' values obtained with the simulator. The reconstruction of the relevant specific growth rates is shown in Fig. 3. The influence of noise is visible in the figures, but it is well dumped by the filter.

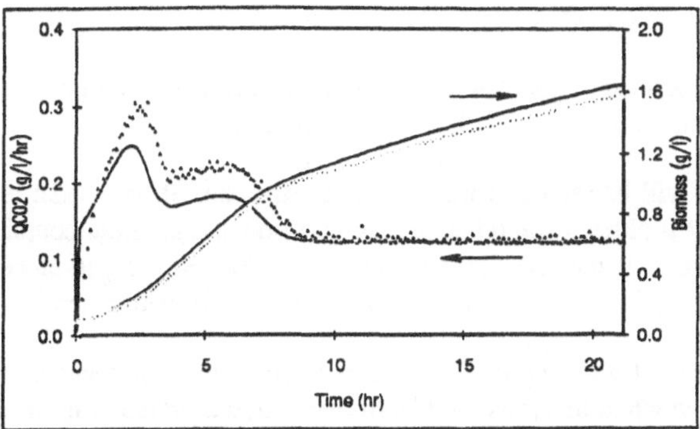

Fig. 2. 'Observed' - (dotted lines) vs. 'true' state properties (continuous line).

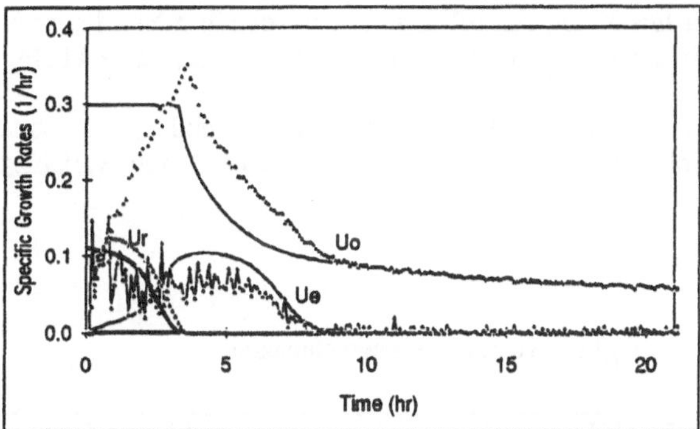

Fig. 3 'Observed' - (dotted lines) vs. 'true' specific growth rates (continuous line).

In the second case study an exponential evolution of the dilution rate [D(t)=0.03 exp(0.1t)] was imposed in MIMOSA. A white noise with variance 0.01 was considered in the measurement of ethanol. Figure 4 shows the observed (dotted lines) and the 'true' values of biomass, glucose and CO_2 outflow obtained from single measurements of dissolved O_2 and ethanol. Here, a simplification had to be performed, having neglected the oxidative specific growth rate on ethanol, μ_e. The 'Observer' is not sensitive to the noise imposed and performs well.

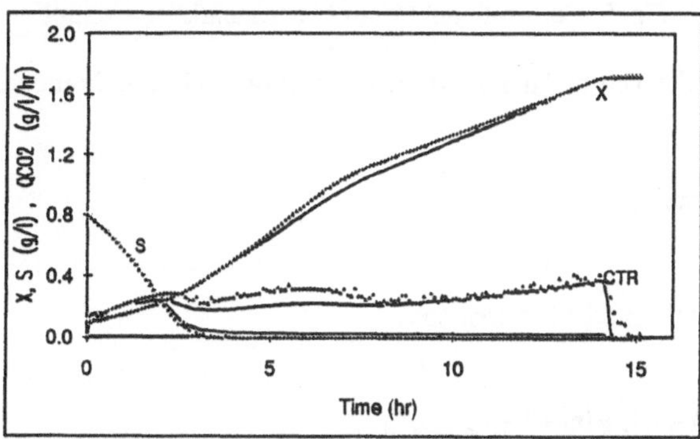

Fig. 4. Estimation of biomass, glucose and outflow of CO_2 from measurements of ethanol and dissolved oxygen.

3.3. EXACT LINEARIZING CONTROL

We shall discuss now the application of linearizing control to bioprocesses. The subject is proposed and extensively detailed by Bastin and Dochain (1990).

In what follows it will be seen that the non-linear structure of the model is fully exploited in solving the control problem. The other remark concerns the possible coupling of this strategy with an observer such as the one described in the previous section, in those cases where state variables are not directly available from measurements and/or parameters are unknown.

3.3.1. Brief theoretical description. The difference between this technique and conventional control lies in the way that linearization is introduced in the problem. In the standard approach, (like a PID control or MV control), we first calculates a linearized approximation of the model, and then we design a 'linear controller' for this approximate model. But the closed loop remains non linear. In exact linearizing control approach we obtain a 'non linear controller' which is precisely designed to achieve a 'linear closed loop'. This is summarised in Fig. 5.

Considering the fermentation process described by the general dynamic model (eq. (19)). The objective is to control a scalar output linear combination of the state variables:

$$y = \sum_{i=1}^{N} C_i \xi_i = C^T \xi \tag{29}$$

where $C^T = [C_1, C_2, ..., C_N]$ is a vector of known constants.

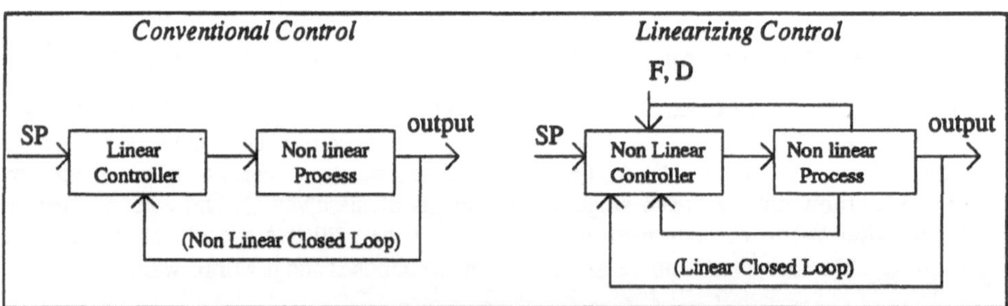

Fig. 5: Conventional Control vs. Linearizing Control

The control input (denoted by "u") is the feed rate of one substrate introduced to the process, from the outside.

$$u = F_i \quad \text{for some i} \tag{30a}$$

$$F = bu + f \tag{30b}$$

With this definition the model is rewritten as

$$\frac{d\xi}{dt} = K\varphi(\xi) - D\xi + bu + f - Q \tag{31}$$

It will be assumed that f and Q are measured on-line and that ξ is known on-line either by measurement or by an asymptotic observer.

The control objective is to track a reference output signal denoted y*(t) (set point is constant in a regulation problem).

The principle of linearization control is to find a control law u(ξ, Q, f, y*) which is a multivariable non linear function of ξ, Q, f e y* such that the tracking error e = (y*-y) is governed by a pre specified stable linear differential equation called a reference model.

Linearizing control design is a three-step procedure:

Step 1: to derive an Input/Output Model (I/O model) by successive differentiation of the general dynamic model:

$$\frac{d^{\delta}y}{dt^{\delta}} = f_0(t) + u(t)f_1(t) \tag{32}$$

where δ the order of the differential equation is called the relative degree.

Note that the I/O model is linear with respect to the control input u(t).

Step 2: A Stable Linear Reference Model of the tracking error is selected as follows:

$$\sum_{i=0}^{\delta} \lambda_{\delta-j} \frac{d^j}{dt^j}\left[y^*(t) - y(t)\right] = 0 \qquad (\lambda_0 = 1) \tag{33}$$

The coefficients $\lambda_{\delta\text{-}j}$ are chosen so that the differential equation is stable.

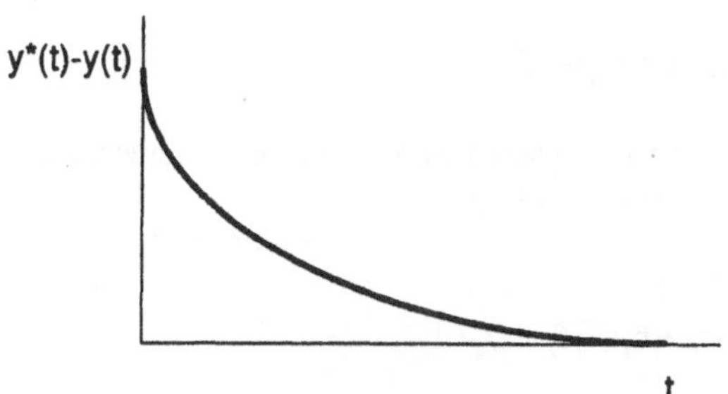

Fig. 6. Tracking error

Step 3: Control Design

To compute the control action u(t) such that the I/O model exactly matches the Reference Model

$$u(t) = \frac{1}{f_1(t)}\left[-f_o(t) + \sum_{j=0}^{\delta-1} \lambda_{\delta-j} \frac{d^j}{dt^j}[y*(t) - y(t)] + \frac{d^{\delta}y*}{dt^{\delta}}\right] \tag{34}$$

3.3.2. Application to Ethanol Regulation in a Yeast Fermentation Process. In the baker's yeast fermentation, the goal is to prevent the fed-batch reactor from ethanol (E) accumulation which is a source of yield decrease and may affect the productivity. Glucose feed rate is chosen as the control action.

$$y = E = C^T\xi \quad \text{with} \quad C^T = [0, 0, 1, 0, 0] \tag{35a}$$

$$u = F_1 = DS_{in} \quad \text{with} \quad b^T = [0, 1, 0, 0, 0] \text{ and } f^T = [0, 0, 0, OTR, 0] \tag{35b}$$

The control problem is solved by the 'Singular Perturbation Technique', which is justified by the following fact - *it is legitimate to assume that the kinetics of some substrates and products are much faster then some limiting ones.* The Singular Perturbation Technique is precisely a method to neglect fast kinetics and dynamics in ordinary differential equations in order to obtain a reduction of the dynamic model.

The following set of "fast" state variables is selected:

$$\xi_f^T = [S, C, G] \quad (M=3) \tag{36}$$

and for "slow" state variables

$$\xi_s^T = [X, E] \tag{37}$$

$$\text{with } \xi^T = [\xi_s^T, \xi_f^T] \tag{38}$$

It may be noticed that ethanol concentration E, which is the state variable to be regulated, does not belongs to ξ_f:

$$E = C_s^T \xi_s \quad \text{with} \quad C_s^T = [0, 1] \tag{39}$$

The general dynamic model is reduced to a set of M (=3) algebraic equations and N (=2) ordinary differential equations as follows:

$$\frac{d\xi_s}{dt} = K_s\varphi - D\xi_s + b_s u + f_s - Q_s \tag{40a}$$

$$K_f\varphi + b_f u + f_f - Q_f = 0 \tag{40b}$$

with

$$K_s = \begin{bmatrix} 1 & 1 & 1 \\ 0 & k_3 & -k_4 \end{bmatrix}, \quad b_s^T = [0, 0], \quad f_s^T = [0, 0], \quad Q_s^T = [0, 0] \tag{41a}$$

$$K_f = \begin{bmatrix} -k_1 & -k_2 & 0 \\ -k_5 & 0 & -k_6 \\ k_7 & k_8 & k_9 \end{bmatrix}, \quad b_f^T = [1,\ 0,\ 0], \quad f_f^T = [0,\ 0,\ 0], \quad Q_f^T = [0, OTR, -CTR]$$

(41b)

Since K_f is a full rank matrix, the vector $\varphi(\xi)$ of growth rates can be written as

$$\varphi(\xi) = K_f^{-1}(Q_f - F_f) = K_f^{-1}(Q_f - b_f u - f_f) \tag{42}$$

with the inverse of K_f computed as:

$$K_f^{-1} = k_2 k_6 \theta \begin{bmatrix} k_6 k_8 & k_2 k_9 & k_2 k_6 \\ (k_6 k_7 - k_5 k_9) & -k_1 k_9 & -k_1 k_6 \\ -k_5 k_8 & (k_1 k_8 - k_2 k_7 & -k_2 k_5 \end{bmatrix} \tag{43}$$

$$\text{where } \theta = \left(k_7 - \frac{k_5 k_9}{k_6} - \frac{k_1 k_8}{k_2} \right)^{-1} \tag{44}$$

Substituting (42) into (40) gives

$$\frac{d\xi_s}{dt} = -D\xi_s + \left[I_{N-M} - K_s K_f^{-1} \right](F - Q) \tag{45}$$

The dynamics of Y (=E) is written as

$$\frac{dY}{dt} = -DY + C_s^T \left[I_{N-M} - K_s K_f^{-1} \right](F - Q) \tag{46}$$

or

$$\frac{dY}{dt} = -DY + C_s^T \left[I_{N-M} - K_s K_f^{-1} \right] bu + C_s^T \left[I_{N-M} - K_s K_f^{-1} \right](f - Q) \tag{47}$$

which leads to the following fully reduced input/output model:

$$\frac{dY}{dt} = -DY + \theta_1 (CTR) + \theta_2 (OTR) + \theta_3 (DS_{in}) \tag{48a}$$

with

$$\theta_1 = \theta\vartheta, \quad \theta_2 = \frac{k_9}{k_6}\theta - \frac{k_4}{k_6}, \quad \theta_3 = \frac{k_8}{k_2}\theta + \frac{k_3}{k_2} \tag{48b}$$

where $\vartheta = \frac{k_1 k_3}{k_2} - \frac{k_4 k_5}{k_6}$ (48c)

Selecting a first order reference model for the tracking error

$$\frac{d}{dt}(y^* - y) + \lambda_1 (y^* - y) = 0 \quad (49)$$

the linearizing control law is readily obtained by substituting (22a) into (23) as follows

$$F_1 (= DS_{in}) = \theta_3^{-1}\{\lambda_1 (E^* - E) + DE - \theta_1 (CTR) - \theta_2 (OTR)\} \quad (50)$$

It is worth noting that the regulation law

i) does not need on-line measurement of the process compounds except that which is regulated (E);
ii) does not require the knowledge of the process kinetics;
iii) makes use of measurement of gaseous outflow rates which are, most often easily accessible on-line;
iv) involves a feedforward compensation of the influent substrate concentration S_{in} (if control action is the dilution rate).

If the yield coefficients which are present in θ_1, θ_2, θ_3 are either badly known or time-varying, an adaptive version of the controller must be implemented. The Adaptive Regulation law is obtained by using eq. (50) but with an on-line parameter estimation of θ_1, θ_2 and θ_3.

4. Closing Comments

Societies are periodically shaken by uncontrolled events which turn clear that 'quality and efficiency' must be in the first line of concern of any production policy, whatever other details are imposed by the market and social forces.

Twenty years ago the oil and energy crises cried for efficiency. Throughout the last decade individual demand and market competition pushed towards product diversity. The more recent collective concern for environmental and health problems is leading to policies of tight pollution control and monitoring of product specification. Labor costs keep increasing, predictably but steadily. The production policies for the next future will only be effectively and economically conducted through the use of computer-based technology and methods for process monitoring and control.

Today we have available tools and theories adapted to the practical application. We have efficient systems for process monitoring. We can implement improved PID-based algorithms and we reached the point where advanced model-based adaptive control algorithms can be used with robust industrial control systems. Other techniques, aiming at employing human experience in control decisions are actively being developed.

The motivation seems strong but we have to be fully aware that there are human, economical and technical difficulties in introducing new plant procedures, viz. -

Computer control systems are expensive and they generate significant running costs, particularly concerning specialised personnel. Also, there is a lack of trained engineers in this field. This often results in a lack of internal motivation to promote changes and in inefficient use of systems already available in the plant. Another problem is that processes often change, leading to the need of modifying the software. Human resources have to be available. In all, there must be a decision on organizational changes to accomodate systems and applications engineers who will become part of the operating personnel.

The pace of changes is the net result of all these factors. But computers are definetely into plant operation.

References

Åström, K.J. and Wittenmark, B. (1984) 'Computer Controlled Systems-Theory and Design', Prentice Hall Inc., N.Y.

Åström, K.J. and Wittenmark, B. (1989) 'Adaptive Control', Addison-Wesley Pub. Comp., Massachusets.

Åström, K.J. and Hagglund, T. (1988) 'Automatic Tunning of PID Controllers', Instruments Society of America, USA.

Aynsley, M.; Hofland, A.G.; Montague, G.A.; Peel, D. and Morris, A.J. (1990) 'A Real Time Knowledge Based System for the Operation and Control of a Fermentation Plant', Proc.ACC, San Diego, USA, pp.1992-1997.

Bastin, G., Dochain, D. (1990) 'On-Line Estimation and Adaptive Control of Bioreactors', Elsevier, Amsterdam.

Beck, M.B. and Young, P.C. (1988) 'An Introduction to System Identification, Parameter and State Estimation', Computer Application in Fermentation Technology: Modelling and Control of Biotechnological Processes, Edts. Fish, N.M.; Fox, R.I. and Thornhill N. F., Cambridge, U.K., 25-29 Sep.

Bequette, B.W. (1991) 'Non-linear Control of Chemical Processes: A Review', Ind. Eng. Chem. Res., 30, 1391-1413.

Box, G.E.P. and Jenkins, G. (1971), 'Time Series Analysis, Forecasting and Control', Halden Day Inc., London.

Bungay, H.R. (1992) 'Introduction to Artificial Intelligence', In NATO ASI on the Use of Computers and Informatic Systems in the Bioprocess Engineering, Eds. A.Moreira, A.Medina, Ofir, Portugal, May, 17-29.

Cadman, T.W. (1991) 'Advances in Bioreactor Control in Recombinant DNA Technology and Applications', Edt. A.Prokop, R. Baijpai, C.Ho, Chapter 16, 477-509, Mc Graw-Hill Inc., N.Y.

Clarke, D. W. (1990) 'Generalized Predictive Control and its Applicaton' Proc. of the CIM-Europe Workshop on Computer Integrated Design of Controlled Industrial Systems, Ed. J. Richalet, Sp. Tzafestas, Paris 26-27 April, pp 57-76.

Clarke, D. W. (1991) 'Implementation of Adaptive Generalized Predictive Control', Intensive Training Course on Model Based Predictive Control, Belgium, 7-8 Oct.

Clarke, D.W., Mohtadi, C. and Tuffs, P.S. (1987) 'Generalized Predictive Control - Part I. The Basic Algorithm', Automatica, 23, 2, 137-147

Clarke, D.W., Mohtadi, C. and Tuffs, P.S. (1987) 'Generalized Predictive Control - Part II. Extensions and Interpretations', Automatica, 23, 2, 148-160

Cooney, C.L., Raju, G.K., O'Connor, G. (1991) 'Expert Systems and Neural Nets for Bioprocess Operation', Int. Symp on Bioproc. Modelling and Control, Ed.J.Morris, Newcastle, U.K., Jan 21-22.

de Keyser, R.M.C. and Van Canwenbergh, A. (1979) 'A Self-tunning Multistep predictor Application', Automatica, 17, 167-174.

de Keyser, R.M.C. (1990) 'Model Based Predictive Control Toolbox', Proc. of the CIM-Europe Workshop on Computer Integrated Design of Controlled Industrial Systems, Ed. J. Richalet, Sp. Tzafestas, Paris 26-27 April, pp 35-56.

Deshpande, P.D. (1992) 'Improve Quality Control On-line with PID Controllers', Chem.Eng.Prog., 88,5,71-76.

Dochain, D. and Bastin, G. (1985) 'Stable Adaptive Algorithm for Estimation and Control of Fermentation Processes', Modelling and Control of Biotechnological Processes, Proc. IFAC Symp., Noordwijkerhout, The Netherlands, 37.

Dochain, D., (1991) 'Design of Adaptive Controllers for Non-linear Stirred Tank Bioreactors: Extension to the MIMO Situation', J. Proc. Cont., Vol 1 , Jan.

Dochain, D., Perrier, M. and Pauss, A. (1991) 'Adaptive Control of the Hydrogen Concentration in Anaerobic Digestion', Ind. Eng. Chem. Res., 30, 129-136.

Dochain, D., Babary, J.P., Tali-Maamar, N. (1992) 'Modelling and Adaptative Control of Non-linear Distributed Parameter Bioreactors via Orthogonal Collocation', to be published in Automatica.

Dochain, D., Perrier, M. and Ydstie, B.E. (1992) 'Asymptotic Observers for Stirred Tank Reactors', Chemical Engineering Science, Vol. 47, pp. 1-11.

Feyo de Azevedo, Pimenta, P., Ferreira, E. and Oliveira, F. (1992a) 'Studies on On-line State and Parameter Estimation through a Real-time Process Simulator', Proc. of the 2nd IFAC Symp. on Modelling and Control of Biotechnical Processes, Keystone, USA.

Feyo de Azevedo, Pimenta, P., Oliveira, F. (1992b) 'Computer Based Studies on Bioprocess Engineering I - Tools for Process Analysis', In NATO ASI on the Use of Computers and Informatic Systems in the Bioprocess Engineering, Eds. A.Moreira, A.Medina, Ofir, Portugal, May, 17-29.

Foss, A.S. (1973) 'Critique of Chemical Process Control Theory', AIChE J., 19, 209-214.

Fowler, G. W., Higgs, R. E., Clapp, D. L., Alford, J. S., Huber, F. M. (1992) 'Development of Real-Time Expert System Applications for the Analysis of Fermentation Respiration data', Proc. of the 2nd IFAC Symp. on Modelling and Control of Biotechnical Processes, Keystone, USA.

Goodwin, G.C. and Sin, K.S. (1984) 'Adaptive Filtering Prediction and Control', Prentice-Hall, New Jersey.

Graupe, D. (1987) 'Control Systems, Identification in Encyclopedia of Physical Science and Technology', Vol. 3, 589-617, Ac. Press Inc., N.Y.

Harmon, P. and Sawyer, B. (1990) 'Creating Expert Systems', J. Wiley, N.Y.

Hydrocarbon Processing (1991), Advanced Process Control Handbook VI, September

Kestenbaum, A.; Shinnar, R. and Thau, F.E. (1976) 'Design Concepts for Process Control', Ind. Eng. Chem. Process Des. Dev., 15, 2-13.

Kosko, G. (1992) 'Neural Networks and Fuzzy Systems', Prentice-Hall Inc., New Jersey.

Landau, I.D. (1987) 'Controls, Adaptive Systems in Encyclopedia of Physical Science and Technology', Vol. 3, 618-626, Ac. Press Inc., N.Y.

Lee, W. and Weekman, V.W. (1976) 'Advanced Control Practice in the Chemical Process Industry: A View from Industry', AIChE J., 22, 27-38.

Lübbert, A. (1991), 'Automation in Biotechnology', in Biotechnology, Edts. Rehm, H.; Reed, G.; Puhler, A. and Stadler, P., Vol. 4, 561-602, VCH, Weinheim.

Lübbert, A.; Beil, S.; Dors, M.; Havlik, I. and Simutis, R. (1992) 'Experiences with Neural Networks in Predicting the Substrate Degradation during a Production-scale Beer Brewery Fermentation', submitted for publication.

Martin-Sanchez, J.M. (1976) 'A new Solution to Adaptive Control', Proc. of IEEE, 64, 8

Martin-Sanchez, J.M. (1986) 'Adaptive Control for Time Variant Processes', Int. J. Control, 44, 315

Martin-Sanchez, J.M., Shah, S.L. and Fisher, D.G. (1984) 'A Stable Adaptive Predictive Control System', Int. J. Control, 39, 215

Morris, A.J.; Montague, G.A.; Tham, M.T.; Ansley, M.; DiMassimo, C. and Lant, P. (1991) 'Towards Improved Process Supervision-Algorithms and Knowlwdge Based Systems', Int. Symp on

Bioprocess Modelling and Control, Ed.J.Morris, New Castle, U.K., Jan 21-22.

Mosca, E.; Zappa, G. and Manfredi, C. (1984) 'Multi Step Horizon Self-tunning Controllers: the MUSMAR Approach', 9th IFAC World Congress, Pergamon Press, Oxford, 155-159.

Pomerleau, Y., Perrier, M. and Dochain, D. (1989) 'Adaptive Nonlinear Control of the Baker's Yeast Fed-batch Fermentation', Proceedings of the 1989 American Control Conference, June 21-23, 1989, Pittsburgh.

Pomerleau, Y. and Perrier, M. (1990) 'Estimation of Multiple Specific Growth Rates in Bioprocesses', AIChE Journal, Vol. 36, n. 2, 207-215.

Pomerleau, Y. and Viel, G.. (1992) 'Industrial Application of Adaptive Nonlinear Control for Baker's Yeast Production', Proceedings of the IFAC Symposium on Modeliing and Control of Biotechival Processes, 29 March-2 April, 1992, Keystone, Colorado, USA.

Richalet, J. (1990) 'Model Based Predictive Control in the Context of Integrated Design', Proc. of the CIM-Europe Workshop on Computer Integrated Design of Controlled Industrial Systems, Ed. J. Richalet, Sp. Tzafestas, Paris 26-27 April, pp 3-34.

Rivera, S. and Karim, M.N. (1992) 'On-line State Estimation in Bioreactors using Recurrent Neural Networks', Proc. of the 2nd IFAC Symp. on Modelling and Control of Biotechnical Processes, Keystone, USA.

Saraiva, P. and Stephanopoulos, G. (1992) 'Continuous Process Improvement through Inductive and Analogical Learning', AIChE J., 38, 161

Schügerl, K. (1988) 'On-line Analysis and Control of Production of Antibiotics, Analytica Chimica Acta, 213, 1-9.

Schügerl, K. (1992) 'Bioreactor Instrumentation and Biosensors', In NATO ASI on the Use of Computers and Informatic Systems in the Bioprocess Engineering, Eds. A.Moreira, A.Medina, Ofir, Portugal, May, 17-29.

Simutis, R.; Havlik, I. and Lübbert (1992) 'A Fuzzy-Supported Extended Kalman Filter: A New Approach to Atate Estimation and Prediction of a Beer Fermentation', submitted for publication.

Sonnleitner, B. and Käppeli, O. (1986) 'Growth of *Saccharomyces cerevisiae* is controlled by its Limited Respiratory Capacity: Formulation and Verification of a Hypothesis', Biotech. Bioeng., Vol. 28, June, pp.927-937.

Stephanopoulos, G. and Stephanopoulos, G. (1986) 'Artificial Intelligence in the Development and Design of Biochemical Processes', Trends in Biotechnology, 241-248.

Stock, M. (1989) 'AI in Process Control', Mc Graw Hill Inc., N.Y.

Twork, J.V., Yacynych, A.M. Edts. (1990) Sensors in Bioprocess of Antibiotics, Analytica Chimica Acta, 213, 1-9.

Acknowledgements - This work was partially supported by JNICT - *Junta Nacional de Investigação Científica e Tecnológica*, under contract numbers BD/224/90-IF and BD/1476/91-RM and INIC - *Instituto Nacional de Investigação Científica*.

MULTIVARIABLE CONTROL OF A CONTINUOUS CULTURE

P. A. Gostomski[1], L.S. Clesceri[2], and H. R. Bungay[1]
Department of Chemical Engineering[1]
Department of Biology[2]
Rensselaer Polytechnic Institute
Troy, NY 12180-3590, USA

ABSTRACT. Multivariable control of both pH and ammonium levels in an aerobic continuous yeast culture was demonstrated. A buffered glucose feed was used for pH control. This control loop was coupled with ammonium control using an on-line ammonia electrode suspended above a pH-adjusted sample stream from the fermentor. The dilution rate of the culture was the sum of the two individual feed rates. The advantages and implications of multivariable feedback control in continuous fermentation are discussed.

1. Introduction

Continuous fermentation is a frequent research tool for studying the effect of environmental changes on an organism's productivity and substrate utilization. Chemostats with a fixed dilution rate are the traditional mode of operation for continuous culture, mainly because of ease of operation and historical equipment limitations. With the availability of inexpensive computers, the possibility for implementing advanced feedback control now exists. These systems permit sophisticated control algorithms, proper signal conditioning and high quality real-time data display. It is also possible to store and manipulate the large amounts of data that a complicated control system can generate. These hardware/software configurations are often user-designed or are commercial products that have the flexibility required by research laboratories. This type of sophistication was too cumbersome and complicated before the advent of personal computers.

For studying microorganisms at steady-state, chemostats have commonly been used, but the advantages for using auxostats have been recognized for sometime (Agrawal and Lim, 1984). Auxostats are now easily computerized for data acquisition and control (DAC) systems. The main advantage of auxostats is the tendency for stable operation near the maximum growth rate of an organism; this is not observed in a chemostat. A second advantage is that auxostats, through feedback control, achieve steady-state more quickly than chemostats. Some experiments with auxostats have been reported. Menchke, et al., (1988) controlled the level of napththalenesulfonic acid in a *Psuedomonas testosteroni* culture using an on-line HPLC. Kleman, et al., (1991) controlled glucose on-line in an *Escherichia coli* culture with a YSI glucose analyzer.

A.R. Moreira and K.K. Wallace (eds.), Computer and Information Science Applications in Bioprocess Engineering, 51-57.

The pH-auxostat is an example of a robust style of auxostat. The pH-auxostat was first reported by Bungay (1972) and developed by Martin and Hempfling, (1976). With fresh medium addition coupled to pH control, the culture is very stable at high growth rates, and the possibility of washout is eliminated. The rate of medium addition is determined by the buffering capacity and not directly by the setpoint (pH) as in a traditional auxostat. Buffering capacity is defined as the equivalents of titrant required to change the medium pH to the reactor pH. By using the glucose feed for pH control, the glucose level can be manipulated through adjustments in buffering capacity. The work reported here expands auxostats to the next level of complexity with multivariable control. This development takes full advantage of the sophistication available with a fermentation system interfaced with computer DAC system.

The advantages of feedback control are diminished at low dilution rates because of the insensitivity of nutrient levels to dilution rate changes. Multivariable control expands the advantages of feedback control over the entire growth rate regime and opens new avenues for research. The ratio of different nutrients, such as carbon to nitrogen, is critical in the production of many bioproducts. These interactions can be explored with a chemostat by manipulating feed ratios, but this is cumbersome. True multivariable control offers the ability to set independently the concentration levels and move from one ratio to another rapidly because of the quick response of a closed-loop system compared to an open-loop system. This type of operation makes a multivariable auxostat superior, compared to a chemostat, for studying the interrelationship between nutrient levels over the entire growth regime of a microorganism.

The two parameters controlled in our system were pH and NH_4^+ – a pH:NH_4^+-auxostat. The feed rate of NH_4^+ medium was manipulated to maintain a desired NH_4^+ concentration in the fermentor and the glucose addition rate controlled the pH. Prior to multivariable control, single variable control was implemented as a NH_4^+-auxostat and a pH-auxostat. An aerobic, glucose-limited *Saccharomyces cerevisiae* culture was used. The defined medium was split into two portions; a glucose reservoir (see Materials and Methods) and an NH_4Cl reservoir. The feed rate of the glucose reservoir was determined by the pH-auxostat portion of the controller and the ammonium feed rate was set by the NH_4^+-auxostat portion. The dilution rate was the sum of these two feed rates divided by the reactor volume. Figure 1 is a simplified diagram of this reactor. An important aspect is that feed concentration of glucose and ammonium are dependent on the total feed rate, so changes in the individual feed rates effect the feed concentrations.

2. Materials and Methods

2.1 Yeast Strain

The strain used in this research was *Saccharomyces cerevisiae* ATCC 42594, maintained on a complex medium consisting of 2% glucose and 2% peptone in agar slants at 5°C.

2.2 Medium

The medium was a based on a medium designed by Fiechter (1984) to be defined and glucose-limited. For this work all major ammonium salts were replaced with equal molar

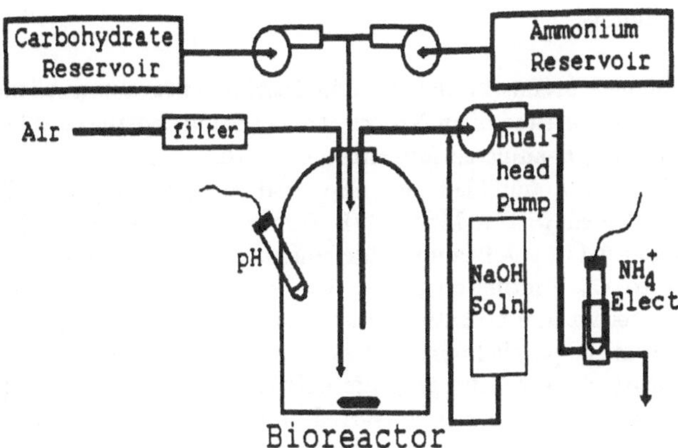

Figure 1 - Bioreactor schematic

amounts of the corresponding potassium and sodium salts. The only source of ammonium was as NH_4^+Cl. The medium consisted of (per liter of deionized water): 5.0g D-glucose, 6.02g K_2SO_4, 5.37g Na_2SO_4, 2.12g K_2HPO_4, 3.25g $Na_2HPO_4 \cdot 7H_2O$, 10.7g NH_4Cl, 0.83g $Na_3C_6H_5O_7 \cdot 2H_2O$. Trace elements: 0.024g $FeCl_2$, 0.7g $CaCl_2 \cdot 2H_2O$, 5ml sulfate stock solution, 5ml vitamin stock solution. The pH was adjusted to 3.00 before addition of the trace elements. The sulfate stock solution consisted of (per ml): 150mg $MgSO_4 \cdot 7H_2O$, 0.78mg $CuSO_4 \cdot 5H_2O$, 3mg $ZnSO_4 \cdot 7H_2O$, and 3.15mg $MnSO_4 \cdot 7H_2O$. The vitamin stock solution consisted of (per ml): 0.01mg biotin, 20mg m-inositol, 10mg Ca-pantothenate, 2mg thiamin, 0.5mg pyridoxin, and 2mg niacin.

For all experiments, the medium was divided into three vessels and appropriately concentrated to maintain the above concentrations. The first feed vessel consisted of all medium constituents except NH_4Cl. The second vessel contained only NH_4Cl. The third vessel was either pure H_2O for chemostat and NH_4^+-auxostat operation or a caustic solution for pH- auxostat and pH-NH_4^+-auxostat operation. A separate dilute caustic solution pumped in proportion to the glucose medium was required for pH-auxostat operation, because the minerals medium would precipitate at the required pH.

2.3 Cultivation Conditions

The reactor was a 1 L 500 Series Research Fermentor (LH Fermentation, Hayword, CA) with a 500 ml working volume. Temperature was maintained at 30°C, and pH was held at 4.0. Air was supplied at a rate of 1 vvm. Medium flow was supplied by peristaltic pumps (Cole-Parmer, Chicago, IL)

During chemostat and NH_4^+-auxostat operation, a triple head was used for pumping fresh medium. A separate pump and concentrated caustic solution were used for pH control. During pH-auxostat operation, the pure water feed reservoir contained instead a dilute caustic solution and the separate pH control loop was removed. During pH:NH_4^+-auxostat operation two feed pumps were used. The first used two heads to supply the glucose solution and the dilute caustic solution. The second pump supplied the NH_4Cl solution.

2.4 Analytical Techniques

On-line ammonium measurement is described in detail in a previous publication (Gostomski and Bungay, 1992). It consisted of an NH_3 electrode (Orion, Boston, MA) suspended above a continuous stream of pH-adjusted fermentation broth.

All analog signals were amplified and converted to digital signals with a Model 140 Analog Interface coupled with a Model 17B Differential Multiplexor (Lawson Lab Inc, Kalispell, MT) mounted in a XT-compatible computer. Pumps were controlled by OAC5 optical relays (Opto 22, Huntington Beach, CA) or with optocoupler chips mounted on Model 7553-05 Masterflex motor control boards (Cole-Parmer, Chicago, IL). Digital signals for control came from the parallel port of the computer. The hardware was controlled by a custom data acquisition and control program written in Turbo Pascal 5.5 (Borland, Scotts Valley, CA). Each analog signal was converted at 1 second intervals and passed through a second-order digital filter. Incremental flow rate adjustments were accomplished with on/off cycles over 10 second intervals. Data were averaged and stored at 5 minute intervals to a 20M harddrive.

3. Experimental Results

Before implementing the dual setpoint auxostat, two preliminary experiments were performed. The yeast culture was first operated in chemostat mode to study the dynamic characteristics of the culture. Step changes in dilution rate were studied to determine the linearity of the system and possibly to fit the responses to a first or second-order lag model. The second set of experiments used single setpoint auxostats i.e., pH-auxostat and NH_4^+-auxostat. These data were needed to determine rough controller settings for multivariable control.

Chemostat results are shown in Figure 2. The dilution rate was changed from 0.21 hr^{-1} to 0.31 hr^{-1}, and the ammonia signal tracked to a higher concentration as expected for an increased dilution rate. In this particular operating regime the system behaved linearly. The response was modeled as a first-order response with deadtime. The time constant was 5.4 hours and the gain was 4.72 with a deadtime of 0.4 hours.

Figure 3 shows the effect of a buffering capacity change while in pH-auxostat mode. The pH was raised from 4.0 to 4.5, which lowered the buffering capacity of the medium. This change elevated the steady state dilution rate as seen in Figure 4. The dilution rate is constrained by the maximum pumping rate of the system. Proportional control was sufficient because of the rapid response of pH compared to other state variables in the system. The ammonia signal is shown rapidly adjusting to a new steady state with the change in dilution rate.

An NH_4^+-auxostat was operated with proportional-integral control. Figure 4 shows the response of the system to a change in the NH_4^+ setpoint from 1.55mM to 2.6mM. The response is characteristic of a tightly tuned controller with rapid response and overshoot of the new setpoint. The drop to a lower steady state dilution rate could not be explained.

These two auxostats were combined to create a pH:NH_4^+-auxostat. The controller was designed as two independent control loops. Figure 5 shows the response of the pH:NH_4^+-auxostat to a step change in the ammonium setpoint from 10mM to 15mM. The ammonium signal rapidly reaches the new setpoint. The interesting result is the dynamics of the

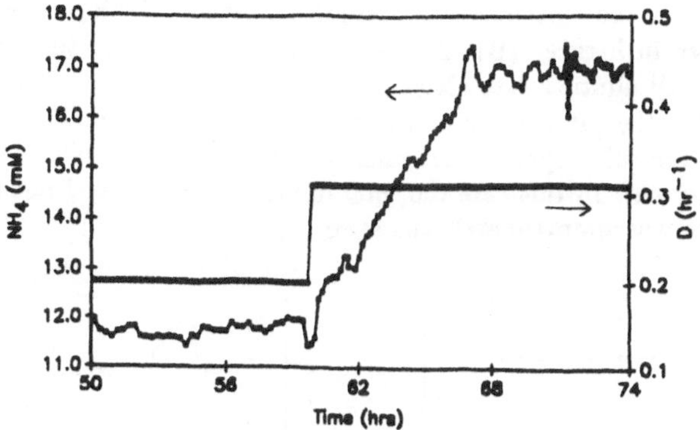

Figure 2 - Open loop response in chemostat mode. A step change in dilution rate from 0.21 hr^{-1} to 0.31^{-1}.

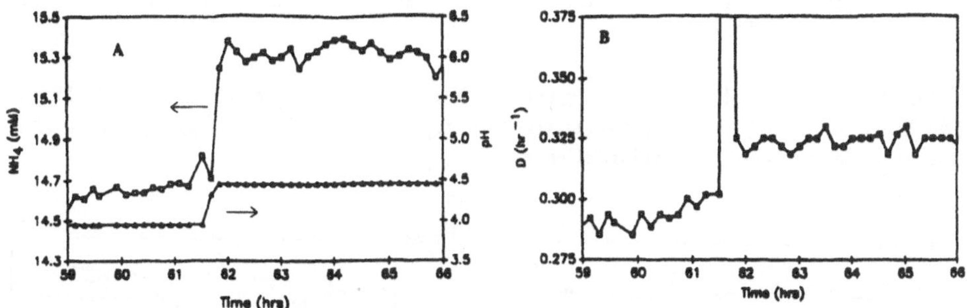

Figure 3 – Closed loop response of a pH-auxostat. Buffering capacity decrease accomplished by raising the pH setpoint from 4.0 to 4.5. (A) Response of pH and NH_4^+. (B) Dilution rate response.

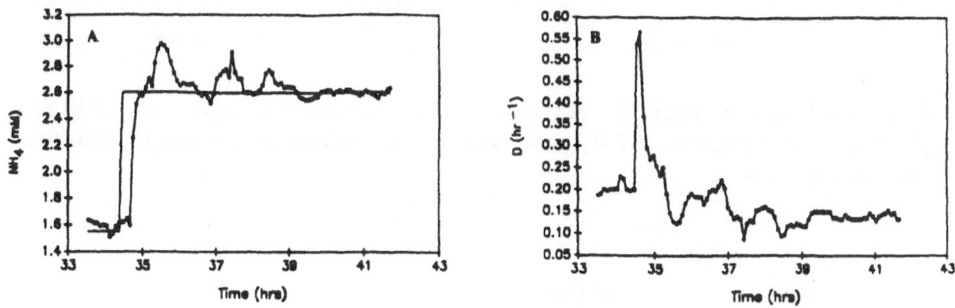

Figure 4 - Closed loop response of an NH_4^+-auxostat. NH_4^+ setpoint change from 1.55mM to 2.6mM. (A) Response of NH_4^+ concentration in the reactor. (B) Dilution rate repsonse.

individual dilution rate contributions from the pH-auxostat and NH_4^+-auxostat portions of the controller shown in Figure 5(B). The NH_4^+ feed rate rises and the glucose feed rate drops, but the overall dilution rate rises. Figure 6 demonstrates a change in buffering capacity. In this case the pH setpoint was lowered, which raised the buffering capacity of the medium. The the pH response was quick and the NH_4^+ controller was successful in maintaining the setpoint. In this case the total dilution rate dropped but the new steady state NH_4^+ feed rate was approximately the same.

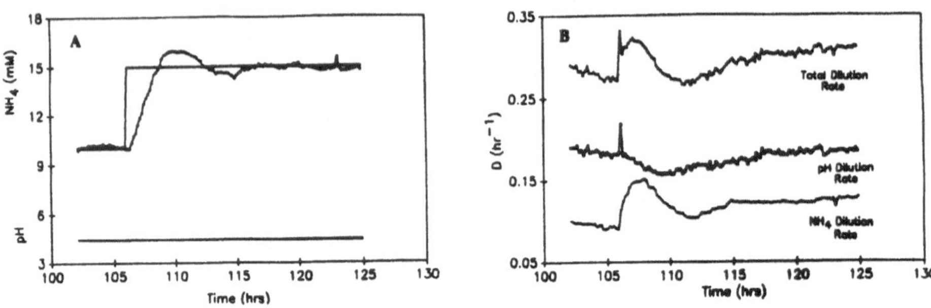

Figure 5 – Response of pH:NH_4^+-auxostat to a NH_4^+ setpoint change from 10mM to 15mM. (A) Tracking of NH_4^+ concentration to the new setpoint and response of pH. (B) Individual flow rate contributions to total dilution rate.

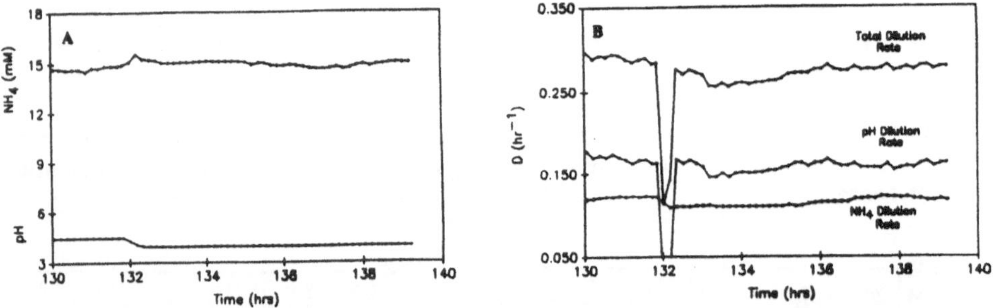

Figure 6 – Response of pH:NH_4^+-auxostat to a buffering capacity change. (A) Shift to new pH setpoint and response of NH_4^+ concentration. (B) Individual flow rate contributions to total dilution rate.

4. Discussion and Conclusions

The multivariable control auxostat shows much promise as a research tool for exploring the relationships between limiting and non-limiting nutrients. Example systems where this could be applied include internal storage materials and exopolysaccarides. Research has shown that the carbon to nitrogen ratio is very important in the production of poly-β-hydroxybutyrate. This material represents a family of biodegradable polymers with many potential uses. The ability to set the C/N ratio directly in the reactor should allow for a quick optimization of growth conditions.

Multivariable control also offers the opportunity to investigate multisubstrate limited growth. This idea has been studied by some groups (Rutgers, et al., 1990), but there is little experimental verification of any of the proposed models. A multivariable auxostat could allow direct experimental study of this area of metabolism.

The multivariable controller design for this system was very straight forward. Two individual control loops were implemented with no attempt to uncouple the control loops to minimize interactions. The success can be attributed to the difference in time constants between the two loops, with pH control having a much smaller time constant than for NH_4^+ control. This would most likely not be the case with other systems and controller design could be more difficult. This represents an interesting area for further work because of the non-linear responses often seen in continuous culture.

5. References

Agrawal, Pramod, Lim, Henry C. (1984) 'Analyses of Various Control Schemes for Continuous Bioreactors' Advances In Biochemical Engineering/Biotechnology, 30, 61-90.

Bungay, H.R. (1972), 'Continuous Culture With Proportional Control Of Substrate Concentration', Proc. IV IFS Ferment. Technol. Today, 117-120.

Fiechter, A. (1984), 'Physical and Chemical Parameters of Microbial Growth', Advances in Biochemical Engineering/Biotechnology, 30, 7-60.

Kleman, Gary L., Chalmers, Jeffrey J., Luli, Gregory W. and Strohl, William R. (1991), 'Glucose-Stat, a Glucose-Controlled Continuous Culture', Applied Environ. Micro., 57, 918-923

Martin, G.A., and Hempfling, W.P. (1976), 'Method for the regulation of microbial population density during continuous culture at high growth rates', Arch. Microbiol., 107, 41-47.

Meschke, J., Bennemann, H., Herbst, H, Dormeier, S. and Hempel, D.C. (1988) 'On-Line HPLC-Measurement and Control of Substrate in a Continuously Operated Biological Tank Reactor', Bioprocess Engineering 3, 151-157.

Rutgers, Michiel, Balk, Peter A. and van Dam, Karel (1990) 'Quantification of Multiple-Substrate Controlled Growth–Simultaneous Ammonium and Glucose Limitation in Chemostat Cultures of Klebsiella Pneumoniae', Arch Microbiol 153, 478-484.

6. Acknowledgement

This work was supported by the New York State Energy Research and Development Authority.

STEADY STATES IN CONTINUOUS CULTURE

H. R. BUNGAY
Department of Chemical Engineering
Rensselaer Polytechnic Institute
Troy, New York 12180-3590, U.S.A.

ABSTRACT. Simulation exercises based on setting the differential equations for cell mass and substrate to zero reveal the effects of various coefficients and conditions on steady states for continuous cultivation of microorganisms.

1. Introduction

1.1. METHODS FOR CONTINUOUS CULTIVATION

Continuous fermentation techniques have been used for years. The advantages include higher productivity and the ability to study cultures and metabolic pathways under steady-state growth conditions. Batch and fed-batch cultures have been used extensively because of the ease of operation and potentially high product concentrations. However, batch results are often difficult to interpret because of constantly changing growth conditions. The traditional methods of continuous culture include the chemostat, the auxostat, and the turbidostat. The chemostat is an open-loop operation with a fixed nutrient flow rate. The auxostat and the turbidostat are closed-loop systems with feedback control of nutrient flowrates.

For feedback controlled methods of the auxostat or the turbidostat, a particular variable is measured in the fermentor and the nutrient flowrate is adjusted to hold it constant. The turbidostat uses turbidity as the control variable. Because of the inherent fouling of optical sensors, this mode of operation is not popular. The auxostat fixes a nutrient concentration. In auxostat mode, the growth rate is determined by the organisms based on the constraints of nutrient availability in the fermentor. This method has been described as using the `internal control system' (idiolimitation) of the culture to stabilize the dilution rate (Rice and Hempfling, 1985). The closed-loop control strategy of the auxostat eliminates the instabilities suffered by the chemostat, and it is possible to operate at dilution rates approaching the maximum growth rate.

Theoretical analysis of auxostat cultures with true proportional control was reported some time ago (Edwards, et al., 1972, Bungay, 1972). Agrawal and Lim (1984) extended these ideas but noted that experimental verification of stability is lacking. Only recently have experimental results been reported with nutrient-controlled cultures using direct feedback control of dilution rate (Menschke, et al., 1988, Fraleigh, et al., 1987).

1.2. LIMITING NUTRIENT

The nutrient in short supply relative to the others will be exhausted first and will thus limit cellular growth. The other ingredients play various roles such as exhibiting toxicity or promoting cellular activities, but there will not be an acute shortage as in the case of the limiting nutrient.

A.R. Moreira and K.K. Wallace (eds.), Computer and Information Science Applications in Bioprocess Engineering, 59-65.

A possible relationship between specific growth rate coefficient and substrate concentration is shown in Figure 1. The curve is nearly linear at low concentrations and levels off for high concentrations. If the nutrient reaches toxic levels, the curve may drop precipitously. Except in very rare cases when all nutrients are in perfectly equivalent proportions, one will be present in low ratio to the others. As cells multiply, this ingredient will run out first and will limit further growth. This concept of a growth-limiting nutrient in shortest supply is crucial to understanding bioprocessesing.

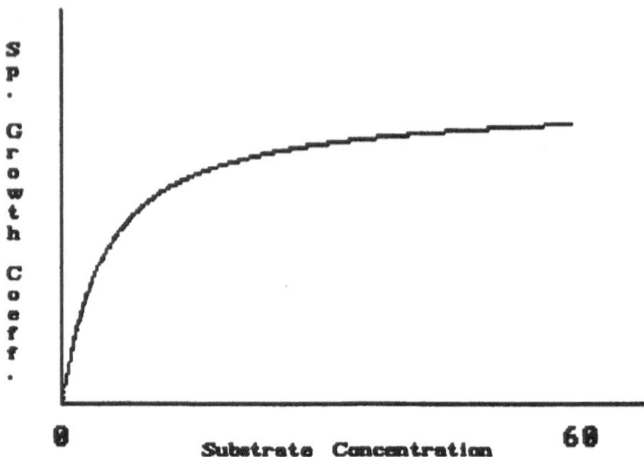

Figure 1. Specific growth rate coefficient versus concentration of limiting nutrient.

2. Well-mixed Continuous Processes

2.1. THEORY

Consider a process with such good mixing that all fluid elements are assumed to be identical. The analysis starts with the mass balance equation:

$$\text{rate of change} = \text{input} - \text{output} \pm \text{reaction}$$

$$V\frac{dx}{dt} = F X_i - F x + \mu X V \tag{1}$$

where
- x = organism concentration at time t
- V = the constant volume of the vessel
- F = feed rate
- μ = specific growth rate coefficient
- X_i = organism concentration in the feed

If we neglect organisms in the feed stream, the $F X_i$ term is not needed. A new term D, the dilution rate, is equal to F/V, thus dividing each term in 1 by V and dropping the zero term gives

$$\frac{dx}{dt} = \mu x - Dx \tag{2}$$

A mass balance for the growth-limiting nutrient gives:

$$\frac{dS}{dt} V = F S_o - F S - \frac{\mu x V}{Y} - M x V \tag{3}$$

where
S = the concentration of limiting nutrient, g/L
S_o = the concentration in the feed stream
Y = the yield coefficient, g of cells/g of limiting nutrient
M = the maintenance coefficient to keep cells alive

Dividing by V, we get:

$$\frac{dx}{dt} = D S_o - DS - \frac{\mu x}{Y} - M x \tag{4}$$

The Monod equation is

$$\mu = \hat{\mu} \frac{S}{K_s + S} \tag{5}$$

where
$\hat{\mu}$ = maximum rate coefficient
K_s = half-saturation constant

At steady state, there is no change thus the derivatives in the differential equations disappear to give:

$$\mu = D \tag{6}$$

and

$$D S_o - DS = \frac{\mu x}{Y} + M X \tag{7}$$

Substituting D for μ in the Monod Equation and solving for S gives:

$$S = \frac{D K_s}{\hat{\mu} - D} \tag{8}$$

Solving Equation 2 for x after substituting D for μ gives:

$$x = DY \frac{S_o - S}{D + MY} \tag{9}$$

2.2. COMPUTER EXERCISE: STEADY STATES

The theory of continuous culture becomes very clear when illustrated by computer graphics. The mathematical basis is the same as presented earlier, and coefficients are substituted into Equations 8 and 9. Note that the values needed are $\hat{\mu}$, K_s, Y, S_o, and M. Productivity, P, equals cell mass times dilution rate. Reasonable values for the parameters when using the scales on the graph for this program on your disk are:

$\hat{\mu}$ 0.1 to 4 hr^{-1}
Y 0.2 to 0.7 g/g
S_o 1 to 200 g/L
K_s 0.2 to 10.0

M 0.001 to 0.2

Computer Instructions:

1. Bring up BASIC on your computer.

2. Type **RUN "MONOD"** [enter]

3. Enter the specifications as prompted.

Entering zero erases the screen for a new start. Entering a minus number aborts the program. Use the shift key and PrtScr key together to get graphs on the printer. You can move the cursor and label the lines before printing.

4. Study the effects of increasing and decreasing $\hat{\mu}$, Y, S_o, and K_s while keeping M very small. Use M = 0.001.

5. Select a set of the other parameters that produces curves that fit the plotting area well and experiment with different values of M.

6. Hold M at a relatively high value and then at a low value while exploring ranges of other parameters.

Some typical results with MONOD are shown in Figure 2. Several plots may be overlayed for easy comparisons, and a cluttered screen can be erased. The most important concepts from the exercise are:

- x goes to zero and S reaches S_o as D approaches $\hat{\mu}$.
- S is not a function of S_o when D is less than $\hat{\mu}$.
- The maintenance coefficient is very important but only at low dilution rates.
- Productivity peaks before D equals $\hat{\mu}$.

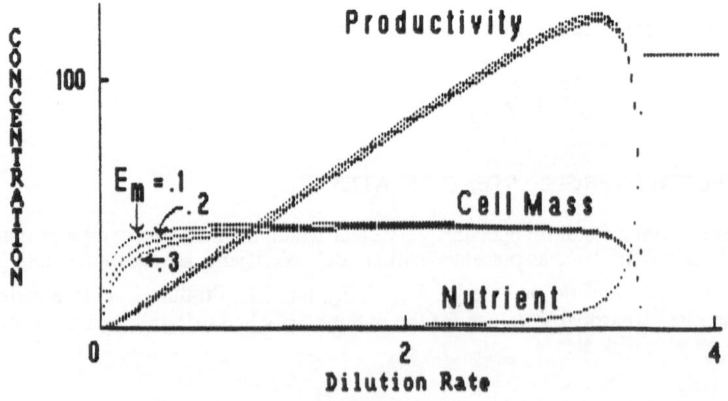

Figure 2. Typical output for MONOD.BAS.

2.3. COMPUTER EXERCISE: RECYCLE

Separation and recycle of cells results in much longer residence times for the cells than for the fluid and permits relatively high cell concentrations. In biological waste treatment, the dilute feed would lead to slow growth rates. More rapid processing is attained by achieving higher populations through cell recycle. High rates of production are also important in industrial fermentations where recycle has the added advantage of saving expensive substrate that would otherwise be needed for new cell growth.

In activated sludge, organisms not in flocs are not collected and they tend to leave the system as recycle increases the proportions of floculating types. Recycle of algae to outdoor ponds has profound influence on the population; seasonal changes can cause small algae to develop so that the concentration of cells declines because the small algae can escape while the large ones die out.

Cells are not the only objective for recycle. Spent medium or the aqueous residue from product recovery may be recycled. Although this recycle stream may save on water and may be nutrittious, a major consideration is less waste treatment. In processes for bioconversion of corn (starch) to ethanol, roughly 93 percent of the still bottoms from alcohol recovery are sent back to the process. This is about the limit on recycle for this process because non-volatile organic compounds accumulate. There may also be two recycle streams: cells collected for recycle and aqueous residue. The cells would be killed or damaged during product recovery, so they are collected just after the bioconversion step.

Mass balances for a continuous culture vessel with recycle are:

$$F_a = (1+\omega)F \tag{10}$$

$$F = F_e + F_{ex} \tag{11}$$

$$F_a = F_e \frac{X_e}{X} + \frac{X_x}{X}(F_{ex} + \omega F) \tag{12}$$

where

F_a = flow rate from vessel
F = fresh feed rate
F_e = liquid from separator
F_x = suspension from separator
F_{ex} = suspension bleed rate
X = cell conc. in bioreactor
X_e = cell conc. in liquid
X_x = cell conc. in suspension
ω = recycle ratio

Assuming perfect mixing:

$$V \frac{dx}{dt} = X_x \omega F - F_a X + V \mu X \tag{13}$$

Combining equations and setting the derivative to 0 for steady state:

$$\mu = (1+\omega) - \omega D \frac{X_x}{X} \tag{14}$$

This leads to:

$$\mu = D + D\omega \left(1 - \frac{1 + \omega - F_e/F - X_e/X}{+1\ \omega/F}\right) \tag{15}$$

Without recycle, washout occurs when D is greater than the maximum specific growth rate. With recycle, D can greatly exceed the maximum specific growth rate. Some of the assumptions may not be rigorous. For example, a constant yield coefficient is assumed although it it known that starving cells at very low dilution rate differ from rapidly dividing cells. Further research with continuous cultivation of microorganisms is needed to provide relationships on which to base improved computer analysis.

RECYCLE.BAS keeps the maximum specific growth rate constant at unity so that you can see the ratio of the washout rate to that without recycle. Of course, you can't recycle more cells than you have. For example, if you recycle 10 per cent of your fluid, the concentration factor must be less than 10 because 0.1 times 10 equals 1, and this is the total fraction of cells. Similarly, a concentration factor of 4 means that the recycle fraction must be less than 0.25.

COMPUTER INSTRUCTIONS:

1. Bring up BASIC on your computer.

2. Type **RUN "RECYCLE"** ENTER
   You will be prompted for the specifications. Entering zero clears the screen.

3. Experiment with permutations of the coefficients.

When concentration factor times recycle fraction approaches 1, the washout can be many times that for no recycle. You may go off the scale of the graph when this happens, so starting with less critical values is recommended. Figure 3 shows typical results.

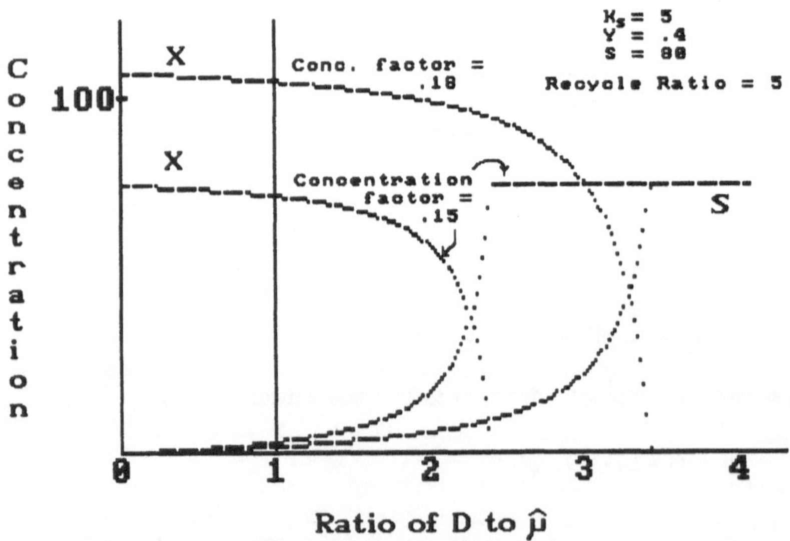

Figure 3. Typical Output for RECYCLE.BAS

3. Additional Reading

Bungay, H.R. (1972) 'Continuous culture with proportional control of substrate concentration', in G. Terui (ed.), Fermentation Technology Today (Proceedings of the 4th International Fermentation Symposium) pp. 117-119, Society for Fermentation Technology, Japan.

Edwards, V.H., Ko, R.C. and Balogh, S.A. (1972) Biotechnol. Bioeng. 15, 939-974.

Fraleigh, S.P., Bungay, H.R., Clesceri, L.S. and Bradley, W.D. (1987) 'The effect of controlled potassium ion concentration on aerobic ethanol production by *Saccharomyces cerevisiae*', in O.M. Neijssel, R.R. van der Meer, R.R. and K.Ch.A.M. Luyben (eds.), Proceedings of the 4th European Congress on Biotechnology, Vol. 3, Elsevier, Amsterdam, pp. 222-225.

Fraleigh, S.P., Bungay, H.R. and Clesceri, L.S. (1989) 'Continuous Culture, Feedback Control and Auxostats', Trends in Biotechnology 7, 159-164

Fraleigh, S.P., Bungay, H.R. and Clesceri, L.S. (1990) 'Aerobic Formation of Ethanol by*Saccharomyces cerevisiae* in a Computerized pHauxostat', J. of Biotechnology 13, 61-72.

Grady, C.P.L. and Lim, H.C. (1980) 'Biological Waste Treatment: Theory and Applications', Marcel Dekker, NY and Basel.

Meschke, J., Bennemann, H., Herbst, H., Dormeier, S. and Hempel, D.C. (1988) Bioprocess Eng. 3, 151-157.

Rice, C.W. and Hempfling, W.P. (1985) Biotechnol. Bioeng. 27, 187-191.

v. Schulthess, R., Bungay, H.R. and Fraleigh, S.P. (1990) 'Competition in a pHauxostat', Biotechnology Letters 12, 93-98.

INTRODUCTION TO ARTIFICIAL INTELLIGENCE

HENRY R. BUNGAY
Department of Chemical Engineering
Rensselaer Polytechnic Institute
Troy, NY 12180-3590, USA

ABSTRACT. A brief overview of artificial intelligence is supplemented with exercises using a commercial neural network shell. The problems are classification of microorganisms and characterization of flow in tanks.

1. Expert Systems

1.1 INTRODUCTION

Artificial intelligence cannot be covered in a brief lecture. Some flavor of A.I. may come from a few comments about expert systems and neural networks. The computer demonstration will be only for a neural network shell that is reasonably priced, quite powerful, and entertaining.

One branch of artificial intelligence called expert systems goes beyond the simple idea of using computer logic and mimics the ways in which humans make decisions. We can interview an expert in order to derive rules that guide decisions, but often the information is **not** yes-no, black-white, or some exact numerical trigger for action. There may be a heirarchy of rules. For example, a physician may base a diagnosis on many factors, one of which is fever. However, absence of fever does not absolutely rule out certain diseases that are usually accompanied by fever. When the information is highly inexact, we make use of "fuzzy logic", but space limitations prevent its coverage here. Weighting of rules so that each contributes in different amounts or in combination with others is in the repertoire of an expert, but this is not easy to code into a computer language. Fortunately, there are programs called expert shells that are not too difficult to use and that systematize logic and rules for rather complicated systems.

A vital element is an *inference engine* that processes IF-THEN statements that may have quite complicated and profound interdependencies. A *production system* uses rules and is also called a pattern-matching inference system. Rules have two parts - a premise and a conclusion. The conclusion may be hard or soft in terms of its application. For example, a hard rule would be "IF an airplane is firing at you, THEN shoot back". A soft rule might be "IF the plane is an enemy AND is not firing, THEN you have a 70 per cent basis for shooting at it". In this case, we need more confirmation. Another applicable rule might be "IF we are at war, THEN shoot at the enemy". This may have exceptions such as "IF there is a flag of truce, THEN don't shoot". We have to be careful in our expert systems to resolve conflicts so that there are not hard rules that insist that we shoot while another rule says don't shoot just as emphatically. Some conflicts are inherent such as quality versus cost, and we can build trade offs into our system.

An inference engine or rule interpretter examines facts and determines the order for processing rules. The facts may encounter semantic difficulties. For example, consider the meaning of "had" in these sentences:

A.R. Moreira and K.K. Wallace (eds.), Computer and Information Science Applications in Bioprocess Engineering, 67-75.

Mary had a baby.

Mary had a little lamb.

I was had by my mother.

The first sentence probably means that Mary gave birth. The second does not mean that at all and denotes possession. To be had by your mother is a very uncommon way to refer to giving birth. A more likely interpretation is that you were outfoxed. This illustrates a problem with a natural language interface. The program may quiz you with true-false, multiple choice, or questions that require answers as words or numbers. Poorly designed questions that elicit ambiguous or incorrect answers cripple your expert system. When possible, the computer should get information free of human error. For example, it could be connected to a device that senses temperature directly instead of asking you to read a thermometer and to type in the value.

The control component in the inference engine must:

- compare a rule to a pattern
- choose the most relevant rule
- implement the rule
- act - e.g., phone the police

It is difficult to anticipate and to program the most efficient path through these procedures.

1.2 ADVANCED FEATURES OF EXPERT SYSTEMS

Mass and energy balancing, energy flux, sources, and sinks deserve special attention for chemical and biochemical processes. The information library for a bioprocess expert should have features such as stoichiometric and thermodynamic data for pathways and for estimation of ATP and reduced pyridine nucleotide production. There can be correlations with measured ratios of reduced to oxidized pyridine nucleotides. Sophisticated instruments such as the flow cytometer could quantify some biochemicals for individual cells. Built in to our expert system should be relationships that interpret diauxie, mass balances, pathways, and energetics. More subtle effects such as mutation, selection, and reversion from finely-tuned mutants back to the wild type may be appreciated qualitatively, but mathematical description may not be possible

1.3 EXPERT SYSTEM SHELLS

Some expert system shells can construct rules if provided with case examples. These are called inductive shells. A state-of-the-art hybrid shell has multiple features for knowledge representation and for inference. The simple rule-based shell is most commonly used at present.

For instruction, we recommend programs from:

EXSYS, Inc., PO Box 75158, Station 14
Albuquerque, NM 97194 phone (505) 836-6676

There is a very large and powerful package for rule-based expert systems, but a truncated version for evaluation is provided at a very modest price. At the time we purchased the small package (soft-cover book and two floppy disks), the price was only $ 15. While there are limits to the number of rules that can be handled by this inexpensive version, we find that is has more than enough power for the expert systems that are appropriate for learning the techniques. There are other commercial programs that are excellent, and most come with computer tutorials and good instructions.

2. Neural Networks

2.1 INTRODUCTION

Just as the nervous system is composed of nerve fibers and junctions called neurons, an artificial neural network has interconnected blocks with information flowing between them. The general idea is pattern recognition. There are input blocks and output blocks with one or more layers of blocks in between. If you know the input and the output, the ways that information flows through the network of blocks can be codified and analyzed. The inputs and outputs for learning are used to improve coefficients through iteration, and eventually the trained system will derive reasonably correct outputs from inputs that it has never encountered before.

Let's examine a very crude example of how a neural network could be useful. Consider a process that has several different feed streams and some process measurements such as concentration, temperature, and pH. There will be some permutations of our input information for which we know the output. This output may be action rules such as add more acid, increase the rate of cooling water, sound the alarm before exiting the premises expeditiously, or the like. As we allow our neural network to iterate and to learn with solid examples of appropriate outputs for given sets of inputs, it will evolve until it recognizes patterns in these inputs. When presented with totally new inputs, it may be able to generate a correct output. The correctness of the output will depend on how well it has been trained, and insufficient examples from which to learn will give us unreliable answers.

The blocks between the input layer and the output layer are called "hidden". There may be several layers, but one hidden layer may be enough. The algorithm develops weights as the system learns. In other words, it may find a pattern by which 10 per cent of the signal from Input Block 1 is summed with 33 per cent of the signal from Input Block 2, and with percentages of other input blocks to be passed on to a hidden node. Different percentages would go to another hidden node.

Some of the most important facts about neural network programs are that learning is slow (medium-sized networks may take a day or more on a slow computer) but decisions with a trained network can be lightning fast. The learning requires iteration, error checking, and testing for convergence. Calculating the output of a trained network is merely once through with multiplications and additions. it makes sense to perform the learning with a very fast computer workstation or even an expensive supercomputer. On the other hand, a very inexpensive computer may be satisfactory when using the trained network. A neural net is very easy to use with no mathematics. Furthermore, the neural network learns automatically; you lay it out but do not have to write any program.

The computer science of neural networks need not be understood to use them, but there are a few concepts that may make neural networks less mysterious. The computational elements called nodes interact locally to process input information to generate output information. The activity of a node is determined by the weighted sum of input signals, but this activity does not lead to a sharp decision. There are many choices for how to relate output to activity, but most neural networks incorporate a sigmoidal (S-shaped, symmetric) relationship because it behaves very nicely for the mathematics. The implication of this comes from asymptotic approach to true or to false; it takes a near absence of activity to generate zero (false) for the output and intense activity to generate one (true) for a sigmoid that rises monotonically with activity. In other words, decisions by the neural network are likely to be something like 0.92 on a scale of 1. This is not the same as as being 92 per cent correct, but it may be helpful to think in percentages. The scores for a set of output blocks are actually relative rankings.

Inputs to a neural network can have greatly different magnitudes. To avoid having large numbers dominate in the calculations, each input is normalized by apportioning it between its highest and lowest values.

2.2 COMMERCIAL SHELLS FOR NEURAL NETWORKS

A number of references and some commercial programs for neural networks were listed in BYTE magazine, p 244-245, August, 1989. We have worked with but one of these programs but find it fully satisfactory for teaching. We like the program NeuroShell from Ward Systems Group, Inc. 245 W. Patrick St., Fredrick, MD 21701.

NeuroShell is user-friendly with menus. There are two versions - one for binary true/false inputs and another that allows a continuous range of inputs. The programs runs much faster if your computer has a math coprocessor. You can input or output data in format accepted by word processing programs and in files compatible with spreadsheets.

The terminology is that data sets for input are called defining characteristics and data sets for output are classifying characteristics. For example, the defining characteristics could be medical symptoms such as headache, blurred vision, high temperature, and the like. Typical classifying characteristics for this system would be the diagnoses such as bacterial infection, blow to the head, etc. Cases of defining characteristics and the corresponding classifying characteristics are presented to the program for learning. The binary version of NeuroShell accepts only inputs that are true/false, black/white, or yes/no. The analog version accepts ranges for the inputs. Both versions assign numerical scores to the classification. All the possible diagnoses are listed and beside each is its relative score.

2.2.1 *Example: Classification of Microorganisms*. This type of problem was suggested by Dr. Lesley Robertson of the Technical University of Delft (Bungay and Bungay, 1991). Convert the following table for keying out bacteria to NeuroShell:

Genus	Rods	Gram	Anaerobic	Micro-aerophillic	Spores	Fix N2	In Pairs
Azotobacter	+	-	-	-	-	+	+
Lactobacillus	+	+	+	+	-	-	-
Bacillus	+	+	-	-	+	-	-
Clostridium	+	+	+	-	+	-	-
Neisseria	-	-	-	-	-	-	+
Veillonella	-	-	+	-	-	-	+
Diplococcus	-	+	+	+	-	-	+

Solution: You must have the programs from Ward Systems Group, Inc. Make a working copy, and put the original in a safe place. Boot your computer system, insert the disk with Neuro Shell, and type **DIR** [**enter**]. Check to be sure that the disk is not too full; you should have at least 50 kilobytes of free space. If there is not much room on the disk, delete one or more of the examples.

1. Type **BINARY** [**enter**]
   Information on the screen guides you.

2. When prompted for the name of a file, type **bacteria** [**enter**]

The program will not find that file and will initiate a new problem. You will soon see the following:

Main Menu Options
Define the characteristics
Enter characteristics for sample cases
Check for duplicate sample cases
Learn the sample cases (develop a model of the problem)
Classify new cases according to the problem model
Print the characteristic definitions

Print the characteristics of classified cases
Select from advanced options menu
Quit this program

3. Use the cursor keys to move up and down through these choices. Select "Define the characteristics" and hit [enter]

4. A screen appears. Position in the top region with the cursor keys and hit the insert key enough times to create seven lines above the dotted dividing line.

5. Position the cursor even with the top line and type **RODS** [enter].

6. Continue entering the characteristics. If you make a mistake, reposition the cursor for the ine with the error, and start typing. Note that the line appears at the top of the screen, and you can use the backspace key to delete.

7. Position the cursor just below the dotted dividing line, and type each bacterial name followed by [enter].

8. When all the names are entered, hit the ESCAPE key to save this information and to return to the menu screen.

9. On the menu screen, select "Enter the characteristics for the sample cases" [enter]

10. The screen with the characteristics appears. A characteristic is selected or deselected as a toggle. Line up the cursor with a characteristic and hit [enter]. If the line was bold (true), it will become normal type (false). If it was normal, it becomes bold. Work through the sample case making the correct pattern for Azotobacter and make only Azotobacter bold among the list below the dotted dividing line. Hit the [rtarrow] key to save this case and to create the next case. Note that the case number is at the top of the screen. You may use Figure 1 as a guide. Note that the selected characteristics are in bold face.

Figure 1
Typical Training Data

Rods
Cocci
Gram
Anaerobic
Micro-aerophillic
Spores
Fix N2
In Pairs

Azobacter
Lactobacillus
Bacillus
Clostridium
Neisseria
Veillonella
Diplococcus

Figure 2
Typical Result for Classification

Rods
Cocci
Gram
Anaerobic
Micro-aerophillic
Spores
Fix N2
In Pairs

Azotobacter	0.20
Lactobacillus	0.77
Bacillus	0.05
Clostridium	0.05
Neisseria	0.28
Veillonella	0.46
Diplococcus	0.05

11. When all the cases have been entered, hit the ESCAPE key to save them and to return to the menu screen.

12. Position the cursor at "Learn the sample cases (according to the problem model)" and hit enter . Some time will elapse. On a slow laptop computer, this takes 3 to 5 minutes. Eventually there will appear a message that learning is over.

13. Position the cursor at "Classify new cases according to the problem model" and hit enter.

14. On the screen that appears, do as before on the top portion to make bold or normal the characteristics. When you have them as you wish (one of the sample cases prior to learning is a good test), use the function key **F3** to classify. A typical classified case is shown in Figure 2.

Note that there are decimals besides the names of each of the bacteria, and the correct organism has a high but not perfect score. Make changes and reclassify to observe the effects. The relative scores will change. This ability to rank decisions is both a great advantage and a drawback of a neural network. When you want absolute scores and firm decisions, a neural network may not be the method of choice. However, a ranking of all the possible decisions is often highly desirable, and further testing can be used to build upon the information from the neural network.

15. When you are finished, hit **ESCAPE** to return to the menu screen. Select "Quit this program" enter.

16. When you see the DOS prompt, type **DIR BACTERIA.*** enter. Observe the files that were created for your problem.

2.2.2 *Example: Flow Characterization.* This problem requires the analog version of NeuroShell. Consider a well-mixed tank with constant volume and feed rate. We make a step input in concentration to the feed steam and observe the output response. From this response, we want our neural net to estimate the dilution rate (Bhagat, 1990). We will train the neural net with ideal cases produced by simulation. The equation is:

$$dC_1/dt = D^* (Ci - C_1)$$

where C_1 = concentration in the tank
Ci = input concentration
D = dilution rate

We will use data at dilution rates of 0.05, .1, .2, .3, and .4 for training and data at dilution rates of 0.15 and 0.22 for testing. Training data are in Figure 3 and testing data are in Figure 4.

Figure 3. Digitized Results for Tank Responses --Training Data

Dilution Rate	=.05	= 0.1	=0.2	=0.3	=0.4
Time					
1	0.48	.951	1.81	2.59	3.29
3	1.39	2.59	4.51	5.93	6.98
5	2.21	3.93	6.32	7.76	8.64
9	3.62	5.93	8.34	9.32	9.72
17	5.72	8.17	9.66	9.93	9.98
25	7.13	9.17	9.93	9.99	9.99
35	8.26	9.69	9.99	9.99	9.99

Figure 4. Digitized Results for Tank Responses -- Testing Data

Dilution Rate	=0.15	= 0.22
Time		
1	1.39	1.97
3	3.62	4.83
5	5.27	6.67
9	7.40	8.61
17	9.21	9.76
25	9.76	9.95
35	9.94	9.99

The steps for using NeuroShell are:

1. With the NeuroShell disk in the proper drive, type **ANALOG** [enter]

2. When the program comes up, type **dilution** [enter]

3. On the selection menu, line up with "define the characteristics" and [enter]

4. Move the cursor to the top and use the INSERT key sufficient times to get seven lines for input characteristics.

5. Check the number of entries below the line. In this case, there is but one output so you do not need to insert more lines.

6. Position the cursor, and type names for each of the lines. It is probably clear enough to use simply r1, r3, r5, r9, r17, r25, and r35 for responses at 1, 3, 5, 9, 17, 25, and 35 minutes and d for the dilution rate.

7. Use the **rightarrow** cursor key at which point you will be prompted to specify the maxima for each input line. Move the cursor and make 10.0 the maximum for each of the responses and 0.4 the maximum for dilution rate.

8. Hit the escape key to save these specs and to return to the menu.

9. Select "Enter the characteristics for sample cases" [enter]

10. Enter the responses for the first data set in Figure 3. You simply move the cursor and type.

11. Use the cursor key (rightarrow) to bring up a new blank case. Continue adding data. You can use the (leftarrow) key to go back and check previous cases.

12. When satisfied with your entry of data, hit ESCAPE to return to the menu.

13. Select "Learn the sample cases" [enter]

    This will take several minutes.

14. Your computer should beep when the algorithm for your new neural net is ready. Select the option "Classify new cases ..." [enter]

15. Enter data from Figure 4 for the test cases. When ready use the function key **F3** to classify. You should get answers very close to the correct answers. If not, check your training data for errors from typing.

    The classifications should be surprisingly good. If they are incorrect, check your data.

3. Additional Reading

3.1 EXPERT SYSTEMS

Anon., "DORIS: An expert system for reactor design", Chem. Engr. Prog. April 1990 pp. 88-90

Basta, N., S. Ushio, and H. Short "Expert systems - Thinking for the CPI" Chemical Engineering, Mar. 14, 1988, pp.26-29

Bensen, R. If/Then, The hands-on introduction to artificial intelligence using Lotus 1-2-3, If/Then Solutions, Palo Alto, CA (1987)

Bielawski, L., and R. Lewand (1988) Expert Systems Development, Building PC-based Applications, QED Information Sciences, Inc., Wellesley, MA

Bungay, H.R. BASIC Environmental Engineering, text and disk with 80 programs, BiLine Assoc., Troy, NY (1988)

DeBernardez, E., P. Dhurjati, and D. Lamb, "A Hybrid Heuristics and Mathematical Model Based Expert System to Diagnose Metabolic State of Cells" paper presented at Am. Chem. Soc. National Meeting, Washington, D.C. (1986)

Hassan, A. and W. Tank, "Expert system for selection of pumps in the chemical industry", Chemie Ing. Technik **61** 838 (1989)

Hohne, B.A., and T.H. Pierce, "Expert Systems Applications in Chemistry", ACS Symposium Series No. 408 (1989)

Hushon, J.M., "Expert systems for environmental problems", Environ. Sci. and Technol. **21** 838-841 (1987)

Hushon, J.M., ed. "Expert systems for environmental applications", ACS Symposium Series 431 (1990)

Jem, K.J., "Spread Sheets & an Expert System for Fermentation Process Optimization" Genetic Engineering News, March 1980 p 12

Joseph, B., H-T. Wu, and B. Allan, "Automation of a batch process with expert system shells", Chem. Engr. Prog. 85: 87-91 (1989)

Lapoint, J., B. Marcos, M. Veillette, G. Laflamme, and M. Dumontier, "Bioexpert - an expert system for wastewater treatment process diagnosis", Computers and Chem. Engr. 13: 619-630 (1989)

Miller, E.J., K.D. Wilson, and C.R. Lewis, "Expert System Shells: Do they deliver what they promise?", Chem. Engr. Prog. 84: 37-44 (1988)

Patry, G.C., and D.T. Chapman, "Dynamic Modelling and expert systems in wastewater engineering" Lewis Publishers (1989)

Raeth, P.G., editor, "Expert Systems --A Software Methodology for Modern Applications", IEEE ISBN 0-8186-8904-8 (1990)

3.2 NEURAL NETWORKS

Bhagat, P., "An introduction to neural nets", Chem. Engr. Prog. August 1990, pp. 55-60

Borgman, S., "Neural network applications in chemistry begin to appear", Chem. Engr. News, April 24, 1989, 24-28

Bungay, H.R. and M.L. Bungay, "Identifying Microorganisms with a Neural Network", BINARY, 13: 51-52 (1991)

Hoskins, J.C., and D.M. Himmelblau, "Artificial neural network models of knowledge Representation in Chemical Engineering", Computers in Chem. Engr. 12: 881-890 (1988)

Obermeier, K.K. and J.J. Barron, "Time to get fired up" Byte, Aug. 1989, pp. 217-224

Petersen, S., H. Bohr, J. Bohr, S. Brunak, R.M.J. Cotterill, H. Fredholm, and B. Lautrup, "Training neural networks to analyse biological sequences", Trends in Biotechnol. 8: 304-308 (1990)

Thibault, J., V. Van Breusegem, and A. Cheruy, "On-line prediction of fermentation variables using neural networks", Biotechnol. Bioengr. 36: 1041-1048 (1990)

Touretzky, D.S. and D.A. Pomerleau, "What's in the hidden layers", Byte, Aug. 1989, pp. 227-233

Wantanabe, K., I. Matsuura, M. Abe, M. Kubota, and D.M. Himmelblau, "Incipient Faule Diagnosis of Chemical Processes via Artifficial Neural Networks", AIChE Journal, 35: 1803-1812 (1989)

NONLINEAR OSCILLATIONS IN BIOREACTION SYSTEMS

R.Y.K. YANG AND J. SU
Bioreaction Engineering Laboratory
Department of Chemical Engineering
West Virginia University
Morgantown, WV 26506-6101, USA

ABSTRACT. An operation strategy of utilizing self-generated limit cycle for performance enhancement of a chemostat system, proposed and demonstrated previously on pure culture with Monod type growth and linearly varying yield coefficient, was applied to pure culture of substrate-inhibition type growth and found to produce performance improvements of similar magnitude.

1. Introduction

The wide-spread use of computers and the development of a range of new concepts concerning non-linear systems, such as deterministic chaos and fractals have made Nonlinear Dynamics a recognized frontier area of research in recent years. This new field, characterized by its unprecedented cross-disciplinary flavor, has already affected the outcome of a wide range of physical, chemical biological, medical and social sciences [1]. However, its potential impacts on engineering research have gradually been recognized only recently [2].

Most bioreaction systems occurring in nature or taking place in industrial bioprocesses are inherently nonlinear and are constantly subject to perturbations from their surroundings. After being perturbed, a majority of them will eventually reach a time-invariant steady state, with or without going through a period of chaotic or non-chaotic oscillations. However, some systems, under certain conditions, may lead instead to a self-sustained oscillation (SSO) known as a limit cycle. Such biological oscillations range in period from milliseconds to days or even months, and may be closely related to biological rhythms, a fundamental property of all plant and animal life. The far-reaching impact of biological rhythms on the work place and thus its significant economic consequences to many industries have gradually been recognized recently.

Nonlinear oscillations in both enzymatic and microbial systems are currently being studied in our laboratory. An example of nonlinear oscillations in an enzymatic system (oxidation of NADH by horseradish peroxidase [3]), simulated numerically in our laboratory [4], is shown in Figure 1, where [A] is NADH concentration and k_1 and k_5 are kinetic parameters. The patterns of oscillation of this system have been extensively studied in recent years (see e.g., [3,5]).

One of the aims of our studies of nonlinear oscillations is to explore the possibility of improving the performance of bioreactors, particularly CSTB (continuous stirred-tank bioreactor, i.e. chemostat) via such oscillations. We have proposed recently a

A.R. Moreira and K.K. Wallace (eds.), Computer and Information Science Applications in Bioprocess Engineering, 77-82.

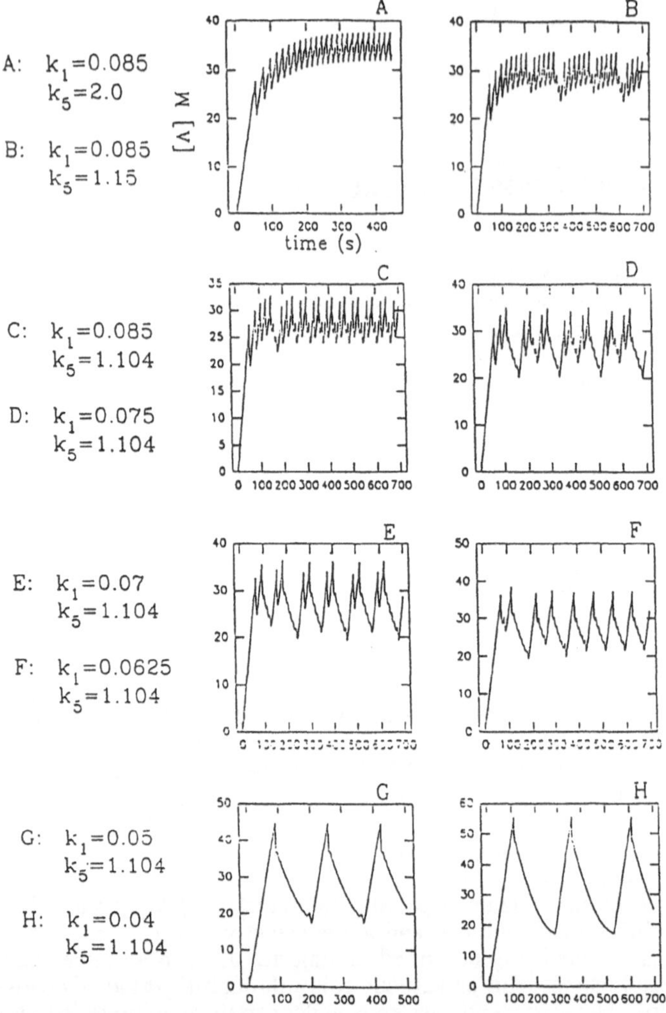

Figure 1. Different patterns of oscillation of NADH concentration in a fed-batch enzyme reactor.

novel operation strategy of employing self-generated SSO to improve chemostat performance without requiring additional input of energy [6]. The process concept was demonstrated via computation on a pure culture with Monod type growth and linearly varying yield coefficient. Several-fold increases in cell-mass concentrations in the outputs were observed [6]. In this paper, the numerical demonstration is extended to pure culture with substrate-inhibition type growth.

2. The Systems

For an objective comparison of the proposed operation strategy two chemostat-in-series systems of equal total volume and volumetric flow rate are considered, one (system 1) with oscillations and one (system 2) without oscillation. In system 1, the operating condition of the first CSTB is carefully chosen to allow SSO to occur, while holding its input constant. The oscillatory output from the first CSTB then causes "forced oscillation" to occur in the second CSTB, resulting in improved overall performance for the system.

The system equations, as listed in Table 1, were put into dimensionless form and then solved numerically for a variety of parameters and conditions [7].

Table 1 System Equations

$$V_1\frac{dS_1}{dt} = F(S_0-S_1) - V_1X_1\frac{\mu(S_1)}{Y(S_1)}, \qquad S_1(0)=S_1^* \tag{1}$$

$$V_1\frac{dX_1}{dt} = F(X_0-X_1) + V_1X_1\mu(S_1), \qquad X_1(0)=X_1^* \tag{2}$$

$$V_2\frac{dS_2}{dt} = F(S_1-S_2) - V_2X_2\frac{\mu(S_2)}{Y(S_2)}, \qquad S_2(0)=S_2^* \tag{3}$$

$$V_2\frac{dX_2}{dt} = F(X_1-X_2) + V_2X_2\mu(S_2), \qquad X_2(0)=X_2^* \tag{4}$$

$$\mu(S)=\frac{\mu_m S}{K_s+S+K_iS^2} \qquad \text{(Specific Growth Rate)} \tag{5}$$

$$Y(S)=\alpha+\beta S \qquad \text{(Yield Coefficient)} \tag{6}$$

3. Results and Discussion

Shown in Figure 2 are the results corresponding to the set of kinetic parameters, operation conditions and initial conditions indicated. Dimensionless substrate and cell mass concentrations in both chemostats are plotted as functions of dimensionless time in Figure 2A. Limit cycle in the first chemostat is clearly illustrated in Figure 2B, while the forced oscillation in the second chemostat is shown in Figure 2C with the region near the oscillatory trajectory further amplified and displayed in Figure 2D. The results indicate that once limit cycle is reached, the frequency of oscillation in both CSTBs are the same, whereas the amplitude of oscillation in the second CSTB is much smaller than that of he first CSTB.

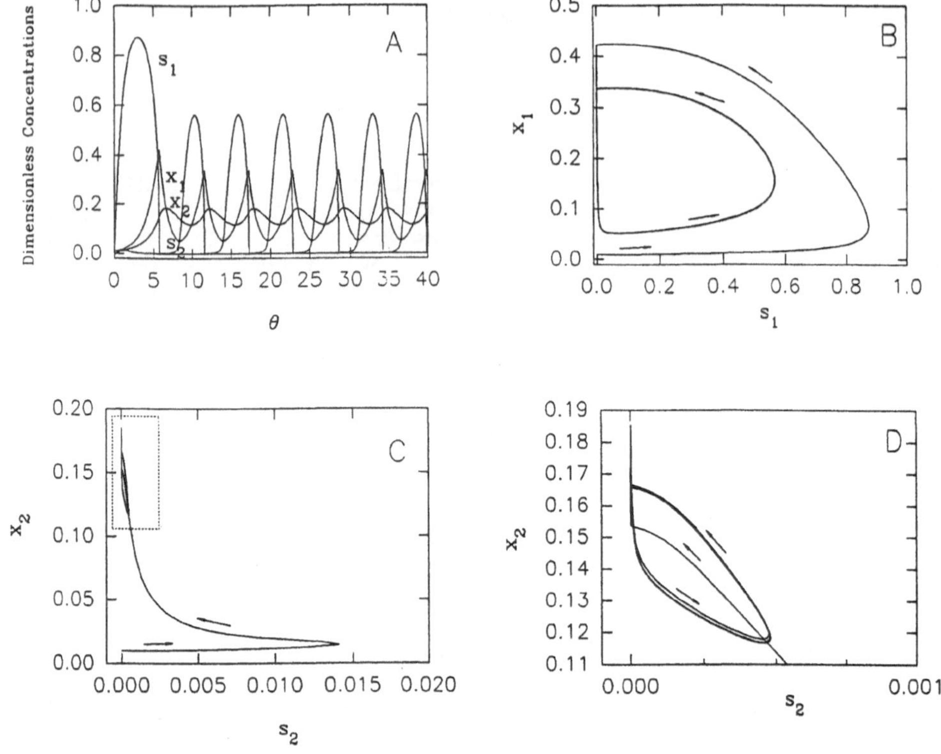

Figure 2. Oscillations of substrate and cell-mass concentrations in a two-chemostat-in-series system with substrate-inhibition type growth. (μ_m = 0.526 hr^{-1}, K_s = 1.824 g/l, K_i = 3.76×10^{-4} l/g, S_0 = 35 g/l, $\alpha = 0.01$, $\beta = 0.03$ l/g, $\tau_1 = 3.5$ hr, $\tau_2 = 10.5$ hr)

The performance of the chemostat system, as measured by the time-averaged cell-mass concentration in the system is plotted against the ratio of the residence time in the first CSTB to that of the second CSTB (see Figure 3). The qualitative behaviors of the chemostat system with substrate inhibition-type growth are similar to that of Monod type growth [6]. Again, three operation regimes, i.e. washout, time-variant (oscillation) and time-invariant (steady state), are observed. As τ_1/τ_2 decreases while keeping $\tau_1+\tau_2$ constant, the operation regime changes from the ordinary time-invariant steady state to the time-variant oscillatory state and finally to washout in the first chemostat. It is observed that the performance in oscillation regime is always better than that of the time-invariant steady-state. Furthermore, in the oscillation regime as τ_1/τ_2 decreases, the amplitude of oscillation increases, but more importantly, the performance of the CSTB system improves dramatically (up to 583% when compared to the steady state reference system).

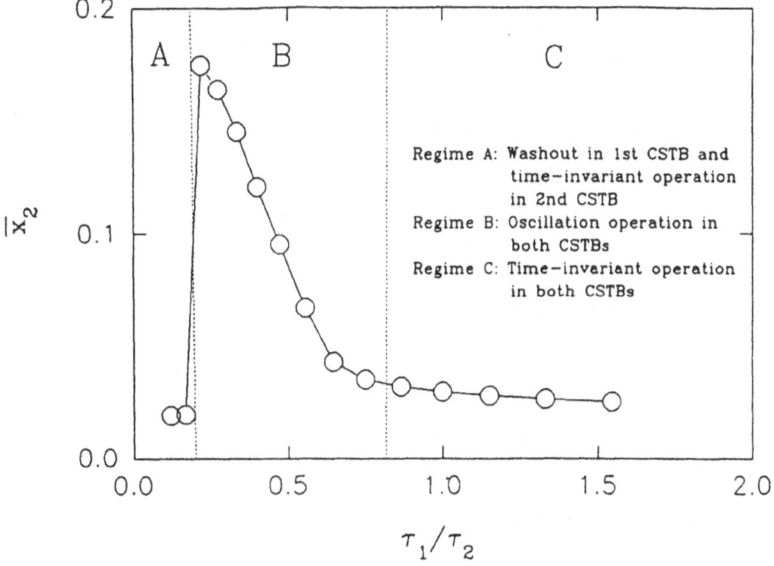

Figure 3. Time-averaged cell-mass concentrations in system output vs. ratios of residence times. ($\mu_m = 0.526$ hr^{-1}, $K_s = 1.824$ g/l, $K_i = 3.76\times10^{-4}$ l/g, $S_0 = 35$ g/l, $\alpha = 0.01$, $\beta = 0.03$ l/g, $\tau_1 + \tau_2 = 14$ hr)

4. Conclusion

The operation strategy of utilizing self-generated limit cycle for performance enhancement of chemostat system [6], when applied to pure culture with substrate-inhibition type growth, produces dynamic behaviors and significant performance improvements similar to that observed for pure culture of Monod type growth.

5. Acknowledgment

Support from the National Science Foundation is gratefully acknowledged. Computation and illustration associated with Figure 1 were performed by G. Ridenour.

6. Notation

F	volumetric flowrate, $l\ hr^{-1}$
K_i	inhibition constant, $l\ g^{-1}$
K_s	Monod constant, $g\ l^{-1}$
S	limiting substrate concentration, $g\ l^{-1}$
s	dimensionless S
t	time, hr
V_i	volume of ith bioreactor, l
X	cell mass concentration, $g\ l^{-1}$
x	dimensionless X
x	time-averaged x
Y	$Y_{x/s}$ (cell mass yield coefficient)
Greek letters	
α	constant in yield coefficient
β	constant in yield coefficient, $l\ g^{-1}$
Θ	dimensionless time, t/τ_1
μ	specific growth rate, hr^{-1}
μ_m	maximum specific growth rate, hr^{-1}
t	mean residence time (V/F), hr
Subscripts	
o	feed stream
i (1 or 2)	ith bioreactor in series
Superscripts	
*	initial conditions

7. References

1. Grebogi, C., "History of Nonlinear Dynamics and Chaos", NIH Workshop on Nonlinear Dynamics in Biological Systems, Bethesda, Md., 1992.

2. Douglas, J., "Seeking Order in Chaos", *EPRI Journal*, June, 1992.

3. Larter, R., C.L. Bush and T.R. Lonis, *J. Chem. Phys.*, 87, 5765, 1987.

4. Ridenour, G., "Dynamic Behavior of Peroxidase Reaction in CSTR Systems", Undergraduate Thesis, West Virginia University, 1992.

5. Steinmetz, C.G. and R. Larter, *J. Chem. Phys.*, 94, 1388, 1991.

6. Yang, R.Y.K. and J. Su, "Improvement of Chemostat Performance via Nonlinear Oscillations, Part 1: Operation Strategy", *Bioprocess Engineering*, in press.

7. Su. J., "Improvement of Chemostate Performances via Nonlinear Oscillations", M.S. Thesis, West Virginia University, 1992.

BIOREACTOR INSTRUMENTATION AND BIOSENSORS

KARL SCHÜGERL
Institut für Technische Chemie
Universität Hannover
Callinstr. 3, D-3000 Hannover, Germany

ABSTRACT. A review on instruments and methods for on-line measurements of cell behavior, medium composition and fluiddynamics of the aerated cultivation medium in bioreactors is presented.

1 Introduction

In bioreactors, complex interrelationships exist between the cells and their micro- and macro-environment. The micro-environment (local concentration profiles of key components) is important for cell aggregates and immobilized cells.
Suspended cells are mainly influenced by their macro-environment (flow pattern and mixing of key components in the bioreactor).
In situ, on-line, and quasi on-line monitoring and control of fluid dynamics, medium composition, and biological state of the cells are requisite for evaluating their interrelationships and to maintain the optimal reactor and process performance.
At universities and research laboratories, several new instruments have been developed during the past few years, which make these measurements feasible. In this review, these new developments will be of central interest.

2 Common Instruments

Temperature, pH value, and dissolved oxygen concentration have considerable effect on cell growth and product formation. Together with the pressure, stirrer speed, aeration rate, and feed rate, they are important control variables. Knowing the reactor weight is significant also. The off-gas composition is often used to control the bioprocess.

Pt-100 resistance thermometers (100 ohm at 0 ^{o}C), which are encased in a protective steel tube that is closed up with a coating of high heat conductivity material, are in great demand. They can be used without calibration with a guaranteed precision of ± 0.26 ^{o}C.

The pH value is measured by a galvanic cell consisting of an Ag/AgCl electrode/reference electrode system which is positioned in internal/external KCl buffer solutions. The measured potential of the outer glass membrane surface is a function of the H^{+}-ion concentration in the medium. On account of the temperature dependence

A.R. Moreira and K.K. Wallace (eds.), Computer and Information Science Applications in Bioprocess Engineering, 83-94.

of the potential, a good temperature control is prerequisite for precise pH measurements.

The dissolved oxygen concentration can be measured by polarographic or galvanic electrodes. In the polarographic electrode the dissolved oxygen is reduced at a constant voltage (650 mV) between the Pt cathode and the Ag/AgCl anode at the surface of the Pt cathode in KCl solution. At the current/voltage plateau (0.6-0.8 V), the current is proportional to the dissolved oxygen partial pressure. In the galvanic electrode, the oxygen is reduced at the noble metal cathode, and the Pb anode is oxidized to $Pb(OH)_2$. On account of the temperature dependence of the current, the temperature must be controlled closely ($\pm$ 0.1°C). With sufficient accuracy of the temperature and pressure measurements, a precision of $\pm$ 5% can be obtained.

The absolute pressure is measured with respect to zero pressure, the gauge pressure with respect to the atmosphere. Strain or capacitance membrane pressure gauges are suitable for aseptic conditions.
The impeller speed is monitored usually by tachometers. The liquid and gas throughputs nowadays are measured with magnetic-inductive flowmeters, the weight of the reactor by load cells.

O_2 and CO_2 are the most important off-gas components; they are monitored by a paramagnetic oxygen analyser and a CO_2 infrared analyser or mass spectrometer. For this purpose, inexpensive quadrupole mass spectrometers with a Faraday cage ion collector are sufficient.

3 Instruments and Methods for Cell Concentration and Behaviour Monitoring

3.1 *IN SITU* METHODS

At low cell concentrations (e.g., in animal cell cultures), the total cell number is counted by a microscope in counting chambers. Microorganism cultures usually have a high cell density. In these cultures, the cell mass often is determined off-line by means of their dry weight or by optical density OD in a photometer at about 600 nm.

For *in situ* monitoring of the optical density, different turbidity probes are used. The instruments on the market employ infrared transmission or retroreflexion for this purpose. Since the transparencies of the light source and detector windows change during cultivation, four beam systems with two light sources and detectors were developed. The photodiodes are turned on and off, while the photodetectors are operated continuously. Two signals with different light paths are received by the detectors and after linearization, they are used for signal correction (e.g., MEX 3 EURO CONTROL, BTG).
The retroreflective turbidity probe measures the intensity of the backscattered light (e.g., Aquasant).

The frequency-dependent permittivity of the cell mass in microbial suspensions is used as well for cell concentration measurements (Bugmeter). In microbial cultivations, however, the ion strength is too high for the measurements. The cells must be resuspended in water for the cell mass determination.

Acoustic methods for determining the liquid density are widely used in the chemical industry. The acoustic resonance densitometer was developed (Blake-Coleman) also for cell concentration measurements, but it can only be used for off-line determinations.

All of these instruments work reliably only if no bubbles are present in the volume to be measured. These instruments, thus, are often installed into a bubble trap containing a recirculation loop for the cultivation medium.

The presented instruments measure the total cell concentration irrespective of the viability of the cells. For measuring viable cell numbers off-line, the trypan blue exclusion technique is combined with the microscopical cell counting.

A heat flux calorimeter can be used for the on-line measurement of the viable cell mass concentration by means of the heat they produce by their metabolic activity (Marison and von Stockar, 1986). In large reactors, the heat loss is negligible compared with the heat produced by the microorganisms, therefore, the viable cell concentration can be determined by means of heat balance, if the heat produced by the stirrer and the heat loss by evaporation are known.

Oxygen utilization rate OUR and CO_2 production rate CPR, determined by the O_2 and CO_2 mass balances, can be used for the viable cell concentration measurement, if the cells are in a balanced growth state.

The same holds true for the measurement of the culture fluorescence. If the cells are irradiated by UV light at 340-360 nm, the reduced adenine dinucleotides (NADH and NADPH) in living cells emit fluorescent light at about 460 nm. The intensity of the fluorescence is affected by the metabolic state of the cells. Only in balanced growth, the viable cell concentration can be related to the flourescence intensity. The culture fluorescence intensity can be influenced by fluorescence-active medium components (casein, peanut flour, etc.), the concentrations of which vary during cultivation. Therefore, two separated measurements are necessary: with cell-free and with cell-containing medium.

Due to the dependence of the intensity of fluorescence emitted by the cells on the biological state of the cells (NADH and NADPH content, which is a function of the specific growth rate, substrate and dissolved oxygen concentration), it can also be used for monitoring the cell behaviour (e.g., change of oxidative to oxidative-fermentative metabolism of yeasts).

The ratio of CPR to OUR, the so-called respiratory quotient RQ, also is a function of the cell metabolism. It is sometimes used, e.g., for the control of the production of baker's yeast.

3.2 *EX-SITU* METHODS

The most widely used instrument for cell components analysis is the flow cytometer. It is well-suited for the analysis of the cell size distribution and the DNA, RNA, protein, and lipid content of the cells as well as (e.g., biopolymer) for products. The analysis rate is extremely high, up to 10 000 cells per sec. Advanced instruments are equipped with a sorter, e.g., for selecting spores of a definite size for inoculation.

Nuclear magnetic resonance NMR spectroscopy is a non-invasive, non-destructive technique to study fundamental physiological and metabolic phenomena *in vivo*. Its major drawback is its low sensitivity and high cost, which explains its infrequent application.

4 Instruments and Methods for on-line Medium Composition Monitoring

Except for temperature, pH, pO_2, and perhaps certain ion concentrations measured *in situ*, the concentrations of all other medium components are measured outside the reactor in a nonsterile area. For on-line analysis, continuous aseptic sampling is necessary.

4.1 *ON-LINE* ASEPTIC SAMPLING

We can sample cell-containing or cell-free medium.
In the case of the sampling of cell-containing medium, we need a steam or chemical barrier for maintaining the aseptic conditions in the reactor. Holst et al. (1988) used a chemical barrier. They applied two coaxially arranged hypodermic syringes for sampling, which were pierced into the reactor. The growth inhibitor was fed through the annular space, and the sample was taken from the inner tubing. The sampling rate was set to be higher than the feed rate of the inhibitor to keep the inhibitor from entering the reactor. Reuss et al. (1987) employed a steam barrier in his sampling system. The cells were separated by a filter and their volume was measured. The filtrate was used for medium analysis.

For sampling cell-free medium, membrane filters are used as barrier. Either separate sterilizable cross-flow filtration modules are used, which are integrated into a medium recirculation loop, or *in situ* sampling filters, which are sterilized with the reactor *in situ*. The only equipment used in industrial practice are *Biopem* (B. Braun Melsungen), a flat membrane cross-flow filtration module combined with a magnetically driven stirrer to reduce fouling effects, and a tubular microfiltration membrane (ABC Biotechnologie/ Bioverfahrenstechnik, Puchheim) that is inserted into the reactor.

After sampling, the specimen must be prepared for analysis. This preconditioning may be done either by dilution, protein precipitation, pH shift, or dialysis.
Since these samples usually contain nutrient salts and substrates, it is possible that microorganisms may grow in them. To avoid microorganism propagation, either growth inhibitors are mixed into the sample, or the analyser system is rinsed occasionally by growth inhibitors and surfactants to remove deposits.

4.2 Continuous flow analysis

Continuous flow analysers are widely used in the chemical industry. The "Autoanalyser" of Technicon (Bran and Lübbe) was the first and most successful air-segmented continuous analyser system employed there. In this equipment, the analyte is converted by a suitable reagent into a product, which can be determined easily, usually in a photometer detector. Continuous air-segmented flow analysers are also applied in biotechnology. However, during prolonged cultivation processes, microorganisms grow in the analyser in spite of all precautions. Thus, they can only be recommended for short cultivation processes. Their advantage is the continuous signal, which is only slightly sensitive to power supply fluctuations caused by the change of the electrical load. Their signal can be used directly for closed loop control.
Their disadvantages are the large amounts of chemicals they require for analysis and the long analysis time, because the product concentration is measured after the chemical equilibrium has been attained.

4.3 Flow injection analysis

The flow injection analysis (FIA), invented in 1974 by Ruzicka and Hansen, has gained in important also in biotechnology during the last few years (Ruzicka and Hansen, 1988).
In this technique, a small amount of sample is injected into the carrier flow, which frequently contains the reagent. The analyte reacts with the reagent, and the product concentration is determined in a detector. Contrary to the continuous flow analysis, which has a steady state signal, the signal in the FIA varies with the time, and the peak height or the surface area under the peak-shaped signal is measured. Since only a small amount of sample is injected into the carrier flow, the preconditioning of the sample for the analysis is easy to obtain by using an optimized carrier composition.
The sample volume required for the analysis is minimal. If the reagent is injected into the carrier as well, the reagent consumption can be kept low. On account of the large dilution of the sample in the carrier, nutrient salts and substrate concentrations are reduced to such low a level, where cell growth is rather improbable. Since the analyte concentration in the carrier passes a bell-type distribution, a controlled dilution is possible by using the signal in different distances from the peak maximum.
The disadvantage of FIA is the discontinuous signal, which is very sensitive to power supply voltage fluctuations. It is difficult to use signal filtering for process control.

4.4 Application of Biosensors as Specific Detectors

In a broader sense, biosensors are instruments in which a biological receiver is combined with a transducer that converts the physical/chemical variation in the receiver into an electric signal, regardless of the type of connection between receiver and transducer.
There is no biosensor as yet on the market, which can be used for *in situ* monitoring and control of medium components. Besides, very few biosensors are used for on-line monitoring and control of bioprocesses. Most published investigations on on-line monitoring and control of medium components are misleading, because actually, the authors merely connected a biosensor to a reactor without validating the biosensor signal. Signal validation takes quite a few months of systematic investigation, otherwise the results are incorrect or even irrelevant.

Several types of biosensors have been developed, but only few were used for on-line monitoring and control of medium components: electrochemical, optical, and calorimetric biosensors.

Two types of electrochemical biosensors are widely used: amperometric biosensors, which frequently use oxygen or H_2O_2 detectors, and potentiometric biosensors, which use pH detectors as transducers.

Optical biosensors (optodes) frequently use the variation of fluorescence dye emission due to the variation of the dissolved oxygen concentration or the pH value caused by the enzymatic reaction or the variation of the NADH (NADPH) emission.
In order to measure the concentration of proteins, immunoassays with fluorescence-tagged antibodies are applied frequently.

Two different types of flow injection analysers with amperometric biosensors are on the market. In one type, the oxidase is immobilized on a membrane, and the H_2O_2 concentration formed during the enzymatic conversion is measured with an amperometric detector (e.g., Yellow Spring Instruments FIA), in the other type, the oxidase is immobilized on carrier particles (e.g., VA Epoxy Biosynth, Riedel de Haen) which are filled into a small (0.8 to 1.5 ml) reactor that is connected with an amperometric oxygen detector (e.g., Eppendorf Variables Analysersystem (EVA) with enzyme cartridges).

Optodes are not as yet on the market. However, protypes have been used for process monitoring.

No calorimetric biosensors are on the market either. Instead, prototypes have also been used for process monitoring.

On account of the wide use of amperometric oxygen electrodes and potentiometric pH electrodes, electrochemical biosensors so far are in greatest demand. Also enzymes are immobilized on the pH-sensitive gate (e.g., $Si/SiO_2/Si_3N_4/Ta_2O_5$) of field effect transistors (FETs) and are combined with a reference (Ag/AgCl) electrode. The drawback of all pH electrodes and FETs is that their signal strongly depends on the buffer capacity of the culture medium. This can be avoided by using fluoride-sensitive (pF) electrodes or gates (e.g., $Si/SiO_2/Si_3N_4/LaF_3$) of FETs together with fluoroaniline and reference electrodes. The H_2O_2 formed during the reaction with oxidase and peroxidase converts fluoroaniline into fluoride ion, which is detected by the pF electrode.

4.5 Chromatographic Analysis

Gas chromatographs (GC) and high performance liquid chromatographs (HPLC) are common instruments for the analysis of complex mixtures. Contrary to continuous flow analysers (CFAs) and flow injection analysers (FIAs) which have specific detectors, GCs are equipped with non-specific detectors as thermal conductivity and flame ionization detectors, and HPLCs are provided with detectors as refractive index detectors, photometers, fluorometers and polarimeters. In several cases, the components must be derivatized to permit detection, e.g., by fluorimetry. The reaction usually is carried out by precolumn derivatization. For amino acids, the most popular derivatization reagent is ortho-phtaldialdehyde (OPA).

In contrast to the large number of publications on GCs and HPLCs, very few papers have been published about their use in an on-line mode.
In liquid chromatography, very high purity and care is required, since the smallest dust particles can stick to the capillary wall or in the pores of the metal filter and cause clogging. The eluents are prepared with twice-distilled water, cleaned through a microfiltration membrane and degassed in an ultrasound bath. To avoid gas absorption, they are gassed with helium during the analysis. The samples are deproteinized by methanol and ultrafiltration for the analysis of low-molecular components.

The advantage of the chromatographic methods is their ability to analyse several components simultaneously. Their disadvantages are the long analysis time and the infrequent discontinuous signal, which makes the use of signal filtering techniques nearly impossible.

5 Local Measurements in Biofilms, in Pellets, and in Immobilized (Gel-Encapsulated) Cells

Since transport limitation frequently occurs in cell aggregates, which can be a rate-limiting process, local measurements are also necessary for system analysis and optimization.
The instruments used for this purpose are miniaturized sensors.

The most common instruments are micro-pH and -pO_2 sensors having diameters of only a few micrometers. Ion-selective sensors can be constructed by applying a thin layer of ion exchangers to the pH sensors. Biosensors consist of pH or pO_2 sensors combined with a tiny enzyme membrane. Such measurements allow the evaluation of the microenvironment of immobilized cells and their influence on cell behaviour and process performance.

6 Instruments and Methods for in situ Monitoring of Bioreactor Fluid Dynamics

6.1 Pressure monitoring

The most common method for the evaluation of the integral gas holdup is the measurement of the volumes of aerated and non-aerated cultivation media. The determination of the volume expansion in aerated stirred tank reactors, however, leads to incorrect results, thus, other methods are preferred. Since the non-aerated volume can be calculated from the weight of the cultivation medium, provided its density has been ascertained, only the volume of the aerated cultivation medium needs to be known. By means of the relative pressure difference measured at different heights in the reactor, the local gas holdup can be evaluated. The mean gas holdup is determined by the relative pressure difference between the bottom and the top of the reactor.

6.2 Optical methods

In non-aerated reactors, the liquid velocity can be measured off-line by a Laser Doppler Anemometer (LDA) if the culture medium is optically transparent. Since this is frequently not the case, the use of optical methods is limited to local light-scattering measurements, whereby solid particles (cells) in non-aerated medium and bubbles and particles in aerated cultivation medium are the determining factors. Separation of light transmission signals caused by solid particles and bubbles is not possible.

Only optical fiber probes based on total light reflexion, can detect bubbles unequivocally. The light beam is guided from the light source through an optical fiber into the culture medium. At the fiber end, where the reflecting coating material is removed, the light can be transmitted into the aerated medium. If the fiber tip is surrounded by a bubble, total reflexion occurs, and the light intensity in the detector, connected to the optical fiber by a light switch, increases.

6.3 Ultrasound methods

The ultrasound Doppler technique can be used for measuring the bubble velocity distributions. Since bubbles act as perfect reflectors for ultrasound, one can compare the frequencies of the original and reflected ultrasound pulses and then calculate the bubble velocity from the Doppler shift.

Similar to light, ultrasound is scattered by bubbles. The specific interfacial area of the bubbles can be calculated from the attenuation of the transmitted power as well as by sound reflection at the surface of the bubbles. The reflected power increases with the reflecting surface area, but decreases due to scattering along the path from the transmitter to the measuring volume and back to the detector. Thus, with increasing bubble density (interfacial area), the signal passes a maximum. By variation of the specific interfacial area by the aeration rate, the maximum of this curve can be determined, which is necessary for the quantitative evaluation of the signal. To avoid this calibration, the reflected power is measured at two different distances from the probe just by varying the predetermined time window by means of the electronic gating. At the same time, the bubble velocity can also be obtained from the Doppler shift. This technique was used for monitoring the bubble velocity distribution and the specific interfacial area during cell growth and product formation with several biological systems.

6.4 Electrochemical methods

Electrical conductivity probes are frequently used for the evaluation of the bubble properties. The electrical conductivity of aerated cultivation medium at the probe tip is measured by means of one, two, or four electrodes. Since cultivation media contain salts, they have a rather high electical conductivity. If a bubble hits the probe tip, the conductivity decreases. Probes with a single electrode measure local holdup, ones with two electrodes evaluate the bubble velocity and the pierced length, probes with four electrodes determine the local gas holdup, the pierce length, and the bubble velocity vector.

6.5 Hot-Film Anemometers (HFA)

The methods mentioned in 6.2 to 6.4 are employed to determine the properties of the dispersed gas phase. In non-aerated liquids, HFAs can be used to determine the flow field of the liquid phase. A thin metal strip on the probe tip is heated. The heat loss is a function of the local liquid velocity. The temperature of the metal strip is kept constant by increasing the heating power. The liquid velocity is calculated from the power necessary to keep the metal strip temperature constant. In a complex flow, the flow direction is not well-defined. Modern versions of the HFA are developed to make the sensor more direction-sensitive. Dual split-film probes allow the determination of the flow direction.

In aerated liquids, the signal is interrupted by the bubbles. It is very difficult to take into consideration the signal variation caused by bubbles.

6.6 Signal-response techniques

The signal-response technique can be employed for measuring the local or global properties of the flow. Local measuring techniques are also called Time-of-Flow (TOF) techniques, because they cause a disturbance in the flow and measure the response time at different distances from the source of disturbance. The most advanced technique is the pseudostochastic heat pulse technique. The pseudostochastic heat pulse sequence is generated by a small high frequency heat transmitter, whereby the liquid temperature is measured downstream from the source by a sensitive resistance thermometer. The local velocity distribution and mixing properties can be evaluated from the cross-correlation function of the test and response signals.

Global signal response techniques are employed for the evaluation of residence times and mixing intensities of the gas and liquid phases in the entire reactor. Helium often is used as tracer in the gas phase, and its concentration is measured with a small and inexpensive quadrupole mass filter.

In clear water dye, salt solutions, acids, or alkaline solutions can be used as tracers. In cultivation media, which are coloured and turbid and have a high buffer capacity and ion strength, none of these are suitable as tracers. Fluorescent dyes applicable at very low concentrations and detectable by fluorometers, can be used if the solid particle (cell) concentration is not too high and the colour of the medium is not too intensive. Again, pseudostochastic test signals are preferred, because the evaluation of the weighting function of the investigated system is much easier, and the accuracy of the measurements are much higher than with deterministic test signals.

6.7 Volumetric Mass Transfer Coefficient K_LA

The volumetric mass transfer coefficient of oxygen is a key parameter of aerobic cultivation processes. They can be evaluated from the oxygen mass balance by means of measuring the oxygen concentration in the off-gas and in the culture medium. For air/oxygen mixtures, the oxygen concentration in the inlet gas must be measured as well. In well-equipped bioreactors, k_La is evaluated in real-time.

7 Conclusion

There has been a remarkable development in the field of bioreactor instrumentation and measuring techniques during the last ten years.
Especially the on-line technique made big advances.
Most of the presented measuring techniques are only employed in laboratories, not as yet in the production areas. Their reliability must be improved considerably before they are suitable for process monitoring and control in industrial production.

8 References

On account of the broad field covered in this review, it would be necessary to refer to a large number of original papers, which would go far beyond the limits of this article. Therefore, instead of citing single papers, review articles and books are referred to.

Flow Injection Analysis and Biosensors:

Ruzicka, Jaromir and Hansen, Elo H. (1988) Flow Injection Analysis, 2nd ed. John Wiley & Sons, New York

No. 1 of Volume 14 (1990) of Journal of Biotechnology, Special Issue on Application of FIA in the Life Sciences.

Schmid, R.D. ed. (1991) Flow Injection Analysis (FIA) Based on Enzymes and Antibodies, VCH, Weinheim

Scheller, F. and Schmid, R.D. (1992) Biosensors: Fundamentals, Technolgies and Applications, VCH, Weinheim

Mattiasson, B. (1991) Biosensors, in "Biotechnology", 2nd ed. (Rehm, H.-J. and Reed, G.) Vol. 4 (Schügerl, K. ed) 75-103, VCH, Weinheim

In situ and On-line analysis of medium components and gases:

Heinzle, E. and Dunn, I.J., (1991) Methods and Instruments in Fermentation Gas Analysis, in "Biotechnology" (Rehm, H.-J. and Reed, G. eds) Vol. 4 (Schügerl, K. ed) 27- 76, VCH, Weinheim

Holst, O., Hakanson, H., Miyabashi, A., Mattiasson, B. (1988) Monitoring of glucose in fermentation process using a commercial glucose analyser, Appl. Microbial. Biotechnol. 28, 32-36.

Reuss, M., Boelcke, C., Lenz, R., Peckmann, U., (1987) A new automatic sampling device for determination of filtration characetistics of biosuspensions and coupling of analysers with industrial fermentation processes, BFT-Biotech-Forum, 4, 2-12.

Schügerl, K. (1991) Common Instruments for Process Analysis and Control and On-line Analysis of Broth (1991) in "Biotechnolgy", 2nd ed. (Rehm, H.-J. and Reed, G. eds) Vol. 4 (Schügerl, K., ed) 1-28 and 149-180, VCH, Weinheim

Measurement of cell concentration and behaviour

Marison, I.W. and von Stockar, U. (1986) The application of a novel heat flux calorimeter for studying growth of *Escherichia coli* W in aerobic batch culture, Biotechnol. Bioeng. 28, 178-1793.

Reardon, K.F. and Scheper, T.H. (1991) Determination of Cell Concentration and Characterization of Cells. in "Biotechnolgy", 2nd ed. (Rehm, H.-J. and Reed, G. eds) Vol. 4 (Schügerl, K. ed) 179-223, VCH, Weinheim.

Fluiddynamic measuring techniques:

Delhaye, J.M. and Cognet, G. eds. (1984)
Mesuring Techniques in Gas-Liquid Two-Phase Flowns, Springer Verlag, Berlin

Lübbert, A. (1991) Characterization of Bioreactors, in "Biotechnology", 2nd ed. (Rehm, H.-J. and Reed, G. eds), Vol. 4 (Schügerl, K. ed), 107-148, VCH, Weinheim

BIOREACTORS FOR IMPROVED PROCESS PERFORMANCE

KARL SCHÜGERL
Institut für Technische Chemie,
Universität Hannover
Callinstr. 3,
D-3000 Hannover, Germany

ABSTRACT. Reactor evaluation is given by means of the oxygen transfer rate, efficiency, and mixing time. The influence of the flow pattern on the reactor performance is discussed.

1 Introduction

Modern bioreactors are closed vessels or towers in which microorganisms, animal, human, or plant cells, or their parts are used for converting raw materials biologically into specific products. Wastewater treatment must be performed in very specific reactors. Many bioreactors for product manufacturing are operated monoseptically. In these bioreactors, the local environment of cells or enzymes is controlled to maintain the optimal conditions for cell growth or product formation. During the past 30 years, several new reactor types have been developed, but only few of them have succeeded in attaining the industrial stage and are used for product manufacturing or for wastewater treatment.
In the present review, only reactor types are considered which work with living cells and are used in the industrial practice.

2 Prerequisites for High Reactor Performance

Most of the known bioreactors are well-suited for the cultivation of non-shear sensitive cells in low viscous cultivation medium at moderate cell concentrations. Cell growth and product formation are only slightly influenced by the reactor type, when compared with the influence of the medium composition.

The specific growth rate and product formation rate, however, (with respect to power input and consumed chemicals) are considerably influenced by the reactor type and operation conditions.

Yet, particular cultivations
- at high cell density,
- in highly viscous cultivation medium,
- of shear-sensitive cells,
- in cultivation media with high solid content, or

A.R. Moreira and K.K. Wallace (eds.), Computer and Information Science Applications in Bioprocess Engineering, 95-107.

- in presence of an explosive gas mixture
  need special reactors.

2.1 OXYGEN TRANSFER RATE AND EFFICIENCY

Aerobic microorganisms require energy sources (substrates) and oxygen for their growth and product formation. Supplying cells with substrates well-soluble in the aqueous phase usually causes no special problems. The substrates can be added to the medium at the beginning of the cultivation process (batch operation), can be fed to the medium gradually (fed-batch operation), or can be added to the continuously fed cultivation medium.

Since the solubility of oxygen in the aqueous cultivation medium, however, is low, it is quickly consumed by the cells and must be replaced continuously. The supply of oxygen usually is performed by its transfer from the dispersed air into the cultivation medium. The oxygen transfer rate Q_{O2} is a function of the volumentric mass transfer coefficient $k_L a$ (h^{-1}) and the driving force (O_2^*-O_2), where O_2^* and O_2 (kg m^{-3}) are dissolved oxygen concentrations at the gas/liquid interface (saturation value) and in the bulk:

$$Q_{O2} = k_L a \, (O_2^* - O_2) \quad (\text{kg m}^{-3}\,\text{h}^{-1})$$

At oxygen limitation (O_2 -> 0):

$$Q_{O2} = k_L a \, O_2^* . \quad (\text{kg m}^{-3}\,\text{h}^{-1})$$

Since the mass transfer coefficient k_L and the dissolved oxygen saturation value O_2^* are practically constant, Q_{O2} is only a function of the the specific gas/liquid interfacial area a (m^{-1}), which is controlled by the specific power input $P\,V_L^{-1}$, where P is the power input (kW) and V_L (m^3) is the volume of the cultivation medium. With increasing specific power input, the specific interfacial area becomes larger and Q_{O2} is enhanced.

However, the efficiency of the oxygen transfer

$$E_{O2} = \frac{Q_{O2}}{P\,V_L^{-1}} \quad (\text{kg } O_2 \, (\text{kWh})^{-1})$$

frequently decreases with increasing specific power input $P\,V_L^{-1}$.
The dependency of Q_{O2} and E_{O2} on $P\,V_L^{-1}$ can be described by:

$$Q_{O2} = C_1 \left(\frac{P}{V_L}\right)^{n+1} \quad \text{and} \quad E_{O2} = C_1 \left(\frac{P}{V_L}\right)^{n}$$

If the Kolmogorov's turbulence theory is applied,

$$a = C_2 \left(\frac{P}{V_L}\right)^{0.4} \qquad n = -0.6 \text{ would hold true.}$$

Yet, the medium properties have considerable influence on exponent n. E.g., with bubble-coalescence-promoting (CP) or -suppressing (CS) media, as well as with highly viscous media (HV), different exponents were found:
In a stirred tank reactor (ST) with a flat-bladed disc turbine (DT),

$n = -0.5$ to -0.6 with CP medium,
$n = -0.1$ to -0.3 with CS medium,
$n = -0.3$ to -0.4 with HV medium.

The same exponents n, but different coefficients C_1 are valid for MIG, INTERMIG, and INTERPROP impellers.

The low negative exponent in CS medium can be explained by the persistence of the primary bubble size controlled by the high energy dissipation rate at the impeller. In CP media, the bubble size quickly increases as soon as the bubbles are transported away from the impeller in regimes with low energy dissipation rates, therefore, n has higher negative values.

In stirred tank loop reactors with a marine propeller (PR),

$n = -0.4$ to -0.6 for CP and CS media.

The medium properties have less of an effect on the oxygen transfer rate in loop reactors than in standard stirred tank reactors, because the energy dissipation rate therein is fairly uniform. In common stirred tanks, the energy dissipation rate decreases from the edge of the impeller to the reactor wall by a factor of 100.

In bubble columns with a static gas distributor in a homogeneous flow regime,

$n = 0.0$ with CP, CS and HV media,

and with static gas distributors in a heterogeneous flow regime,

$n = -0.3$ to -0.4 with CP, CS and HV media.

In a homogeneous flow regime, E_{O2} is independent of the specific power input, but in the heterogeneous regime it drops with increasing specific power inmput $P\ V_L^{-1}$. In both of these systems, the medium properties have a slight effect on the exponent, because the primary bubbles frequently are larger than the dynamic equilibrium bubble size, and the energy dissipation rate is uniform in bubble columns in both flow regimes.

The difference between the exponents found in homogeneous and heterogeneous regimes is due to the lower energy loss in the homogeneous regime because of the absence of the turbulence.

In bubble columns with two-phase nozzles (injector 8/14 and slot nozzle of Bayer Co),

n = -0.6 to -0.8 with CP,CS media.

These relatively high negative exponents are a result of the high local energy dissipation rate in the nozzle throat. Unexpected is the slight difference between CP and CS media.

In surface aerators,

n = 0.0 to +0.2 for CP and CS media.

The positive values of n are unexpected. With inceasing specific power input its efficiency increases, probably due to the increasing volume of the agitated liquid in the pool.

By increasing the aeration of the cultivation medium, the specific power input increases. The dependence of E_{O2} on the specific aeration rate $q_G^* = q_G V_L^{-1}$ can also be represented by

$$E_{O2}^* \alpha\, q_G^{*m},$$

where q_G = gas throughput ($m^3 \, min^{-1}$).

In a stirred tank with DT,

m = -0.4 to -0.7 with CP, CS, and HV media.

The slight dependence of m on the medium properties can be explained by the uniform energy dissipation in aerated liquid.

But in stirred tank loop reactors with DT and fully filled,

m = -0.1 to -0.2 with CP, CS media,

and with DT and overflow,

m = -0.3 to -0.4 with CP, CS media.

The exponent in a fully filled reactor is lower than at overflow operation. This was unexpected.

In stirred tank loop reactor with PR and fully filled,

$$m = -0.3 \text{ to } -0.4 \text{ with CP medium,}$$

and PR and overflow,

$$m = -0.1 \text{ to } -0.2 \text{ with CP medium.}$$

The low exponent in the overflow is probably caused by the surface aeration.

Table 1
Comparison of reactors with low viscosity media: water (W), cultivation medium (C) as well as salt (SA), sulfite (S), and alcohol (A) solutions, in different specific power input P/V_L ranges with respect to their oxygen transfer rate, Q_{O2}, and efficiency E_{O2} (Schügerl, 1991)

	P/V_L = 0-2.5 kW m^{-3}		P/V_L = 4-8 kW m^{-3}		P/V_L = 10-50 kW m^{-3}		
	Q_{O2}	E_{O2}	Q_{O2}	E_{O2}	Q_{O2}	E_{O2}	
reactor	kg/m^3h	kg/kWh	kg/m^3h	kg/kWh	kg/m^3h	kg/kWh	
stirred	0.7-2.5	0.7-2.5	---	---	3	0.3	W
tank	1.0-3.6	0.8-2.6	5.5	1.2	7.1	0.75	C, SA
stirred	1.7	0.8-0.9	3.0	0.51	---	---	W
loop	1.9-2.2	0.9-1.1	3.7-4.5	0.6-0.8	6.2	0.27	W
surface aerator		1.5-2.4	---	---	---	---	W
bubble	0.3-2.7	3.2-9.2	---	---	---	---	W
column	2.3-20	27-73	---	---	---	---	A
	0.3-4.8	2.3-12	---	---	---	---	SA
airlift	0.7-1.8	3.2-9.2	---	---	---	---	W
tower external loop	1.6-3.2	10-16	---	---	---	---	C
airlift	1.0	6	10	2.5	---	---	A
tower	1.2-14	1.3-7	---	---	---	---	S
internal loop	5-8	2-4	10	1.5	---	---	C

These examples indicate that the reactor performance with regard to the oxygen transfer rate and efficiency frequently depend on the medium properties and the operation conditions (specific power input and specific aeration rate of the reactor).

Some Q_{O2} and E_{O2} values evaluated in low viscosity media are compiled in Table 1.

Stirred tank and stirred loop reactors can be operated at high specific power inputs, at which high oxygen transfer rates can be obtained. However, under these conditions, the energy efficiency of the oxygen transfer is very low (below 1 kg $(kWh)^{-1}$).

In bubble columns, in which the power input is maintained by aeration alone, only low specific power input can be attained, mainly because of the foaming problems occurring at high aeration rates.
Yet, high efficiencies of the oxygen transfer can be obtained, i.e., averaging 3 to 10 kg $(kWh)^{-1}$ in cultivation (coalescence suppressing) media.

In airlift tower reactors with an external loop, the allowable specific power input is higher than in bubble columns, because, due to the circulating liquid, the transition of the homogeneous to the heterogeneous regime occurs at higher aeration rates. The efficiency of oxygen transfer is high (3 to 10 kg $(kWh)^{-1}$). The medium effect is less than in a bubble column reactor.

In airlift reactors with an internal loop, the allowable specific power input is higher than in one with an external loop (up to 6 kW m^{-3}), where very high oxygen transfer rates (10 kg m^{-3} h^{-1}) can be attained, yet only at low efficiency (1.6 kg $(kWh)^{-1}$). On an average, Q_{O2} = 5 to 8 kg m^{-3} h^{-1} at E_{O2} = 2 to 4 kg $(kWh)^{-1}$ can be obtained. Again, the medium effect is less than in bubble column reactors. With increasing viscosity, Q_{O2} and E_{O2} diminish.

In plunging jet reactors with two-phase jets high oxygen transfer rates (3.5 to 6.2 kg m^{-3} h^{-1}) at high specific power input (5 kW m^{-3}) can be obtained, but at low efficiency (1.1 to 1.25 kg $(kWh)^{-1}$) can be obtained. With single phase jets the reactor performances are much lower.

The oxygen transfer performances of packed bed reactors with biofilms are moderate (E_{O2} = 2.0 to 3.6 kg $(kWh)^{-1}$ at $P\ V_L^{-1}$ = 1.1 to 2.6 kW m^{-3}).

The oxygen transfer performances of membrane reactors are rather low (0.1 to 0.2 kg m^{-3} h^{-1}). They are used for cultivation of animal cells at low stirrer speeds (50 to 100 rpm), at which the Q_{O2} values with direct aeration of the cultivation medium are in the same range.

2.2 MIXING PERFORMANCE

In small laboratory reactors and in low viscous cultivation media, the mixing times are sufficiently low to maintain uniform medium composition in the reactors. With increasing reactor volume, medium viscosity and cell concentration, the medium segregation becomes more and more pronounced. In large (e.g., 100 m^3) reactors, considerable nonuniformity of dissolved oxygen concentration also prevails with low cell concentration in low viscous media.

With increasing reactor diameter D_R, the impeller diameter D_I also increases, because the D_I/D_R ratio is usually kept constant (0.33 at low viscosity and 0.60 at moderate viscosity of the cultivation medium). The power input in the turbulent regime of non aerated liquid is given by

$$P = k\, N_R^3\, D_I^5\, d,$$

where k = constant
N_R = stirrer speed (min^{-1})
D_I = impeller diameter (m)
d = density of the cultivation medium (kg m^{-3})

Scale-up is usually based on keeping the specific power input constant. Therefore, with increasing reactor/impeller diameter, the stirrer speed must be reduced. In the turbulent regime, the mixing number N_M is usually constant. Since

$$N_M = N_R\, \theta_R,$$

where θ_R = the mixing time (min),

the mixing time increases with decreasing stirrer speed.

Impellers which impart axial motion (propeller) are more efficient in mixing than ones which produce tangential/radial motion (flat-bladed disc turbine).
At high Reynolds numbers ($Re_N = N_R\, D_I^2\, v^{-1} = 10^5$), where v is the kinematic viscosity of the cultivation medium, the power uptake of a flat-bladed disc turbine is by a factor of 25 higher than that of a propeller in the same reactor. The mixing time θ_R for a propeller is, however, only by a factor of 1.5 higher, i.e., the specific mixing time with regard to the power uptake θ_R/P is by a factor of 16 less for a propeller than for a flat-bladed disc turbine.

In bubble column (BC) and airlift tower loop (ATL) reactors, still higher mixing times prevail: E.g.,
θ_{R95} = 18 s in 2.0 m^3 BC and 28 s in 20 m^3 BC reactors,
θ_{R95} = 63 s in 0.4 m^3 ATL, 85 s in 2.0 m^3 ATL and 101 s in 27 m^3 ATL reactors.

In packed beds, the mixing time of liquids is extremely high. It, therefore, is necessary to have a high recirculation flow rate to reduce the mixing time in these reactors.

In fluidized bed reactors, the mixing time is in the same order of magnitude as in bubble columns.

In large reactors, distributed aeration and substrate feeding are necessary to reduce the medium separation.

2.3 CELL MASS PRODUCTIVITY

Adequate substrate and oxygen supply are necessary, but not the only prerequisites for high cell growth rate and product formation in aerobic cultivations.

Cell cultivation with substrate limitaton is common in biotechnological practice to avoid oxygen transfer limitation, and as a consequence, the reduction of the yield coefficient of the cell growth with respect to the substrate consumption $Y_{X/S}$, e.g., by formation of ethanol (*Saccharomyces cerevisiae)* or acetate (*Escherichia coli*), or foaming causes the starvation of the cells. Under this condition, the Monod relationship between cell growth rate R_X, the concentrations of substrate S, and cell X is

$$R_X = \mu_m \frac{S\,X}{S + K_S},$$

where μ_m = the maximum specific growth rate
K_S = the model parameter, or saturation constant, of the substrate.
At substrate limitation, μ_{max} is reduced to the actual specific growth rate

$$\mu = \frac{R_X}{X} = \mu_m \, S \, K_S^{-1}.$$

The specific growth rate μ is proportional to S. Nonuniformity of the substrate concentration impairs the cell productivity.

At oxygen transfer limitation, the growth rate is given by

$$R_X = \mu X = Q_{O2} \, Y_{X/O2} = k_L a \, O_2^* \, Y_{X/O2},$$

where $Y_{X/O2}$ = yield coefficient of growth with respect to the oxygen consumption, i.e., R_X is proportional to the oxygen transfer rate Q_{O2}.

This relationship frequently is used to determine Q_{O2} in cultivation media containing microorganisms which do not change their metabolism in absence of oxygen (e.g., *Trichosporon cutaneum)* and have constant yield coefficients $Y_{X/O2}$.

Nonuniformity of the specific interfacial area or reduction of O_2^* caused by the consumption of oxygen in the air bubbles reduces the cell productivity.

The local concentrations of the substrate in the liquid as well the oxygen in the gas phase depend on the reactor geometry, stirrer type, impeller speed (in stirred tanks), aerator type, and aeration rate as well as on the flow patterns of the liquid and gas phases.

3 Flow Pattern in Different Bioreactors

In non-aerated clear water, several measurements were carried out with lased Doppler anemometers (LDA) to evaluate the liquid flow pattern in stirred tank reactors. In aerated cultivation media, however, LDA fails on account of the turbidity of the medium and the presence of the bubbles. Recently, Lübbert at al. (Lübbert, 1991) developed new techniques for the determination of the velocity vectors of the liquid and gas bubbles in multiphase systems. Other groups used a flow follower to measure the circulation flow in bioreactors (Reuss and Bajpai, 1991).

The determination of the liquid velocity vectors in multiphase flows is based on the signal-response technique. Pseudostochastic heat pulses are produced by a high frequency heat transmitter, where the temperature variation is measured at some cm distance from the heat source by a resistance thermometer. The time-of-flow distribution is determined by means of the cross-correlation of the test and response signals, which allows the determination of the local velocity and the local mixing of the liquid. By shifting this probe in radial and axial direction, the complete velocity profile can be determined.

Figure 1 shows velocity distributions measured in a airlift tower loop reactor 23 m in height. At low aeration rates, the velocity profile is uniform. The mixing intensity is low. With increasing aeration rate, the velocity profiles become more and more non-uniform and approach the velocity profiles of bubble column reactors. At the same time, the intensity of mixing increases. Small bubbles follow this velocity profile. Large bubbles mainly rise in the column center at high aeration rates.

The determination of the bubble velocity vectors is based on the ultrasound Doppler technique. By means of a trasmitter-receiver, the frequencies of the emitted and reflected ultrasound waves are compared, and the local bubble velocities are determined from the frequency shift. This is possible, because the bubbles act as perfect reflectors for the ultrasound. By using ultrasound pulses, the local specific interfacial area can be determined from the power reflected back into the detector. By varying the electronic gating, the distance from the probe and the local volume investigated can be varied. By

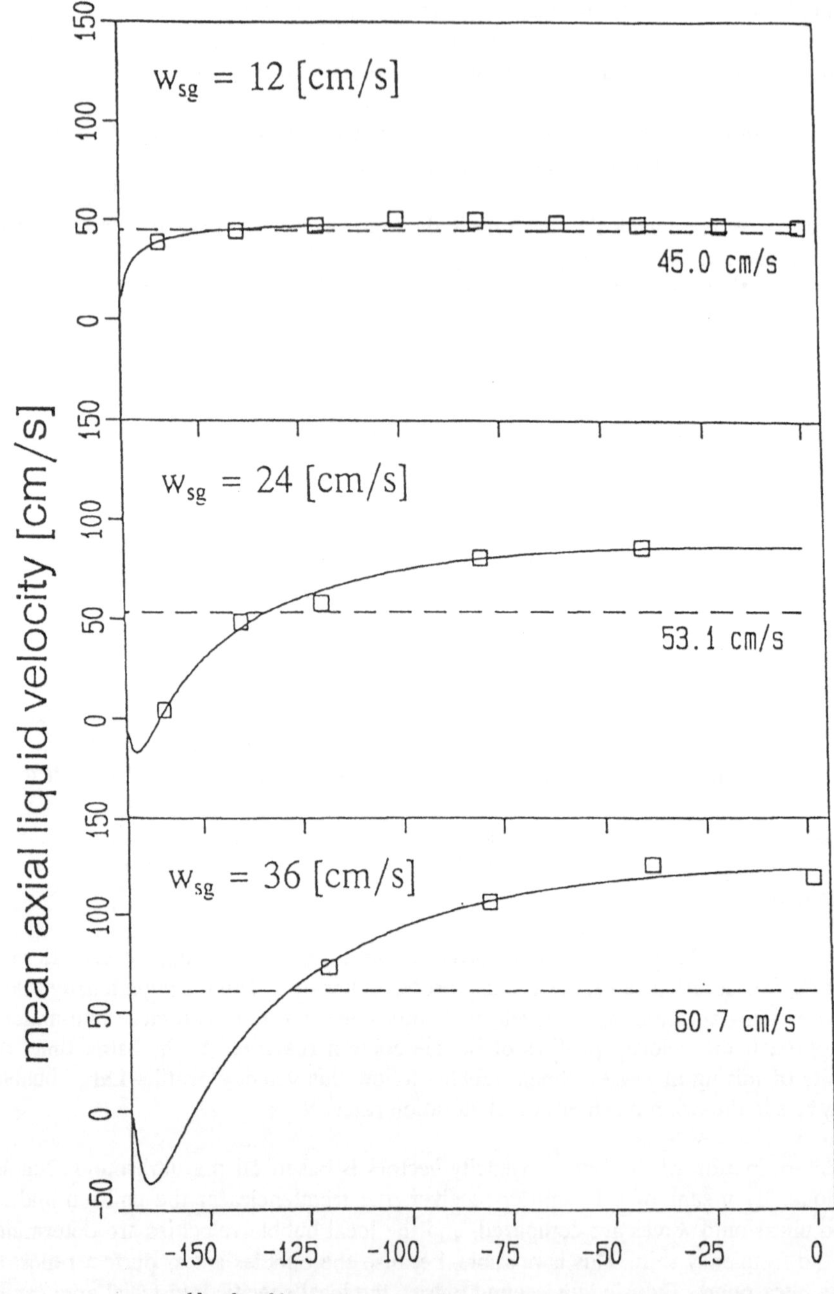

Figure 1. Liquid velocity profiles in the riser of an airlift tower loop reactor, 23 m in height, at different aeration rates.

this technique, axial and radial velocity profiles of bubbles in stirred tanks were evaluated with different stirrers and at different impellers speeds.

Figure 2 shows such bubble velocity profiles measured in a 300-l stirred tank reactor equipped with three-stage Rushton turbines.

These measurements clearly show the circulation patterns between the impellers. Such measurements allow the optimization of the applied stirrer type and impeller speed to maintain the optimal flow pattern.

3.1 STIRRED TANK REACTORS

Short loops around the impeller are advantageous for the mixing of substrate and dissolved oxygen in the liquid and oxygen in the bubbles. Poorly mixed regions are formed between the impellers if they have too large a clearence. The circulation velocity around the impellers decreases, if there is too small a clearance between the impellers.

By means of these flow patterns, optimal clearence, optimal positions for substrate feeding, gas distributior and sensors for *in situ* or on-line monitoring of key parameters for process control can be determined. Also local dead water regions may be identified, at which wall growth or depositions can occur.
Furthermore different impeller types can be compared and their optimal speed can be evaluated as a function of the cultivation conditions.

3.2 BUBBLE COLUMN AND AIRLIFT TOWER LOOP REACTORS

In bubble column and airlift tower loop reactors, the bubble and liquid velocity profiles can be determined. By means of the cross-section ratio of the riser and downcomer and the aeration rate, the mixing intensities in a wide range between those in bubble column and airlift tower loop reactors can be set. Real tower reactors for fed-batch operation usually have a complex construction to assure the optimal operation at the batch phase (with a low amount of cultivation medium) during the fed-batch operation (with variable medium volume) and during the final batch operation (with maximum medium volume).

These reactors frequently consist of an airlift tower loop body and a bubble column head. The construction, especially the combination of these reactor parts, are critical for optimal operation. By means of the flow pattern, the reactor construction can be optimized.

4 Conclusions

Cells can be supplied with a sufficient amount of oxygen if the specific power input P/VL is increased. However, at high P/V_L values, the efficiency of oxygen transfer with respect to the power input is low. Stirred tanks can be operated at high specific power

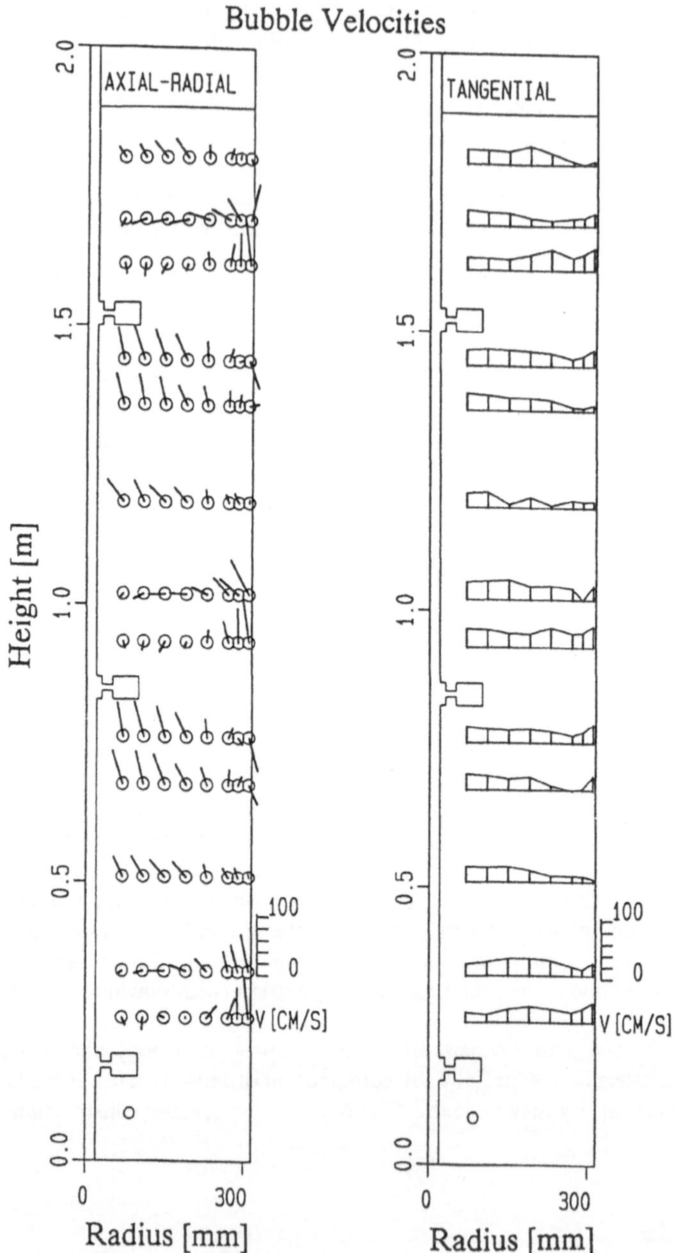

Figure 2. Bubble velocity in a 600 L working volume stirred tank reactor with a flat-bladed disc turbine in water.

input and high oxygen transfer rate/cell concentration, yet only at low efficiency. Bubble column and airlift tower loop reactors can be operated only at rather low specific power supply/oxygen transfer rate, but at a high efficiency.
With increasing bioreactor size in the culture medium, segregation occurs, which impairs the reactor performance. Determination of the flow pattern allows the optimization of the reactor construction.

5 References

Lübbert, A. (1991) Characterization of Bioreactors, in "Biotechnology" 2nd Ed. (H.-J. Rehm, G. Reed, eds), Measuring, Modelling and Control, Vol. 4 (K. Schügerl, ed.) VCH, Weinheim, pp. 107-148.

Reuss, M. and Bajpai, R. (1991) Stirred tank reactors, in "Biotechnology" 2nd Ed. (H.-J-. Rehm, G.Reed, eds.), Measuring, Modelling and Control, Vol. 4 (K. Schügerl, ed.) VCH, Weinheim, pp. 297-348

Schügerl, K. (1991) Bioreaction Engineering, Vol. 2 John Wiley & Sons, Chicester, Chapter 4.

BIOREACTOR FAULT DETECTION USING DATA RECONCILIATION

John J. Prior and Charles L. Cooney
Biotechnology Process Engineering Center
Department of Chemical Engineering
Massachusetts Institute of Technology
Cambridge, MA 02139 USA

ABSTRACT: We summarize experience using a data reconciliation technique to detect faults and inconsistencies in process data collection from lab and industrial scale fermentations.

1. INTRODUCTION

Although limited by available monitoring technology, industrial companies often collect data from hundreds of sensors installed on reactors and associated sub-systems throughout their plants. The questions the operators and engineers using this data need answered are: (1) is the data accurate and consistent, (2) is the process operating correctly, and (3) how can the process be improved. The nonlinear, time-varying, and imprecisely modelled nature of these processes makes answering these questions challenging. Our goal is to develop and evaluate a general methodology for assessing bioreactor data, assumption, and model validity.

Our approach to answering these questions is to combine data from a variety of sources to perform an *assumption reconciliation*. We focus on: (1) the process stoichiometry, (2) the sufficiency of medium nutrients, (3) the gas train material balances and correlations, (4) the water balance, and (5) the process kinetics, to provide this information. This paper describes experiences using the process stoichiometry as a reconciliation tool.

This method for data reconciliation has three phases. The first step is determining what benefits can the analysis provide. The problem is ultimately one of estimation of either unmeasured parameters or measurement errors. If there are not sufficient measurements and constraints, part of the system will be unobservable. The second step is application of low-level analyses to estimate parameters and isolate inconsistencies. The data can be considered on a cumulative or rate basis to differentiate among possible problems. The final step is assignment of inconsistencies to potential faults and communication of this information to the operator or engineer.

An industrial *Bacillus licheniformis* fermentation producing α-amylase served as the model system for this study (Figure 1). 5000 hours of data from 46 batches were collected manually by plant operators on an hourly basis. This dataset provided realistic scenarios for testing diagnostic techniques. While the data contains numerous errors and fault episodes, it is sometimes difficult to resolve exactly what occurred after the fact since limited analytical data was collected. To compensate for this, and to gain experience with other strains, lab-scale fermentations have been conducted to investigate deliberate fault conditions and collect data on additional variables. In this paper, experiments using baker's yeast are described.

A.R. Moreira and K.K. Wallace (eds.), Computer and Information Science Applications in Bioprocess Engineering, 109-114.

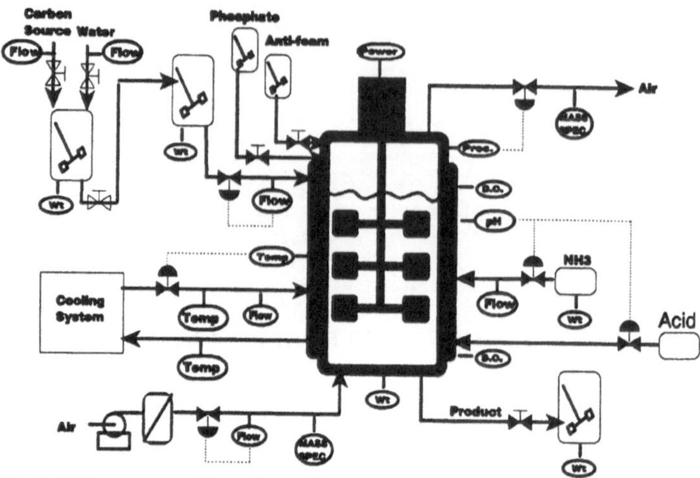

Figure 1 Typical bioreactor configuration and sensors.

1.1 Stoichiometric Model

It is often difficult to define a simple stoichiometric model for a process. There may be several possible sources of carbon, nitrogen, and energy in the medium and several by-products can be formed. Furthermore the stoichiometric constants may vary with time. The question is whether a simplified model consisting of the major material flows is representative enough for the analysis to be useful.

The simplified stoichiometric model for the industrial process used in this study is

$$CH_2O+O_2+NH_3 \rightarrow Biomass+CO_2+H_2O \tag{1}$$

where biomass is a lumped parameter representing cell mass and enzyme product. The composition of these two components is not different enough to be differentiated by the available measurements. For the laboratory scale baker's yeast experiments, ethanol is added to the stoichiometric model[1]:

$$CH_2O+O_2+NH_3 \rightarrow Biomass+CO_2+H_2O+EtOH \tag{2}$$

1.2 Data Reconciliation

When more measurements and assumptions are present than necessary to observe all a system's parameters, the extra (redundant) information can be used to check the consistency of the measurements. The approach was first introduced in 1961[2] and has been refined and applied to bioprocesses by several researchers[3-4]. Assumed process constraints such as material balances, energy balances, and empirical correlations can be represented mathematically as

$$f+Ax+By = 0 \tag{3}$$

for linear relations, or more generally in the form

$$F(x,y)=0 \tag{4}$$

for non-linear relationships. Here **x** represents a vector or measurements and **y** represents estimated parameters. In general, these relationships will not be satisfied when redundant measurements are present:

$$x^{+}=x+e \tag{5}$$

Here **x** is the true value and **e** represents the error due to sensor noise or imprecise assumptions. The first step of the reconciliation process is adjustment of the measured values to obtain a consistent set of data. This adjustment is made so as to minimize the sum of the squares of accuracy-weighted adjustments.

$$\hat{x}=x^{+}+v \tag{6}$$

$$h=\min\ v^{T}\Sigma^{-1}v \tag{7}$$

where $\hat{x}$ is new estimated measurement value, **v** is the adjustment, and Σ is the estimate of the covariance matrix.

1.3 Fault Detection

If there are gross errors present, the statistical basis for the adjustment is not valid. Potential causes of gross errors include miscalibrated sensors, failed sensors, process leaks, unknown products, and dynamic episodes.

The null hypothesis is that no gross errors are present and that inconsistency is due to random noise and assumption uncertainty. The adjustments are compared against the expected chi-square distribution to determine the likelihood that the inconsistency is due to the expected random errors.

$$h < \chi^{2}_{(1-\theta)} \tag{8}$$

This information can be represented as the p-value at which the null-hypothesis would be rejected. This p-value can be plotted graphically against time to monitor the consistency of the process data being analyzed. Ideally, introduction of a fault will result in a sudden increase in the p-value. A tolerance can be assigned to decide when further investigation is necessary (Figure 2).

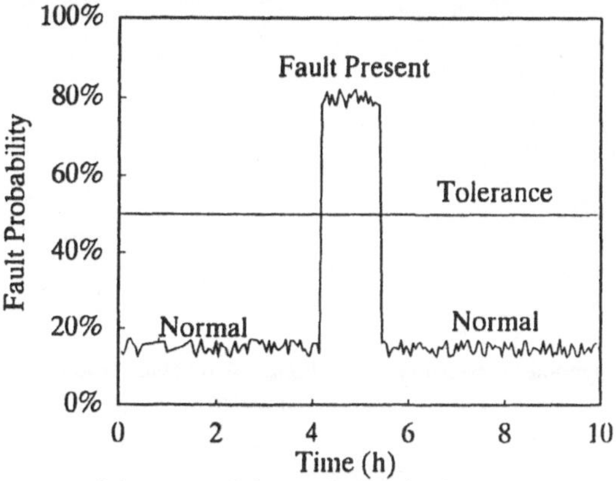

Figure 2 Ideal response of the test statistic to a fault episode.

If the dataset is found to exceed the tolerance and be unreasonably inconsistent, an attempt is made to locate the cause of the problem. If sufficient redundancy exists, each measurement is removed from the dataset in a serial fashion. If the remaining data are significantly more self-consistent, that measurement is suspected as faulty.

2. RESULTS AND DISCUSSION

2.1 Batch Analysis of Industrial Fermentation Data

Hourly measurements from 46 batches of an industrial fed-batch fermentation of *Bacillus licheniformis* producing α-amylase were gathered. Four elemental balance equations may be written from the stoichiometry. Online measurements of sugar addition rate, ammonia addition rate, oxygen consumption, and carbon dioxide evolution can be reconciled against the four elemental balance relationships. In this situation, there are two more measurements than necessary to observe the system. These extra measurements provide the opportunity to detect inconsistencies in the data and to locate the most likely cause of disagreements when they occur.

The overall balances for each batch were considered first. Figure 3 shows the p-value for rejection of the no-fault null hypothesis for each run. In contrast with the idealized behavior show in Figure 2, the level of inconsistency in the data is high for most of the runs. Serial deletion of measurements for the batches with p-values higher than 90% indicated that the ammonia addition measurement was faulty. In general, the reconciled measurement value was lower than the measurement.

Plant operators investigated and found the sensor had a calibration error that resulted in the reported measurement 50% being above actual. This correction was made retroactively to the dataset and the analysis repeated. Figure 4 shows the resulting improvement in dataset consistency. The p-values with the ammonia fault corrected were reduced an average of 27%.

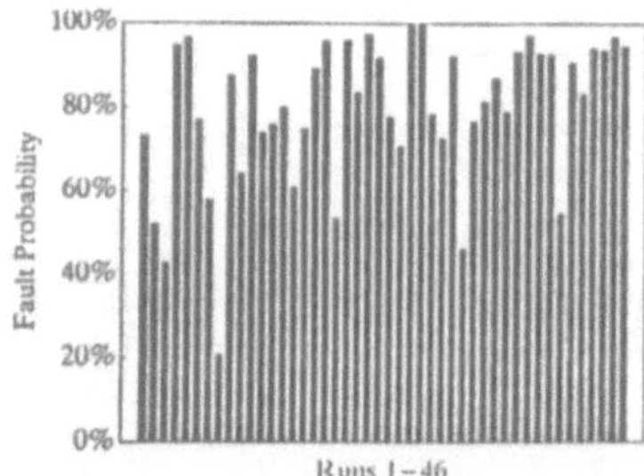

Figure 3 p-values for null-hypothesis rejection for each of the 46 batches.

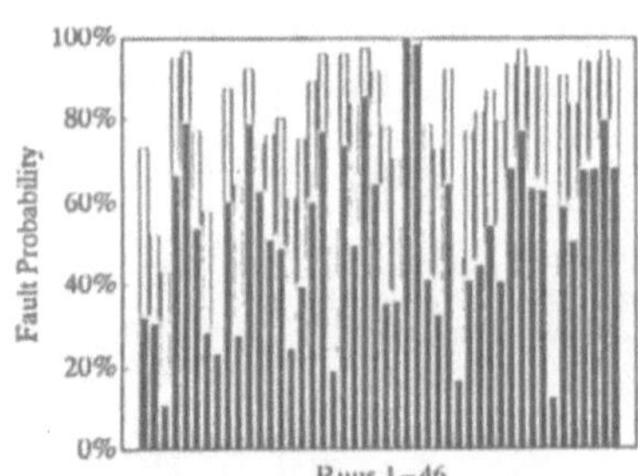

Figure 4 p-values before □ and after ■ NH3 correction.

The remaining inconsistency is quite high. The next leading cause of inconsistency was found to be drifting and miscalibrated gas analysis equipment in the early batches. Other improvements were made by isolating individual fault episodes in some batches. In many cases more analytical data or equipment would be needed to differentiate between possible problems.

2.2 Fault Identification at Laboratory Scale

Fermentations were done at the laboratory scale to generalize the approach to other strains and to better monitor reactor conditions. Data from a fed-batch fermentation of baker's yeast was reconciled against the stoichiometric balances in real-time.

Oxygen, carbon dioxide, ammonium, and sugar rates were measured. The presence of the unmeasured ethanol product reduced the redundancy by one. In this situation, the degree of inconsistency can be measured but the cause cannot be isolated. To overcome this limitation, we assumed that ethanol production was negligible until inconsistency indicates that assumption might not be valid.

Figure 5 shows the p-values for hourly data from the first 20 h of a run. The peaks at 8 and 18 h were traced to perturbations in the rate of ammonia addition. At 8 h there was a sudden change in pH as the poorly tuned controller stabilized (Figure 6). At 18 h there was an increase in the weight of the ammonium addition tank. This increase is often observed when the pressure in the reactor drives flow back through a defective pump (Figure 7).

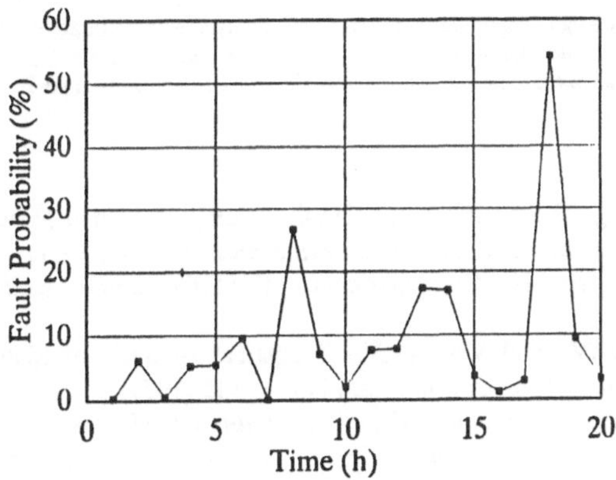

Figure 5 p-values for hourly data from a lab-scale baker's yeast fermentation.

3. CONCLUSIONS

Experience gained in these and other case studies made it clear that there is rarely true redundant information available in bioprocesses. Careful assumption, simplification, and uncertainty assignment is necessary in order to create a formulation that will identify and isolate important disturbances and inconsistencies.

We have had most success by starting with large windows of analysis (i.e., whole batches) and high inconsistency tolerances. This allows major, long term faults to be identified and corrected before more subtle problems are identified in smaller time frames using a smaller fault tolerance level.

In general, it is important to relate inconsistencies back to primary measurements. Each "measurement" in the elemental balance is calculated based on as many as six primary measurements. To be useful in an industrial environment, particular sensors and assumptions must be flagged for investigation.

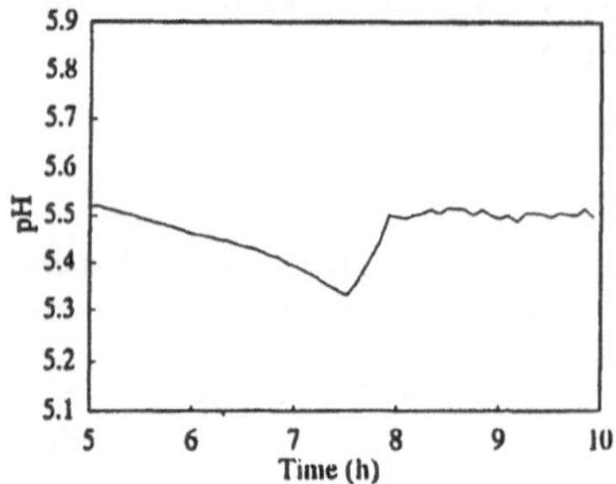

Figure 6 Sudden pH shift from base addition above demand rate.

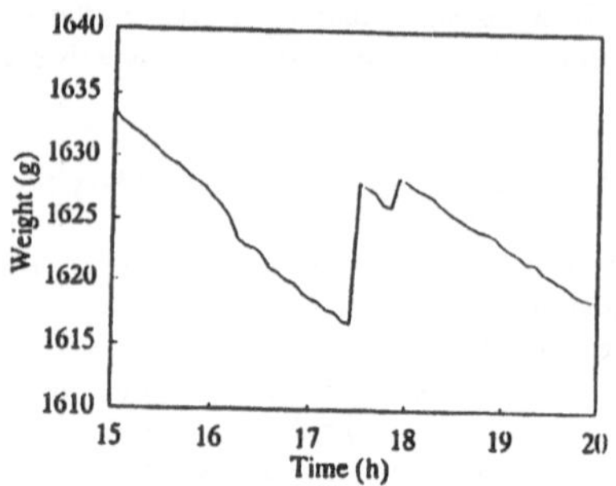

Figure 7 Increase in base tank weight causing inconsistency at 18h.

The authors wish to acknowledge support from International Biosynthetics (Kingstree, SC), and also from the National Science Foundation under the engineering research center initiative to the MIT Biotechnology Process Engineering Center (Cooperative agreement ECD8A-03014) that made this research possible.

REFERENCES

1 Cooney, C.L., Wang, H.Y., and Wang D.I.C., 'Computer-Aided Material Balancing for Prediction of Fermentation Parameters', *Biotechnol. Bioeng.*, **19**, (1977)

2 Kuehn, D.R. and Davidson, H., 'Computer control II. Mathematics of control', *CEP*, **56**, No. 6, 44 (1961)

3 Madron, F., Veverka,V., and Vanecek,V., 'Statistical Analysis of Material Balance of a Chemical Reactor', *AIChE Journal*, **23**, No. 4, 482 (1977)

4 Wang, N.S., and Stephanopoulos, G., 'Application Of Macroscopic Balances To The Identification of Gross Measurement Error', *Biotechnol. Bioeng.*, **25** (1983)

THE DESIGN APPROACH OF MULTIPRODUCT PLANTS

J. Mata-Alvarez*, and C. Rolz**
*Department d'Enginyeria Química. Universitat de Barcelona.
Martí i Franquès 1, 6p. 08028 Barcelona (Spain)
** Central American Research Institute for Industry (ICAITI)
P.O.Box 1552, Guatemala 01901, Central America

ABSTRACT. Methods for designing multiproduct batch plants are very important for biotechnological process, because the size of production fits quite well to this approach. In this paper an introduction to this topic is carried out, pointing out the basic methods and the computer design resolutions. An example plant representing a biotechnological process is used as a basis for the application of the methods.

1. Introduction

Batch processes are typical for fine chemicals production, characterized by their relatively small production and their high added value. This high value comes mainly from the rather complicated synthesis procedures.

A field of application of biochemical reactions is precisely the production of these kind of products, as for example, for the pharmaceutical industry. This type of industry and others related (cosmetics, for example) often require the manufacture of several products, which, essentially, require the same equipments. This is the basis of a multiproduct plant. Normally in this plant only one product is produced at a time during the so called "campaign". After several campaigns the whole production is obtained.

As described in Sparrow et al. (1975) there is another type of plants, called multipurpose plants, which are like an engineering "job shop". There, there is not a common production pattern for the different products and even different routes can be taken for the product in different batches. The approaches for both types of plants are obviously different.

In this lecture, the principles of this type of plants will be outlined giving some examples elaborated using the available software. First of all, the model of Loonkar and Robinson (1970) for a single product will be examined. Second an extension of this model will be considered, taking into account the operation with overlapping. Finally, the approach of multiproduct plants of Sparrow et al. (1975) together with the more complex model of Knopf et al. (1982) will be examined.

2. Loonkar and Robinson formulation

Loonkar and Robinson (1970) formulation is applied to design a plant for a single product. Later, these authors also studied the multiproduct case (Loonkar and Robinson, 1972). Batch plants can be

A.R. Moreira and K.K. Wallace (eds.), Computer and Information Science Applications in Bioprocess Engineering, 115-124.

operated basically in two modes: a) Without overlapping, in which the new batch is not started until the previous one has left the last operation in the sequence; b) With overlapping operation, in which the new batch is begun in accordance with the bottle-neck equipment. This equipment operates continuously, that is, once the batch has been unloaded, a new batch is immediately loaded (see Figure 1). In the approach of Loonkar and Robinson (1970) only the non-overlapping case was considered. However, both cases will be solved for illustrative purposes.

2.1 NON OVERLAPPING OPERATION

The procedure will be illustrated with the following example of a biotechnological process. The discontinuous equipment of it consist of a sterilizer, a fermentor and a drier. The rest of the equipment, which operates semicontinuously, is formed by three pumps and an ultrafiltration unit. Figure 2 shows the scheme of this simple process.

In this approach the production per batch W kg prod./h is defined and, if V is the amount of product per batch, then:

$$W = V/T \tag{1}$$

where T is the batch time (without overlapping):

$$T = t_j + \Theta_k \tag{2}$$

t_j, the operation time of discontinuous equipment, will normally be fixed, thus only Θ_k, the operation time of semicontinuous equipment, can be changed. Once the required plant capacity is defined and the individual t_j and Θ_k are known, the sizes of the equipment become known and the plant is completely specified.

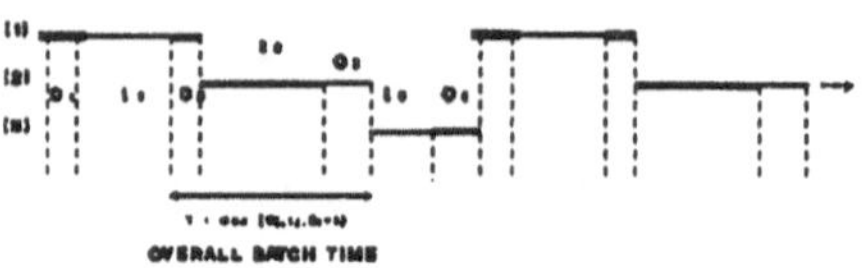

Figure 1. Basic modes of operation of a batch plant

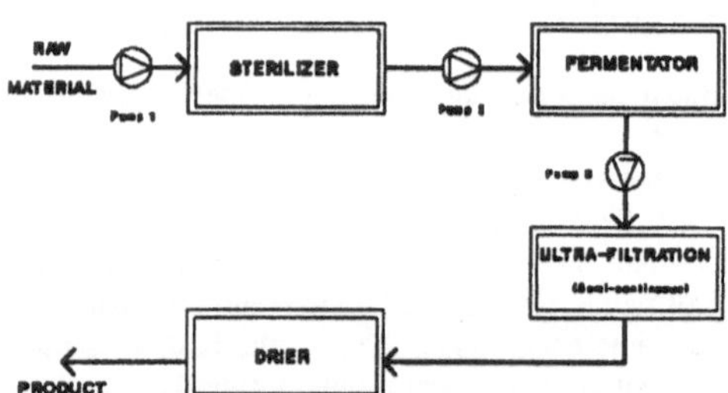

Figure 2. Flow diagram of a simple batch plant to carry out a biotechnological process

Accordingly with the formulation of Loonkar and Robinson, the major item in this kind of industry is the capital cost, the rest of costs and process efficiency can be considered fixed. As the size of V_j is proportional to V, the capital costs, I, of the plant, can be estimated according with the following equation:

$$I = \Sigma_j \; a_j V^{\alpha j} + \Sigma_k \; b_k (V/\Theta_k)^{\beta k} \tag{3}$$

TABLE 1
Loonkar and Robinson approach for the example of Figure 2, using a spreadsheet. Values represent the optimal solution found applying a simple manual procedure.

Production per batch W = 10 kg prod./h
V_j is proportional to V

	a_j	$Alpha_j$	t_j	a'_j	Cost
Steril.	9000	0.6	0.5	35829.64	225078
Ferm.	22000	0.55	12	78058.94	420732
Drier	50000	0.9	8	397164.1	6253237

	b_k	$Beta_k$	$Theta_k$	b'_k	Cost
Pump 1	700	0.4	0.032	1758.320	23719
Pump 2	500	0.4	0.025	1255.943	18701
Pump 3	600	0.4	0.83	1507.131	5528
U.Filtr	10000	0.65	0.83	44668.35	369145

Batch time T =	21.387 hr.		C.TOTAL	7316139
Variables	Theta 1	Theta 2	Theta 3	
	0.032	0.025	0.83	

The summation is over k, the number of semi-continuous trains. In accordance with Loonkar and Robinson, these costs are the most significative. As a consequence, only them will be considered for an economical optimization of the plant.

The problem consist in finding the appropriate values of the semi-continuous equipment residence times to minimize the overall equipment costs. If residence times are low, the cost of semi-continuous equipments are high, but the batch equipment is filled faster and, as a consequence, with a lower volume is possible to achieve the desired production.

Taking into account equation (1), equation (3) can be written as:

$$I = \Sigma_j \; a_j \, (W.T)^{\alpha j} + \Sigma_k \; b_k(W.T/\Theta_k)^{\beta k} \tag{4}$$

If:

$$a'_j = a_j \, W^{\alpha j} \quad \text{and} \quad b'_k = b_k . W^{\beta k} \tag{5}$$

then, the objective function can be formulated in terms of the batch time:

$$I = \Sigma_j \; a'_j \, T^{\alpha j} + \Sigma_k \; b'_k(T/\Theta_k)^{\beta k} \tag{6}$$

This optimization problem is a simple one, as it does not contain any constrain. Thus, simple optimization methods can be used to solve it. The resolution will be carried out using the example plant depicted in Figure 2. Data on coefficient costs of the equipment are presented in Table 1, which has been extracted from a spreadsheet. As can be seen, this problem has only three variables, the semi-continuous operation times Θ_1, Θ_2 and Θ_3 (this one, common for the pump and the ultra-filtration unit, which form a semi-continuous subtrain). This uncomplicated problem can be handled by an elementary optimization procedure such as the classical one-variable-at-a-time. It consist in optimizing one of the factors, varying it step by step, whereas the rest of the factors remain constant. After this factor has been optimized, it is maintained at the optimum level found, and the second one is optimized. The procedure is repeated cyclically until no improvement of the objective function is observed. Applying this method the residence times which minimize the cost of the plant can easily be obtained. By means of a spreadsheet, the procedure can be carried out even manually in a very fast way. The results are also showed in Table 1. Using the GAMS software for non linear programming cases, which is based in a generalized reduced gradient (GRG) method, the same solution is also found (Brooke et al., 1988).

TABLE 2
Results of problem presented in Table 1, using the overlapping approach and GAMS software.

Semicontinuous time Θ_1	12.144
Semicontinuous time Θ_2	0.021
Semicontinuous time Θ_3	0.644
Overall batch time T	12.665
Capital cost I	4714697

2.2. OVERLAPPING OPERATION

If overlapping is allowed, a reduction of the time to complete successive batches is possible. In fact, the batch time T, will be the maximum of:

$$T = \max(\Theta_{j-1}, t_j, \Theta_{j+1}) \tag{7}$$

that is, the maximum time of discontinuous equipment operation, which includes the loading and unloading times. A more easy-to- implement mathematical form of equation 7, consist in expanding it into inequality constrains:

$$T \geq \max(\Theta_{j-1}, t_j, \Theta_{j+1}) \tag{8}$$

For the particular example considered, this implies an optimization problem of 3 variables and 3 inequality constrains. Of course, the problem is not very complex, but it cannot longer be handled with the spreadsheet and a simple optimization procedure such as that used before. GAMS solution of the problem is presented in Table 2. As can be seen there, semicontinuous time Θ_1, with overlapping situation has increased until achieving the maximum allowed (Θ_2+t_2+Θ_3). The increase was in equipment 1 (a pump), the more expensive, so that the impact on the overall cost is the greatest. A considerable reduction on capital investment costs I can also be observed comparing Tables 1 and 2.

3. The Sparrow et al. approach

A more complex problem is when several products are to be produced a batch plant such as that represented in Figure 2. Sparrow et al. (1972) proposed a methodology to optimize the economics of the plant considering only equipment specified by their volume and not by their processing rate. Equipment specified by their throughput like pumps, is considered to be chosen in accordance with past experience and, normally, it will represent a minor percentage of capital cost.

TABLE 3
Spreadsheet formulation of the multiproduct plant depicted in Figure 2, using the Sparrow et al. approach.

S_{ij}	Prd. 1	Prd. 2	Prd. 3	T_{ij}	Prd. 1	Prd. 2	Prd. 3
Steriliz.	10	5	20	Steriliz.	0.5	0.5	0.5
Fermentor	12	6	21	Fermentor	20	40	12
Drier	7.5	4.5	20.4	Drier	15	8	17
				TLi	20	40	17
COST ESTIM. ($)		a_j	$Alpha_j$		Prd. 1	Prd. 2	Prd. 3
Steriliz.		9000	0.6	Prod.Mt/y	1650	2200	3000
Fermentor		22000	0.55	Bi	24.30	48.60	10
Drier		50000	0.9	Req.time	1358	1811	4729
		SIZE	COST				
Steriliz.		243	243000				
Fermentor		292	499325				
Drier		220	6414317	HOURS/YEAR ------------>			7900
OBJECTIVE FUNCTION -->			7156642	Employed Total Time --->			7898

3.1 MATHEMATICAL MODEL

The criterion in selecting equipment sizes will be the process economics. As before, the sum of the costs of all the equipment will be used as the function to optimize:

$$I = \Sigma_j \, a_j \, (V_j)^{\alpha j} \tag{9}$$

where V_j is the characteristic size of equipment j.

Each product i requires to spend some time to be processed in the equipment j. Let this time to be T_{ij}. Each equipment has a characteristic production factor S_{ij}, which specifies the required volume (or generalizing, size) of equipment j to produce a mass unit of product i.

A batch process will normally be operated with overlapping operation which is more economical as was previously shown. Thus, the longest T_{ij} over j will be the necessary time to complete a batch of product i (T_i)(see figure 1). Thus:

$$T_i \;=\; \mathrm{Max}\ (T_{ij}/n_j) \ (j = 1,M) \tag{10}$$

On the other hand, in each batch, a characteristic amount of product i will be manufactured. Let B_i to be this batch size. Then:

$$B_i \;= \min\ (V_j/S_{ij}) \ (j = 1,M) \tag{11}$$

Finally, if there is H time available in the year, and Q_i mass units are required of product i during this period of time, the following constrain can be formulated:

$$\Sigma_i\ Q_iT_i/B_i\ \leq H \qquad (12)$$

Given this model, an optimization problem exists when the size of the equipments have to be decided. If large sizes V_j are used, large B_i will result and some time will be wasted (total time used < H). Thus, a right choice of volumes V_j has to be done, so that to use the whole time H. Several options may exist in matching this requirement, thus allowing an optimization study to be done.

3.2 EXAMPLE SOLUTION

An example of the manufacture of three products in the example-plant of Figure 2 will be considered. Data of batch times, factor sizes, required annual production for products together with the cost factors are shown in Table 3. This table has been extracted from another spreadsheet file. In order to better appraise the problem, the spreadsheet can be operated, so that the effect on the cost of the different volumes can be observed. The solution presented is one that fits the required time period, but others can exist with better results. In fact, this problem involving several constraints must be solved with an optimization program such as that presented before. GAMS2 optimization results are presented in Table 4. As can be seen, the solution is quite different from that obtained with the spreadsheet.

4. The Knopf et al. approach

The design problem of a multiproduct plant with consideration of both, batch and semicontinuous units was carried out by Knopf et al. (1982). Some new definitions are used in this approach. Thus, the semicontinuous equipment is characterized by a given rate R_k. Moreover, in a given multiproduct plant, the following variables will be known:

S_{ji} mass balance factor (volume or mass of material which must be processed at stage j to result in one unit volume or mass of final product i)

S'_{ki} mass balance factor (volume or mass of material which must be processed at stage k to result in one unit volume or mass of final product i)

T_{ji} processing time at stage j used to produce i.

In addition, the required production Q_i of each product over the planning horizon and the length T of this planning horizon will also be known. Variables to be fixed by the design engineer will be the following:

V_j characteristic volume of batch equipment j

TABLE 4 Solution of the Sparrow approach using GAMS program

SIZE OF EQUIPMENT J	
V_1	210.119
V_2	105.060
V_3	367.709
PRODUCTION PER BATCH	
B_1	18.385
B_2	17.510
B_3	18.025

OPTIMAL CAPITAL COST: 1,069,100

R_k characteristic rate of semicontinuous equipment k
Θ_{ki} processing time at semicontinuous unit k to produce i.
B_i Batch size for product i
T_i Batch time to produce i
Of the latter the first two, that is, the characteristic sizes of the equipment will affect directly to the process economy.

4.1. MATHEMATICAL MODEL

The mathematical model is formed by the restrictions inherent to the variables, together with the economic objective function. The problem consist in minimizing the capital investment function:

$$I = \Sigma_j a_j V^{\alpha j} + \Sigma_k b_k(R_k)^{\beta k} \tag{13}$$

which will be subjected to a series of constraints. First of all, the volume of batch equipment must be selected to accommodate the largest batch size. This means that:

$$V_j = \mathrm{Max}\ (S_{ji}\ B_i) \tag{14}$$

On the other hand, the operating time and the rate of semicontinuos equipment are related in accordance with the expression:

$$\Theta_{ki} = \frac{S'_{ki}\ B_i}{R_k} \qquad k=1,...,N;\ i=1,...,P; \tag{15}$$

Considering two consecutive semicontinuous units, in the processing sequence the following equation must be accomplished:

$$\Theta_{k-1,i} \geq \Theta_{ki} \tag{16}$$

For the overlapping case, the cycle time for product i has to be that for which:

$$T_i = \max\ (\Theta_{k-1,i} + t_j \geq \Theta_{ki}) \tag{17}$$

where k-1 and k are preceding and following respectively the batch equipment j. Finally, it is necessary to consider the time availability. There is a limited time T to complete the required production:

$$T \geq \Sigma_i \frac{Q_i\ T_i}{B_i} \tag{18}$$

4.2 EXAMPLE SOLUTION

As an illustration of the application of this methodology, Table 5 shows the source code written with GAMS language, to solve the plant depicted in Figure 2. As can be seen, values for the mass balance factors are expressed as tables. More details about the program can be found in Brooke et al. (1988).

Although the problem formulation is a little bit longer, the solution is found quickly. Batch volumes are respectively 32.6, 33.5 and 23.9; Batch sizes are 23.9, 21.7 and 23.9 respectively for products 1,2 and 3. Units are in accordance with those given in Table 5. Capital cost is 1,193,600 $.

5. Conclusions

Three approaches to solve the design problem of single product and multiproduct plant have been presented. All of them are based on the minimization of the capital investment costs, the most significative for this kind of processes. GAMS software has been used to solve the tree problems. In all of the cases the solution was found very quickly. Although the mathematical model is more complex, the results are more significative using the Knopf et al. approach.

Finally, to point out that this is an introduction to the design of multiproduct batch plants and that more complex and therefore, complete, approaches are available. For instance, the possibility of parallel equipment and the introduction of the scheduling problem are difficulties that often must be taken into account and that complicates the mathematical model. References given at the end of this chapter, are a good starting point to consider all these aspects.

6. Nomenclature

a_j Cost coefficient for equipment j
b_k Cost coefficient for equipment k
B_i Batch size for product i
I Capital cost (investment)
M Number of batch units
P Number of products
R_k Characteristic rate of semicontinuous equipment k
S_{ji} Mass balance factor (volume or mass of material which must be processed at stage j to result in a unit volume or mass of final product i)
S'_{ki} Mass balance factor (volume or mass of material which must be processed at stage k to result in a unit volume or mass of final product i)
T Batch time
T_i Batch time to produce i
T_{ji} Processing time at stage j used to produce i.
V_j Characteristic volume of batch equipment j
V Volume of product produced per batch
W Production per batch

TABLE 5 Listing of the program, using GAMS software, to solve the example plant depicted in Figure 2, in accordance with the Knopf et al. approach.

```
SETS
  J  batch equipment/ 1, 2, 3 /
  K  semicont. equipment  /1,2,3,4/
  I  products /1,2,3/;
PARAMETERS
  ALFA (J)  cost exponent of equipment k
   /  1    0.65, 2 0.55, 3  0.90/
  A(J) cost coefficient for equipment j
   /  1  9000, 2  22000, 3  50000/
  BETA (K)  cost exponent of equipment k
   /1   0.40, 2  0.40, 3  0.40,  4 0.65/
  B(K)  cost coefficient for equipment k
   /1   700,  2  500,  3  600,  4  10000/
  Q(I)  required annual production of i
    /   1    1650, 2  2200, 3  3000 /
BINI(I) initial values for B(I)
   / 1   10, 2   10, 3   10 /
RINI(K)  Initial values for s.c.
         equipment rates
   / 1  1.2, 2  1.2, 3  1., 4   1.2/
VINI (J) initial values for volume
         of equipment j
   /  1   150,  2   150, 3  150 /
TINI(I) Initial cycle times
   / 1    20, 2    20, 3    20 /;
TABLE SB(J,I)  mass balance factor
              batch equipment
            1      2      3
   1       1.2    1.5    1.3
   2       1.4    1.2    1.
   3       1.0    1.1    1.0;
 TABLE TB(J,I)  times for batch
              equipment
            1      2      3
   1       0.5    0.5    0.5
   2       20     40     12
   3       15     8      17;
TABLE SC(K,I) mass balance factor
              for s.c. equipment
            1      2      3 .
   1       1.2    1.5    1.1
   2       1.2    1.5    1.1
   3       1.2    1.5    1.1
   4       1.4    1.5    1.2;
```

```
VARIABLES

      V(J)  equipment j capacity
      R(K)  production rate m3_h
      BAT(I) batch production of p
      CC  capital cost of the plant
      T(I) batch time to produce I;

      R.L(K)=RINI(K);
      BAT.L(I)=BINI(I);
      V.L(J)=VINI(J);
      T.L(I)=TINI(I);

      BAT.LO(I)=1;
      R.LO(K)=1;
      V.LO(J)=1;
      T.LO(I)=1;

EQUATION
 COST    define objective function
 TIMEMAX     available time
 VOL(J,I)   volume big enough
 TIME1(I)   enough time to produce I
 TIME2(I)   enough time to produce I
 TIME3(I)   enough time to produce I
 UNIT34(I)  consecutive units 2,3;  COST..
CC =E= SUM(J,A(J)*V(J)**ALFA(J)) +
SUM(K, B(K)*R(K)**BETA(K));
TIMEMAX..
SUM(I, Q(I)*T(I)/BAT(I)) =L= 8000;
VOL(J,I)..  V(J) =G= SB(J,I)*BAT(I);
TIME1(I)..  T(I) =G= SC('1',I)*BAT(I)/R('1')
+ TB('1',I) + SC('2',I)*BAT(I)/R('2');
TIME2(I)..  T(I) =G= SC('2',I)*BAT(I)/R('2')
+ TB('2',I) + SC('3',I)*BAT(I)/R('3');
TIME3(I)..  T(I) =G= SC('4',I)*BAT(I)/R('4')
+ TB('3',I);
UNIT34(I).. SC('3',I)/R('3') =G=
SC('4',I)/R('4');

MODEL KNOPF /ALL/ ;
SOLVE KNOPF USING NLP MINIMIZING
CC;
```

Subindexes:
i subindex for the product p
j subindex for the batch equipment
k subindex for the semicontinuous equipment

Greek letters:
α_j Cost exponent for equipment j
β_k Cost exponent for equipment k
Θ_{ki} processing time at semicontinuous unit k to produce i.

7. References

Brooke A., Kendrick, d. and Meeraus, A. (1988). GAMS: A user's guide. G. Smith, Eds. The Scientific Press. Redwood City, (CA).

Knopf, F.C., Okos, M.R. and Recklaitis, G.V. (1982). 'Optimal design of batch/semicontinuous processes' Ind. Eng. Chem. Proc. Des. Develop., 21, 79-86.

Suhami, I and Mah, R.S.H. (1982) 'Optimal design of multipurpose batch plants'. Ind. Eng. Chem. Proc. Des. Dev., 21, 94-100.

Loonkar, Y,R, abd Robinson, J.D. (1970) 'Minimization of capital investment for batch processes'. Ind. Eng. Chem. Proc. Des. Develop., 9, 625-629.

Grossmann, I.E. and Sargent, R.W.H. (1979) 'Optimum design of multipurpose chemical plants'. Ind. Eng. Chem. Proc. Des. Develop., 18, 343-348.

Sparrow, R.E., Forder, G.J. and Rippin, D.W.T. (1975) 'The choice of equipment sizes for multiproduct batch plants. Heuristics vs. branch and bound'. Ind. Eng. Chem. Proc. Des. Develop., 14, 197-203.

STRATEGIES FOR BIOREACTOR SCALE-UP

F. MAVITUNA
Department of Chemical Engineering
University of Manchester Institute of Science and Technology
Manchester, M60 1QD, U.K.

ABSTRACT. Scale-up has been one of the major problems in bioprocess technology. Earlier methods of trial-and-error have now been replaced by those which involve the understanding of the fundamental phenomena, at least for the simpler cases. For more complicated heterogeneous systems the employment of various rules of thumb is still widely practised. A summary of various strategies for bioreactor scale-up is given below.

1. Introduction

Scale-up is a procedure for designing and building a large scale system on the basis of the results of experiments with small-scale models. Different scale-up methods have been reviewed by Kossen and Oosterhuis (1985) and Sweere et al. (1987). Other texts on scale-up include Brown (1982), Ho and Oldshue (1987), Johnstone and Thring (1957), Khang and Levenspiel (1976), Oldshue (1966) and Tatterson (1971).

There are differences in the performance of bioprocess equipment at various scales. The three different kind phenomena important for process design are : thermodynamic phenomena, microkinetic (intrinsic) phenomena and transport phenomena (momentum, heat and mass). Thermodynamic and microkinetic phenomena are independent of scale. Transport processes on the other hand are very dependent on scale and this is the only reason for the scale-up problems. Scale-up problems in industrial fermentations are more pronounced for aerobic than for anaerobic processes and more for continuous than for batch processes. Transport phenomena in bioprocesses are governed by two mechanisms: flow (convection) and diffusion (conduction). Other phenomena directly related to flow and diffusion are : shear, mixing, mass transfer, heat transfer, and macrokinetics (apparent kinetics as a result of coupling microkinetics and mass transfer as in immobilized systems). Further complications in scale-up arise from microbial properties such as growth, adaptation, decay and shear sensitivity. Table 1 summarises some of the general effects of scale-up for a range of bioreactors from bench to production scale.

2. Scale-up Methods

Scale-up methods in use include :

- Fundamental methods
- Semi-fundamental methods
- Dimensional analysis/regime analysis
- Rules of thumb
- Trial and error

A.R. Moreira and K.K. Wallace (eds.), Computer and Information Science Applications in Bioprocess Engineering, 125-142.

TABLE 1. General effects of scale-up factors at different levels

Criteria	Bench top (10 l)	Pilot plant (2000-5000 l)	Production (50000 l)
Reproducibility	Variable, can be good	Reasonable, can be good	Good to not-so-good
Sterility			
Continuous nutrient sterilizer, addition vessels, piping valves,	Somewhat difficult	Less difficult	Less difficult
nozzles	Requires proper steam seals, jackets, etc		
Heat transfer	Easily controlled	Well controlled	Often limiting
Aeration/power input	Easily controlled	Fairly easily controlled	Compressor capacity may be limiting
Agitation (depending upon viscosity and plasticity)	Fairly uniform	Uniform	Nonuniform
Process control response time	Small	Small	Large, nonuniform
Sampling errors	Small, may affect volume	Small	Significant, nonuniform
Risk tolerated	High	Moderate	Low

These methods are often used in combination with each other.

2.1. FUNDAMENTAL METHODS

These methods involve solving the microbalance equations for momentum, mass and heat transfer (Bird et. al., 1960). Some complications arise when these balances are used for process design; for example the balances have to contain terms for transport in three dimension and the boundary conditions are very complicated. Furthermore, the balances for momentum transfer are coupled with heat and mass transfer. Lastly, the momentum balances are usually set up for a homogeneous fluid which is not often a realistic representation of bioprocess systems such as aerated fermentation broth. Therefore, the fundamental methods of scale-up are only used for very simple cases, for example the laminar flow of a liquid film, or in the absence of flow.

2.2. SEMI-FUNDAMENTAL METHODS

These methods involve the use of simplified equations, particularly for the flow. The model describing flow are generally one or a combination of the following types: plug flow, plug flow with dispersion and well mixed. Some examples of the application of semi-fundamental methods can be found in Shaftlein and Russel (1968), Norwood and Metzner (1960), Khang and Levenspiel (1976), Ovaskainen et al. (1976), Laine and Kuoppamaki (1979), Mann (1983) and Harnby et al. (1985). Semi-fundamental methods together with rules of thumb are the most widespread in the design of bioprocesses.

2.3. DIMENSIONAL ANALYSIS

Dimensional analysis applied to scale-up requires the values of dimensionless group of parameters to be kept constant. The physical meaning of these dimensionless groups, often called dimensionless numbers, concerns the ratio of the time constants for the different mechanisms involved. Therefore, keeping the dimensionless groups constant means that the relative importance of the mechanisms influencing the process does not change during scale-up. Many books and articles have been published on dimensional analysis (e.g. Johnstone and Thring, 1957; Becker, 1976).

Although this method is very powerful it has severe limitations. Often, it is not possible to keep all the dimensionless groups constant during scale-up. Therefore, one has to determine the most important groups and neglect the rest. In order to do this a regime analysis is necessary. If a change in regime takes place during scale-up the dimensional analysis method breaks down. The formal application of the dimensional analysis sometimes leads to technically unrealistic values, for example for power consumption, stirrer speed, etc. The choice of parameters involved in the dimensional analysis is not always obvious and sometimes rather arbitrary.

The basic concept of formal dimensional analysis is the geometric similarity. Geometric similarity is often maintained in going from bench to pilot plant scale; however, in the industrial size process the configurations of equipment are often changed and this presents a real design challenge.

TABLE 2. Some rules of thumb to scale-up (Jordan, 1968)

Process	Scale-up rule	Remarks
1. Blending of liquids		
(a) Continuous, low viscosity	Geometrical similarity P/V constant τ/t_m = constant or τ = constant if $\tau > t_m$	t_m depends on reactor scale; tm = constant requires too much energy on a large scale; a change of τ/t_m may cause a change of regime during scale-up
(b) Continuous, high viscosity	Geometrical similarity P/V = constant t_m = constant	Only reasonable mixing times can be achieved with helix-shaped impellers Laminar flow: Nt_m = constant (not checked on a large scale)
(c) Continuous, non-newtonian liquid		Only rapid mixing with relatively big impellers (high-energy input or special shaped impellers, e.g. helix)
2. Dispersion of non-miscible liquids	Geometrical similarity P/V = constant τ or τ/t_m = constant	
3. Suspension of solids	Geometrical similarity P/V = constant	Propeller is most effective; if the agitator has also another function, e.g. gas dispersion, then another agitator can be chosen

TABLE 3. Scale-up Criteria in Fermentation Industries

% of Industries	Scale-up Criterion Used	
30	constant	P/V
30	constant	$k_l a$
20	constant	υ_{tip}
20	constant	pO_2

2.4. RULES OF THUMB

Rules of thumb for scale-up involve 'translation equations'; which are actually performance indices such as power input per unit volume (P/V), torque per unit volume (T_Q/V), or speed ratio (N_2/N_1). Table 2 lists some rules of thumb which can be used for the scale-up of process equipment. Scale-up rules generally used in the fermentation industries are given in Table 3.

Various scale-up rules for mixing are presented in Tables 4 and 5. The earliest approach by Tatterson (1971) which uses P/V, covers seven different operations for turbulent mixing and five for viscous mixing as shown in Table 4. In Table 5, in addition to P/V, two other scale-up parameters, T_Q/V and agitator speed, N, are used for the five operations in the turbulent regime.

Values of the scale-up rules or criteria, such as P/V in Tables 2 - 5, can be calculated for different scales using correlations available from the literature. In fact, the existence of many forms of these correlations, often for the same phenomenon, has been the source of some confusion. Tables 6 and 7 list some of the correlations useful for the calculation of scale-up criteria for stirred tanks and bubble columns, respectively.

Table 8 shows the results of such calculations for two geometrically similar stirred tank systems with the model system volume, $V_m = 10$ l and the production system volume, $V_P = 10\ m^3$ (linear scale-up factor of 10). In Table 8, the values in every row give, for a particular scale-up criterion, the ratio of the values for prototype and model of the variables mentioned at the top of the table. This shows that different scale-up criteria result in entirely different process conditions on a production scale.

TABLE 4a. Scale-up recommendations (Tatterson, 1971) Laminar or viscous mixing, impeller Reynolds number <300

Criteria	Scale-up procedures
1) Equal heat transfer per unit volume	$\frac{(P/V)_2}{(P/V)_1} \sim \left(\frac{V_2}{V_1}\right)^1 \sim \left(\frac{D_2}{D_1}\right)^3$
2) Equal heat transfer coefficients	$\frac{(P/V)_2}{(P/V)_1} \sim 1$
3) Equal blend time	$\frac{(P/V)_2}{(P/V)_1} \sim 1$
4) Equal tip speed	$\frac{(P/V)_2}{(P/V)_1} \sim \left(\frac{V_2}{V_1}\right)^{-0.66} \sim \left(\frac{D_2}{D_1}\right)^{-2}$
5) Equal impeller Reynolds Number	$\frac{(P/V)_2}{(P/V)_1} \sim \left(\frac{V_2}{V_1}\right)^{-1.32} \sim \left(\frac{D_2}{D_1}\right)^{-4}$

TABLE 4b. Scale-up recommendations (Tatterson, 1971) Turbulent mixing, impeller Reynolds number >300

Criteria	Scale-up procedures
1a) Equal mass and heat transfer coefficients to particles, bubbles or drops	$\frac{(P/V)_2}{(P/V)_1} \sim 1$
1b) Equal bubble or drop diameter	
2) Equal heat transfer coefficients stationary surfaces	$\frac{(P/V)_2}{(P/V)_1} \sim \left(\frac{V_2}{V_1}\right)^{0.15} \sim \left(\frac{D_2}{D_1}\right)^{0.45}$
3) Equal blend time	$\frac{(P/V)_2}{(P/V)_1} \sim \left(\frac{V_2}{V_1}\right)^{0.66} \sim \left(\frac{D_2}{D_1}\right)^{2}$
4) Equal tip speed	$\frac{(P/V)_2}{(P/V)_1} \sim \left(\frac{V_2}{V_1}\right)^{-0.33} \sim \left(\frac{D_2}{D_1}\right)^{-1}$
5) Equal impeller Reynolds number	$\frac{(P/V)_2}{(P/V)_1} \sim \left(\frac{V_2}{V_1}\right)^{-1.32} \sim \left(\frac{D_2}{D_1}\right)^{-4}$
6) Equal Froude number	$\frac{(P/V)_2}{(P/V)_1} \sim \left(\frac{V_2}{V_1}\right)^{0.15} \sim \left(\frac{D_2}{D_1}\right)^{0.45}$
7) Equal solids suspension	$\frac{(P/V)_2}{(P/V)_1} \sim \left(\frac{V_2}{V_1}\right)^{-0.19} \sim \left(\frac{D_2}{D_1}\right)^{-0.55}$

TABLE 5a. Scale-up procedures for impeller Reynolds number >300 (Penney, 1971)

Criteria	Scale-up procedures
Equal blend time	$\frac{(P/V)_2}{(P/V)_1} \sim \left(\frac{D_2}{D_1}\right)^2$
Equal Froude number	$\frac{(P/V)_2}{(P/V)_1} \sim \left(\frac{D_2}{D_1}\right)^{0.45}$
Equal mass transfer coefficients to particles, drops or bubbles; equal bubble or drop diameter	$\frac{(P/V)_2}{(P/V)_1} \sim 1$
Equal solids suspension	$\frac{(P/V)_2}{(P/V)_1} \sim \left(\frac{D_2}{D_1}\right)^{-0.55}$
Equal tip speed	$\frac{(P/V)_2}{(P/V)_1} \sim \left(\frac{D_2}{D_1}\right)^{-1}$

TABLE 5b. Scale-up procedures for impeller Reynolds number >300 (Uhl and von Essen, 1986)

Criteria	Scale-up procedures
Equal blend time	$\frac{(T_Q/V)_2}{(T_Q/V)_1} \sim \left(\frac{D_2}{D_1}\right)^2$
Equal surface effects	$\frac{(T_Q/V)_2}{(T_Q/V)_1} \sim \left(\frac{D_2}{D_1}\right)^1$
Equal dispersion	$\frac{(T_Q/V)_2}{T_Q/V)_1} \sim \left(\frac{D_2}{D_1}\right)^{0.66}$
Equal dilute solids suspension	$\frac{(T_Q/V)_2}{(T_Q/V)_1} \sim \left(\frac{D_2}{D_1}\right)^{0.5}$
Equal fluid motion (velocity)	$\frac{(T_Q/V)_2}{(T_Q/V)_1} \sim 1$

TABLE 5c. Scale-up procedures for impeller Reynolds number >300 (Rautzen et al., 1976)

Criteria	Scale-up procedures
Equal blend time	$\frac{N_2}{N_1} \sim 1$
Equal surface behaviour	$\frac{N_2}{N_1} \sim \left(\frac{D_2}{D_1}\right)^{-0.5}$
Equal mass transfer (rate)	$\frac{N_2}{N_1} \sim \left(\frac{D_2}{D_1}\right)^{-0.66}$
Equal solids suspension	$\frac{N_2}{N_1} \sim \left(\frac{D_2}{D_1}\right)^{-0.75}$
Equal liquid motion (velocity)	$\frac{N_2}{N_1} \sim \left(\frac{D_2}{D_1}\right)^{-1}$

TABLE 6. Correlations for a stirred vessel with one standard Rushton-turbine impeller (Kossen and Oosterhuis, 1985)

Power input	$P = 6\rho N^3D^5$	Ungassed
	$P_g = 0.4\ P$	Gassed
	$P_g = 0.312\ P\ Fr^{-0.16}Re^{0.064} \left(\frac{Q}{ND^3}\right)^{-0.38} (T/D)^{0.8}$	Gassed, Reuss et al., 1980
Mass transfer	$k_La = 2.6 \times 10^{-2}\ (P_g/V)^{0.4}(\upsilon_s)^{0.5}$	Coalescing systems
	$k_La = 2.0 \times 10^{-3}(P_g/V)^{0.7}(\upsilon_s)^{0.2}$	Non-coalescing systems
Mixing/circulation	$t_{circ} = V/\theta_{circ}$	
	$\theta_{circ} = 2\theta_p \qquad Nt_{circ} = \frac{V}{2.6D^3}$ (Re>5 x 10^3	
	$\theta_p = 1.3\ ND^3$	
	$t_m = 4t_{circ}$	
Shear	$\sim\eta N$	Laminar flow
	$\sim\rho(ND)^2$	Turbulent flow
	$\sim N/D$	Wang and Fewkes, 1977
	$\sim\rho(ND)^2N\sim P/V$	Turbulent flow and fatigue

TABLE 7. Correlations for bubble column (Kossen and Oosterhuis, 1985)

Power input	$P = \upsilon_s \rho_1 g V$	(H<2 m)
	$P = \dfrac{\theta_{m,gas} RT}{M_{gas}} \ln \left(\dfrac{P_1}{P_2}\right)$	(H>2 m)
Mass transfer	$k_L a = 0.32 \upsilon_s^{0.7}$ 0.55% decrease of oxygen concentration in air per metre column	for coalescing systems
	For non coalescing systems $k_L a$ depends entirely on the initial bubble size	
Mixing	$D_e^1 = 0.35\ (\upsilon_s g T^4)^{1/3}$ $t_m \sim H^2/D_e^1$	
Shear	In bubble columns shear is usually not important	

TABLE 8. Different scale-up criteria and their consequences (Kossen and Oosterhuis, 1985)

Scale-up criterion	Value at 10 m^3 scale (V_m=10 l)						
	P	P/V	N (or t_m^{-1})	ND	Re	N/D	
Equal P/V	10^3	1	0.22		2.15	21.5	0.022
Equal N (or t_m^{-1})	10^5	10^2	1		10	10^2	0.1
Equal tip speed	10^2	0.1	0.1		1	10	10^{-2}
Equal Re number	0.1	10^{-4}	10^{-2}		0.1	1	10^{-3}
Equal shear to flow ratio	10^8	10^5	10		10^2	10^3	1

2.5. REGIME ANALYSIS

The regime concept was introduced by Johnstone and Thring (1957). The word 'regime' is used for the 'dominance' of a particular mechanism in the performance of a system. For instance, the kinetic regime for a bioreactor means that the performance of the bioreactor is dominated by kinetic phenomena rather than transport phenomena. A regime can be pure, with only one mechanism dominating, or mixed, with two or more mechanism playing a comparable part in the performance of the system.

Regime analysis is important for the design of scale-up experiments because as mentioned earlier, not all dimensionless groups or scale-up criteria can be kept at a constant value during scaling-up. Regime analysis shows whether one regime is rate determining, which regime it is and whether this regime changes during a change of scale. Dimensional analysis combined with regime analysis and small-scale experiments is much used to solve scale-up problems, especially those involving mass, heat and momentum transfer.

The experimental methods of regime analysis are based on the variation of a quantity such as velocity, concentration, particle size, and temperature, which has a significant influence on one of the mechanisms involved in the behaviour of the system. Some examples of the use of experimental regime analysis are given by Levenspiel (1972). The experimental technique can indicate whether one regime is rate determining and which regime it is but it cannot tell whether the regime will change during scale-up. This can be answered by theoretical regime analysis.

Theoretical regime analysis based for example on the numerical solution of the balance equations, involves parameter sensitivity assessment. Parameter sensitivity, P_i can be defined as :

$$P_i = \frac{\Delta Y / Y}{\Delta X_i / X_i}$$

where Y = the value of the important variable at the original value of the parameter
ΔY = the variation of this variable due to the variation of the parameter
X_i = the original value of the parameter
ΔX_i = the variation of this parameter

The analytical approach to theoretical regime analysis is often based on a comparison of time constants, of transfer and conversion steps, of forces, of pressures, or dimensionless groups.

In comparing rate phenomena, time is the most convenient characteristic parameter. Characteristic time is a measure of the rate of a mechanism and can be considered as the time needed by that mechanism to smooth out a change. In the literature, terms like time constant, process time (constant) and relaxation time are also used. The term time constant is commonly used for first-order processes only, and is equal to the time needed for a mechanism to proceed to 63 percent conversion.

Characteristic time is used to characterise the rate of mechanisms. Therefore, it is possible to characterise non-linear processes and processes of a higher order with only one characteristic time. A low value of a characteristic time means a fast mechanism; a

TABLE 9. Definitions of various characteristic times

Mixing time	$t_m \equiv \dfrac{V}{2.6ND^3}$ (Re>5 x 10^3 Rushton turbine)
Conversion time	$t_c \equiv \dfrac{C}{r}$
Diffusion time	$t_D \equiv \dfrac{L^2}{D}$
Residence time	$\tau \equiv \dfrac{V}{\theta_v}$ or $= \dfrac{L}{v}$
Mass transfer time	$t_{mt} \equiv \dfrac{1}{k_L a}$

high value means of a slow mechanism. Characteristic times can be determined experimentally or theoretically. Examples of experimental determination include liquid mixing time and liquid circulation time in mechanically stirred vessels.

The theoretical determination of characteristic times may be based on literature correlations, the solution of differential equations of transport phenomena, or rules of thumb. Table 9 lists the relationships for the calculation of various characteristic times and Table 10 of various characteristic pressures and tensions (Kossen and Oosterhuis, 1985).

In addition to physical mechanism, the performance of a bioreacter may also be influenced by physiological mechanisms. Characteristic times for substrate consumption or product formation can be calculated by means of the integrated Michaelis-Menten kinetics for batch growth.

TABLE 10. Definitions of different pressures and tension for regime analysis of flow mechanisms

Shear stress (laminar)	$\frac{v}{\eta_L}$	N m^{-2}
Shear stress (turbulent)	ρv^2	N m^{-2}
Buoyancy	$\Delta \rho g L$	N m^{-2}
Yield stress	τ_o	N m^{-2}
Surface tension	σ	N m^{-1}

3. Conclusion

Although only bioreactor scale-up has been the focus of attention here, scale-up is equally important for other upstream and downstream unit operations in bioprocess technologies. For instance, scale-up of the preparation of inocula and media sterilisation can be crucial for a successful run. Although process engineers are more used to scaling-up conventional downstream processing operations, new separation methods and equipment, as well as the difficulties pertinent to the new biological products and stricter product purity requirements pose new challenges.

In any study and application of scale-up, scale-down is equally important. In order to obtain reliable results at laboratory or pilot scale which will be used for scale-up, growth and production have to be investigated under conditions similar to the production scale. In such scale-down practises, regime analysis and time constants are useful, For the study of flow and mixing, geometrical similarity is essential.

Computers and computational procedures can be used not only in monitoring, data logging and data analysis but also in performing 'computer experiments' using real time simulations. For instance, such applications of computers are becoming increasingly popular for flow visualisation. 'Computer experiments', if successfully applied, can reduce the time and cost of scale-up and scale-down by focusing on the most important phenomena, parameters and the range of their values.

NOMENCLATURE

a	specific surface area	m^{-1}
C	concentration	$kg\ m^{-3}$
D	stirrer diameter	m
D	diffusion coefficient	$m^2\ s^{-1}$
D_e	dispersion coefficient	$m^2\ s^{-1}$
Fr	Froude number	-
g	acceleration due to gravity	$m\ s^{-2}$
H	height of column	m
k_la	mass transfer coefficient	s^{-1}
L	length	m
M_{gas}	molecular weight of gas	
N	stirrer speed	s^{-1}
P_1	pressure at the bottom	$N\ m^{-2}$
P_2	pressure at the top	$N\ m^{-2}$
P	power	W
P_g	gassed power	W
pO_2	partial pressure of oxygen	$N\ m^{-2}$
Q	volumetric flow	$m^3\ s^{-1}$
r	reaction rate	$kg\ m^3\ s^{-1}$
R	gas constant	$N\ m\ kmol^{-1}\ K^{-1}$
Re	Reynolds number	-
t_c	conversion time constant	s
t_{circ}	circulation time	s
t_D	diffusion time constant	s
t_m	mixing time	s
t_{mt}	mass transfer time constant	s
T	temperature	^0K
v	velocity	$m\ s^{-1}$
v_s	superficial velocity	$m\ s^{-1}$
v_{tip}	stirrer tip velocity	$m\ s^{-1}$
V	volume	m^{-3}
η	dynamic viscosity	$N\ s\ m^{-2}$
ρ	density	$kg\ m^{-3}$
σ	surface tension	$N\ m^{-1}$
τ	residence time	s
τ_0	yield stress	$N\ m^{-2}$
θ_{circ}	circulating flow rate	$m^{-3}\ s^{-1}$
θ_m	mass flow rate	$kg\ s^{-1}$
θ_p	pumping capacity of stirrer	$m^3\ s^{-1}$
θ_v	volumetric flow rate	$m^3\ s^{-1}$

REFERENCES

Becker, H. A. (1976) Dimensionless Parameters: Theory and Methodology, Applied Science Publishers, London.

Bird, R. B., Stewart, W. E. and Lightfoot, E. N. (1960) Transport Phenomena, Wiley, New York.

Brown, D. E. (1982) 'Industrial scale operations of microbial processes', J. Chem. Technol. Biotechnol., 32, 34-46.

Harnby, N., Edwards, M. F. and Nienow, A. W. (eds.) (1985) Mixing in the Process Industries, Butterworth, London.

Ho, C. S. and Oldshue, J. Y. (eds.) (1987) Biotechnology Processes: Scale-up and Mixing, AIChE, New York.

Johnstone, R. E. and Thring, M. W. (1957) Pilot Plants, Models and Scale-up Methods in Chemical Engineering, McGraw-Hill, New York.

Khang, S. J. and Levenspiel, O. (1976) 'New scale-up and design method for stirrer agitated batch mixing vessels', Chem. Engng. Sci., 31 (7), 569-577.

Kossen, N. W. F. and Oosterhuis, N. M. G. (1985) 'Modelling and scaling-up of bioreactors', in H. -J. Rehm and G. Reed (eds.), Biotechnology, Vol. 2, H. Brauer (Vol. ed.) Ch. 24, 572-605, VCH, Weinheim.

Laine, J. and Kuoppamaki, R. (1979) 'Development of the design of large scale fermenters', Ind. Engng. Chem. Process Des. Dev., 18, 501-506.

Levenspiel, O. (1972) Chemical Reaction Engineering, Wiley, New York.

Mann, R. (1983) Gas-liquid Contacting in Mixing Vessels, IChemE, Rugby.

Norwood, K.W. and Metzner, A. B. (1960) 'Flow patterns and mixing rates in agitated vessels, A.I.Ch.E. J., 6, 432-437.

Ovaskainen, P., Lundell, R. and Laiho, P. (1976) 'Engineering of fermentation plants. Part 2. Fermenter design and scale-up', Process Biochem., 11, 37-55.

Penney, W. R. (1971) Chem. Engr., 787, 86.

Reuss, M., Bajpai, R. K., Lenz, R., Niebelschutz, H. and Papalexiou, A.(1980) 'Scale-up strastegies based on the interaction of transport and reaction'VIth Int. Ferment. Symp., July 20-25, London, Ontario, Canada, F-7.2.1.

Shaftlein, R. W. and Russel, T. W. F. (1968) 'Two-phase reactor design, tank-type reactors', Ind. Engng. Chem., 60, 12-27.

Sweere, A. P. J., Luyben, K. Ch. A. M. and Kossen, N. W. F. (1987) Regime analysis and scale-down: tools to investigate the performance of bioreactors, Enzyme Microb. Technol., 9 (July), 386-392.

Tatterson, G. B. (1971) 'Scale-up procedures and power consumption in agitated vessels', Food Technol., May, 65-70.

Uhl, V. W. and von Essen, J. A. (1986) in Mixing: Theory and Practice, V. W. Uhl and J. B. Gray (eds.), Vol. III, Ch. 15, Academic Press, New York.

Wang, D. I. C. and Fewkes, R. C. J. (1977) 'Effect of operating variables and geometric parameters on the behavior of non-Newtonian, mycelial antibiotic fermentations', Dev. Ind. Microbiol., 18, 39-56.

STATISTICAL DESIGN AND ANALYSIS OF EXPERIMENTS

C.E. ROLZ
Center for Scientific and Technological Studies
Central American Research Institute for Industry, ICAITI
P.O. Box 1552
01901 Guatemala, Central America

ABSTRACT. A problem solving strategy employing experimental designs is illustrated in optimizing the yield of a secondary metabolite in solid substrate culture. The strategy employs a screening Plackett-Burman design that selects two of six factors. Then a two-level factorial design is carried out in order to define a search direction. The optimizing search is done employing the Simplex algorithm. A surface response quadratic model in three factors is generated to predict optimum conditions.

1. Introduction

Statistical experimental design is an effective problem solving technique, especially in those cases where response variables are influenced by many independent factors and the theoretical base is not well understood. It has been used in process experimentation in order to screen and identify important process variables, and to find optimum settings of chosen factors by response surface methodology. The reader is referred to the texts by Box et al. (1978) and Mason et al. (1989). Lately, it has also been an important statistical tool for quality improvement (Barker, 1985). Haaland (1989) has written an excellent monograph of its application to solve complex biological problems: enzyme-immunoassays, bioactive surfaces and monoclonal antibodies

The proliferation of PC's in research laboratories has stimulated the development of commercial software for the generation of experimental designs and the analysis of data. The reader is referred to Directories published by several journals, i.e. Chemical Engineering Progress, Quality Progress.

The present illustration has been planned in order to exemplify a possible strategy in problem solving employing experimental designs. The chosen example will be the yield optimization of a cell secondary metabolite. The data will be fictitious but logical. In the screening step the software *Screen** will be employed. A special type of fractional factorial design proposed by Plackett and Burman in 1946 will be employed. Out of this exercise, two independent variables will be selected as most influential. For the optimization step, the following software will be used: *Simplex-V*, *X-Stat* and *Echip**. The optimization algorithm will be the Simplex, either on its fixed-size original option or the variable-size one of Nelder and Mead. The model will be generated by response-surface methodology employing central-composite (or Box-Wilson), Box-Behnken and computer generated designs. In some instances, graphical methods will be used to identify "bad" data, with support from the software *Systat/Sygraph* or *Stata**.

A.R. Moreira and K.K. Wallace (eds.), Computer and Information Science Applications in Bioprocess Engineering, 143-156.

2. Screening Phase

Plackett-Burman designs, a special class of fractional factorial experiments are useful for picking one or two most important factors (independent variables) from a list of factors, which the experimenter believes have influence on the response (dependent variable). They can be generated in a multiple of four runs. Each design generator can be used to construct an experiment having up to one fewer factor than the number of test runs. They are usually resolution III designs (all main effects are partially confounded with two factor interactions). However, they can be "folded-over" in order to assure that main effects are not confounded. This implies more runs but avoids selecting the wrong factors for further study.

The selected problem for maximizing the yield of a secondary metabolite involves a solid substrate culture process employing a selected strain of a *Streptomyces*. The experimenters believe six factors are important: time, temperature, amount of initial protein, substrate pretreatment (thermal autohydrolysis), culture agitation and the presence of an inducer. Initial culture pH has been already explored and from this knowledge an optimum interval defined between 5.8 and 6.6, hence it was decided to fix it for the screening phase at its lowest value and then include it as a third variable in the response surface design.

Twelve and a "fold-over" of 27 runs are defined by Haaland (1989) in his Design Digest for seven factors, identified as PB0712 and PB0724. One more "dummy" factor can be added to the problem, which can be used to estimate pooled uncertainty and interaction effects.

The Screen software generates a 19 run Plackett-Burman, "foldover", with a single experimental repetition at a specific one factor combination. This is illustrated in Table 1. A "plus sign" indicates the higher level of the factor. The level values chosen for each factor are displayed in Table 2. The runs are performed and yield values obtained are reported in the last column of Table 1. The results of the analysis are illustrated in Table 3, as they are actually printed. It shows clearly that two factors affect yield the most: temperature and time. Substrate pretreatment, the amount of initial protein and the culture agitation exert minor effects upon yield. The effect of the inducer is confounded with the statistical noise.

3. Region Optimization Step

With the two most important variables, time and temperature, a full factorial at two levels design incorporating a center point repeated three times, was defined. The idea behind this experimentation was to explore a surface and look for a direction of movement.

The design, the yield values and the data analysis obtained employing the software *X-Stat* are found in Table 4. The linear model fits adequately the data, the coefficients are statistically significant and increasing both factors seems to be the direction to look for.

The software *Simplex-V* has been designed for process experimentation employing the Simplex algorithm in various options. Table 5 presents the results of a search with a fixed-size Simplex, starting with three of the previously known points. The strategy of the Simplex is straightforward: to achieve improved response and move away from bad response. The search has been bounded in a region limited by time: from 12 to 120 hours and temperature: 15 to 55 deg C. Six experiments were requested, although two of them were out of bounds. The three best experiments improved the yield significantly and defined the last vertex. The search was stopped due to eventual cycling (not shown) originated by the fixed-size of the Simplex.

TABLE 1. Plackett-Burman seven factor, 19 runs, "fold-over" screening design generated by the software screen.

Experiment	time	Temp	Prot	Pretr	Agitat	Inducer	Dummy	Yield
1	+	+	+	−	+	−	−	6.4
2	−	+	+	+	−	+	−	4.7
3	−	−	+	+	+	−	+	1.8
4.	+	−	−	+	+	+	−	3.6
5	−	+	−	−	+	+	+	4.4
6	+	−	+	−	−	+	+	3.1
7	+	+	−	+	−	−	+	6.7
8	−	−	−	−	−	−	−	1.6
9	−	−	−	+	−	+	+	1.7
10	+	−	−	−	+	−	+	2.9
11	+	+	−	−	−	+	−	6.0
12	−	+	+	−	−	−	+	4.2
13	+	−	+	+	−	−	−	3.6
14	−	+	−	+	+	−	−	4.5
15	−	−	+	−	+	+	−	1.5
16	+	+	+	+	+	+	+	6.9
17	−	−	−	−	−	−	−	1.5
18	−	−	−	−	−	−	−	1.6
19	−	−	−	−	−	−	−	1.7

TABLE 2. Factor levels for the screening design.

Factor	Level (+)	Level (−)
Time, h	48	24
Temperature, °C	40	18
Initial protein, %	2.5	0.5
Substrate pretreatment, °C	210	180
Agitation, rpm	50	Static

TABLE 3. Analysis of the screening design.

```
SCREEN 150160
Analysis of RESPONSES for CARLOS ROLZ
Project name: METSSC
----------------------------------------------

EFFECT ON RESPONSE 1, "Yield %"

     FACTOR        EFFECT           MSQ                 (-1)          (+1)
--------------------------------------------------------------------------------
2    Temp          3.00             36.00               18 deg C      40 deg C
1    time          1.85             13.69               24 h          48 h
4    Pretr         4.25E-01         7.23E-01            180 deg C     210 deg C
3    Prot          1.00E-01         4.00E-02            0.5 %         2.5 %
5    Agitat        5.00E-02         1.00E-02            0 rpm         50 rpm
6    Inducer       2.50E-02         2.50E-03            0 mg %        100 mg %
7    Dummy         -2.50E-02        2.50E-03            1             2
================================================================================
SCREEN 150160
Analysis of UNCERTAINTY for CARLOS ROLZ
Project name: METSSC
----------------------------------------------

Foldover          Yes
Replicate Whole   No
Replicate Single  Yes

     RESPONSE      SQUARES          D.F.     VARIANCE
----------------------------------------------------------
1    Yield %       2.00E-02         3        6.67E-03
----------------------------------------------------------
```

TABLE 4. Analysis of a 22 factorial with center point repeated three times employing *X-stat*

```
                        METSSC
                        ======

        Controlled FactorsCharacteristic
RUN  *---------*---------*  *---------*
 #       TIME      TEMP         YIELD

 1      24.00     18.00         2.000
 2      48.00     18.00         3.500
 3      24.00     40.00         4.400
 4      48.00     40.00         6.800
 5      36.00     29.00         3.400
 6      36.00     29.00         3.200
 7      36.00     29.00         3.600
YIELD:
  standard deviation about the regression = 0.5727
  explained variation about the mean (R-squared) = 90.09%
  condition of design matrix = 11.30
  model = LINEAR
     Regression Coefficients for YIELD
     =================================

                                 Standard                Confidence
Coefficient       Term           Error       T-Value    Coef <> 0

  -2.839        1 (constant)     1.164       2.439       92.1%
   0.08125      TIME             0.0239      3.405       96.2%
   0.1295       TEMP             0.0260      4.976       98.1%

Confidence figures are based on 4 degrees of freedom
```

TABLE 4 (continued)

Analysis of Variance for YIELD

Source	df	SS	MS	F-Ratio
Total (corrected)	6	13.23714		
Regression	2	11.92500	5.96250	18.18 (1)
Residual	4	1.31214	0.32804	
Lack of fit	2	1.23214	0.61607	15.40 (2)
Pure error (3)	2	0.08000	0.04000	

(1) Implies 97.6% confidence regression equation is nonzero.
(2) Implies 10.6% confidence pure error explains lack of fit.
(3) Determined from replicated runs:
(6, 7, 5)

Table of Residuals for YIELD

RUN #	Controlled Factors: TIME	Controlled Factors: TEMP	Characteristic YIELD: Observed	Characteristic YIELD: Predicted	Characteristic YIELD: Residuals
1	24.00	18.00	2.000	1.443	0.557
2	48.00	18.00	3.500	3.393	0.107
3	24.00	40.00	4.400	4.293	0.107
4	48.00	40.00	6.800	6.243	0.557
5	36.00	29.00	3.400	3.843	-0.443
6	36.00	29.00	3.200	3.843	-0.643
7	36.00	29.00	3.600	3.843	-0.243

```
      Residuals
|---------|--------|
MIN       0      MAX
:         :        X
:         : X      :
:         : X      :
:         :        X
:  X      :        :
X         :        :
:     X   :        :

            X      X
X  X  X     X      X
|---------|--------|
MIN       0      MAX
```

MAXIMUM YIELD

A maximum of 6.243 was achieved under the following conditions.

Value at Maximum		Lower Limit	Upper Limit
	Factors		
48.0	TIME	24.00	48.00
40.0	TEMP	18.00	40.00
	Characteristics		
6.24	YIELD		

Starting factor values: 36.00, 29.00

YIELD

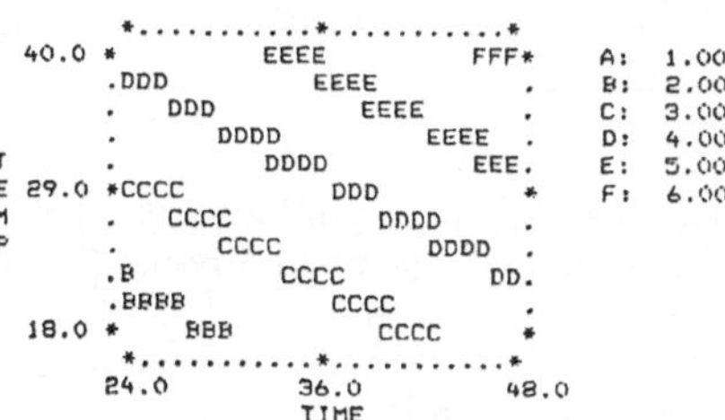

TABLE 5. Fixed-size Simplex algorithm search employing the software *Simplex-V.*

[#110533] FACTOR INFORMATION FOR PROJECT MSSCFS

NO	FACTOR	UNITS	LOWER	UPPER	START	STEP
1	TIME	HOURS	12.00	120.00	None	None
2	TEMP	DEG C	15.00	55.00	None	None

[#110533] SIMPLEX HISTORY FOR PROJECT MSSCFS

VERTEX	TIME HOURS	TEMP DEG C	RESPONSE
1) I	36.00	29.00	3.40
2) I	24.00	40.00	4.40
3) I	48.00	40.00	6.80
4) R	36.00	51.00	7.50
5) R	60.00	51.00	13.20
6) R	48.00	62.00	BV
7) R	72.00	62.00	BV
8) R	84.00	51.00	11.30
9) R	72.00	40.00	9.20

[#110533] CURRENT SIMPLEX FOR PROJECT MSSCFS

VERTEX	TIME HOURS	TEMP DEG C	RESPONSE
5) R - 3	60.00	51.00	13.20
9) R - 1	72.00	40.00	9.20
8) R - 2	84.00	51.00	11.30
10) R - 0	48.00	40.00	0.00

TIME TEMP RESPONSE

Table 6 shows much better results, obtained employing a variable-size Simplex, starting at the center point of the factorial design, using steps of 5 hours (5% of the span for that variable) and 4 deg C (10% of the span for temperature). These two figures can be varied at will. Twenty-two experiments were requested, five were out of bounds and not done. In seventeen experiments, the yield has been increased to a value between 14 and 15 g of metabolite per kg of final moist product at conditions of temperatures between 52 and 55 deg C after 60 hours of solid substrate culture.

TABLE 6. Variable-size Simplex search employing the software *Simplex-V*.

[#110533] SIMPLEX HISTORY FOR PROJECT MSSCVS

VERTEX	TIME HOURS	TEMP DEG C	RESPONSE
1) I	36.00	29.00	3.40
2) I	41.00	29.00	4.00
3) I	36.00	33.00	4.20
4) R	41.00	33.00	4.90
5) E	43.50	35.00	5.20
6) R	38.50	39.00	5.30
7) E	37.25	44.00	6.30
8) R	44.75	46.00	7.60
9) E	49.12	52.50	9.10
10) R	42.87	61.50	BV
11) Cr	43.03	54.87	8.90
12) R	54.91	63.37	BV
13) Cr	50.49	58.53	BV
14) Cw	41.66	48.84	7.70
15) R	47.76	46.47	8.10
16) R	55.22	50.12	10.90
17) E	62.00	50.77	13.10
18) R	63.36	56.80	BV
19) Cr	59.46	54.21	14.00
20) R	72.33	52.48	12.40
21) Cr	66.53	52.49	14.90
22) R	64.00	55.93	BV
23) Cr	63.50	54.64	14.20

[#110533] FACTOR INFORMATION FOR PROJECT MSSCVS

NO	FACTOR	UNITS	LOWER	UPPER	START	STEP
1	TIME	HOURS	12.00	120.00	36.00	5.00
2	TEMP	DEG C	15.00	55.00	29.00	4.00

[#110533] CURRENT SIMPLEX FOR PROJECT MSSCVS

VERTEX	TIME HOURS	TEMP DEG C	RESPONSE
21) Cr - 2	66.53	52.49	14.90
23) Cr - 1	63.50	54.64	14.20
19) Cr - 3	59.46	54.21	14.00
24) R - 0	70.57	52.91	0.00

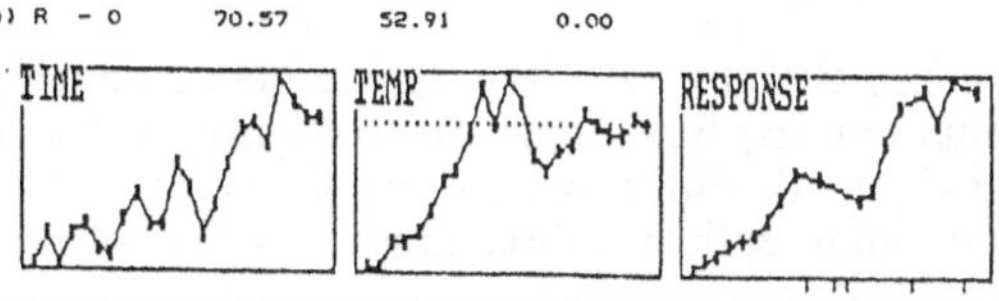

TIME TEMP RESPONSE

4. Model Development

A statistically based model relating yield with time, temperature and initial culture pH was constructed using response surface methodology. Three experimental designs were generated as detailed in Figure 1. The factor levels are given in Table 7. The analysis of the computer generated design (A) is given in Table 8, as obtained using *X-stat*. The fitted model was quadratic and statistically significant. Significant factors were pH and time square (outside points from linear correlation of the probability plot of the coefficients, Figure 2). Although the analysis shows that, at alpha=0.05, also temperature and the time*temperature interaction were important. The model residuals (difference between observed and predicted values) were normally distributed: linear correlation in probability plot and skewness and kurtosis tests significant, see Figure 3. The only point in doubt is run 10 (a residual equal to 1.093), a duplicated experiment. In fact, at those factor levels, the largest difference between duplicates exists. The maximum yield is 15.6 predicted by the model at an initial pH of 5.8, a temperature of 51.5 deg C and 64.7 hours of culture. The predicted surface response is plotted for a constant pH of 5.8 as a function of time and temperature in Figure 4. The fitting of the three models are summarized in Table 9. In terms of cost/benefit ratio, the Box-Behnken design is best: in fewer runs it provides a better fit identifying the important effects. However, the predicted optimum conditions are rather similar for the three models.

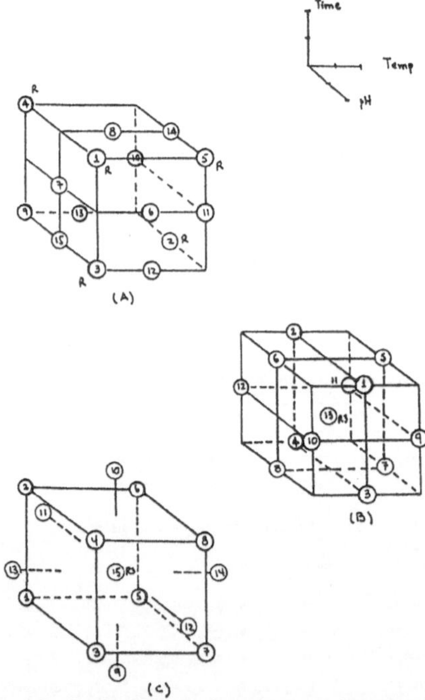

Figure 1. Response surface designs. (A) 20 experiments computer generated (ECHIP) with five replicated points (R); (B) Box-Behnken, 16 experiments, center point replicated three times (BB0316 of Haaland, 1989); (C) 18 runs Central Composite Design or CCD (Box-Wilson) with center point replicated three times (CC0318 of Haaland, 1989)

TABLE 7. Factor levels for response surface designs

Factor	(-1.68)	(-1)	(0)	(+1)	(+1.68)
pH	5.5	5.8	6.2	6.6	6.9
Time	56.6	60	65	70	73.4
Temperature	48.3	50	52.5	55	56.7

TABLE 8. Analysis employing X-stat of response surface design experiment A.

DESIGN A

RUN #	Controlled Factors: TIME	TEMP	PH	Characteristic: YIELD
1	1.000	-1.000	1.000	7.20
2	1.000	-1.000	1.000	7.60
3	-1.000	1.000	0.000	10.30
4	-1.000	1.000	0.000	10.00
5	-1.000	-1.000	1.000	6.80
6	-1.000	-1.000	1.000	7.40
7	1.000	-1.000	-1.000	13.10
8	1.000	-1.000	-1.000	13.20
9	1.000	1.000	1.000	4.00
10	1.000	1.000	1.000	5.60
11	0.000	0.000	1.000	9.30
12	0.000	-1.000	0.000	14.20
13	1.000	0.000	0.000	9.70
14	-1.000	-1.000	-1.000	12.60
15	0.000	1.000	-1.000	13.50
16	0.000	1.000	1.000	7.00
17	-1.000	0.000	1.000	7.70
18	-1.000	0.000	-1.000	14.20
19	1.000	1.000	0.000	8.40
20	-1.000	-1.000	0.000	9.50

YIELD:
standard deviation about the regression = 0.7711
explained variation about the mean (R-squared) = 96.58%
condition of design matrix = 8.008
model = QUADRATIC

Regression Coefficients for YIELD

Coefficient	Term	Standard Error	T-Value	Confidence Coef <> 0
13.03	1 (constant)	0.6140	21.22	99.9%
-0.3890	TIME	0.2173	1.790	89.9%
-0.8160	TEMP	0.2223	3.671	99.6%
-2.982	PH	0.2401	12.42	99.9%
-0.5999	TIME*TEMP	0.2333	2.572	97.6%
0.09028	TIME*PH	0.2489	0.3626	30.9%
-0.1026	TEMP*PH	0.2685	0.3820	31.9%
-2.283	TIME^2	0.4465	5.113	99.9%
-0.8479	TEMP^2	0.4400	1.927	92.1%
-0.5930	PH^2	0.4023	1.474	82.5%

Confidence figures are based on 10 degrees of freedom

TABLE 8 (continued)

Analysis of Variance for YIELD

Source	df	SS	MS	F-Ratio
Total (corrected)	19	174.085		
Regression	9	168.139	18.6822	31.420 (1)
Residual	10	5.946	0.5946	
Lack of fit	5	4.356	0.8712	2.740 (2)
Pure error (3)	5	1.590	0.3180	

(1) Implies 99.9% confidence regression equation is nonzero.
(2) Implies 15.5% confidence pure error explains lack of fit.
(3) Determined from replicated runs:
(1, 2)
(5, 6)
(8, 7)
(9, 10)
(3, 4)

MAXIMUM YIELD

A maximum of 15.58 was achieved under the following conditions.

	Value at Maximum	Factors	Lower Limit	Upper Limit
(64.7)	-0.0465	TIME	-1.000	1.000
(51.5)	-0.392	TEMP	-1.000	1.000
(5.8)	-1.00	PH	-1.000	1.000
		Characteristics		
	15.6	YIELD		

Starting factor values: 0.000, 0.000, 0.000

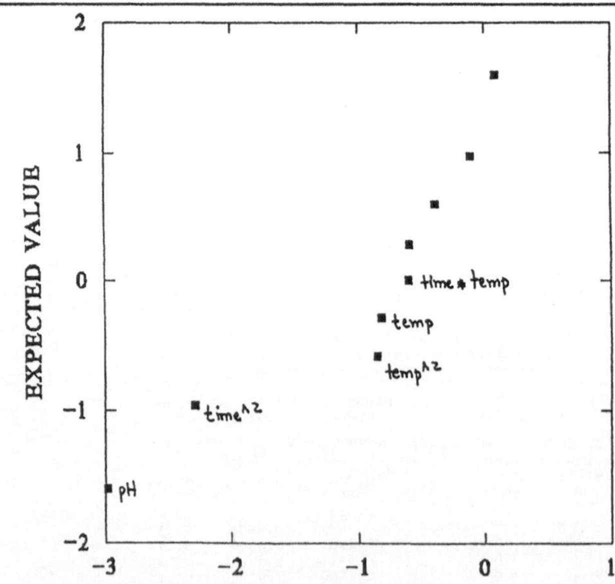

Figure 2. Probability plot of response surface model (A) coefficients obtained with SYSTAT/SYGRAPH

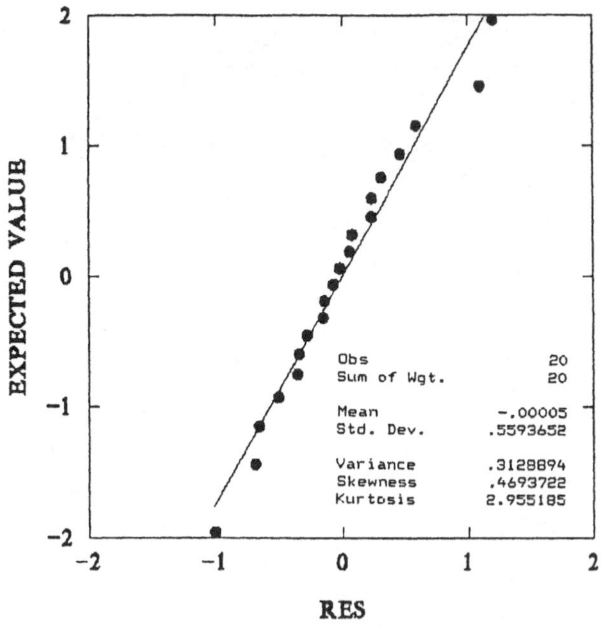

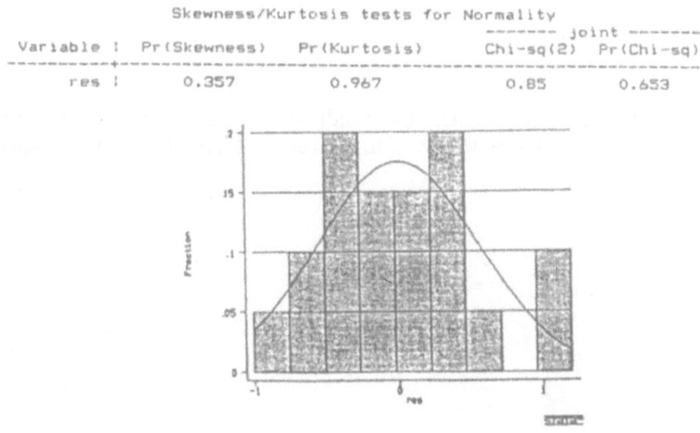

Skewness/Kurtosis tests for Normality

Variable	Pr(Skewness)	Pr(Kurtosis)	joint Chi-sq(2)	joint Pr(Chi-sq)
res	0.357	0.967	0.85	0.653

Figure 3. Linear probability plot of residuals. Results of skewness and kurtosis tests and histogram, as done by *STRATA*.

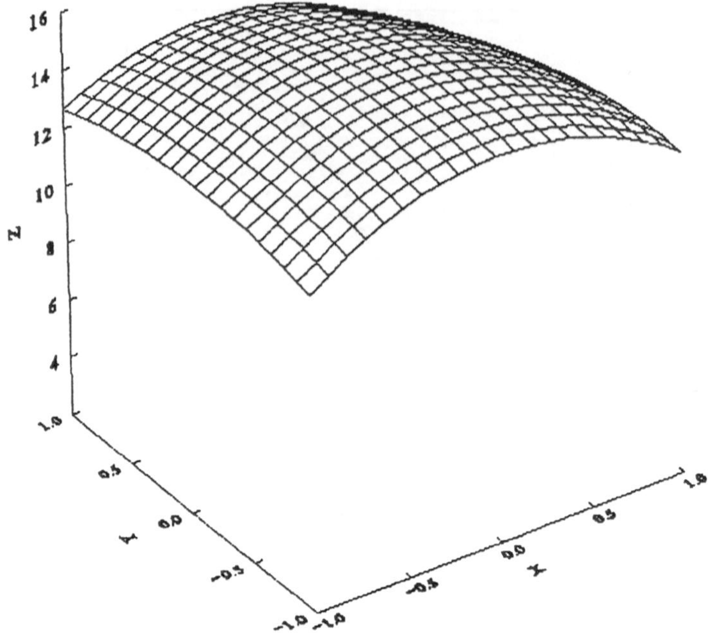

Figure 4. Response surface generated by quadratic model of data from experiment A. The yield is z; x is time and y is temperature in units defined by their levels of Table 7 (*SYSTAT/SYGRAPH*)

TABLE 9. Summary of response system models.

Model	Runs	Residual sum of squares	Mean square	Significant factors
Computer generated (A)	20	5.95	18.68	pH, time^2, temp, time*temp
Box-Behnken (B)	16	1.17	10.62	pH, time, time^2
Central Composite (C)	18	15.10	12.50	pH

TABLE 9. Summary of response surface models (continued)

Model	Maximum yield predicted	pH	Optimum conditions Temperature	Time
Computer generated (A)	15.6	5.8	51.5	64.7
Box-Behnken (B)	14.5	5.8	52.7	62.5
Central Composite (C)	13.1	5.8	54.2	61.8

5. Was it a Good Experimental Strategy?

The yield was certainly increased although it took 59 experiments. This figure is much less if from the start we would have chosen a factorial experiment (without replicates) of nine factors at two levels, which would have required 512 runs. Besides there was no assurance that the optimum region would have been included. Was it correct to leave a very significant variable constant for the screening and search phases? Not necessarily. In this case, although pH was the most significant factor at the end, it had no significant interactions with the other two important independent variables. This fact was obtained by chance.

As defined in the introduction, the data was simulated, although it does represent what could happen in solid substrate culture where, according to the experience of the author, surprises abound.

**Screen* and *Simplex-V* are registered names from Statistical Programs, Houston; *X-Stat* from John Wiley & Sons, New York; *Echip* from Echip, Inc., Hockessin; *Stata* from Computing Resource Center, Santa Monica; *Systat/Sygraph* from SYSTAT, Inc., Evanston, all in the USA.

6. References

Barker, T.B. (1985) Quality by Experimental Design, Dekker, Inc., New York and Basel.

Box, G.E.P., Hunter, W.G., and Hunter, J.S. (1978) Statistics for Experimenters, Wiley and Sons, Inc., New York.

Haaland, P.R. (1989) Experimental Design in Biotechnology, Dekker Inc., New York and Basel.

Mason, R.L., Gunst, R.F., and Hess, J.L. (1979) Statistical Design and Analysis of Experiments, Wiley and Sons, Inc., New York.

SIMULATION OF A COMPLEX BIOTECHNOLOGICAL PROCESS USING AN OBJECT-ORIENTED KNOWLEDGE-BASED SYSTEM

M. POLAKOVIČ and C. F. MANDENIUS
University of Linköping
Department of Physics and Measurement Technology
S-581 83 Linköping
Sweden

ABSTRACT. A novel approach is presented for the use of object-oriented knowledge-based systems as a framework for building computer integrated bioprocess engineering software. The features of object-oriented programming for the simulation of complex systems are demonstrated on an example of the simulation of ethanol plant using the real-time expert system shell G2.

1. Expert systems

Expert system technology is the most successful application of artificial intelligence research. Bowerman, Clover (1988) define expert system as a system of software or combined software and hardware capable of competently executing a specific task usually performed by human experts. They became widespread after the so-called knowledge-based expert systems appeared in mid- and late 1970s. Knowledge-based expert systems possess specific inference or reasoning mechanisms for processing knowledge expressed in a form of rules or frames.

The limitations of static expert systems in some applications led to the development of real-time expert systems in mid- 1980s. Real-time expert systems execute reasoning over time. At the same time the emergence of commercially available tools for building expert systems accelerated the application of both static and real-time expert systems. Some real-time expert systems were built on static expert shells (Krebs, Respondek, 1991), but it is generally preferable to build real-time expert systems on real-time expert system shells (Lalka, Weber, 1991, Leitch, et al., 1991). G2 from Gensym Corporation (Moore, 1991) is the leading software of this type on the market. Although G2 was designed for intelligent process control, it has many built-in capacities, which make it a powerful object oriented graphical environment of wider use.

2. Computer integrated bioprocess engineering - state-of-the-art of and perspectives

Simulation, or process flowsheeting is the centrepiece of computer-aided engineering (CAE) software. The foundations of CAE of chemical processes were laid in 1960s. During this period two different approaches to process flowsheeting evolved:

- sequential-modular approach used only for steady-state simulation;

A.R. Moreira and K.K. Wallace (eds.), Computer and Information Science Applications in Bioprocess Engineering, 157-163.

- equation-oriented approach, which enables also dynamic simulation.

Commercial CAE software packages like ASPEN PLUS, HYSIM, or SPEEDUP have been doing a good service to chemical engineers in process design and retrofit in recent years. Nevertheless, the trend today is towards computer integrated process engineering (CIPE) which should integrate the following software (Winter, 1992):

- CAE,
- CAD (computer-aided design),
- database,
- control.

With the quick development in the area of knowledge-based systems, it is reasonable to expect they will also become a part of CIPE. The integrated systems should bring benefits in improved process design, better plant operations and increased engineering productivity. Graphics will play much bigger role not only as an output medium but also for the definition of knowledge. Simulation will also be used for more diverse purposes (Winter, 1992) like:

- new and revamped design;
- on-line process optimization;
- control system design;
- model-based control;
- data reconciliation;
- plant diagnostics;
- operator training;
- safety studies;
- business applications.

The information given above applies to conventional chemical processes, but one can expect the architecture of computer integrated bioprocess engineering (CIBPE) and its purpose should be very similar. There was a substantial progress made in the computerization of bioprocesses in the last decade. Nevertheless, the most important hint on the way of bioprocess software integration is that CAE of bioprocesses is practically not existent. There are several reasons for that:

- mathematical models of many important operations especially purification steps are not available; mainly because not enough research effort was paid to the physical operations in biotechnological systems;
- it is difficult to design general purpose models because of very specific behaviour of biological systems even in similar applications,
- more common steady-state flowsheeting is of limited use for bioprocess simulation as most of the biotechnological processes are run in discontinuous mode;
- specific regulations in pharmaceutical industry impose a burden for engineering activities.

Above mentioned fact could invoke a feeling that it is preliminary to speak about CIBPE. Our intention is to show that object-oriented knowledge-based system tools (OOKBST) could be a feasible framework for building CIBPE packages. One reason is that there has been a considerable interest in these systems in the recent period and many biotechnological companies own OOKBSTs. However, the more important reason is that these tools have an open-system architecture with built-in capacities for interfacing with the most common databases and control systems. If they also posses some flowsheeting capabilities, a user can easily develop bioprocess flowsheeting software in a form of a knowledge-base.

3. Description of ethanol fermentation process

The experiences reported here come from a project aimed on building an expert system for the supervisory control of an ethanol fermentation plant. Building a simulation module of the process is the first step. Fig. 1 shows the process flowsheet. It is a continuous process with six overflow fermentors in series, splitted after the second fermentor into three parallel lines. After fermentation the suspension of cells and fermented liquor proceed to continuous centrifuges where cells are separated and partly recycled. The centrifuged liquor continues to further purification in the distillation part. The raw material is spent-sulphite liquor from pulp plant, which contains 5 different sugars (three of them are fermentable by yeast) in a total concentration of 4 %. Baker's yeast is used for the conversion of sugars to ethanol.

The basic elements of simulation module are the material balances of following process-equipment:

- fermentor,
- container,
- centrifuge,
- flow mixer,
- flow splitter,
- pump,
- valve.

Except of the total material balance, the material balances of biomass, substrate (total fermentable sugars) and ethanol were composed for each unit. The material balances were represented for fermentor and container by ordinary differential equations and for the other units by algebraic equations. The kinetic equation for the biomass growth was expressed by a Monod-type equation including product inhibition. The substrate consumption and ethanol formation were related to the biomass growth by simple equations through constant yield factors.

The second most important part of the simulation module was formed by the simulation of sensors, controllers and control valves.

4. Simulation in G2

The elementary unit carrying information in a G2 knowledge-base is called item. Some examples of items are objects, connections, workspaces, rules, procedures, formulas, simulation formulas, relations, displays, buttons, functions, etc. All items in G2 knowledge-base are represented graphically. Whereas rules, formulas are only rectangles containing some defined text expression, objects have the most sophisticated user-defined graphical representations, which are called icons. Object icons are created and modified by icon editor.

Figs. 1a and 1b show the G2 workspaces of process schematics, on which several types of objects can be recognized, e.g.: fermentor tank, final-line fermentor, centrifuge, yeast container, pump, flow mixer, flow splitter, on-off valve, manual valve, control valve, sugar analyser and discrete PID controller. These object types are called object classes. Each object class is characterized by an object definition. The object class definition includes the specific information like parameters and variables describing the unit, icon description, connection site specifications, for example.

A very useful feature in G2 is that the classes can be organized into an object hierarchy. In our prototype most of the above mentioned classes are the subclasses of the class process-equipment,

ETHANOL PLANT - OVERVIEW

PRETREATMENT
PRIMARY-FERMENTATION
FINAL-FERMENTATION
FROM PULP PLANT
TO EXPEDITION
DISTILLATION
SEPARATION

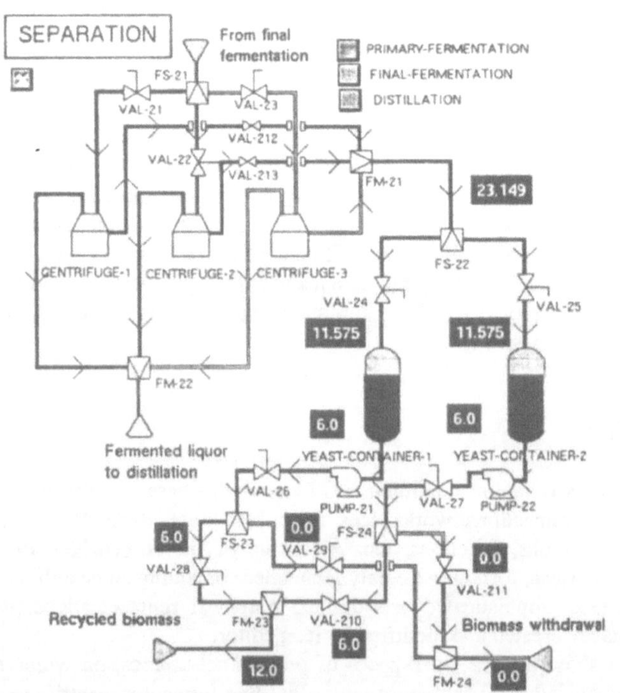

Figure 1a. Process flowsheet.

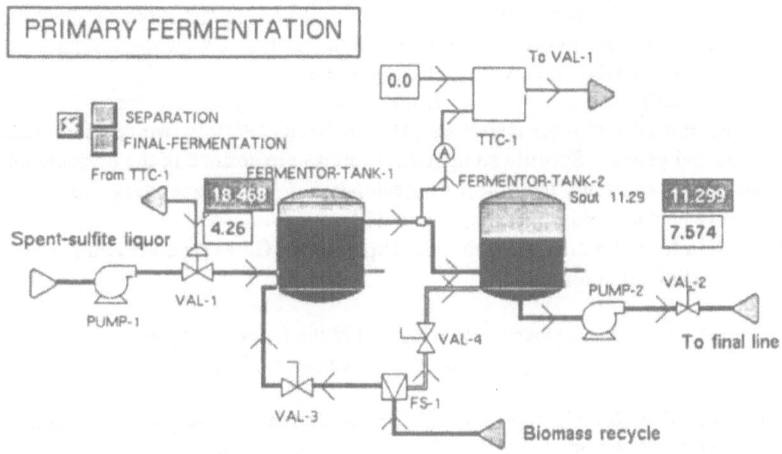

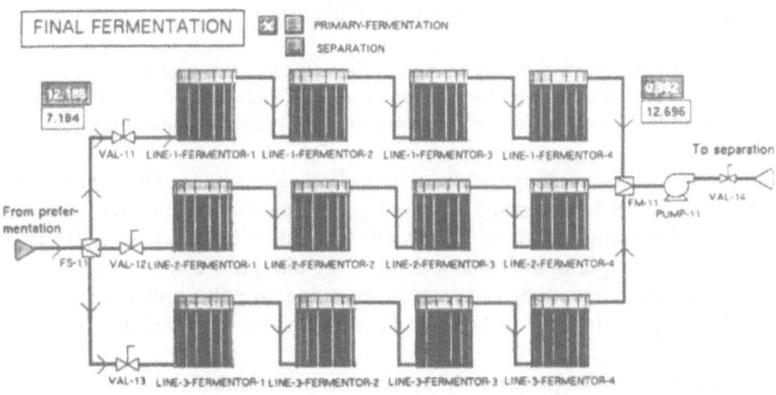

Figure 1b. Process flowsheet.

or there is a class fermentor with two subclasses fermentor-tank and final-line-fermentor. Figure 1b shows that there are 2 instances of the fermentor-tank and 12 instances of the final-line-fermentor. The introduced object-oriented representation combined with the option of the so-called generic knowledge enables a very efficient way of expressing knowledge.

The concept of specific and generic knowledge is very clearly distinguished in simulation formulas. G2 simulator evaluates the simulated values of variables from two types of simulation formulas: specific and generic. Specific simulation formulas are defined in the attribute tables of an object instance. For example, the object fermentor-tank-1 has a simulation subtable for the attribute initial-volume where this attribute is set to 850.

Generic simulation formulas are on the contrary expressed in G2 as separate items. Here is one example of generic simulation formula:

> the inflow of any process-equipment PE = the first of the following expressions that has a value (the outflow of the process-equipment connected at an input of PE, 0)

This generic simulation formula applies to all process units that have process-equipment as a superior class, have attributes inflow and outflow and are connected. It is noteworthy that the class process-equipment does not have attributes inflow and outflow in its object definition. Quite different is the case of the class fermentor and its two subclasses whose object definitions differ mainly in the icon descriptions. From that reason all process variable attributes are defined for the higher class fermentor and are automatically inherited for the subclasses fermentor-tank and final-line-fermentor. Similarly, practically all generic simulation formulas are defined for the class fermentor, as for example the material balance of biomass:

> state variable: d/dt (the xout of any fermentor F) = (if the volume of F <= 0 then 0 else (the liquor-inflow of F * the xin of F - the liquor-outflow of F * the xout of F + the recycle-inflow of F * the xrec of F + the volume of F * the growth-rate of F) / the volume of F))

The usefulness of generic knowledge declaration is clearly shown here. One generic simulation formula is sufficient for 14 fermentors. The use of generic simulation formulas is preferable in G2. Our knowledge-base employs about 100 generic simulation formulas but only few specific simulation formulas. We performed simulations to obtain information about the dynamic behaviour of our system in open-loop and closed-loop mode. We also tested the process control performance (Fig. 2) simulating different on-line and off-line sensors placed in different positions in the system.

The use of generic knowledge not only spares the programmer's time but it makes the knowledge-base much more flexible. Without much effort can any part of the knowledge-base be modified and recompiled for a similar application. We tried this for the simulation of baker's yeast process and of a recombinant fermentation.

The process flowsheeting is also efficient and easy with G2. It is possible with object menu commands create, delete, clone objects, place them on the workspace, connect them with another objects, transfer the group of objects to another workspace. If the generic simulation formulas are defined in a robust way, they will not require modifications after the changes in the flowsheet.

The two examples of generic simulation formulas demonstrated the structured natural language of G2. Its use enabled to implement a text sensitive editor in G2, which significantly speeds up the compilation of knowledge-base. Another advantage of the language is that formulas, rules,

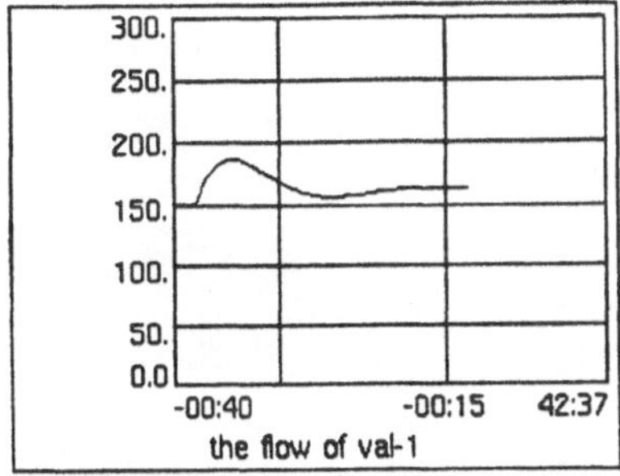

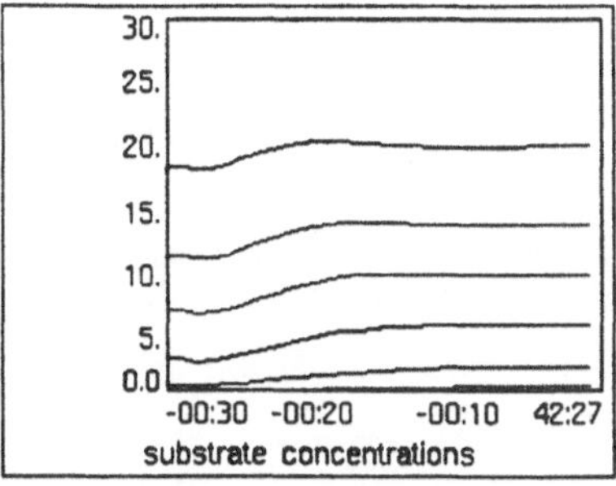

Figure 2. Illustration of system response to the change of the setpoint of a PID controller. Controlled variable was the sugar concentration after the fermentor-tank-1 measured off-line.

etc. can be much more easily understood by an outsider. However, we must make a comment here that it takes some time to get accustomed to writing mathematical equations in this way.

Finally, we would like to briefly mention the time scheduling in G2. This feature is very useful from the point of view of bioprocesses as these are mostly discontinuous or semicontinuous systems that are also sequenced in time. Three different time-modes are available: real time, simulated time, as fast as possible. This fact makes it possible to use the same knowledge-base for different purposes, like for example real-time supervisory control of processes, design and operational studies and operator training.

Acknowledgment

We wish to thank Lars Ahlenius, MoDo Papers AB, Sweden, for valuable discussions. The project was financially supported by the Swedish Ethanol Foundation and the National Board for Technical Development.

References

Bowerman, R. G., Glover, D. E. (1988) Putting expert systems into practice, Van Nostrand Reinhold Company, New York

Krebs, V., Respondek, T. (1991) Automatization of complex chemical processes using real-time expert systems, Automat.-tech. Praxis 33, 82-86 (in German)

Lalka, C.J., Weber, R. (1991) Real-time versus static shells for real-time expert systems, AIChE Spring National Meeting, Houston, Texas, April 7-11, 1991, Paper No. 14A

Leitch, R., Kraft, R., Luntz, A. (1991) RESCU: a real-time knowledge based system for process control, IEE Proceedings-D 138, 217-227

Moore, R.L. (1991) G2: a software platform for intelligent process control, Proceedings of the 1991 IEEE International Symposium on Intelligent Control, 13-15 August 1991, Arlington, Virginia, 1-5

Winter, P. (1992) Computer-aided process engineering: the evolution continues, Chem. Eng. Progr. 88, 76-83

PART 2:

ENZYME TECHNOLOGY

THE IMPLICATION OF pH ON THE DESIGN OF AN ENZYMATIC REACTOR FOR PENICILLIN HYDROLYSIS.

GOMEZ-AGUIRRE Alfonso (*),
QUINTERO R. Rodolfo
LOPEZ-MUNGUIA Agustin.

Instituto de Biotecnologia, Universidad Nacional Autonoma de Mexico. Apartado Postal 510-3, Cuernavaca, Morelos, MEXICO.
(*) Instituto de Biotecnologia, Universidad Nacional de Colombia. Apartado Aereo 14490, Bogota, COLOMBIA.

ABSTRACT. In this paper penicillin acylase is used to exemplify the effect of pH on the design of enzymatic reactors. Simulation programs are used to describe the batch, continuous stirred tank, packed bed and recirculated reactors using a triple inhibition model for enzyme kinetics, an electrochemical model to describe the pH changes with conversion and an Arrhenius type equation to account for the pH effect on enzyme stability.

1. Introduction

Several design criteria for enzymatic reactors have been published [1,2]. It has been pointed out that in the design of an immobilized enzyme process, three important factors must be considered: the source and extent of purification of the enzyme, the size and chemical nature of the support and the method of immobilization. These factors have a direct effect on fundamental properties of the biocatalyst, such as the specific activity and the effectivity of the union between enzyme and support. Also, the particle size is of considerable importance in the design of the reactor as well as in the eventual diffusional limitations in the reaction rate. Another consequence resulting from these factors is the long term microbial and mechanical stability of the biocatalyst.

Two additional factors are highly important: enzyme kinetics and enzyme stability. In the first case, as industrial reactions are run at high levels of conversion and high substrate concentration, the enzymes may be subjected to severe inhibition effects. Therefore, optimum operation policies have to be defined and in some extreme cases, the inhibition may even be the factor defining the type of reactor [3,4].

In general, biocatalyst stability is one of the most important factors from the economical point of view, directly related with the productivity. Hence, the selection of the operating temperature is critical and should be done considering not only its effect on the reaction rate, but also on other factors such as the operational half life and the risk of contamination. In the majority of industrial enzymatic processes, pH does not represent an inconvenience and its effect on enzyme stability is regulated through the use of buffers.

Nevertheless, there are a few cases where strict control of pH is of crucial importance and will therefore be the main factor to consider in the selection of the reactor configuration. This is the case of penicillin acylase (PA), an enzyme responsible of the specific cleavage of penicillin G to produce 6-aminopenicilanic acid (6-APA). This is one of the few immobilized enzymes in use at the industrial scale, with annual sales only below those of glucose isomerase. For this process, the packed bed column option is excluded [4], as buffers stronger than

A.R. Moreira and K.K. Wallace (eds.), Computer and Information Science Applications in Bioprocess Engineering, 167-177.

50mM phosphate interfere with the purification process. Although some research has been carried out with the continuous stirred tank reactor (CSTR) [5], the only actual alternatives at industrial scale are the batch and recirculated packed bed reactor [6,7]. In this chapter we will examine the behavior of penicillin acylase in various reaction configurations with the aid of simulation programs, taking into account the effect of pH.

2. Biocatalyst stability

The production process of a penicillin acylase biocatalyst from purified extracts of *E. coli* has been described elsewhere [8,10]. It has an activity of 160-180 Ug^{-1} catalyst (wet weight), particle size of 100-200 μm, density of 1.02 g cm^{-3} and a porosity of 0.5. One unit is defined as the amount of enzyme producing 1 μmol min^{-1} of 6-APA at pH 7.8 and 37 ^{o}C using 2% (w/v) PGK. The enzyme displays an optimum activity at 49° C, but due to the combined detrimental effect of temperature and pH on the enzyme stability it is often used at 37 ^{o}C. At this lower temperature the activity is reduced in 40% but its half life is increased from 28 h to 2880 h at pH 7.5. The half life at 37 ^{o} C as a function of pH has been described with an Arrhenius type model as follows.

$$-\frac{dE}{dt} = kE \qquad (1)$$

with:

$$k = A \exp [B / pH] \qquad (2)$$

where:

E = enzyme activity (U g-1 catalyst)

$A = 1.54 \times 10^{-6}$	B = 37.1	for pH < 7.5
$A = 3.26 \times 10^{6}$	B = 173.98	for pH > 7.5

Equations 1 and 2 may be used simultaneously with kinetic equations to take account of enzyme deactivation with pH in reactors. In this case the reaction rate is considered independent of pH as shown by the pH activity profile at 37° C.

3. Relationship between penicillin conversion and pH

If we consider the reaction catalyzed by penicillin acylase as:

$$[HP]_T \longrightarrow [PAA]_T + [APA]_T \qquad (3)$$

where $[HP]_T$, $[PAA]_T$ and $[APA]_T$ are the total concentration of penicillin, phenyl acetic acid and 6-aminopenicillanic acid respectively. Considering the dissociation equations of all the chemical species present in the reaction and their relation with penicillin conversion (defined as X= So-S/So), it is possible to write:

$$[HP]_T = [HP] + [P^-] = So\,(1 - X) \quad (4)$$

$$[PAA]_T = [HPAA] + [PAA^-] = So\ X \quad (5)$$

$$[APA]_T = [H_2APA^+] + [HAPA^+] + [APA^-] = So\ X_o \quad (6)$$

$$[PO4]_T = [H_3PO_4] + [H_2PO_4^-] + [HPO_4^{2-}] + [PO_4^{3-}] = C_1 + C_2 \quad (7)$$

$$[Ammonia]_T = [NH_4^+] + [NH_3] = CA \quad (8)$$

$$[H_2O]_T = [H_2O] + [OH^-] \quad (9)$$

where C_1 and C_2 are the molar concentrations of KH_2PO_4 and K_2HPO_4. In order to construct the model, the fraction of each dissociated specie is defined, introducing the definition of the equilibrium constant. For instance, the fraction of dissociated penicillin from equation 4 is given by:

$$\frac{[P^-]}{[HP]_T} = F_{10} = \frac{[P^-]}{[P^-]+[HP]} = \frac{[P^-]}{So\,(1-X)} \quad (10)$$

and

$$Ka_1 = \frac{[H^+]\,[P^-]}{[HP]} \quad (11)$$

or

$$[HP] = [H^+][P^-]/Ka_1 \quad (12)$$

substituting equation 12 on equation 10 and introducing the definition of pH and pK, it is possible to obtain the fractions (in this case F_{10}) only in terms of pH and the pKs of the chemical species:

$$F_{10} = \frac{1}{1 + 10^{(pK_{a1}-pH)}}$$

Similar expressions may be obtained for fractions F_{20} (PAA^-), F_{30} (APA^-), F_{32} (H_2APA^+), F_{40} (PO_4^{3-}), F_{41} (HPO_4^{2-}), F_{42} ($H_2PO_4^-$) and F_{50} (NH_3). Finally all the expressions are substituted in an electrochemical balances to give:

$$X = \frac{F_{10}\,So + (C_1 + C_2)\,(3F_{40}+2F_{41}+F_{42}) - K + M + CA\,(F_{50} - 1)}{So(F_{10} + F_{32} - F_{20} - F_{30})} \quad (13)$$

or

$$CA = \frac{F_{10}So+(C_1+C_2)(3F_{40}+2F_{41}+F_{42})-X\ So\ (F_{10}+F_{32}-F_{20}-F_{30})-K+M}{(1 - F_{50})} \quad (14)$$

where CA is the amount of ammonia required to maintain the pH at a given value, K is the concentration of potassium ions (= $S_o + C_1 + C_2$) and M is obtained from the water equilibrium constant:

$$M = 10^{(pH-14)} - 10^{-pH}$$

A detailed description of the model has been published elsewhere [9].

4. Enzyme kinetics

There is a general agreement that PA is inhibited by excess substrate in addition to competitive inhibition by phenylacetic acid and non competitive inhibition by 6-APA. The kinetic equation may be written as:

$$vi = \frac{k_2 E S}{Km \left\{1+\frac{PAA}{Kpaa}+\frac{APA}{Kapa}+\frac{PAA\ APA}{Kpaa\ Kapa}\right\} + S\left\{1+\frac{APA}{Kapa}+\frac{S}{Ks}\right\}} \qquad (15)$$

with:

k_2 = 170 μmol penicillin min^{-1} $gcat^{-1}$
Km = 4.17 mM
Kpaa = 68.6 mM
Kapa = 100.7 mM
Ks = 413 mM

5. Simulation of penicillin acylase reactors.

In the following sections the analysis of penicillin acylase in the basic reactor configurations is examined (batch reactor, CSTR, packed bed reactor and recirculated reactor). In each case the reactor dimensions correspond to experimental reactors constructed in the laboratory. The simulation consists in the solution of the mass balance equation for each reactor configuration, simultaneously with equations 13 and 14 to take account of pH changes during the reaction and equations 1 and 2 to predict the stability of the biocatalyst in each system.

5.1 THE BATCH REACTOR.

If a reactor is well mixed, the pH variation around the set point, will be mainly due to the efficiency of the control system. A 50 ml batch reactor was used for batch reactions at 37°C and with an enzyme load of 120 U per gram of penicillin, with pH regulation with 2N NH_4OH in a range of ± 0.1 pH units around 7.5. Data from 150 reactions was produced. On the other hand, a simulation program contained the following sequence was made:

a) evaluation of one batch defining a time interval (Δt) and initial conditions (So, initial biocatalyst activity and pH= 7.5).
b) evaluation of reaction conversion by numerical solution of equation 15 for each Δt.
c) evaluation of the pH change according to equation 13.

d) when the pH reaches 7.4 or lower (lower regulation limit) the pH is reset to 7.6 (upper regulation limit), calculating the ammonia required using equation 14.
e) evaluation of the residual activity according to equations 1 and 2.
f) end the simulation when conversion reaches 95%.

The simulation from steps a) to e) may be repeated to evaluate the effect of the number of batches on biocatalyst activity. Also, the values of pH in step d) may be set in a wider range to simulate the effect of deficient mixing and/or pH regulation systems in the stability of the biocatalyst. The results of this simulation is shown in figure 1, where it may be observed that the prediction of the experimental data is adequate and the deactivating effect of wider pH oscillations is evident.

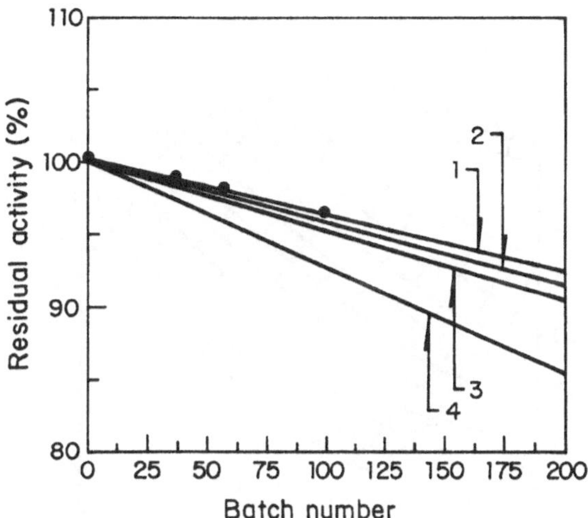

Figure 1. Residual activity of a penicillin acylase biocatalyst in a batch reactor after 200 reactions. Continuous lines represent the simulation for the following conditions: 10% penicillin, 95% conversion, 120 units per gram of penicillin, 37°C, in a 50 ml reactor with pH regulation with 2N NH_4OH at pH=7.5 ±Δ with Δ=0.1 (1), Δ=0.2 (2), Δ=0.3 (3) and Δ=0.4 (4). Experimental results are also shown.

5.2 THE PACKED BED REACTOR

A simulation program of a packed bed reactor requires the integration throughout the reactor length and also throughout time. The simulation is carried out as follows:

a) definition of initial conditions: inlet substrate concentration, initial activity throughout the bed and inlet pH= 7.5
b) definition of a time interval Δt
c) definition of a reactor length interval Δz
d) for each Δz evaluation of outlet conditions by the solution of the plug flow mass balance

equation, with the consumption term given by equation 15. Input conditions of each new Δz correspond to outlet calculations in the previous one. Proceed until the exit of the reactor (z=L), evaluating substrate concentration, conversion and pH (equation 13) through the reactor.

e) repetition of the simulation for the next Δt, evaluating the residual activity profiles as a function of time in each Δz using equations 1 and 2.

No experimental data was generated for the continuous packed bed reactor. In figure 2 an example of the simulation results is shown. It is evident that shorter residence times results in higher stabilities but also in low outlet conversions.

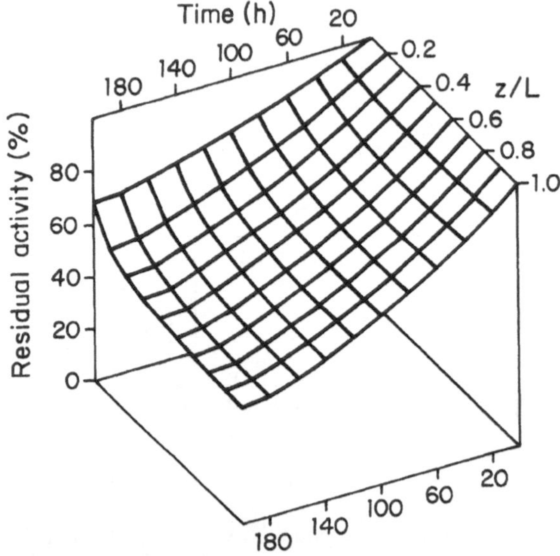

Figure 2. Simulation results for a packed bed reactor with penicillin acylase biocatalyst. Residence time= 294 sec fwith 3.5 g of catalyst, or 95% conversion. 37oC, 10% penicillin.

5.3 THE CONTINUOUS STIRRED TANK REACTOR (CSTR).

The strongest inhibition to which PA is subjected comes from the products, phenylacetic acid in particular. It is therefore not recommended to use this configuration, where the reaction rate takes place at the exit conditions. In figure 3 the experimental data of conversion obtained at steady state for various flow rates, when 10% penicillin were used in the reactor , is presented. It is clear that conversions are low even for high residence times. Therefore, in terms of productivity, the CSTR is not an adequate option when compared to the batch reactor.

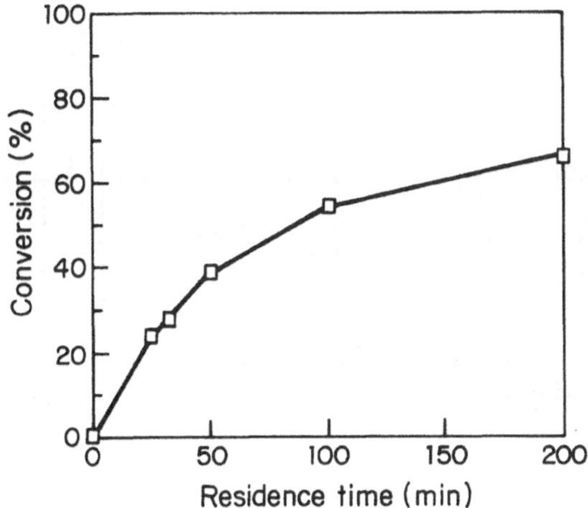

Figure 3. Steady states conversions obtained in a continuous stirred tank reactor at different residence times. Reactor volume 50 ml, 37°C, 10% penicillin, 120 units per gram of penicillin, pH regulation at 7.5 with 2N NH_4OH. (Conversion was 95% in 2 hours in a batch reaction under identical reaction conditions).

5.4 THE RECIRCULATED PARALLEL PACKED BED REACTORS (RPPBR).

It is important to observe that the configuration of the majority of industrial penicillin acylase reactors is the agitated batch type. One exemption is the process by Toyo Jozo using multiple recirculated packed beds. It has been shown that the CSTR is not adequate due to inefficient use of the biocatalyst in the operating conditions. It has been pointed out [9], that in reactors where the biocatalyst is in direct contact with the alkali solution used to regulate the pH, important deactivation effects may result from non efficient addition and distribution systems. On the other hand, the packed bed reactor is useful only if low conversions without significant changes in pH are mantained. As the process requires high conversions, there is an evident need for recirculation. This in turn is limited by the pressure drop in the reactor and the obvious solution is the distribution of the catalyst in a certain number of parallel columns as shown in figure 4. It is the best system for penicillin acylase in terms of stability, avoiding handling of the biocatalyst and direct contact with the pH regulation system. Experiments to define the minimum Reynolds number to avoid external diffusion control is essential.

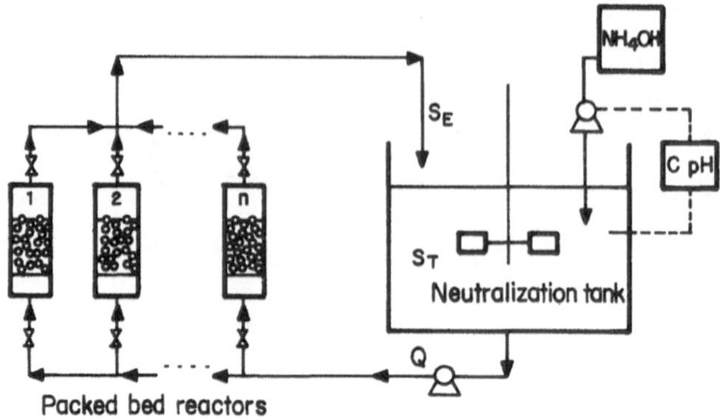

Figure 4. Layout of the recirculated parallel packed bed reactor system.

The simulation program for this configuration consists of the following steps:

a) definition of initial conditions of a batch run t=0, X = 0, So = 100 g l^{-1}
b) definition of global recirculation flow rate (Q)
c) definition of the number of parallel reactors (NR) of the same size
d) definition of reactor geometry L / D and the recirculation flow rate for each reactor QR = Q / NR. Verification of minimum Reynolds number.
e) calculation of pressure drop (Ergum equation)
f) determination of the substrate, conversion and pH profiles in the reactor each Δz as in the packed bed. This will result in the outlet substrate concentration (Se) equal in all the reactors (Se).
g) evaluation of the average residual activity in the reactor, integrating the residual activity profile.
h) mass balance in the neutralization tank, assuming that in the time interval (Δt), the input (Se) and output (S_T) concentrations are constant.

$$S_T(t+\Delta t) = [S_T(t)V_T + Se\, Q\, \Delta t - S_T(t)\, Q\, \Delta t] / V_T$$

$S_T(t+\Delta t)$ = substrate concentration in the tank at time t + Δt
$S_T(t)$ = substrate concentration in the tank at time t

i) calculate pH and NH_4OH concentration in the tank (equations 13 and 14)
j) evaluation for the next time interval (Δt): stop when conversion reaches 95% and start a new batch.

In figure 5 the residual activity of the biocatalyst is shown when two columns are used in parallel for more than 100 hours of use in recirculation mode at various flow rates. The simulation program does not predict differences in activity between the 3 studied flow rates during the first 100 hours. The three flow rates are above the minimum required for non external diffusional limitations (47.2 cm^3 min^{-1}), but the highest flow rate resulted in an excesive pressure drop (14.2 psi).

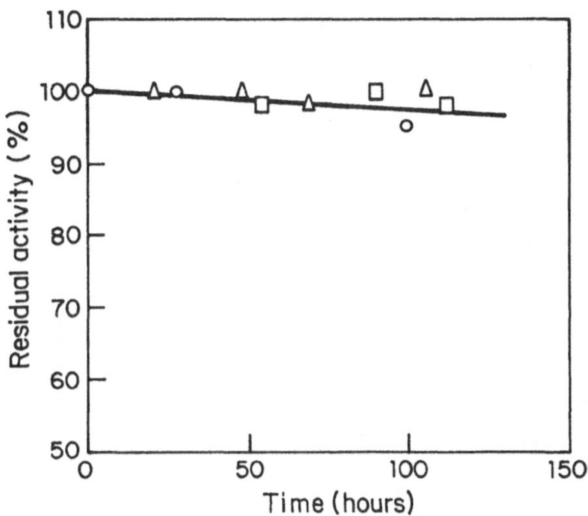

Figure 5. Residual activity of a penicillin acylase biocatalyst in two parallel packed bed reactors of 780 μl each recirculating a total volume of 30 ml of 10% penicillin, with 44.5 units per gram of penicillin and operating at 37°C. pH is regulated at 7.5 with 2N NH_4OH. (recirculating flow rate in cm^3 min^{-1}: - o - 273 ; - Δ - 85.5 and - - 47.2)

5.5 STRATEGY FOR CONTINUOUS PRODUCTIVITY.

In the batch process, as the biocatalyst losses activity two options are followed: either the reaction time is increased to reach a given conversion or the biocatalyst load is increased to account for the activity lost. In the recirculation mode with several columns (NR) another alternative is possible by redefinition of the equation proposed by Pitcher [1]:

$$Rp = \exp \left[\frac{H}{NR} \ln 2 \right] \qquad (16)$$

Rp = relation between the lowest and the highest reaction time desired in a batch.
H = number of half-lives of biocatalyst use
NR = number or recirculated beds required in parallel

As an example, for the case of 10 columns used for one half life (140 days) and a final conversion of 95%, according to equation 16, the variation in reaction time will be of ±3.1% (Rp = 0.93). The start up program is illustrated in figure 6. During this period, each 14 days a new column in parallel is started until the 140th day, when the oldest column has reached one half life and is stopped. By this time 10 columns will be operating in parallel with small variation in the batch reaction time.

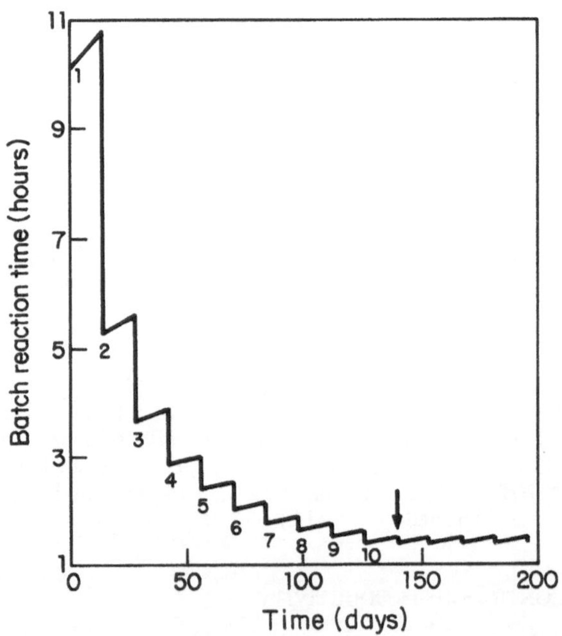

Figure 6. Start up program for a penicillin acylase 10 parallel recirculated packed bed reactors system, based on equation 16, for a continuous productivity strategy.

6. Conclusions.

It has been shown through the aid of simulation programs that a penicillin acylase biocatalyst, due to its stability behavior with pH, requires specific considerations when designing an industrial reactor. For this particular case it is shown that the recirculated parallel packed bed columns is the best arrangement in terms of stability.

7. References

1. Pitcher W.H., 1980. Immobilized Enzymes for food processing C.R.C. Press., Boca Raton, Florida.

2. Weetall H.H. & Pitcher W.H., 1986. Scaling up an Immobilized Enzyme System. Science, 232, 1396-1402.

3. Martinez Espindola J.P. & Lopez-Munguia C.A., 1985. On the kinetics of dextransucrase and dextran synthesis in batch reactors Biotechnol. Lett. 7,7, 483-486.

4. Klein J., Vorlop K.D.J. Wagner F., 1984. The implication of reaction kinetics and mass transfer on the design of biocatalytic processes with immobilized cells. Ann. N.Y. Acad. Sci. 434, 437-449.

5. Carleysmith S.W. & Lilly M., 1979. Deacylation of benzylpenicillin by immobilized penicillin acylase in a continuous four stage stirred tank reactor. Biotechnol. Bioeng., 21, 1057-1073.

6. Mollgaard H.,1987. Choice of reactor for semacylase. Ann. N.Y. Acad. Sci. 501, 473-476.

7. Cardoso J.P. & Bento Correia M., 1986. Modelling of the hydrolysis of benzylpenicillin to 6-APA and PAA by an immobilized penicillin acylase in a small pilot plant bench recirculated reactor. British Polym. J., 18, 5, 323-332.

8. Ospina S., Lopez-Munguia A., Gonzalez R.L. & Quintero R., 1992. Characterization and use of a penicillin acylase biocatalyst. J. Chem. Technol. & Biotechnol., 53. In press.

9. Gomez-Aguirre A., Ospina S., Quintero R. & Lopez-Munguia A, 1992. Modelling and simulation of a pH dependent bioprocess enzymatic conversion of Penicillin G to 6-APA. In: Bioprocess commutations in Biotechnology Vol. II, Ed. T.Ghose. Ellis Horwood, UK. In Press.

10. Ospina S., 1991. Characterization and application of a penicillin acylase biocatalyst. M Sc Thesis. Inst. of Biotechnology, UNAM, Mexico D.F.

SIMULATION OF BATCH ENZYME REACTORS.

LOPEZ-MUNGUIA, Agustin.
Instituto de Biotecnologia, Universidad Nacional Autonoma de Mexico.
Apartado Postal 510-3, Cuernavaca, Morelos, MEXICO.

ABSTRACT. Several enzyme systems are presented in order to demonstrate the use of simulation in the description and characterization of enzyme reactions. The importance of the batch reactor, both in research and industry is demonstrated, providing examples of situations where simulation may be a useful tool.

1. Introduction

Different type of reactors can be used for process scale operations with enzymes in their free or immobilized forms. Based upon the mode of enzyme use, reactors are classified as batch and continuous, being the former, the most frequently used in industrial processes and in laboratory experiments. In spite of the enormous work done on enzyme immobilization during the last two decades, most of the applications of enzymes in industry still use the catalyst in its soluble form. As exemplified in Table 1, such is the case for the food industry applications.

Table 1. Examples of enzymes used in soluble form in industrial applications.

Enzyme	Application	% of total market
α-amylase and glucoamylase	glucose syrups production	16.7
rennin	cheese manufacture	10
pectinases	fruit processing	6.7
alkalin protease	detergents, protein hydrolysates,...	25
papain and other proteases	beer chillproofing, pharmaceuticals,...	30
β-galactosidase	lactose hydrolysis in milk	< 1.5

The use of free enzymes is generally restricted to batch reactors. In this case, the free enzyme is loaded into the reactor and the reaction is carried out to the desired degree of conversion. At the end of the process no attempt is made to recover the enzyme. There are also some immobilized enzymes that are used in batch reactions, specially when production requirements do not justify a continuous operation. In other cases, enzymes are used in batch reactors because of kinetic and stability contraints. For instance, β-galactosidase from *K. lactis* is an enzyme that has been proved difficult to immobilize for long time operation; penicillin acylase, due to the need of pH regulation, cannot be used in packed bed reactors and the continuous stirred tank reactor is not recommended due to strong product inhibition [1]. If it is also considered that enzymes are studied in the laboratory in batch reactors (test tubes, beakers, formal reactors) it may therefore be concluded that the batch reactor is a valuable tool and a frequent alternative in the integration of enzymes to industrial processes.

A.R. Moreira and K.K. Wallace (eds.), Computer and Information Science Applications in Bioprocess Engineering, 179-189.

From the kinetic point of view, the study of the behavior of enzymes in batch reactors is an unavoidable step in their characterization for industrial applications. The effect of products on the reaction rate, the reversibility of the reaction, alternative reactions and the operational stability of the enzyme are a few examples of enzyme properties, not observed in the classical "initial rate" experiments performed for their kinetic characterization. Some of these properties are increased under industrial operating conditions. Therefore, models obtained from initial rate experiments, have to be verified in batch reactions for a proper design of enzymatic processes in any type of reactor. Once a kinetic model is defined, simulation may be very helpful providing data on the most efficient operating policy and the effect of varying model parameters under different reaction conditions.

In this chapter, the use of computers as an aid to the experimental description of enzymes in batch reactors is illustrated.

2. Simulation software.

There are many packages of simulation software commercially available for the numerical solution of differential equations. They are mainly based on the modern, primarily expression based languages, such as DSL and CSSL. ISIM is an interactive simulation language developed in Salford University [2] available for PC use. It is divided in three regions (INITIAL, DYNAMIC and TERMINAL) together with an optional control region. The initial region is used to define the value of constants as well as the initial conditions. The Dynamic region contains the differential equations and the Terminal region may be used for end or run calculations. It has two useful commands: OUTPUT and PLOT to obtain the data values in tabular or graphical form. The data can also be stored with the command PREPARE and might be later shown with the command GRAPH. It is particularly useful that the program may be interrupted at any moment to modify the value of the parameters. It provides a choice of four integration methods, with a default option for a Runge-Kuta 5th order, variable step explicit method.

3. Batch reactors

A material balance for the substrate (S), of uniform composition and constant density throughout the reactor at any instant of time without any fluid entering or leaving the reaction mixture is given by:

rate of loss of S due to enzymatic reaction	=	-	rate of accumulation of S within the reactor

$$- r_s = \frac{dS}{dt} \qquad (1)$$

or if we define conversion as X = (So-S)/So then:

$$- r_s = So\,\frac{dX}{dt} \qquad (2)$$

In these equations r_s is the reaction rate and So is the initial substrate concentration. When the reaction rate expression is simple (0, 1st, 2nd order) or even for Michaelis-Menten kinetics, the integration is straight forward and a simple equation describing the batch reactor may be obtained. For instance, the substitution of the Michaelis-Menten model in equation 1 gives:

$$-\frac{dS}{dt} = \frac{k_2 E\ S}{Km + S} \tag{3}$$

which may be integrated to give:

$$k_2\, E\, t = So\, X - Km \ln(1 - X) \tag{4}$$

Equation 4 is often referred to as the integrated Michaelis-Menten rate equation and is used in the design of batch reactors. It may also be used to estimate the parameters of the model from data of a single run. This of course is not recommended and instead, equation 4 should be used to verify the model obtained from initial rate experiments.

Very often, the kinetic model is far more complex than the Michaelis-Menten equation. In these cases the numerical solution of equation 2 is required. This is also the case of cumbersome equations that result from the combination of models such as product or substrate inhibition. In table 2, some examples of these type of rate equations followed by common enzymes are presented.

Table 2. Examples of rate equations describing the kinetics of commercial enzymes.

Rate equation		
$-r_s = \dfrac{k_2E\ [S - P/Keq]}{Kmf + S + \dfrac{Kmf}{Kmr}P}$	reversible reactions	glucose isomerase
$-r_s = \dfrac{k_2 E\ S}{Km + S + S^2/Ki}$	excess substrate	invertase
$-r_s = \dfrac{k_2 E\ S}{Km\left[1 + \dfrac{P_1}{Ki_1} + \dfrac{P_2}{Ki_2}\right] + S\left[1 + \dfrac{P_2}{Ki_2}\right]}$	two products inhibition	lactase
$-r_s = \dfrac{k_2 E\ S_1\ S_2}{Km_1Km_2 + Km_1S_2 + Km_2S_1 + S_1S_2}$	two substrates	glucose oxidase

4. Simulation Examples.

4.1 DEFINING A RATE EQUATION: THE CASE OF PENICILLIN ACYLASE.

Penicillin acylase is an enzyme used in industry for the hydrolysis of penicillin. It is an enzyme inhibited by its two products, phenylacetic acid (competitive) and 6-

aminopenicilanic acid (non competitive), as well as by excess of substrate. When determining initial rates, the concentration of both products is negligible and if a low substrate concentration is used, then a Michaelis-Menten behavior will be obtained: Km and Vmax may then be determined by classical graphical procedures.

However, when the reaction proceeds to high conversions, equation 4 fails to describe the data. The independent effect of products on initial rates have to be analyzed in order to determine the inhibition parameters. For an immobilized enzyme obtained from *E.coli*, these values are Km = 4.17 mmol dm^{-3} , Kpaa = 68.6 mmol dm^{-3}, Kapa = 100.7 mmol dm^{-3} and Ks = 413 mmol dm^{-3} for a catalyst with an specific activity of 170 U g^{-1} [1]. This triple inhibition phenomena is described by the following rate equation:

$$-r_s = \frac{k_2 E S}{Km \left[1 + \frac{P_1}{Kpaa} + \frac{P_2}{Kapa} + \frac{P_1 P_2}{Kpaa\ Kapa}\right] + S\left[1 + \frac{P_2}{Kapa} + \frac{S}{Ks}\right]} \qquad (5)$$

In figure 1 the effect of substrate concentration is studied in batch reactions and described by a simulation program that solves equations 1 and 5 using an enzyme concentration equivalent to 120 units per gram of penicillin.

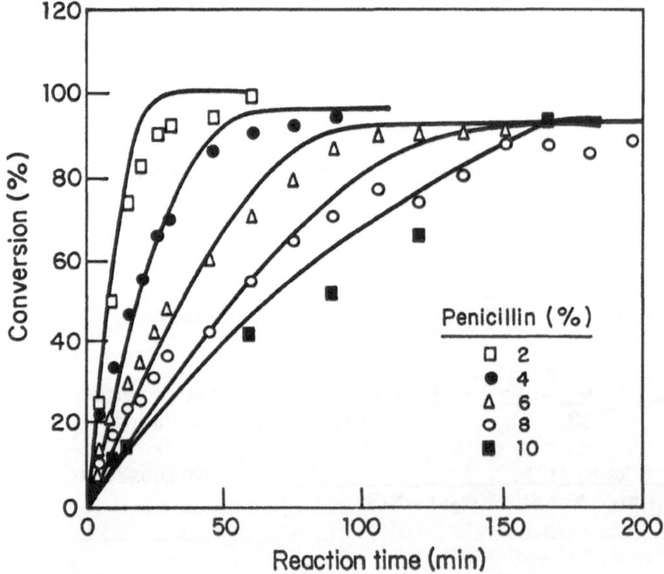

Figure 1. Penicillin hydrolysis in batch reactors with a penicillin acylase catalyst at various substrate concentrations, as described by a simulation program constructed with equation 5.
(From ref. [1], courtesy of SCI, Elsevier)

The same program is modified to carry out succesive batch reactions by repeating the reaction with 10% penicillin and including a first order deactivation model to consider the stability of the enzyme:

$$-\frac{dE}{dt} = k\,E \qquad \text{with } k = f(T) \qquad k = 6 \times 10^{-4}\ h^{-1} \text{ at } 37^{0}\,C \tag{6}$$

This simulation and other useful parameters in the evaluation of the catalyst is presented in Figure 2.

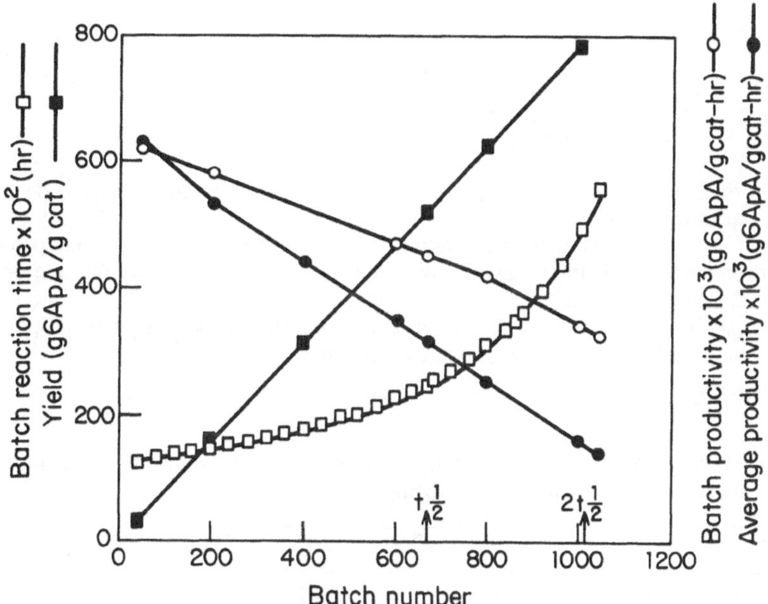

Figure 2. Simulation of the effect of succesive batch reactions on the yield, productivity and stability of a penicillin acylase biocatalyst. Batch reaction time to reach 95% conversion. (From ref. [1], courtesy of SCI, Elsevier)

4.2 AN ENZYME ACTING ON TWO SUBSTRATES: THE CASE OF PHENYLALANINE AMMONIA LYASE.

Phenylalanine ammonia lyase (PAL) is an enzyme whose forward and reverse reactions have industrial interest. It has been used in the synthesis of phenylalanine from *trans*-cinnamic acid and ammonia, and it has been proposed for the deamination of phenylalanine in protein hydrolysates in the production of phenylalanine-free products for phenylketonurics [3]. However, tyrosine, which is present in the hydrolysates is also a substrate for the enzyme. For cells of *Sporiodobolus pararoseus* each substrate was studied independently to determine the kinetic parameters. It was found that $K_{mT} = 0.28$ mM and $K_{mP} = 0.53$ mM for tyrosine and phenylalanine respectively, for cells having 11.5 U g^{-1} cell with phenylalanine as substrate. The maximum initial reaction rate with tyrosine as substrate was 4.03 times lower.

When the enzyme is applied to a protein hydrolysate, both substrates are present and their deamination rate is described by the simultaneous solution of the Michaelis-Menten equation for each substrate while the other acts as a competitive inhibitor:

$$- r_{Phe} = \frac{k_{Phe}\, E\, [Phe]}{K_{mP} \left\{ 1 + \frac{[Tyr]}{K_{mT}} \right\} + [Phe]} \tag{7}$$

$$- r_{Tyr} = \frac{k_{Tyr}\, E\, [Tyr]}{K_{mT} \left\{ 1 + \frac{[Phe]}{K_{mP}} \right\} + [Tyr]} \tag{8}$$

The solution of equations 7 and 8 is presented in Table 3 for various enzyme doses. It is shown that the time required for 90% conversion of phenylalanine is considerably increased if tyrosine is present. Obviously, the amount of tyrosine deaminated is independent of enzyme concentration (65% in all cases shown in Table 3). If we consider that the objective is to eliminate the maximum phenylalanine and the minimum tyrosine, an essential aminoacid that if lost has to be replaced, alternative strategies have to be proposed. For instance by studying the effect of temperature on the relative reaction rates.

Table 3. Phenylalanine deamination by *S. pararoseus* phenylalanine ammonia lyase in the presence and absence of tyrosine

Enzyme dose (U/ml)	0.5	0.10	0.20	0.50	1.0	5.0
(Time required to reach 90% conversion of phenylalanine)						
with tyrosine	2285	1140	570	228	114	22.6
without tyrosine	605	300	150	60	30	6

Reaction conditions: the substrate was a casein hydrolysate containing 5.24 g l^{-1} of phenylalanine and 5.82 g l^{-1} of tyrosine. Temperature = 40 ° C, pH = 8.8,

4.3 OLIGOSACCHARIDE PRODUCTION BY β-GALACTOSIDASE.

It has been shown that many carbohydrases also behave as transglucosidases depending on reaction conditions. This is the case of *Aspergillus niger* β-galactosidase which produces a trisaccharide during lactose hydrolysis. One possible mechanism of this reaction might be:

$$
\begin{array}{c}
\qquad\qquad\qquad\qquad\qquad\qquad G \\
E + L \rightleftharpoons E*L \rightleftharpoons E*Ga \dashrightarrow E + Ga \\
\qquad\qquad\qquad\qquad + \\
\qquad\qquad\qquad\qquad L \\
\qquad\qquad\qquad\qquad \Updownarrow \\
\qquad\qquad\qquad\qquad E + Tri
\end{array}
$$

where L= lactose, G= glucose, Ga= galactose and Tri= trisaccharide. If we neglect the formation of enzyme substrate complexes and deal only with overall reactions, the complicated kinetic equation resulting from these kind of mechanisms is simplified to more practical equations. This simplification is made for the mechanism shown above for the formation of the trisaccharide:

$$E + L \rightleftharpoons E*L \dashrightarrow E + Ga + G$$

$$L + Ga \underset{k_2}{\overset{k_1}{\rightleftharpoons}} Tri$$

As the first equation is of Michaelis-Mente type the differential equations describing the production of galactose, lactose and the trisaccharide in a batch reaction are:

$$-\frac{dL}{dt} = \frac{k\,E\,L}{Km + L} + k_1\,L\,Ga - k_2\,Tri \qquad (9)$$

$$-\frac{dGa}{dt} = -\frac{k\,E\,L}{Km + L} + k_1\,L\,Ga - k_2\,Tri \qquad (10)$$

$$-\frac{dTri}{dt} = -k_1\,L\,Ga + k_2\,Tri \qquad (11)$$

From Yang and Okos [4], the value of Km is taken as 0.0656 mM (at 50°C) with 0.146 M of lactose and a kE value of .0058 M min $^{-1}$. There are no batch reaction data in their results but the relation between conversion and trisaccharide conversion (fig. 3) may be adequately described with equations 9-11. It is interesting to observe the effect of k_1 and k_2 in the simulation: a ratio of $k_1/k_2 = 1.2$ provided $k_2 < 1$, predicts 3.74% of oligosaccharide when conversion reaches 52%, which corresponds to the data presented in figure 3. Also, Km values may be modified together with k E reductions to favor the transglucosidase activity in the overall reaction.

4.4 BATCH REACTION WITH INTERNAL DIFFUSIONAL LIMITATIONS.

Immobilization of microorganisms for the application of a single enzymatic activity is an alternative to enzyme immobilization that often results in biocatalysts of lower cost and higher estability.

Cell entrapment in gels such as alginate, carrageenan or gelatin, is the most frequent technique used for this purpose, due to its simplicity and versatility. However, one of the major problems of this system is the limitation produced by internal mass transfer. Very often, an experimental effectiveness factor (η) is used to correct the reaction rate for diffusion. It is defined as the ratio of the actual reaction rate to the rate predicted by the kinetic model. In immobilized cells by entrapment, it is often obtained experimentally by measuring the activity of the biocatalyst before and after mechanical desintegration. However, the factor measured by this procedure is only valid during initial rate conditions, when substrate concentration is high and changes are negligible. The following example is used to illustrate this situation.

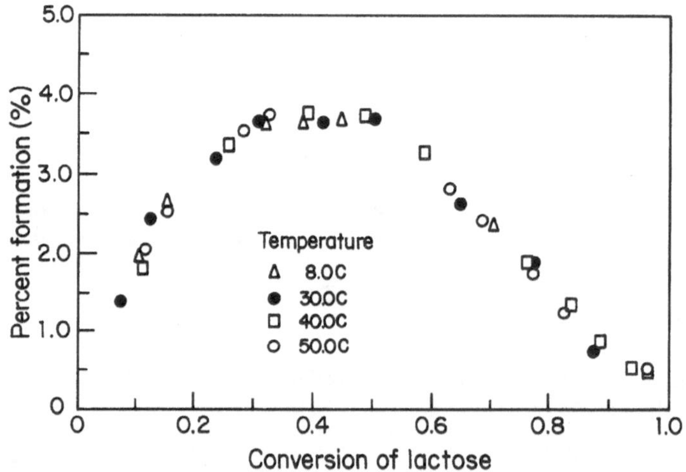

Figure 3. Effect of conversion and temperature on the oligosaccharide formation during lactose hydrolysis by β-galactosidase.
(From ref.[4], courtesy of John Wiley & Sons Inc.).

Consider a β-galactosidase catalyst consisting of *K.fragilis* cells immobilized on gelatin. The enzyme in the cells follows Michaelis-Menten kinetics with Km = 5.33 $g\ l^{-1}$; the biocatalyst has an observed activity of 510 $U\ g^{-1}$, but after desintegration an activity of 1050 $U\ g^{-1}$ is found. (η= 0.486) [5]. For a batch reaction the following models are possible:

a) kinetic control:

$$-\frac{dS}{dt} = \frac{k2\,E\,S}{Km + S} \tag{12}$$

b) diffusional control:

$$-\frac{dS}{dt} = \eta\,\frac{k2\,E\,S}{Km + S} \tag{13}$$

However, equation 13 has to take into account the effect of substrate concentration on the effectiveness factor. This is done by solving equation 13 simultaneously with the mass balance equation in the catalyst, which for a spherical particle is given by:

$$\frac{d^2S}{dr^2} + \frac{2}{r}\frac{dS}{dr} - \frac{k\,E\,S}{Km+S}\,\frac{1}{De} = 0 \tag{14}$$

and predicting the overall reaction rates by integration of the local rates.

For a particle diameter of 1 mm and an initial lactose concentration of 50 g l^{-1}, it was found by an iterative procedure that the effective diffusivity was 4.21×10^{-8} $cm^2\,sec^{-1}$.

Another alternative is to use equation 13 with the experimental description of the relation between substrate concentration and the effectiveness factor. In this case the experimental data obtained was described by a polynomial correlation:

$$\eta = a + bS + cS^2 + dS^3 \tag{15}$$

with $a = 0.1552$
$b = 1.4327 \times 10^{-2}$
$c = -2.5951 \times 10^{-4}$
$d = 2.25 \times 10^{-6}$

In figure 4 the results of the simulation in a batch reactor is presented and compared with the experimental results. It is clear that equation 13 taking into account the increasing effect of internal diffusional limitations with reaction time, gives the best description of the experimental data.

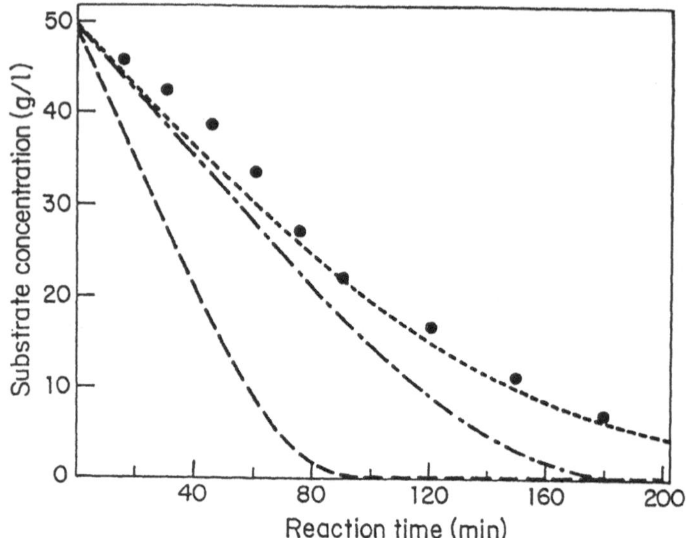

Figure 4. Evolution of a batch reaction limited by internal diffusion with an immobilized cell β-galactosidase biocatalyst. (_ _ _) equation 12; (_ . _ . _) equation 13 with constant effectiveness factor; (------) equations 13 and 15. (From ref.[5], courtesy of Butterworth-Heinemann)

5. Conclusions

Simulation procedures to solve the differential equations involved in the description of batch enzymatic reactors are very helpful under various circumstances. In this text this has been demonstrated with various examples: the veification of models for the description of enzyme kinetics, the study of the effect of inhibitors, alternative substrates and alternative activities on the enzyme and the analysys of internal diffusion effects on the observed reaction rate.

6. References

1. Ospina S., Lopez-Munguia, A., Gonzalez R.L. and Quintero R. (1992). Characterization and use of a penicillin acylase biocatalyst. J. Chem. Tech. Biotechnol.,. 53. In press.

2. ISIM, Salford University. Business Services Limited. P.O. Box 50, Salford, M66BY, UK.

3. Kawasaki T.S., Hernandez Velasco G., Iturbe C.F. and Lopez-Munguia, C.A. (1992). Phenylalanine ammonia lyase from *Sporidiobolus pararoseus* and *Rhodosporidium toruloides*. World J. Microbiol. Biotechnol., 8 , 4, In Press.

4. Yang S. and Okos M. (1989). Effects of temperature on lactose hydrolysis by immobilized β–galactosidase in plug flow reactor. Biotechnol. Bioeng. 33, 873-885.

5. Castillo E., Rodriguez, M., Casas L., Quintero R. and Lopez-Munguia A. (1991). Design of two immobilized cell catalyst by entrapment on gelatin: internal diffusion aspects. Enzyme & Microb. Technol. 13, 127-133.

MODELING OF MEMBRANE BIOREACTORS

D. M. F. Prazeres, F. Lemos, J. M. S. Cabral
Departamento de Engenharia Química
Instituto Superior Técnico
1096 Lisboa Codex
Portugal

ABSTRACT. A theoretical analysis of membrane bioreactors is presented and the use of systematic simulation for experimental data analysis is exemplified. A computational model was developed to describe the lipase catalysed hydrolysis of olive oil in a reversed micellar media, carried out in a CSTR recycle type membrane reactor. Simulations were carried out using a fourth-order Runge-Kutta integration method and with varying kinetic expressions. As in other reported cases, the dynamic of this reactor is mainly controlled by the kinetics of the reaction.

1. Introduction

In most chemical processes, the main procedure is some sort of chemical transformation, either catalysed or not, in homogeneous or heterogeneous systems. The separation of the products from the unreacted reagents and, in homogeneous catalytic systems, the separation of the catalyst from the rest of the reaction mixture, is a necessary step, which adds to the general complexity of the unit.

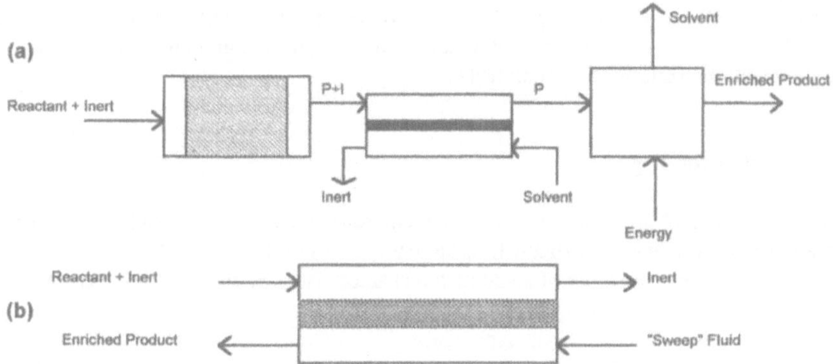

Figure 1. Comparison of the classical process, involving reaction + separation + enrichment **(a)**, with the membrane bioreactor, integrating all steps into a single unit **(b)**.

A.R. Moreira and K.K. Wallace (eds.), Computer and Information Science Applications in Bioprocess Engineering, 191-200.

Enzyme catalysed reactions fall in this later category, and a classical system could be represented by the scheme in figure 1a. Membrane bioreactors constitute an attempt to integrate catalytic conversion, product separation and product enrichment into a single operation, resulting in greater productivity and leading to a much simpler overall process (see figure 1b).

Most of the enzymatic conversion processes used in widely different areas, such as food, pharmaceutical and chemical industries, are conducted in traditional batch type reactors. Despite its broad use, batch processes have several distinct disadvantages that decrease their efficiency and raise operation costs, relative to continuous type processes [1]. Batch to batch oscillations, high labour costs, frequent start-up and shut-down procedures and the need to inactivate or separate the enzyme from the process outlet stream are some of the critical aspects in the performance of these reactors.

The immobilisation of the biocatalyst in a reactor, with retention of its catalytic activity, has emerged as a practical solution to overcome some of the disadvantages of batch processes using free biocatalysts in solution. This technique not only enables the recovery and reuse of the enzyme, but also offers the opportunity to carry out continuous processes. Further advantages include better process control, better productivity and more uniform products [1]. Apart from the classical methods of immobilising enzymes, by chemical or physical attachment to solid surfaces, an alternative and/or complementary strategy has been developed by the use of membrane reactors.

The basic and general concept of membrane reactors is the separation of enzyme and products by a semi-permeable membrane that creates a selective physical/chemical barrier. Due to size exclusion, the enzyme is confined in one side of the membrane where reaction with substrate occurs. The resulting products, which must have adequate dimensions and/or chemical compatibility with the membrane, permeate through the membrane, either by diffusion or convection. Substrate molecules may be either freely permeable or impermeable. In this later case, membrane reactors offer an additional advantage, by the selective removal of the product and the consequent displacement of chemical equilibrium [2].

Ultrafiltration membranes are the most adequate to the retention of the majority of enzymes (10,000 - 100,000 daltons). There is a large variety of synthetic membranes found in the market, with suitable specifications for use in membrane bioreactors. Important characteristics, to be taken into account, include morphology, porosity, pore size distribution, molecular weight cut-off, chemical resistance and biological inertness, temperature, pH and pressure tolerance and, of course, price [3].Several membrane modules can be found in the market, each of them having particular advantages and disadvantages. Plate and frame, tubular, capillar, hollow fibers, spiral wound and dynamic membrane modules are some of the commonly used devices in ultrafiltration and microfiltration operations [4].

2. Membrane Reactor Types

The existence of several applications, types of membrane devices and possible configurations and operation modes, makes it difficult to establish a well-defined classification of membrane bioreactors. Table 1 presents a list of some of the characteristics to be taken into account when classifying these reactors, according to [5].

The three examples of continuous type reactor in figure 2 show the complexity of the classification of this type of reactors. In all three cases the enzyme is in the liquid phase (although it could reasonably be in or on the membrane),. Nevertheless their operating mode is very different. In cases (a) and (c) the flow relative to the membrane is tangential, while in the dead end cell it is perpendicular.

Table 1. Characteristics for the classification of membrane bioreactors.

Characteristic	Typical Alternatives
Membrane Type	UF, Dialysis or Liquid membrane
Biocatalyst Type	Enzyme, Whole cells
Number of Liquid Phases	
Location of Biocatalyst	in Liquid Phase, on or in Membrane
Geometry of the Membrane	Flat, Tubular, Spiral, etc.
Driving Force for Transport	ΔP or ΔC
Flow Relative to Membrane	Tangential or Perpendicular
Type of Operation for Individual Liquid Phases	Batch or Continuous

Case (c) differs from the other two since the enzyme is in a different compartment from the one where the main flow takes place. This means that, for the reaction to occur, the substrate has to diffuse across the membrane to the compartment where the enzyme is located. This limits the application of these type of reactors to low molecular weight substrates.

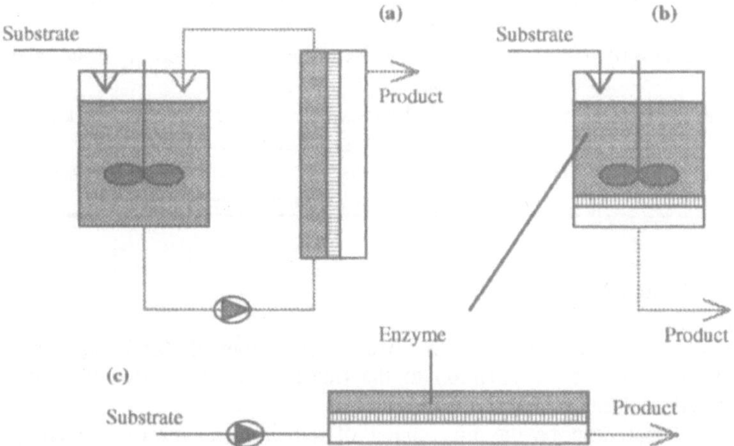

Figure 2. Different arrangements and modes of operation for membrane bioreactors: Continuous Stirred Tank Reactor (CSTR) with recirculation arrangement **(a)**, dead-end cell **(b)**, tubular with entrapped enzyme **(c)**.

The membrane reactor concept can be usually applied to any biocatalysed reaction with the aim of developing high productivity continuous processes. A great emphasis has been put especially in the hydrolysis of macromolecules such as proteins, carbohydrates (starch and cellulose) and lipids. Table 2 lists some representative enzymatic membrane reactor applications published in the literature in the past decade.

Table 2. Some recent applications of enzymatic membrane bioreactors

References	Application description
Deeslie and Cheryan (1982) [6]	Protein hydrolysis in an ultrafiltration membrane system
Jandel *et al* (1982) [7]	L-alanine production from fumarate in a bi-enzymatic system with two hollow fiber membranes
Hausser *et al* (1983) [8]	Two enzyme system for continuous dextrose production from starch
Ohlson *et al* (1984) [9]	Continuous enzymatic hydrolysis of cellulose using an ultrafiltration bioreactor
Leuchtenberger *et al* (1984) [10]	Production of L-amino acids by enzymatic resolution in a hollow fiber system
Hoq *et al* (1985) [11,12]	Enzymatic reactions of lipids (hydrolysis [11] and synthesis [12]) in a three phase membrane bioreactor system
Kinoshita *et al* (1986) [13]	Hydrolysis of cellulose by cellulases in an ultrafiltration membrane bioreactor
Nakajima *et al* (1989) [14]	Sucrose inversion in a ceramic membrane bioreactor
Lozano *et al* (1990) [15]	Pectin hydrolysis in a cross flow membrane bioreactor
Malcata *et al* (1991) [16]	Hydrolysis of butter oil glycerides in a three phase flat-sheet membrane bioreactor
Prazeres *et al* (1992) [17]	Hydrolysis of olive oil in a reversed micellar ultrafiltration membrane bioreactor

3. Fundamental Equations

In this paper our interest is focused in the dynamic behaviour of membrane bioreactors, and therefore we will pay a close attention to the transient equations that describe the time-dependent operation of these systems.

The equations that describe the functioning of a membrane reactor are not particularly different from the ones that describe any other reactor. The fundamental difference is, as it is obvious, the additional transport phenomena that occur at the membrane. The main phenomena to be taken into account are:

- *Reactor dynamics*
- *Chemical kinetics*
- *Transport phenomena across the membrane*

In the case of multiphase reactors, additional problems may arise and transport phenomena between the various phases may have to be taken into consideration.

Reactor dynamics can be described by starting with the general continuity equation for a chemical reactor, but due to the tubular geometry that is usually assumed by the reactors, a dispersion type model will generally be applicable [18].

$$\frac{\partial C}{\partial t} = D_z \frac{\partial^2 C}{\partial z^2} - u\frac{\partial C}{\partial z} + r \qquad \text{equation 1}$$

Chemical kinetics of biological systems are usually rather complex, but a Michaelis Menten type equation (for enzyme catalysed processes) or a Monod expression (for cases involving cellular growth) will usually be adequate as a starting point.

Transport phenomena will always be significant in membrane type reactors, namely across the membrane. Mass transport across a membrane can be due to several types of gradient, depending on the membrane separation process, the component being transported and the media in which the transport occurs. It can usually be due to chemical potential gradient, a pressure gradient or even an electrical field gradient, and can, in a first approach, be written as[19]:

$$J_n = L_{nn}\nabla\mu + L_{nv}\nabla P + L_{ne}\nabla\varphi \qquad \text{equation 2}$$

where ∇ operator refers to the gradient of the indicated quantities, and represents the driving forces associated with each of these quantities (chemical potential, pressures and electrical field). The quantities L are the phenomenological coefficients associated with the transport of n due to the corresponding gradient, which can usually be related to known physical properties. For instances, L_{nn} can be substituted by the molecular diffusivity if a standing dilute solution is considered (in this case the concentration will also be a measure of the chemical potential). In ultrafiltration the driving force is also a pressure gradient and intraparticle convection must be taken into account. This is particularly important when the flux across the membrane is significant, as in dead end cells. As in most membrane bioreactors only concentration (chemical potential) and pressure gradients are important, the electrical field gradient can be considered as null.

In multi-phase systems, mass transport between the various phases can also become limiting, namely in systems with live cells or immobilised biocatalysts, especially in the case of very fast reactions.

4. Modeling of Membrane Bioreactors - General Considerations

Although the diversity of membrane bioreactors is large, we think that most cases can be simplified, by taking into account the particular characteristics, so that only a reasonable amount of work is required.

The first problem likely to be encountered is that this type of systems often have very different time constants associated with the different parts of the system. The flux through the membrane is usually much lower than the main flow across the reactor, giving rise to a much longer mean residence time in the permeate section comparatively to the main flow section. These very different time constants imply a stiff system of differential equations.

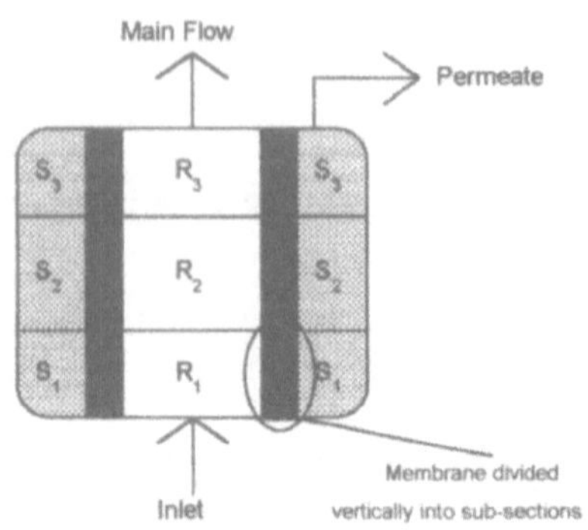

Figure 3. Subdivision of a membrane type reactor for simulation purposes.

This can be further reinforced by the chemical reaction taking place in the reactor, which will still have another time constant, often limiting in terms of the global operation of the reactor [2]. In order to accurately describe what occurs in the reactor the integration of the necessary differential equations will have to be carried out with a step size that is regulated by the shortest time constant, but the number of integration steps required will be ruled by the largest of these time constants [20]. This gravely limits the obtainable computation time.

If significant gradients develop inside the membrane, it must be radially divided into segments (see figure 3) so that an accurate account of the transport phenomena is achieved.

Another important issue in biological systems is the complexity of the reaction schemes. It is important to consider a good description of the reaction kinetics, especially if the reactor arrangement reveals a high sensitivity to the kinetic equation being used. Oversimplified kinetic approaches will result in a poor model, with a short application range. This fact was, indeed, observed in the case study presented below.

5. Olive Oil Hydrolysis - A Case Study

A great number of lipase catalysed reactions are carried out in biphasic media due to poor reactant and/or product solubilities in the conventional aqueous media. An alternative approach has been the encapsulation of lipase in reversed micellar media, where those reactions can be carried out in a "pseudo" one phase system [21]. Other advantages of these systems include greater interfacial area, lower mass transfer limitations, enhancement of catalytic activity, and increased stability. The development of reactor design enabling continuous reaction and product separation is one of the critical demands in reversed micelles technology in present time.

The enzymatic hydrolysis of olive oil (see figure 4a) using *Chromobacterium viscosum* lipase B encapsulated in reversed micelles of dioctyl sodium sulfosuccinate (AOT) in isooctane has been investigated in a membrane bioreactor [17]. A tubular ultrafiltration ceramic membrane was used in a CSTR recycle configuration (see figure 4b).

A preliminary study using a batch reactor supplied a first approximation to the kinetics of the reaction, indicating that a strong product inhibition existed (up to third order), and the collected data was consistent with the generally accepted idea that lipase act at interfacial surfaces [22], provided in this case by the reversed micelles shells.

In the case of the membrane reactor, lipase is, at least, partially adsorbed on the membrane, and most of the reaction occurs in its vicinity. Both substrate (olive oil) and products (oleic acid) permeate through the membrane. To obtain transport (rejection) coefficients for the most important components of the reaction mixture, individual transmission experiments were carried out in the reactor system. Although a slight dependence of the transport coefficients on the composition of the mixture was observed, it was not considered relevant and constant rejection coefficients were used in the rest of the work for the substrate. The rejection coefficient for oleic acid was measured during the reaction experiment and used accordingly.

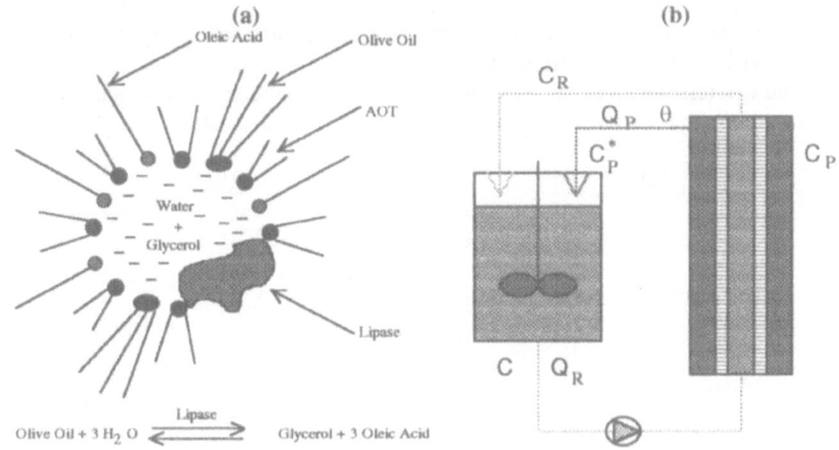

Figure 4. Schematic representation of the reversed micelle where the reaction takes place **(a)**, and reactor arrangement used in this work **(b)** - V_R=17.6 ml, V_P=3 ml, V=80 ml.

For modeling purposes the bioreactor system was divided into subsections. Considering the fact that we were working with a CSTR type reactor, concentration was supposed uniform in each of the subsystems: reactor core (V_R), reactor shell (V_P) and external vessel (V). Due to the length of the tube connecting the reactor shell and the external vessel (recirculation of the permeate stream), a time delay (θ) was explicitly considered in this model. The equations describing the system are shown below (see figure 4b for variable assignment).

$$\frac{dC}{dt} = (Q_P C_P^* + (Q_R - Q_P)C_R - Q_R C)/V \qquad \text{equation 3a}$$

$$\frac{dC_P}{dt} = (Q_P C_R (1 - \sigma) - Q_P C_P)/V_P \qquad \text{equation 3b}$$

$$\frac{dC_R}{dt} = (Q_R C - (Q_R - Q_P)C_R - Q_P C_R(1 - \sigma))/V_R + r \qquad \text{equation 3c}$$

$$C_P^*(t) = C_P (t - \theta) \qquad \text{equation 3d}$$

To integrate these equations a program was written in Turbo Pascal (Version 6 - from Borland Int.) and run on an IBM compatible computer equipped with 386SX/25 MHz and a Cyrix Fasmath arithmetic coprocessor. Several integration schemes were tried, including Euler, fourth-order Runge-Kutta and Adams-Moulton predictor-corrector multistep method. Due to the low residence time of the reaction mixture inside the reactor, the low permeate flux across the membrane and comparatively low reaction rate, the Adams-Moulton scheme revealed itself liable to instability and was abandoned. Fourth-order Runge-Kutta method was selected as the most efficient algorithm for the solution of this problem.

The program allowed the choice of several kinetic expressions and of the integration method itself. Additionally it compared the experimental results with the simulated data that was being obtained. Due to the length of the simulation runs and the final number of parameters to be estimated, only visual fitting was performed.

The program was tested with the above mentioned transmission runs and was seen to describe accurately the global flow scheme of the apparatus, as can be seen in figure 5.

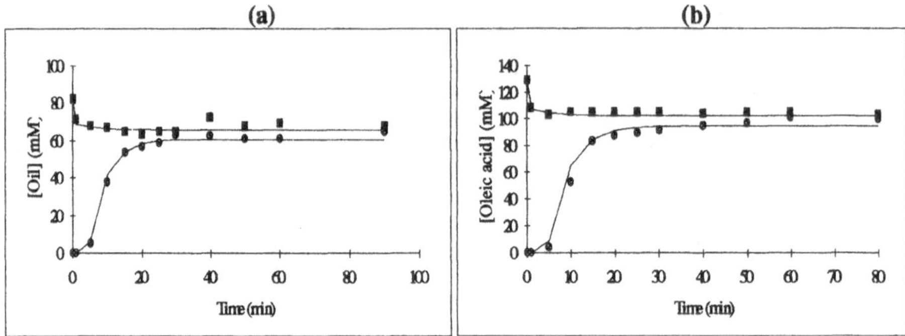

Figure 5. Experimental data (symbols) obtained during the experiments to measure rejection coefficients of oil **(a)** and oleic acid **(b)**, compared to the simulated curves (lines) obtained with the model.

A series of kinetic equations was used, according to literature, but none of the published reaction schemes was able to fit the data, neither in the batch reactor nor in the membrane one. The model that best fitted the batch reactor data included, as mentioned above, an inhibition term for the product up to third order, with a Michaelis-Menten type dependence on the substrate concentration.

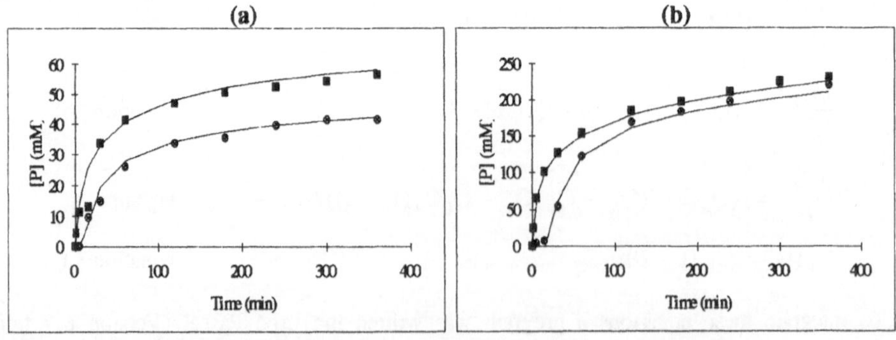

Figure 6. Comparison of experimental (symbols) and simulated data (lines) for time profiles of product concentration in the permeate stream (circles) and concentrate (squares) for two typical experiments (Q_R=1080 ml/min, [E]=0.02 mg/ml, [S]=25 mM **(a)**; [S]=134 mM **(b)**).

However, for the membrane reactor the first order in relation to the substrate could not explain the global change in reaction rate as the substrate concentration increased. A best fit was obtained using a second order kinetics (see figure 6).

This change in the observed kinetics indicated that, in order for lipase to function on the membrane, a lipidic "coating" should exist, and that this coating was supplied by the substrate itself, raising the observed order for the substrate from one (in "pseudo" homogeneous system) to two. The kinetic expression obtained through the simulation of this system is given in equation 4, and figure 7 depicts the change in rate with both the substrate and the product concentrations.

$$\frac{r}{E} = \frac{kS^2}{K_M(1 + k_1P + k_2P^2 + k_3P^3) + S + k_{S2}S^2} \qquad \text{equation 4}$$

The mechanism subjacent to this kinetic law is consistent with the accepted scheme for lipase action, involving an activation step followed by the catalytic transformation [23]. The peculiarity in this case is the fact that the activating lipid is the substrate itself, instead of the AOT shells as in the "pseudo" homogeneous system.

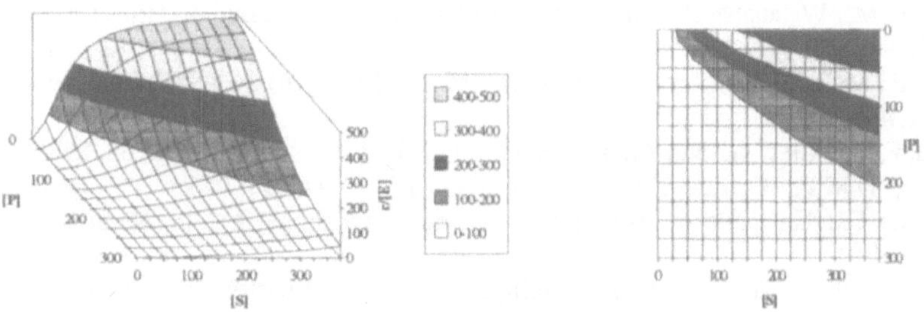

Figure 7. Surface representation of the change of the reaction rate (μmol.ml^{-1}.min^{-1}) as a function of substrate and product concentrations (μmol.ml^{-1}).

6. Acknowledgements

A fellowship to D.M.F. Prazeres from JNICT/Portugal (Ciência BD/51/90-IS) is acknowledged.

7. References

1. Cheryan, M. and Mehaia, M.A. (1986) "Membrane Bioreactors" in W.C. McGregor (Ed.), Membrane Separations in Biotechnology, Marcel Dekker Inc., New York, pp. 255-302.

2. Van der Padt, A. and Van't Riet, K. (1991) "Membrane Bioreactors" in C.A. Costa and J.S. Cabral (Eds.), Chromatographic and Membrane Processes in Biotechnology, Kluwer Academic Pub., Dordrecht, pp. 443-448.

3. Hildebrandt, J.R. (1991) "Membranes for Bioprocesses: Design considerations" in C.A. Costa and J.S. Cabral (Eds.), Chromatographic and Membrane Processes in Biotechnology, Kluwer Academic Pub., Dordrecht, pp. 363-378.
4. Santos, J.A.L., Mateus, M. and Cabral, J.M.S. (1991) "Pressure Driven Membrane Processes" in C.A. Costa and J.S. Cabral (Eds.), Chromatographic and Membrane Processes in Biotechnology, Kluwer Academic Pub., Dordrecht, pp. 177-205.
5. Prenosil, J.E., Dunn, I.J., and Heinzle, E. (1987) "Biocatalyst Reaction Engineering" in J.F. Kennedy (Ed.), Biotechnology: A Comprehensive Treatise in 8 Volumes, Vol. 7A, VCH, Weinheim, pp. 489-545.
6. Deeslie, W.D. and Cheryan, M. (1982) Biotechnol. Bioeng., 24, 69-82.
7. Jandel, A.S., Hustedt, H. and Wandrey, C. (1982) Eur. J. Appl. Microbiol. Biotechnol., 15, 54-63.
8. Hausser, A.G., Goldberg B.S. and Mertens, J.L. (1983) Biotechnol. Bioeng., 25, 525-532.
9. Ohlson, I., Trägårdh G. and Hahn-Hägerdal, B. (1984) Biotechnol. Bioeng., 26, 647-653.
10. Leuchtenberger, W., Karrenbauer, M. and Plöcker, U. (1984) Ann. N. Y. Acad. Sci., 434, 78-86.
11. Hoq, M.M., Yamane, T. and Shimizu, S. (1985) J. Am. Oil Chem. Soc., 62, 1016-1021.
12. Hoq, M.M., Yamane, T. and Shimizu, S. (1985) Agric. Biol. Chem., 49, 335-342.
13. Kinoshita, S., Chua, J.W., Kato, N., Yoshida, T. and Taguchi, H. (1986) Enzyme Microb. Technol., 8, 691-695.
14. Nakajima, M., Watanabe, A., Jimbo, N., Nishizawa, K. and Nakao, S. (1989) Biotechnol. Bioeng., 33, 856-861.
15. Lozano, P., Manjón, A., Iborra, J.L., Cánovas, M. and Romojaro, F. (1990) Enzyme Microb. Technol., 12, 499-505.
16. Malcata, F.X., Hill, C.G. and Amundsen, C.H. (1991) Biotechnol. Bioeng., 38, 853-868.
17. Prazeres, D.M.F., Garcia, F.A.P. and Cabral, J.M.S. (1992) Paper presented in the symposium on Fundamentals of Biocatalysis in Non-Conventional Media, Noordwijkerhout, The Netherlands.
18. Calo, J.M. (1981) "The Modeling of Multiphase Chemical Reactors" in A.E. Rodrigues et al. (Eds.), Multiphase Chemical Reactors. Volume II - Design Methods, Sijthoff & Noordhoff, Alphen aan den Rijn, pp. 3-63.
19. Strathmann, H. (1991) "Fundamentals of Membrane Separation Processes" in C.A. Costa and J.S. Cabral (Eds.), Chromatographic and Membrane Processes in Biotechnology, Kluwer Academic Pub., Dordrecht, pp. 153-175.
20. Constantinides, A. (1987) "Applied Numerical Methods with Personal Computers", McGraw Hill, New York.
21. Prazeres, D.M.F., Garcia, F.A.P. and Cabral, J.M.S. (1992) J. Chem. Technol. Biotechnol., 53, 159-164.
22. Sarda, L. and Desnuelle, P. (1958) Biochim. Biophys. Acta, 30, 513-521.
23. Brady, L., Brzozowski, A.M., Derewenda, Z.S., Dodson, E., Dodson, G., Tolley, S., Turkenburg, J.P., Christiansen, L., Huge-Jensen, B., Norskov, L., Thim L. and Menge, U. (1990) Nature, 343, 767-770.

MEMBRANE BIOREACTOR WITH IMMOBILIZED LIPASE: MODELLING AND COMPUTATIONAL CONSIDERATIONS

F. XAVIER MALCATA
Escola Superior de Biotecnologia
Universidade Católica Portuguesa
Rua Dr. António Bernardino de Almeida
4200 Porto
Portugal

ABSTRACT. Several problems pertaining to the simulation of a three-phase hollow fiber bioreactor for the controlled accelerated hydrolysis of butterfat are addressed encompassing mathematical modelling, statistical fitting, and computational procedures. Common sense considerations and order of magnitude analyses are employed to simplify the general mass balances to substrate and enzyme forms whereas the rate expressions are derived on the basis of reasonable mechanisms postulated for the enzyme-catalyzed reaction and the enzyme deactivation/rearrangement.

1. Introduction

During the last two decades, enhanced consumer interest in the relationship of diet to good health (and concomitant increased demand for low fat dairy products) has led to a decline in consumption of whole milk and butter, which has created a situation of worldwide surplus of milkfat [1]. Hence, the industry has urged researchers to find pragmatic and palatable uses for surplus milkfat. The increased availability of lipases (glycerol ester hydrolases, EC 3.1.1.3) from microbial sources in the recent past has made it possible for technologists to employ the selectivity and relatively high catalytic activity of these enzymes in innovative ways. One application in which the use of lipases has become well established is the production of lipolyzed flavors from milkfat feedstocks by hydrolysis of their triglycerides [2]. The controlled release of the short chain fatty acid moieties of these triglycerides (e.g., C_4, C_6, and C_8) in their free form can impart sensations of richness, creaminess, buttery flavor, and a variety of cheese aromas [3]. The resultant lipolyzed flavor bases have an expanding number of uses in confections, bakery goods, snack foods, condiments, salad dressings, and pet foods [4].

Despite the trend for an increasing number of applications of lipases, these enzymes do not currently account for more than 3% of all enzymes produced commercially [5]. An obvious way to overcome their relatively high cost is their reuse, which can easily be achieved via immobilization. In spite of a multitude of protocols for enzyme attachement to a variety of organic and inorganic carriers, immobilization of lipases by adsorption on hydrophobic polymeric supports has the potential to (i) preserve, and in some cases enhance, the activity of lipases (as compared with their free counterparts), (ii) increase their thermal stability, (iii) avoid contamination of the lipase-modified product with residual activity, (iv) allow regeneration of the catalytic activity of the reactor (via sequential steps of

A.R. Moreira and K.K. Wallace (eds.), Computer and Information Science Applications in Bioprocess Engineering, 201-220.

desorption of deactivated lipase and adsorption of fresh lipase), (v) increase system productivity per unit of lipase employed, and (vi) permit the development of continuous food-grade processes. One the best reactor configurations using immobilized lipase consists of a bundle of hollow fibers made from microporous polypropylene potted at both ends and contained in a suitable casing (see Figure 1). This configuration takes advantage of the high effective diffusivity, high chemical stability, high membrane surface area to volume ratio, and high hydrophobic character of the membrane. The present communication focuses on the simulation methodology associated with this type of reactor for the release of free fatty acids within a phase of melted butterfat in the presence of an independent aqueous phase.

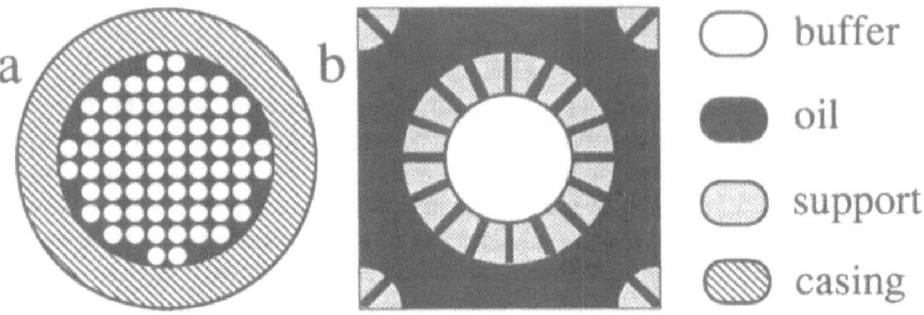

Figure 1. Schematic cross sectional views of (a) the hollow fiber reactor and (b) a single hollow fiber.

The aqueous and oil phases flow cocurrently in an upward fashion; the aqueous buffer is pumped through the lumen of the fibers, whereas the melted butterfat is pumped through the shell side of the reactor. The pressure differential between the two phases is adjusted so as to avoid percolation of either phase through the membrane walls.

2. Theoretical considerations

For a bioreactor which effects an enzyme-catalyzed reaction in the absence of electric and magnetic fields, the balance to any chemical species i can be written in its most general form as

$$\frac{\partial C_i}{\partial t} + \vec{\nabla}.\left(C_i \vec{v}\right) - \vec{\nabla}.\left(L_C \vec{\nabla}\mu_i + L_P \vec{\nabla}P + L_T \vec{\nabla}T\right) + \ldots$$

$$\ldots\ a\left(K_C \Delta\mu_i + K_P \Delta P + K_T \Delta T\right) - \sum_j v_{ij} r_j = 0 \qquad \text{equation 1}$$

where C_i denotes the concentration of species i, t is the time elapsed since start-up of the reactor, $\vec{\nabla}$ is the nabla vector operator (defined as $\vec{i}_x \frac{\partial}{\partial x} + \vec{i}_y \frac{\partial}{\partial y} + \vec{i}_z \frac{\partial}{\partial z}$), $\vec{v}$ is the local velocity vector, L_C, L_P, and L_T are the phenomenological conductances associated with the transport of species i by chemical potential, pressure, or temperature differential driving forces, respectively, K_C, K_P, and K_T are the mass transfer coefficients associated with the transport of species i by chemical potential, pressure, or temperature finite driving forces, respectively, across the boundary layers in the vicinity of interfaces, a is a specific area of mass transfer, ν_{ij} is the stoichiometric coefficient of species i in reaction j, and r_j denotes the rate expression for the j-th chemical reaction. In general, L_C, L_P, L_T, K_C, K_P, K_T, and r_j are functions of the activities of all chemical species present (substrate and enzyme forms) in addition to pressure and temperature. Hence, the first term in equation 1 is accounted for by the local variation of the concentration of species i, the next three terms represent the fluxes of i by convection, molecular transport, and transfer through an interface, respectively, and the last term corresponds to the sources/sinks of species i. In the absence of significant enthalpy changes of the reactions, heat transfer through the reactor wall, shaft work on the reacting fluid, and viscous dissipation, the energy balance (similar in form to equation 1) is not relevant, and so will not be considered here.

In the case of interest, the microporous membrane in hollow fiber shape separates the aqueous and oil phases. The chemical equation which describes the hydrolysis process can be written as

$$S_1 + S_2 \longrightarrow S_3 + S_4 \qquad \text{equation 2}$$

where S_1, S_2, S_3, and S_4 denote a glyceride moiety, a water molecule, an alcohol moiety, and a free fatty acid, respectively; the oil phase supplies species S_1 and dissolves species S_3 and S_4, whereas the aqueous phase supplies species S_2.

If M different types of fatty acid residues are considered, the general form of the rate expression describing the change with time of the concentration of reactant or product species must include the concentrations of all M free fatty acids, water, M monoglycerides of each of three different positional types, M^2 diglycerides of each of three different positional types, and M^3 triglycerides. Hence the total number of species that should be included in the rate expression is given by $(M+1)[1+(M+1)^2]$ [6]. For the case of butterfat, the number of fatty acid residues existing in non-negligible molar concentrations (say, larger than 0.5%) is 10 (i.e., C_4, C_6, C_8, C_{10}, C_{12}, C_{14}, C_{16}, C_{18}, $C_{18:1}$, and $C_{18:2}$, where the digit after the colon indicates the number of double bonds), which leads to 1342 different species to be included in the rate expression! The search for a rate expression containing so many terms presents an intractable problem because enormous amounts of experimental data would be required and the numerical work necessary to conduct the nonlinear regression analyses would be prohibitive. Hence, it will hereafter be assumed that the attack on each labile ester bond associated with a given type of fatty acid moiety is equally probable irrespective of the position of esterification in the glycerol backbone and presence (or absence) of other residues esterified in the other positions of the same backbone. This assumption implies that for uniresponse lumped models one should use simply the total molar concentration of fatty acids esterified in the positions susceptible to catalytic action of the lipase in question (termed hereafter as *glycerides*), and that for multiresponse lumped models one should use, in a similar fashion, the total concentrations of each type of fatty acid moiety.

The chemical equations which describe the deactivation/rearrangement processes can be written as

$$E_1 \rightarrow E_2 \qquad \text{equation 3i}$$

$$E_1 \rightarrow E_3 \qquad \text{equation 3ii}$$

$$E_2 \rightarrow E_4 \qquad \text{equation 3iii}$$

where E_1 and E_2 denote the native and rearranged active forms of lipase, respectively, and E_3 and E_4 denote deactivated forms of lipase.

Based on inspection of available thermodynamic information and *a priori* order of magnitude analyses, the following simplifying assumptions were made: (i) the concentrations in the aqueous phase of species S_1, S_3, and S_4 are negligible compared with their concentrations in the oil phase (i.e., only a balance to the oil phase is germane); (ii) the flow of oil is essentially unidirectional (i.e., $\vec{v} \sim v_z\{x,y,z\}\vec{i_z}$, where z denotes the longitudinal coordinate); (iii) the longitudinal velocity vector is essentially constant along the reactor (i.e., $\vec{\nabla}.(C_i\vec{v}) \sim v_z\frac{\partial C_i}{\partial z}$); (iv) the oil solution behaves ideally (i.e., $\vec{\nabla}\,\mu_i \sim \vec{\nabla}\,C_i$); (v) the membrane wall material is inert and insoluble and the rate of transport of water molecules from the aqueous phase through the oil phase filling the inner pores of the membrane is very high compared with their consumption by chemical reaction (i.e., $\Delta\mu_i{\sim}0$); (vi) the linear velocities are small and there is no percolation of either liquid phase through the membrane (i.e., $\vec{\nabla}\,P \sim 0$ and $\Delta P \sim 0$); (vii) the operation is isothermal (i.e., $\vec{\nabla}\,T \sim 0$ and $\Delta T \sim 0$); (viii) the diffusivity is essentially independent of the position within the reactor (i.e., $\vec{\nabla}\,.\left(L_C\,\vec{\nabla}\,C_i\right) \sim L_C\,\vec{\nabla}^2 C_i$); (viii) the time scale of diffusion in directions other than the longitudinal is very small (i.e., $L_C\,\vec{\nabla}^2 C_i \sim L_C\frac{\partial^2 C_i}{\partial z^2}$); and (ix) the time scale of diffusion in the longitudinal direction is very large compared with the reactor space time (i.e., $L_C\frac{\partial^2 C_i}{\partial z^2} \sim 0$). Under the above circumstances, the set of partial differential equations (together with associated initial and boundary conditions) denoted as equation 1 can be simplified in the case of substrate S_i to yield

$$\frac{\partial C_{S_i}}{\partial t} + v_z\frac{\partial C_{S_i}}{\partial z} - \nu_{S_i}\, r_{ht}\left\{\overline{C_S},\overline{C_E}\right\} = 0 \;,\; i=1,2,3,4 \qquad \text{equation 4i}$$

subject to the initial condition

$$t = 0\,,\, 0 \leq z \leq L\,,\, \overline{C_S} = \overline{C_{S,0}} \qquad \text{equation 4ii}$$

and the boundary condition

$$t \geq 0 \ , z = 0 \ , \overline{C_S} = \overline{C_{S,0}} \qquad \text{equation 4iii}$$

Here $\overline{C_S}$ is the vector of concentrations of all substrates, $\overline{C_E}$ is the vector of concentrations of all enzyme forms, r_{ht} is the observed rate of hydrolysis at time t, and subscript o denotes inlet stream conditions.

In the case of enzyme, the assumption that the enzyme phase is stationary with respect to the flow of oil implies that $\vec{\nabla}.(C_i \vec{v}) \sim 0$, $\vec{\nabla}.\left(L_C \vec{\nabla}\mu_i + L_P \vec{\nabla}P + L_T \vec{\nabla}T\right) \sim 0$, and $a\left(K_C \Delta\mu_i + K_P \Delta P + K_T \Delta T\right) \sim 0$); so, the following simplification of equation 1 results:

$$\frac{dC_{E_i}}{dt} - \nu_{E_i} r_{d,i}\left\{\overline{C_E}\right\} = 0 \ , \ i=1,2,3,4 \qquad \text{equation 5i}$$

subject to the initial condition

$$t = 0 \ , 0 \leq z \leq L \ , \ \overline{C_E} = \overline{C_{E,0}} \qquad \text{equation 5ii}$$

where subscript o denotes initial conditions. Integration of equation 5 eventually leads to $C_{E_i} \equiv C_{E_i}\{t\}$, so equation 4 can be reformulated to give

$$\frac{\partial C_{S_i}}{\partial t} + v_z \frac{\partial C_{S_i}}{\partial z} - \nu_{S_i} a_t\{t\} r_h \left\{\overline{C_S}\right\} = 0 \ , \ i=1,2,3,4 \qquad \text{equation 6}$$

where r_h is the rate of hydrolysis corrected for time zero and a_t is the activity of enzyme at time t normalized by the activity of the enzyme at time zero.

Preliminary fits have indicated that more complex models for enzyme deactivation and/or rearrangement than that depicted in equation 3 can not be fit on the 5% significance level; this model can be viewed as a step of irreversible unimolecular rearrangement (described by kinetic constant k_r) of the native active enzyme form (E_1) to an active, more stable form (E_2) accompanied by parallel steps of irreversible unimolecular deactivation (described by kinetic constants k_{dn} and k_{dr}, respectively) of both the native and the rearranged forms to yield as many deactivated enzyme forms (E_3 and E_4, respectively). This mechanism (hereafter denoted as Model 3), which is a simplified form of the general mechanism proposed by Henley and Sadana [7], is justified by the fact that the microenvironment of the lipase is changed from macroaqueous during the immobilization steps to microaqueous during regular operation; hence, the existence of two alternative forms of enzyme with different activities and lability to thermal deactivation is anticipated (E_1 should be the dominating active form in the former situation and E_2 should the the dominating active form in the latter situation). Two nested mechanisms were also considered for the thermal deactivation of the immobilized lipase: Model 2, corresponding to irreversible unimolecular deactivation of E_1 to E_3 accompanied by a parallel irreversible unimolecular rearrangement of E_1 to E_2; and Model 1, corresponding to an irreversible unimolecular deactivation of E_1 to E_3. The associated rate expressions for the formation of E_1 ($r_{d,1}$) and formation of E_2 ($r_{d,2}$) can be written as

$$r_{d,1} = - k_{dn}\, C_{E_1} \qquad \text{equation 7i}$$

$$t = 0 \;,\; C_{E_1} = C_{E_1,0} \qquad \text{equation 7ii}$$

for Model 1,

$$r_{d,1} = - \left(k_{dn} + k_r\right) C_{E_1} \qquad \text{equation 8i}$$

$$r_{d,2} = k_r\, C_{E_1} \qquad \text{equation 8ii}$$

$$t = 0 \;,\; C_{E_1} = C_{E_1,0} \;,\; C_{E_2} = C_{E_2,0} \qquad \text{equation 8iii}$$

for Model 2, and

$$r_{d,1} = - \left(k_{dn} + k_r\right) C_{E_1} \qquad \text{equation 9i}$$

$$r_{d,2} = k_r\, C_{E_1} - k_{dr}\, C_{E_2} \qquad \text{equation 9ii}$$

$$t = 0 \;,\; C_{E_1} = C_{E_1,0} \;,\; C_{E_2} = C_{E_2,0} \qquad \text{equation 9iii}$$

for Model 3. Combination of equations 7, 8, and 9 with equation 5 followed by adequate integration leads to

$$a_{t,1} \equiv \frac{C_{E_1}}{C_{E_1,0}} = \exp\left\{- \lambda_{11}\, t\right\} \qquad \text{equation 10}$$

$$a_{t,2} \equiv \frac{C_{E_1} + \alpha\, C_{E_2}}{C_{E_1,0} + \alpha\, C_{E_2,0}} = \lambda_{21} \exp\left\{- \lambda_{22}\, t\right\} + \left(1 - \lambda_{21}\right) \qquad \text{equation 11}$$

$$a_{t,3} \equiv \frac{C_{E_1} + \alpha\, C_{E_2}}{C_{E_1,0} + \alpha\, C_{E_2,0}} = \lambda_{31} \exp\left\{- \lambda_{32}\, t\right\} + \left(1 - \lambda_{31}\right) \exp\left\{- \lambda_{33}\, t\right\} \qquad \text{equation 12}$$

respectively, where α is the specific activity of form E_2 relative to form E_1. (In all the above situations, the lumped kinetic parameters must be all nonnegative and of the general form $\lambda_{ij} \equiv \lambda_{ij}\{k_r, k_{dn}, k_{dr}, C_{E_1,0}, C_{E_2,0}\}$ [7].)

According to several sources of physical evidence [8], the mechanism of lipase-catalyzed hydrolysis of glycerides follows a Ping Pong Bi Bi mechanism [9] which, owing to the presence of excess water, can be viewed as essentially irreversible. Using Cleland's nomenclature, and denoting as $v_{max,f}$ and $v_{max,r}$ the maximum rate of reaction in the forward and reverse directions, respectively, as $K_{m,j}$ (j=1,2,3,4) the Michaelis-Menten constant for substrate S_j, as $K_{i,j}$ (j=1,4) the inhibition constant for substrate S_j, and as K_{eq} the equilibrium constant of the hydrolysis reaction, the associated form of the rate expression reads

$$r_{h,3} = \frac{(v_{max,f}\, v_{max,r}\, C_{S_2})\, C_{S_1}}{v_{max,r} K_{m,2} C_{S_1} + v_{max,r} K_{m,1} C_{S_2} + \frac{v_{max,f} K_{m,4}}{K_{eq}} C_{S_3} + \ldots} \qquad \text{equation 13}$$

$$\ldots \frac{v_{max,f} K_{m,3}}{K_{eq}} C_{S_4} + v_{max,r} C_{S_1} C_{S_2} + \ldots$$

$$\ldots + \frac{v_{max,f} K_{m,4}}{K_{eq} K_{i,1}} C_{S_1} C_{S_3} + \frac{v_{max,r} K_{m,A}}{K_{i,4}} C_{S_2} C_{S_4} + \frac{v_{max,f}}{K_{eq}} C_{S_3} C_{S_4}$$

Using the stoichiometry of equation 2, the mass balance equations to alcohol moieties and free fatty acids can take the form

$$C_{S_1,0} - C_{S_1} = C_{S_3} - C_{S_3,0} = C_{S_4} - C_{S_4,0} \qquad \text{equation 14}$$

whereas the oil phase can be assumed to be saturated at all times with a $C_{S_2,sat}$ concentration of water. Combination of equations 13 and 14 results in

$$r_{h,3} = \frac{\theta_{31} C_{S_1}}{1 + \theta_{32} C_{S_1} + \theta_{33} C_{S_1}^2} \qquad \text{equation 15}$$

which will hereafter be denoted as Model 3 for hydrolysis. Two nested simplifications thereof can be devised, viz.

$$r_{h,2} = \frac{\theta_{21} C_{S_1}}{1 + \theta_{22} C_{S_1}} \qquad \text{equation 16}$$

denoted as Model 2, which corresponds to a Ping Pong Bi Bi mechanism controlled by the rate of deacylation of the enzyme, and

$$r_{h,1} = \theta_{11} C_{S_1} \qquad \text{equation 17}$$

denoted as Model 1, which corresponds to a Ping Pong mechanism controlled by the rate of acylation of the enzyme if it exhibited very low affinity for the substrate. (In all the above situations, the lumped kinetic parameters can be either positive or negative and are of the form $\theta_{ij} \equiv \theta_{ij}\{v_{max,f}, v_{max,r}, K_{m,1}, K_{m,2}, K_{m,3}, K_{m,4}, K_{i,1}, K_{i,4}, C_{S_1,0}, C_{S_2,0}, C_{S_3,0}, C_{S_4,0}\}$.)

Combining equations 6 and 15-17, one obtains

$$\frac{\partial C_{S_1}}{\partial t} + v_z \frac{\partial C_{S_1}}{\partial z} + a_t\{t\}\, r_h\{C_{S_1}\} = 0 \qquad \text{equation 18i}$$

$$C_{S_2} = C_{S_2,sat} \qquad \text{equation 18ii}$$

$$C_{S_3} = C_{S_1,0} + C_{S_3,0} - C_{S_1} \qquad \text{equation 18iii}$$

$$C_{S_4} = C_{S_{1,0}} + C_{S_{4,0}} - C_{S_1} \qquad \text{equation 18iv}$$

Recalling the chain differentiation rule, equation 18i can be transformed into

$$\frac{\partial\left(\displaystyle\int \frac{dC_{S_1}}{r_h\{C_{S_1}\}}\right)}{\partial t} + v_z \frac{\partial\left(\displaystyle\int \frac{dC_{S_1}}{r_h\{C_{S_1}\}}\right)}{\partial z} + a_t\{t\} = 0 \qquad \text{equation 19i}$$

coupled with the initial and boundary conditions:

$$t = 0\,,\ 0 \leq z \leq L\ ,\ C_{S_1} = C_{S_{1,0}} \qquad \text{equation 19ii}$$

$$t \geq 0\,,\ z = 0\ ,\ C_{S_1} = C_{S_{1,0}} \qquad \text{equation 19iii}$$

Equation 19 has the general solution

$$\Phi_i\left\{C_{S_1},C_{S_{1,0}},\bar{\theta}\right\} = \Psi_j\left\{t,\bar{\lambda}\right\}\ ,\ t \leq \tau \qquad \text{equation 20i}$$

$$\Phi_i\left\{C_{S_1},C_{S_{1,0}},\bar{\theta}\right\} = \Psi_j\left\{t,\tau,\bar{\lambda}\right\}\ ,\ t \geq \tau \qquad \text{equation 20ii}$$

For each model of lipase-catalyzed hydrolysis, the functions Φ_i are defined as:

$$\Phi_1\left\{C_{S_1},C_{S_{1,0}},\bar{\theta}\right\} \equiv \frac{1}{\theta_{11}} \ln\left\{\frac{C_{S_{1,0}}}{C_{S_1}}\right\} \qquad \text{equation 21i}$$

for Model 1,

$$\Phi_2\left\{C_{S_1},C_{S_{1,0}},\bar{\theta}\right\} \equiv \frac{1}{\theta_{21}} \ln\left\{\frac{C_{S_{1,0}}}{C_{S_1}}\right\} + \frac{\theta_{22}}{\theta_{21}}\left(C_{S_{1,0}} - C_{S_1}\right) \qquad \text{equation 21ii}$$

for Model 2, and

$$\Phi_3\left\{C_{S_1},C_{S_{1,0}},\bar{\theta}\right\} \equiv \frac{1}{\theta_{31}} \ln\left\{\frac{C_{S_{1,0}}}{C_{S_1}}\right\} + \frac{\theta_{32}}{\theta_{31}}\left(C_{S_{1,0}} - C_{S_1}\right) + \ldots$$

$$\ldots \frac{\theta_{33}}{2\,\theta_{31}}\left(C_{S_{1,0}}^2 - C_{S_1}^2\right) \qquad \text{equation 21iii}$$

for Model 3. For each model of deactivation/rearrangement of lipase, the functions Ψ_j are defined as:

$$\Psi_1\left\{t,\bar{\lambda}\right\} \equiv \frac{1}{\lambda_{11}}\left(1 - \exp\left\{-\lambda_{11}\, t\right\}\right) \qquad \text{equation 22i}$$

for Model 1,

$$\Psi_2\left\{t,\bar{\lambda}\right\} \equiv \frac{\lambda_{21}}{\lambda_{22}}\left(1 - \exp\left\{-\lambda_{22}\, t\right\}\right) + \left(1-\lambda_{21}\right) t \qquad \text{equation 22ii}$$

for Model 2, and

$$\Psi_3\left\{t,\bar{\lambda}\right\} \equiv \frac{\lambda_{31}}{\lambda_{32}}\left(1 - \exp\left\{-\lambda_{32}\, t\right\}\right) + \frac{1 - \lambda_{31}}{\lambda_{31}}\left(1 - \exp\left\{-\lambda_{33}\, t\right\}\right) \qquad \text{equation 22iii}$$

for Model 3 during the transient period, as well as

$$\Psi_1\left\{t,\tau,\bar{\lambda}\right\} \equiv \frac{1}{\lambda_{11}}\left(\exp\left\{-\lambda_{11}\,(t-\tau)\right\} - \exp\left\{-\lambda_{11}\, t\right\}\right) \qquad \text{equation 23i}$$

for Model 1,

$$\Psi_2\left\{t,\tau,\bar{\lambda}\right\} \equiv \frac{\lambda_{21}}{\lambda_{22}}\left(\exp\left\{-\lambda_{22}\,(t-\tau)\right\} - \exp\left\{-\lambda_{22}\, t\right\}\right) + \left(1-\lambda_{21}\right) \tau \qquad \text{equation 23ii}$$

for Model 2, and

$$\Psi_3\left\{t,\tau,\bar{\lambda}\right\} \equiv \frac{\lambda_{31}}{\lambda_{32}}\left(\exp\left\{-\lambda_{32}\,(t-\tau)\right\} - \exp\left\{-\lambda_{32}\, t\right\}\right) + \ldots$$

equation 23iii

$$\ldots \frac{1 - \lambda_{31}}{\lambda_{31}}\left(\exp\left\{-\lambda_{33}\,(t-\tau)\right\} - \exp\left\{-\lambda_{33}\, t\right\}\right)$$

for Model 3 during the post-transient period.

The functions $\Psi_j\{t,\tau\}$ play the role of a reactor corrected space time, i.e., the mean residence time in the reactor that would lead to the same degree of hydrolysis if the enzyme activity remained constant at its initial value throughout time. It is interesting to note that the relationships $\Phi_i\{C_{S_1}\}=\Psi_j\{t,\tau\}$ defined in equations 21 and 23 are also the solutions of the ordinary differential equation (ODE)

$$\frac{dC_{S_1}}{dt^*} + a_{t,j}\left\{t^*\right\} r_{h,i}\left\{C_{S_1}\right\} = 0 \qquad \text{equation 24i}$$

where the modified time is defined as $t^* \equiv t+\tau$, provided that the following initial condition is employed:

$$t^* = t - \tau\,,\; C_{S_1} = C_{S_{1,0}} \qquad \text{equation 24ii}$$

For the case of a multisubstrate approach, equation 24i takes the form

$$\frac{dC_{S_1^i}}{dt^*} + a_t\left\{t^*,\overline{\lambda}\right\} r_{h_i}\left\{\overline{C_{S_1}},\overline{\theta_1},\overline{\theta_2},\overline{\theta_3}\right\} = 0 \quad , \quad i=1,2,...,10 \qquad \text{equation 25i}$$

$$t^* = 0\, ,\, \overline{C_{S_1}} = \overline{C_{S_{1,0}}} \qquad \text{equation 25ii}$$

where S_1^i denotes a glyceride containing the i-th type of fatty acid moiety and $\overline{C_{S_1}}$ is the vector of concentrations of all types of glycerides. The above set of ordinary differential equations can be rewritten as

$$\frac{dC_{S_1^i}}{dt^*} + a_t\left\{t^*,\overline{\lambda}\right\} \frac{\theta_1^i\, C_{S_1^i}}{r_h^*\left\{\overline{C_{S_1}},\overline{\theta_2},\overline{\theta_3}\right\}} = 0 \quad , \quad i=1,2,...,10 \qquad \text{equation 26i}$$

Hence,

$$\frac{dC_{S_1^i}}{\theta_1^i\, C_{S_1^i}} = - \frac{a_t\left\{t^*,\overline{\lambda}\right\}}{r_h^*\left\{\overline{C_{S_1}},\overline{\theta_2},\overline{\theta_3}\right\}}\, dt^* \quad , \quad i=1,2,...,10 \qquad \text{equation 27i}$$

which can be integrated with the help of the following set of initial conditions:

$$C_{S_1^j} = C_{S_{1,0}^j}\, ,\, C_{S_1^i} = C_{S_{1,0}^i} \quad ; \quad i=1,2,...,j-1,j+1,...,10 \qquad \text{equation 27ii}$$

(where j denotes an arbitrarily chosen reference substrate). This implies for a situation where 10 different responses (i.e., the overall concentrations of butyric, caproic, caprylic, capric, lauric, myristic, linoleic, palmitic, oleic, and stearic acid residues in butterfat) are measured in each experiment that the set of ten ODEs denoted as equation 26 can be replaced by the following set of one ODE and nine algebraic equations, viz.

$$\frac{dC_{S_1^j}}{dt^*} + a_t\left\{t^*,\overline{\lambda}\right\} \frac{\theta_1^j\, C_{S_1^j}}{r_h^*\left\{C_{S_1^j},\overline{C_{S_{1,0}}},\overline{\theta_2},\overline{\theta_3}\right\}} = 0 \qquad \text{equation 28i}$$

$$C_{S_1^i} = C_{S_{1,0}^i}\left(\frac{C_{S_1^j}}{C_{S_{1,0}^j}}\right)^{\theta_1^i/\theta_1^j} \quad ; \quad i=1,2,...,j-1,j+1,...,10 \qquad \text{equation 28ii}$$

Sets of mixed algebraic and ordinary differential equations as the one above can be easily handled by the DDASAC program [27]. This algorithm calculates the states and first order sensitivity coefficients as functions of time using an implicit integrator especially designed to handle stiff systems; the predictor-corrector integration is based on Gear's variable-order, variable-step method.

As seen below, the uniresponse analyses indicate that the best fit is obtained when Model 2 is used to describe the lipase-catalyzed hydrolysis; hence, this model will be considered hereafter. Since even in this situation the number of parameters (10 values of θ_1^i, 10 values of θ_2^i, and 2 values of λ) is excessively large for an efficient fitting strategy, a dramatic reduction must be effected otherwise most of the parameters will remain statistically indeterminate. One way to achieve this goal is to assume that the distribution of the θ_1^i and θ_2^i is given by

$$\theta_1^i = \frac{\alpha}{\sigma} \exp\left\{-\frac{(n_{C,i} - \chi\, n_{D,i} - \mu)^2}{2\,\sigma^2}\right\} \qquad \text{equation 29i}$$

$$\theta_2^i = \beta \qquad \text{equation 29ii}$$

where $n_{C,i}$ and $n_{D,i}$ are the number of carbon atoms and double bonds, respectively, of the hydrocarbon backbone of the fatty acid moiety in question. Equation 29 corresponds physically to assuming that the values of parameter θ_1^i exhibit a bell-shaped distribution centered at μ with variance σ, where, for the purpose of fitting, each double bond is equivalent to χ extra carbon atoms (α and β are constants). This postulated Gauss-type distribution can be explained in terms of the induced fit theory [11] in the following way: the essentially inert form of the enzyme in the absence of substrate acquires a catalytically active form upon binding of a substrate molecule with the optimal degree of hydrophobicity; the probability that substrates with lower or higher degrees of hydrophobicity will be able to induce optimal fits on the enzyme will be lower as the deviations with respect to optimality increase (in either direction).

3. Statistical considerations

A general nonlinear uniresponse regression model can in our case be formulated as

$$\bar{y} = \bar{\eta}\left\{\bar{t^*}, \bar{\lambda}, \bar{\theta}\right\} + \bar{z} \qquad \text{equation 30}$$

where $\bar{y}$ is the vector of N experimental observations, $\bar{\eta}$ is the vector of N expected responses, and $\bar{t^*}$ is the vector of modified times. The disturbance vector, $\bar{z}$, is assumed to have a spherical normal distribution, viz. $E\{\bar{z}\}=0$ and $E\{\bar{z}.\bar{z}^T\}=\sigma^2$ (where σ denotes a variance). The criterion for the best estimates of the parameters corresponds to minimizing the residual sum of squares, $||\bar{z}||^2$.

If M responses are obtained in each experiment, the nonlinear multiresponse regression model can be written as

$$\bar{Y} = \bar{H}\left\{\bar{t^*}, \bar{\lambda^*}, \bar{\theta^*}\right\} + \bar{Z} \qquad \text{equation 31}$$

where $\bar{Y}$ is the NxM observation matrix, $\bar{H}$ is the NxM expected response matrix, $\bar{Z}$ is the disturbance matrix, $\bar{\lambda^*}$ is the vector of parameters λ_{21} and λ_{22}, and $\bar{\theta^*}$ is the vector of parameters α, β, μ, σ, and χ.

The criterion for the best estimates of the parameters in multiresponse models will depend on the assumptions about the disturbance. Following Bard [12], the model used to describe the disturbance term is a normal distribution with $E\{\bar{Z}\}=\bar{0}$ and $E\{\bar{Z}\,\bar{Z}^T\}=\sigma^2\,\bar{I}$ ($\bar{I}$ is the NxN identity matrix). In this case, the least squares criterion is appropriate and one should minimize the sum of the squared residuals of all NM responses, i.e., minimize $tr\{\bar{Z}^T\bar{Z}\}$. Following Box and Draper [13], the model used to describe the disturbance term is a normal distribution with $E\{Z_{nm}\}=0$ and $E\{Z_{nm}Z_{ri}\}=\{\Sigma\}_{mi}$ for n=r, or $E\{Z_{nm}Z_{ri}\}=0$ for n≠r (where $\bar{\Sigma}$ is a fixed MxM covariance matrix). That is, the measurements from different experiments are assumed to be independent but measurements from the same experiment are assumed to be correlated with each other. In this case, the maximum likelihood estimates are obtained by minimizing $|\bar{Z}^T\bar{Z}|$. In practice, the inference regions associated with the parameters using Box and Draper's method of multiresponse estimation are usually narrower than those obtained using Bard's method; however, this advantage is often offset by the fact that the former method is more prone to optimization problems than the latter (e.g., achievement of convergence depends critically on the initial estimates selected; the Hessian must be positive definite; and convergence to spurious optima may occur as a result of implicit mass balances constraining the data).

The estimation of the parameters can proceed to advantage using GREG, a software package for General REGression analysis of uni- and multiresponse data to given models [14] via minimization of the appropriate statistical objective function, S. The minimization of such objective function, i.e., $||\bar{z}||^2$ for uniresponse data, and $tr\{\bar{Z}^T\bar{Z}\}$ (or $|\bar{Z}^T\bar{Z}|$) for multiresponse data, is done by a modified Newton method, starting from the user's initial guess for the parameters. In each iteration, S is expanded as a local quadratic function S* over a feasible predefined region of the parameter vector in order to obtain a set of normal equations for the minimization of S*, followed by the search for a local minimum of S*. To complete an iteration, the resulting parameter corrections are tested, and reduced when necessary to obtain a descent of the true objective function S. These steps recur until a convergence criterion is met or until the quota of a maximum number of iterations is expended. In case linear dependencies arise, the system responds by using a linear combination of the responses as proposed elsewhere [15].

Traditionally, the information on the dependence of the rate of an enzyme-catalyzed reaction on the substrate concentration(s) is decoupled from the information on the rate of change of the enzyme activity with time; hence, data concerning substrate conversion via an enzyme-catalyzed reaction under conditions of negligible deactivation are fitted to a postulated rate expression prior to fitting data on the maximum rate of reaction obtained in long term experiments to appropriate deactivation models. In the present case, such approach is not feasible because, since the time scale associated with the conformational change in the lipase is of the order of magnitude of the time scale associated with the lipase-catalyzed modification of the substrate, these two phenomena can not be decoupled from a statistical standpoint. Therefore, the data on the release of free fatty acids were fitted to rate expressions derived from combinations of suitable mechanisms of enzyme deactivation and/or rearrangement with enzyme catalytic mechanisms.

The results of the uniresponse statistical fit as provided by the postconvergence report generated by GREG are tabulated in Tables 1i-iii as the best estimates for the parameters, the 95% marginal inference intervals associated thereof, and the corresponding normalized covariance matrix.

Table 1i. Results of the uniresponse regression analyses for the parameters involved in Models 1-3 for deactivation/rearrangement of the lipase coupled with Model 1 for the lipase-catalyzed hydrolysis.

ID[a]	value	95%MII[b]	NCM[c]			
λ_{11}	0.098 hr^{-1}	0.018	1			
θ_{11}	0.0052 hr^{-1}	0.0011	0.721	1		
λ_{21}	0.863	0.053	1			
λ_{22}	0.0109 hr^{-1}	0.0043	-0.426	1		
θ_{11}	0.118 hr^{-1}	0.024	0.174	0.678	1	
λ_{31}	0.47	0.32	1			
λ_{32}	0.0046 hr^{-1}	0.0012	0.075	1		
λ_{32}	0.19 hr^{-1}	0.42	-0.655	0.365	1	
θ_{21}	0.18 hr^{-1}	0.12	-0.944	0.196	0.792	1

Table 1ii. Results of the uniresponse regression analyses for the parameters involved in Models 1-3 for deactivation/rearrangement of the lipase coupled with Model 2 for the lipase-catalyzed hydrolysis.

ID[a]	value	95%MII[b]	NCM[c]				
λ_{11}	0.0329 hr^{-1}	0.0057	1				
θ_{21}	0.00616 hr^{-1}	0.00088	0.325	1			
θ_{22}	-0.486 $l.mol^{-1}$	0.026	0.643	-0.306	1		
λ_{21}	0.911	0.024	1				
λ_{22}	0.0107 hr^{-1}	0.0019	-0.457	1			
θ_{21}	0.0420 hr^{-1}	0.0059	-0.100	0.518	1		
θ_{22}	-0.473 $l.mol^{-1}$	0.021	-0.250	-0.035	0.626	1	
λ_{31}	0.89	0.17	1				
λ_{32}	0.0112 hr^{-1}	0.0043	-0.899	1			
λ_{32}	0.0005 hr^{-1}	0.0036	-0.987	0.874	1		
θ_{21}	0.0423 hr^{-1}	0.0063	-0.297	0.495	0.287	1	
θ_{22}	-0.474 $l.mol^{-1}$	0.021	0.058	-0.100	-0.101	0.561	1

Table 1iii. Results of the uniresponse regression analyses for the parameters involved in Models 1-3 for deactivation/rearrangement of the lipase coupled with Model 3 for the lipase-catalyzed hydrolysis.

ID[a]	value	95%MII[b]	NCM[c]					
λ_{11}	0.037 hr^{-1}	0.029	1					
θ_{31}	0.00616 hr^{-1}	0.00091	-0.128	1				
θ_{32}	-0.38 $l.mol^{-1}$	0.87	0.981	-0.211	1			
θ_{33}	-0.06 $l^2.mol^{-2}$	0.45	-0.977	0.202	-1	1		
λ_{21}	0.911	0.025	1					
λ_{22}	0.0107 hr^{-1}	0.0019	-0.438	1				
θ_{31}	0.0422 hr^{-1}	0.0023	-0.303	0.128	1			
θ_{32}	-0.47 $l.mol^{-1}$	0.55	-0.297	-0.008	0.972	1		
θ_{33}	0.00 $l^2.mol^{-2}$	0.28	0.288	0.006	-0.966	-0.999	1	
λ_{31}	0.89	0.18	1					
λ_{32}	0.0113 hr^{-1}	0.0046	-0.899	1				
λ_{33}	0.0005 hr^{-1}	0.0035	-0.985	0.876	1			
θ_{31}	0.043 hr^{-1}	0.025	-0.257	0.257	0.208	1		
θ_{32}	-0.453 $l.mol^{-1}$	0.58	-0.186	0.128	0.135	0.970	1	
θ_{33}	-0.01 $l^2.mol^{-2}$	0.30	0.188	-0.132	-0.139	-0.965	-0.999	1

[a] ID: Identification of Parameter; [b] MII: Marginal Inference Interval; [c] NCV: Normalized Covariance Matrix

The results of the multiresponse statistical fit generated in a similar way are tabulated in Table 2 as the best estimates for the parameters, the 95% marginal inference intervals associated thereof, and the corresponding normalized covariance matrix. In this case, and since the total concentrations of each type of fatty acid residue are quite dissimilar (they range from ca. 0.02 for C_8 to ca. 0.32 $mol.l^{-1}$ for C_{16}), the concentration of each fatty acid residue was normalized by its total concentration prior to fitting (i.e., the regression was performed in such a way as to force the model fit all the responses equally well in percent terms).

To decide which is the simplest nested model to fit a data set adequately, one may proceed as in the linear case and use a likelihood ratio test [16]; because of the spherical normal assumption, this leads to an assessment of the extra sum of squares due to the extra parameters involved in going from the partial to the full model. Letting S denote the sum of squares, N the number of observations, and P the number of parameters, with subscripts f and p for the full and partial (nested) models, respectively, one evaluates the F-ratio, F_r, defined as

$$F_r \equiv \frac{\frac{S_p - S_f}{P_f - P_p}}{\frac{S_f}{N - P_f}} \qquad \text{equation 32}$$

and compares it to Fisher's distribution at the α significance level, viz. $F\{P_f-P_p, N-P_f; \alpha\}$. The partial model should be accepted if the calculated mean square ratio is lower than the tabulated value of the *F*-distribution; otherwise, the extra terms of the full model should be retained. Examples of the application of this technique are illustrated in Tables 3i-ii. Although this type of analysis is only approximate, the distribution of F_r is only affected by intrinsic nonlinearity and not by parameter effects nonlinearity; the former is usually very small (except for highly inadequate models, which would be rejected anyway on the basis of general inadequacy of the fit) [17].

Table 2. Results of the multiresponse regression analyses for the parameters involved in Model 2 for deactivation/rearrangement of the lipase coupled with Model 2 for the lipase-catalyzed hydrolysis and a Gauss distribution for the θ_1^i parameters.

ID[a]	value	95%MII[b]	NCM[c]						
α	0.466 hr^{-1}	0.051	1						
μ	10.13	0.48	-0.266	1					
σ	10.3	1.1	0.803	-0.363	1				
χ	5.65	0.84	0.122	-0.337	0.089	1			
β	-0.4801 $l.mol^{-1}$	0.0082	0.313	-0.008	-0.032	-0.002	1		
λ_{21}	0.9185	0.0098	-0.040	-0.004	-0.009	-0.001	-0.268	1	
λ_{22}	0.01173 hr^{-1}	0.00093	0.320	0.000	-0.001	0.000	-0.047	-0.387	1

[a] ID: Identification of Parameter; [b] MII: Marginal Inference Interval; [c] NCV: Normalized Covariance Matrix

Inspection of Table 3i indicates that the models which provide the statistically best fit are Models 12, 22, and 32 (which correspond to Model 2 for the deactivation/rearrangement reaction coupled with every model for the lipase-catalyzed hydrolysis reaction). Inspection of Table 3ii indicates that, among Models 12, 22, and 32, the model which provides the statistically best fit is model 2 for deactivation/rearrangement coupled with model 2 for the lipase-catalyzed hydrolysis reaction.

The experimental data overlaid with the theoretical results as predicted by this combination of models are plotted in Figure 2 as the percent conversion of the lumped glycerides versus the reactor corrected space time.

The experimental data overlaid with the theoretical results as predicted by a similar combination of models for multiresponse data (on the assumption of a Gauss distribution of parameters θ_1^i) are plotted in Figure 3 as the fractional conversion of the glycerides containing caprylic (i) and lauric (ii) acid moieties versus the reactor corrected space time. (These acids were selected because the first has been implicated with the typical flavor of Cheddar cheese whereas the second has been associated with soapy off-flavors in cheese.)

Table 3i: Extra residual sum of squares analysis between the various rate expressions for the lipase-catalyzed hydrolysis (first digit) and deactivation/rearrangement of the lipase (second digit).

Nested models	Source	SSR[a]	DF[b]	MSSR[c]	F_r	*p*-value[d]
	extra parameters	0.357	1	$3.57x10^{-1}$	11.3694	0.001918
	rate expression 12	1.036	33	$3.14x10^{-2}$		
11⊂12						
	rate expression 11	1.393	34			
	extra parameters	0.011	1	$1.10x10^{-2}$	0.3438	0.561758
	rate expression 13	1.025	32	$3.20x10^{-2}$		
12⊂13						
	rate expression 12	1.036	33			
	extra parameters	0.1775	1	$1.775x10^{-1}$	40.488	<0.000001
	rate expression 22	0.1403	32	$4.384x10^{-3}$		
21⊂22						
	rate expression 21	0.3178	33			
	extra parameters	0.0004	1	$4.000x10^{-4}$	0.0886	0.767950
	rate expression 23	0.1399	31	$4.513x10^{-3}$		
22⊂23						
	rate expression 22	0.1403	32			
	extra parameters	0.1769	1	$1.769x10^{-1}$	39.1458	<0.000001
	rate expression 32	0.1401	31	$4.519x10^{-3}$		
31⊂32						
	rate expression 31	0.3170	32			
	extra parameters	0.0005	1	$5.000x10^{-4}$	0.1075	0.745286
	rate expression 33	0.1396	30	$4.653x10^{-3}$		
32⊂33						
	rate expression 32	0.1401	31			

The predicted curves plotted in Figures 3i-ii coincide because the substrates in question, C_8 and C_{12}, have the same n_D values, and n_C values virtually equidistant from μ.

Table 3ii: Extra residual sum of squares analysis between the various rate expressions for the lipase-catalyzed hydrolysis (first digit) and deactivation/rearrangement of the lipase (second digit).

Nested models	Source	SSR[a]	DF[b]	MSSR[c]	F_r	*p*-value[d]
12⊂22	extra parameters	0.8957	1	8.957×10^{-1}	204.3111	<<0.000001
	rate expression 22	0.1403	32	4.384×10^{-3}		
	rate expression 12	1.036	33			
22⊂32	extra parameters	0.0002	1	2.000×10^{-4}	0.0443	0.834674
	rate expression 32	0.1401	31	4.519×10^{-3}		
	rate expression 22	0.1403	32			

[a] SSR: sum of squares of residuals; [b] DF: number of degrees of freedom; [c] MSSR: mean sum of squares of residuals; [d] *p*: probability that the given value of the *F*-ratio happens by pure chance: e.g., 0.0443 = *F*(1, 31; 0.834674).

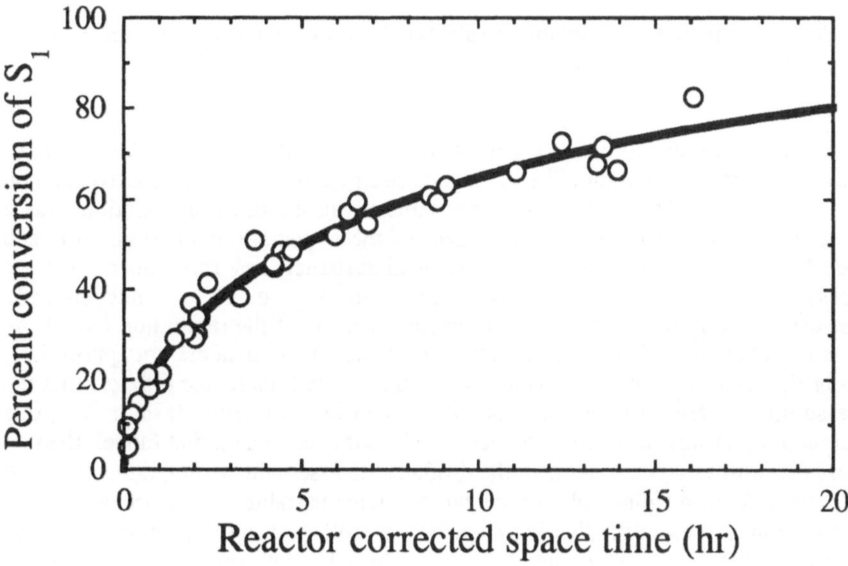

Figure 2. Plot of the percent conversion of substrate S_1 versus the reactor corrected space time.

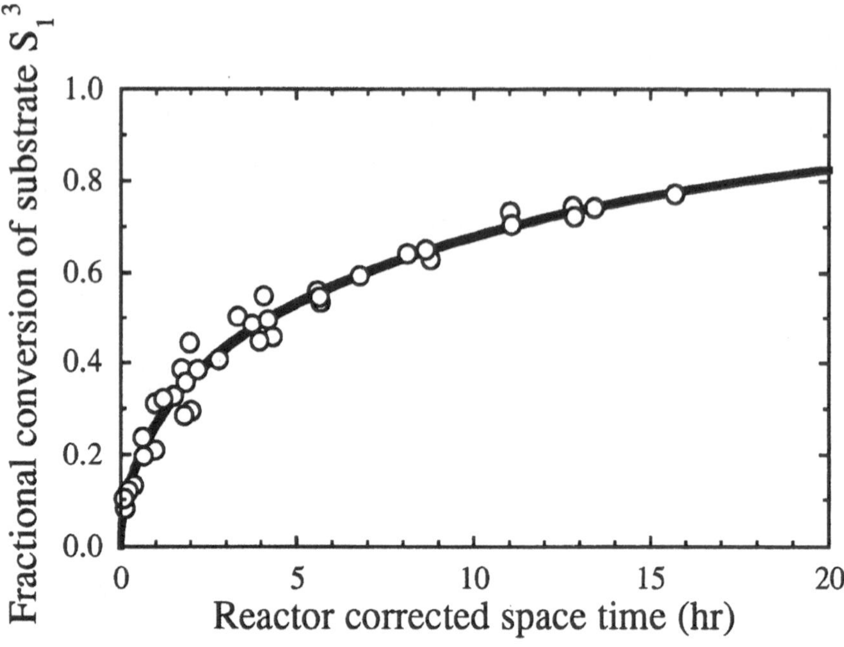

Figure 3i. Plot of the percent conversion of substrate S_1^3 versus the reactor corrected space time.

There is a number of statistical assumptions which lead to the use of the least squares analysis as a basis for the maximum likelihood parameter estimates, viz. (i) the expectation function is correct, (ii) the response is exprimable as an expectation function plus a disturbance, (iii) the disturbance is independent of the expectation function, (iv) each disturbance has a normal distribution, (v) each disturbance has zero mean, (vi) the disturbances have equal variances, and (vii) the disturbances are distributed independently [17]. These assumptions encompass several different aspects of the regression model. As with any statistical analysis, if the assumptions on the model and data are not appropriate, the results of the analysis will not be valid. Since one can not guarantee *a priori* that the different assumptions are all valid, one should proceed in an iterative fashion; i.e., one should entertain a plausible model for the data, analyze the data using that model, then go back and use diagnostics (such as plots of the residuals) to assess the assumptions. Since all diagnostics made (e.g., residuals plotted vs. the predictor variable, vs. the quantiles of a normal distribution, and vs. each other in a sequence, as well as visual inspection of the fits) did not indicate failure of the assumptions in either the deterministic or stochastic components, the models employed to obtain the predicted curves in Figures 2 and 3 should then be accepted as statistically correct.

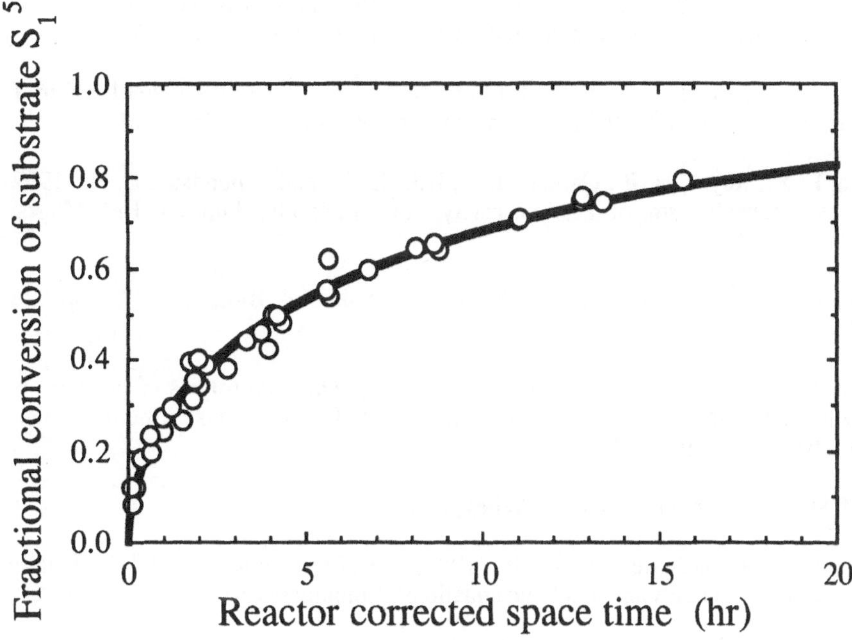

Figure 3ii. Plot of the percent conversion of substrate S_1^5 versus the reactor corrected space time.

4. Acknowledgements

Fruitful discussions on the reasonings reported above with Profs. Charles G. Hill, Jr. and Clyde H. Amundson (c/o University of Wisconsin - Madison, USA) are gratefully acknowledged.

5. References

1. Keniston, M., Novakovic, A., and Cosgrove, T. (1990). 'A complex game of supply and demand', Dairy Foods 91(11): 80-82.

2. Kilara, A. (1985) 'Enzyme-modified lipid food ingredients', Process Biochem. 20: 33-45.

3. Nelson, J. H. (1972) 'Enzymatically produced flavors for fatty systems', J. Am. Oil Chem. Soc. 49: 559-562.

4. Lindsay, R. C. (1988) 'Milkfat as a source of concentrated dairy flavors', paper presented in the CDR Cheese Research Conference, Madison, Wisconsin, USA.

5. Godfrey, T. and Reichelt, J. R. (1983) 'Introduction to industrial enzymology' in T. Godfrey and J. Reichelt (eds.), Industrial Enzymology, Nature Press, New York, p.1

6. Malcata, F. X., Reyes, H. R., Garcia, H. S., Hill, C. G., and Amundson, C. H. (1992) 'The kinetics and mechanisms of reactions catalyzed by immobilized lipases', Enz. Microb. Technol. 14: 426-446.

7. Henley, J. P. and Sadana, A. (1986) 'Deactivation theory', Biotechnol. Bioeng. 28: 1277-1285.

8. Malcata, F. X., Hill, C. G., and Amundson, C. H. (1992) 'Hydrolysis of butteroil by immobilized lipase using a hollow fiber reactor: Part II: Uniresponse kinetic studies', Biotechnol. Bioeng. 39: 984-1001.

9. Segel, I. H. (1975) Enzyme Kinetics, Wiley, New York.

10. Caracotsios, M. and Stewart, W. E. (1985) 'Sensitivity analysis of initial value problems with mixed ODE's and algebraic equations', Computers Chem. Eng. 9: 359-365.

11. Jencks, W. P. (1969) Catalysis in Chemistry and Enzymology, Dover, New York.

12. Bard, Y. (1974) Nonlinear Parameter Estimation, Academic Press, New York.

13. Box, G. E. P. and Draper, N. R. (1965) 'The Bayesian estimation of common parameters from several responses', Biometrika 52: 355-365.

14. Stewart, W. E., Caracotsios, M., and Sørensen, J. P. (1992) GREG Software Package Documentation, Department of Chemical Engineering, University of Wisconsin, USA.

15. Box, G. E. P., Hunter, W. G., MacGregor, J. F., and Erjavec, J. (1973) 'Some problems associated with the analysis of multiresponse models', Technometrics 15: 33-51.

16. Draper, N. R. and Smith, H. (1981) Applied Regression Analysis, Wiley, New York.

17. Bates, D. M. and Watts, D. G. (1988) Nonlinear Regression Analysis and its Applications, Wiley, New York.

PROTEIN EXTRACTION BY REVERSED MICELLAR SYSTEMS

M.R. AIRES-BARROS and J. M. S. CABRAL
Laboratorio de Engenharia Bioquimica
Instituto Superior Tecnico
1000 Lisboa-Portugal

ABSTRACT. This chapter describes the concepts involved in the solubilization of proteins in organic solvents via reversed micelles. The characterization of reversed micellar systems is described and the parameters that affect protein solubilization and phase transfer are discussed. Applications of reversed micellar systems as a bioseparation technique for proteins are analysed and three case studies are described: selective separation of microbial lipases, recovery of an alkaline protease from whole fermentation broth and extraction of recombinant cytochrome b_5.

1. Introduction

Liquid-liquid extraction has been employed in many different sectors of the chemical industries but has found a limited application for separation, concentration and purification of proteins. One reason for this is the lack of suitable solvents having the desired selectivity and capacity, and which are not harmful to the proteins.

These problems can be overcome using liquid-liquid extraction techniques with reversed micelles solutions as extractants. In these systems the proteins are solubilized inside the polar core of surfactant aggregates and stabilized in the organic solvent by a surfactant shell layer, that protects the biomaterials from denaturation against the organic phase. Reversed micellar systems have shown particular promise in the recovery of bioproducts by solvent extraction (1-6).

A.R. Moreira and K.K. Wallace (eds.), Computer and Information Science Applications in Bioprocess Engineering, 221-235.

2. Characterization of the reversed micellar system

Reversed micelles are aggregates of surfactants in apolar media. The surfactant molecules are amphiphiles, that can be arranged in a way that the hydrophobic parts are in contact with the apolar bulk solution and the polar head groups are turned towards the interior of the aggregate, forming a polar inner core. Most of the work done with proteins in reversed micelles, uses the anionic surfactant AOT (sodium bis [2-ethylhexyl] sulphosuccinate), which aggregates spontaneously in hydrocarbon solvents, forming water pools with radii over 170Å (7).

Usually, the reversed micellar system is a ternary system of water, organic solvent and surfactant. Extra components are added when it is advantageous or necessary. In the case of cationic surfactants, a cosurfactant is introduced, usually an aliphatic alcohol, essential for the stabilization of the reversed micelles, that partitions among the micelle interphase and the continuous phase. Depending on the relative concentrations of the various components of the system, phase structure can change dramatically, including normal and reversed micelles, liquid crystals and lamellar phases, protein extraction being performed when phase separation occurs. Figure 1 shows the phase diagram for the ternary system AOT/isooctane/water.

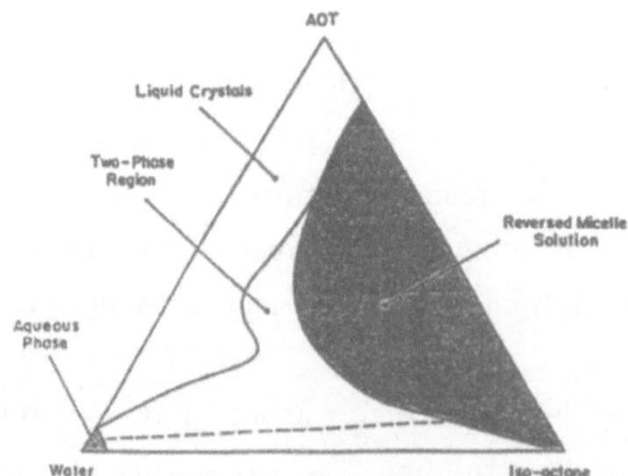

Figure 1- Phase diagram of AOT/water/isooctane system

Usually, in the forward transfer of proteins, the micellar phase does not contain water, except for the residual water. The amount of water transfered into reversed micelles during the mixing of the two phases is low (8), (w_0 < 20; w_0= $[H_2O]/[surfactant]$), depending primarily on the salt concentration of the aqueous phase.

3. Protein solubilization in reversed micelles

Although the phase transfer of proteins in biphasic system with reversed micelles is being exploited as a means of recovering proteins from aqueous media, the kinetic and thermodynamic aspects of the process have not been clarified. The driving forces responsible for the phase transfer, as well as the structure of the protein-containing micelles are still not completely understood. Apparently, there is not a defined limit for the size of the hosted molecules, as suggested by the solubilization in reversed micelles of large biologic entities such as nucleic acids (9), plasmids (10) and bacterial cells (11). Experimental evidence seems to support the prediction that hydrophilic proteins are hosted within the water core of he reversed micelles (12), while lipophilic macromolecules can be located in the interphase or partially exposed to the organic solvent (5).

3.1 Factors affecting phase transfer

The distribution of proteins between a micellar organic phase and an aqueous solution is largely determined by the conditions in the aqueous bulk phase, namely pH, ionic strength and type of salt. Other parameters related to the organic phase, also influence the partition pattern of a protein, like the concentration and type of surfactant, presence of co-surfactant and type of solvent. Changes in temperature can also affect the solubilization of biomolecules.

3.1.1 pH

The pH of the aqueous solution determines the net charge of proteins. Electrostatic interactions between the protein and the surfactant head groups can favour the transfer of protein into the organic phase. This trend has been observed using anionic (2) and cationic surfactants (3,13), solubilization occurring only when the protein and the surfactant head groups have oposite charges. However, this is not always observed, and Luisi et al (14) obtained identical pH-profiles for α-chymotrypsin (pI=8.3) and pepsin (pI<1.1) using TOMAC, a cationic surfactant in cyclohexane. Similar results were reported by Goklen (15) with another cationic surfactant, didodecyldimethyl ammonium bromide (DDAB) and a mixture of proteins with isoelectric points ranging from 5.5 to 7.8. In both studies using quaternary ammonium salts, the transfer of proteins into the reversed micellar phase is more effective in the alkaline pH range (less protonated residues on the protein), which could be explained by ion-pairing between the anionic surfaces of the protein and the positively-charged surfactant. Luisi et al (14) found sharp decreases in the percentages of transfer at pH higher than 13 for α-chymotrypsin and pepsin, attributed to competition of the biopolymers with OH^- ions. Protein denaturation and changes in the ionization state of the surfactant at extreme pH values make additionally difficult the interpretation of the phase-transfer pattern.

3.1.2 Ionic strength

Increasing ionic strength of the aqueous phase will reduce electrostatic interactions as a result of the Debye screening effect, stronger for larger ions. Thus, at a higher ionic strength, the interactions between hydrophilic biomolecules and the surfactant polar groups are reduced, smaller micelles being formed. As a consequence of this effect of increasing the ionic strength of the aqueous solution, the solubilization capacity of the organic phase for water and biomolecules will decrease. Complete solubilization of three proteins, cytochrome-c, lysozyme and ribonuclease-a, using an AOT/isooctane system, was obtained at low ionic strength (0.1 M KCl) and no solubilization occurs at high ionic strength (1.0 M KCl). The inversion of the partition coeficients is achieved in a very narrow range of pH, that depends on the protein species and does not correlate with the isoelectric point of the protein. Ribonuclease-a is the protein with the lower isoelectric point, but the sharp change of solubilization behaviour occurs at a KCl concentration (0.5 M) lower than the case of lysozyme (0.7 M). The importance of the particular electric characteristics on the protein surface is suggested by these results. Participation of electrostatic interactions in phase transfer is supported by the narrowing of the pH peaks, when the ionic strength is increased, as observed for the extraction of cytocrome-c (15). These results, found with anionic surfactant AOT, are consistent with the phase transfer behaviour of α-amylase (13) in a system including TOMAC. Additional effects of the ionic strength in the solubilization of biomolecules in reversed micellar biphasic systems, are competition with ionic species for transfer into the reversed micelles and changes in the electrostatic state of the micelles and/or proteins, more likely to occur at high ionic strength.

It has been notice that when the ionic strength of the aqueous phase is below a certain limit, reversed micelles and phase separation do not occur, a stable macroemulsion being formed. Therefore, the transfer of proteins between phases requires a minimum value for the ionic strength of the aqueous solution. A minimum concentration of KCl around 0.1 M was found to achieve quantitative transfer of cytocrome-c into AOT/isooctane system (8). However, if the aqueous solution is buffered, the buffer can supply enough amount of electrolyte.

3.1.3 Type of electrolyte

The salt composition of the aqueous phase can also affect the efficiency of transfer of proteins between phases, apart the expected different contribution to the ionic strength of the solution. Specific interactions of the protein and/or the surfactant with the ions in solution can account for a change in the solubilization pattern of proteins at given conditions, in the presence of different ionic species (15-17). The buffering system itself can influence the solubilization behaviour of proteins (15).

3.1.4 Surfactant concentration

The solubility capacity of the organic phase changes directely with the surfactant concentration, possibly due to changes in electrostatic interactions. Increasing the surfactant concentration favours protein solubilization in the organic phase, resulting in broader pH-solubilization peaks (15). On the other hand, higher surfactant concentrations difficult the backward transfer of proteins into a second aqueous phase (18). Therefore, the optimum surfactant concentration in a double-phase transfer corresponds to the minimum limit for achieving maximum transfer into the organic phase.

3.1.5 Other factors affecting reversed micelle structure

The type of organic solvent affects the size of the reversed micelles and consequently the water solubilization capacity (15). A different micellar structure resulting from the solvent effect can be exploited to significant changes in the solubilization of α-chymotrypsin with different solvents. When comparing the results obtained with several proteins, in a system including cyclohexane as organic solvent, the transfer efficiences were also very different.

The introduction of a co-surfactant can improve the solubility capacity of the organic phase (3), probably due to an increase in the micelle size, when using cationic surfactants.

Changes in temperature have drastic effects on the physico-chemical properties of the reversed micellar system. Increasing the temperature can enhance the solubility of proteins in the organic phase. Luisi et al (14) obtained an increase of 50% in the transfer yield of α-chymotrypsin into a NH_4^+/chloroform phase.

4. Use of biospecific ligands

Significant enhancements in the selectivity of the protein extraction process with reversed micellar systems can be achieved by introducing affinity ligands in the organic phase (19). It was obtained an increased partitioning in the transfer of concanavalin-a into reversed micelles if the biosurfactant octyl β-D-glycopyran (2-20% of total surfactant concentration) was included in the organic phase. The transfer of the control protein, ribonuclease-a was not affected by the presence of the biological detergent, while glucose inhibited the solubilization of concavalin-a due to competition with the ligand for two binding sites in the protein. High values for protein/ligand binding constants in the organic phase, ligand partition coefficient between the organic and the aqueous phases and number of binding sites in the protein favour the solubilization of proteins in reversed micelles, whereas a low protein/ligand binding constant in the aqueous solution is desired.

5. Stability of recovered proteins

The separation of proteins using reversed micelles requires that they are efficiently recovered, with retention of biological function. A criterious choice of the components of the reversed micellar system, namely surfactant, co-surfactant and organic solvent, is necessary to achieve maximal stability of the extracted proteins. The sufactant nature can have a strong effect on stability due to direct interactions with the hosted molecules. Surfactant concentration and temperature determine the rate of denaturation (20). Interactions between the proteins solubilized inside the reversed micelles and the co-surfactant can also be important.

Some proteins are not completely stripped from the organic phase by changing the ionic strength and the pH of the aqueous phase. An evident disadvantage of this method is that the organic phase can not be reused without solvent regeneration. If addition of polar solvents is used, the polar solvent should also be carefully chosen. Eremin and Metelitsa (20) investigated the stability of several enzymes after recovery of mixed reversed micelles of AOT and Triton X-45, with organic solvents, mixtures of acetone and water or with aqueous solutions containing urea, glycerol, tween 20 and ammonium sulphate. It was observed that the method and medium of extraction affect the retention of activity dramatically and that each enzyme retains the highest activity following a different procedure.

The stability of the micelle also affects the stabilization of the proteins. Unstable micellar aggregates promote the contact between the molecules solubilized in the inner aqueous space, simultaneously causing a higher contact between the proteins and the organic solvent, increasing the denaturation rate. Increased agitation during the extraction process, while decreasing mass transfer effects during the phase transfer, also contributes to increase the denaturation rate of biomolecules. The investigation of the factors that affect the cohesion of the micellar aggregate is extremely important to optimize the extraction efficiency.

6. Case Studies

6.1 Separation and purification of two lipases from *Chromobacterium viscosum* using reversed micelles

Solubilization and selective separation of a lipolytic preparation from *Chromobacterium viscosum* containg two lipases, is described.

The separation and recovery of the two lipases consists of two steps: the extraction step in which equal volumes of the micellar solution and the buffered aqueous solution containing the lipases, were mixed; and the second step involved the back-extraction of the extracted lipase, in which the lipase-loaded micellar solution from the first step was mixed with a fresh aqueous buffered solution, with addition of ethanol to the mixed phases.

6.1.1 Effect of pH on lipases solubilization

The effect of pH on the protein transfer from the aqueous to the micellar solution, for the two lipases is shown in Figure 2.

At high pH values aproximately 50% of lipase B is solubilized, when the pH of the system is lower than the isoelectric point of lipase B (pI=7.3), the solubilization increases and 100% of lipase B is extracted into the micellar solution, at the lower pH values (5.2 to 6.5). At these pH values lipase B is positively charged and is easily solubilized in the AOT reversed micelles. The amount of solubilized lipase B decreases again as pH was reduced to 4.7. At this pH some denaturation of the protein occurred, and that the unfolded protein is not solubilized in reversed micelles. The water content for the resulting micellar solution containing lipase B was determined to be $w_0=20$.

Lipase A was not extracted into the reversed micellar solution at all the pH values tested as it is negatively charged (pI=3.7).

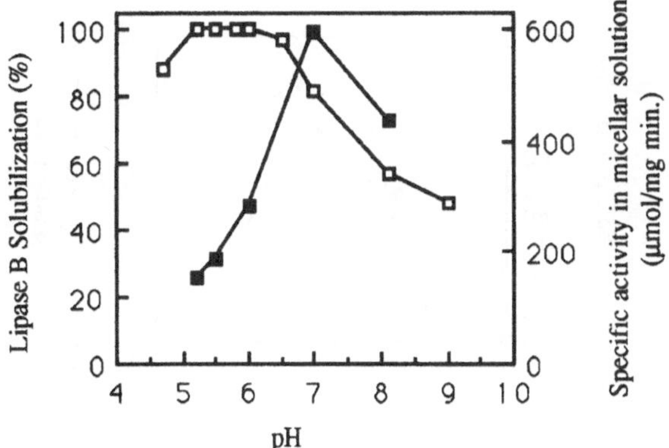

Figure 2- Effect of pH on the solubilization of lipase B from *Chromobacterium viscosum* in AOT (250 mM)/isooctane reversed micelles, at w_o=20, 50 mM phosphate buffer (pH 6.0) with 50 mM KCl and a phase ratio of 1.0: (□) lipase B extracted and (■) specific activity of lipase B in the micellar phase.

Protein solubilization seems to require an attractive electrostatic interaction between the micelle and the protein, and therefore pH values lower than the pI of a given protein are required to solubilize it in a micellar solution of an anionic surfactant. However, approximately, 50% of lipase B is solubilized at pH values higher than its pI, indicating that hydrophobic interactions between the lipase and the surfactant and/or organic solvent are also important. Lipase A is a relatively large protein and the size exclusion effect must be more important than the protein-surfactant interactions. Figure 2 also shows that lipase B remained active in the micellar phase, and the maximum specific activity was obtained at pH 7.0.

6.1.2 Recovery of lipase B from micellar solution

Lipase B was not recovered from the micellar solution into a new aqueous phase at all used pH and ionic strength variations, indicating a very strong-surfactant interactions. However, the addition of 2.5% by volume of ethanol, 85% of lipase B could be recovered into a fresh bufered aqueous phase (50 mM K_2HPO_4 with 50 mM KCl) at pH 6.0 (Figure 3).

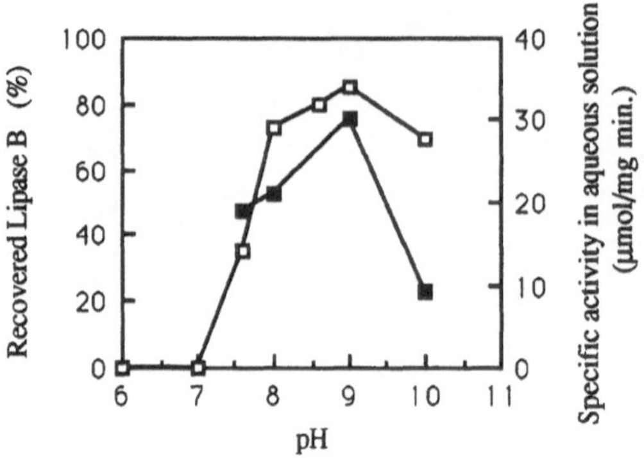

Figure 3- Recovery of lipase B as a function of pH of the aqueous phase (50 mM K_2HPO_4 with 50 mM KCl), at a wo value of 15, phase ratio of 1.0, and 2.5% (v/v) ethanol: (□) protein recovered and (■) specific activity in the aqueous phase.

At pH values lower than the isoelectric point of lipase B (pI=7.3), no recovery was achieved indicating that electrostatic interactions between the protein and the surfactant are dominant. At higher pH values the back-extraction of lipase B increased, reaching a maximum at pH 9.0. Without addition of ethanol, no recovery occurred as mentioned above. This suggests that other type of interactions, namely hydrophobic interactions between the protein and the hydrophobic tails of the surfactant and/or the solvent, play an important role on the protein solubilization (see Figure 2) and recovery. Figure 3 also shows the specific activity of the back-extracted lipase B in the aqueous solution, and the maximum specific activity (30 µmol/min mg) was obtained at pH 9.0. The results of the purification procedure are summarized in Table 1.

Table 1- Purification of lipases from *Chromobacterium viscosum* using AOT reversed micelles (6).

Lipases	Solubilized protein (%)	Back-extracted yield (%)	Recovery yield (%)	Purification factor
A	0	0	86	2.6
B	100	85	76	1.5

6.2 Recovery of an extracellular alkaline protease from whole fermentation broth using reversed micelles

Alkaline protease from *Bacillus* sp. was extracted (4) from a complex fermentation broth using an organic solution of AOT in isooctane. The enzyme has a molecular weigth of 33 KDa and a pI of 10.

The effect of the equilibrium forward transfer pH (the pH of the aqueous solution in contact with the micellar solution after the forward transfer) on the recovery of protein mass and activity is shown in Figure 4. These experiments were carried out at a pH of 8.0 for the back-transfer stripping solution.

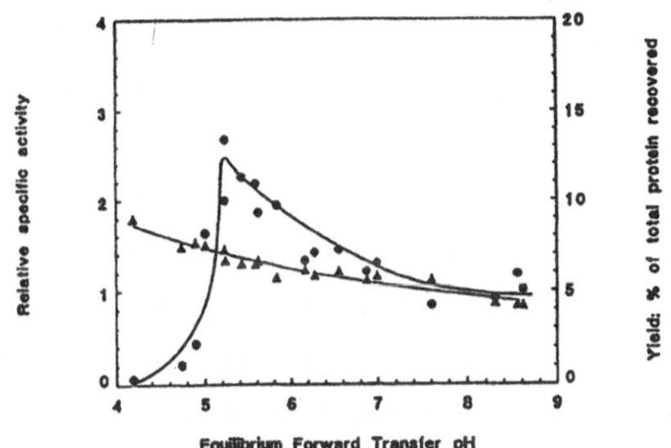

Figure 4- Recovery of alkaline protease from broth as a function of forward transfer pH: (●) relative specific activity and (▲) total protein yield (4).

As the forward transfer pH was lowered, the favourable interaction between the increasingly positively-charged proteins and the negatively charged surfactant heads was enhanced. This enhanced the protease transfer into the micellar phase and thus greater recovery of the protein in the extraction/stripping cycle. The relative specific activity also increased as the pH was lowered, reaching a maximum of 2.7 at pH 5.4. Below this pH, activity fell off sharply due to protein denaturation. As the pH increased and the protease charge decreased, the selectivity of the extraction for the protease was lost, the relative specific activity asymptotically approached a value of unity, and no purification was obtained.

The effect of variations in the equilibrium back transfer pH, using the optimum forward transfer pH of 5.4, is shown in Figure 5.

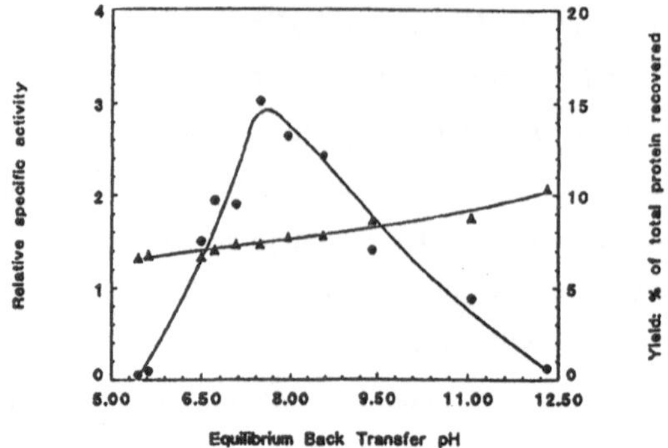

Figure 5- Recovery of alkaline protease from broth as a function of back-transfer pH: (●) relative specific activity and (▲) total protein yield (4).

At low back-transfer pH, the favourable electrostatic interaction between the protein and the micellar shell makes it less likely that the protease will leave the micellar phase, and as seen in Figure 5, the protein yield increased steadily as the equilibrium pH for the back-transer was increased. A maximum purification factor of 3.0 was obtained at a back transfer pH of 7.5. At low back-transfer pH, the pH was essentially that used for the forward-transfer, which lead to the transfer of the protease into and not out of the micellar phase, so the active protein remained in the micellar phase. At very high back-transfer pH values, protease activity again decreased. This decrease appears to be due to protein instability in the presence of the organic solvent at high pH values. Agitation of the protease alone, at very high pH, and for the time used for back-transfer, did not result in a decrease in activity. Agitation with either isooctane solvent alone, or the AOT/isooctane micellar system, at high pH, resulted in almost complete loss of activity.

6.3 Liquid-liquid extraction of a recombinant protein with a reversed micellar phase

Cytocrome b_5 from rat is a small protein (13,600 Da) with a low isoelectric point (pI 4.4). This protein was cloned into the plasmid pUC 13 and expressed in *Escherichia coli* TB1 (21). The recombinant protein can be extracted from an aqueous solution to a reversed micellar phase of 100 mM CTAB in cyclohexane/10% (v/v) decanol. The extraction depends on three main parameters: ionic strength, pH and temperature. If this process is mainly due to an electrostatic interaction between the protein and the reversed micelle, it should be expected a strong dependence on the pH of the aqueous solution, as it controls the overall charge of the protein, and on the ionic strength, since it controls the magnitude of its interaction. By controling these three parameters it was possible to observe a shift in the type of forces that are responsible for the solubilization of cytocrome b_5 in the micellar phase.

Figure 6 shows the extraction results obtained for cytocrome b_5 at ranges of pH between 4.4 and 7.0, and ionic strength between 0.1 and 1.0 M KCl.

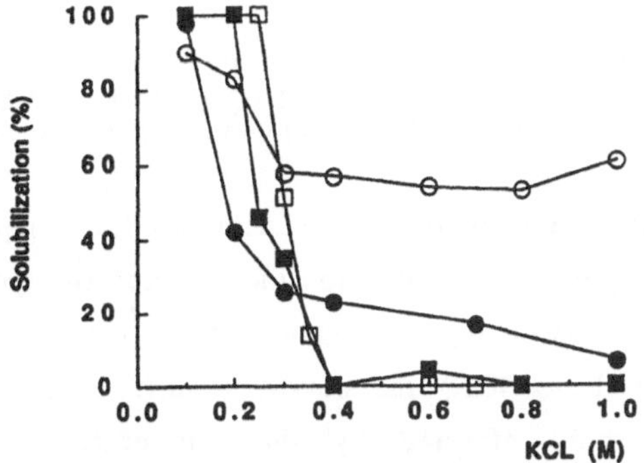

Figure 6- Extraction of cytochrome b_5 to a micellar phase of 100 mM CTAB/cyclohexane/10% (v/v) decanol at several pH and ionic strengths: (○) pH 4.4, (●) pH 5.0, (■) pH 6.0 and (□) pH 7.0.

It can be seen that at pH far above the pI (6.0 and 7.0) the extraction is mainly governed by an electrostatic interaction with the positive surfactant. At high ionic strength this interaction is damped and the extraction does not occur. At a pH close to the pI the extraction seems to be due to hydrophobic interactions between the protein and the surfactant, as above 0.4 M KCl it is independent on the ionic strength of the aqueous phase. The initial decrease in the extraction can be attributed to the dampening of electrostatic interactions that still exists at low ionic strength.

Since hydrophobic interactions are strongly dependent of the temperature (becoming less important at lower temperatures) the extraction of cytocrome b5 was tested at several temperatures, between 10 to 40 °C, using a pH of 4.4 (close to the pI) and 7.0 (above the pI), and different ionic strengths, 0.1 and 0.8 M KCl. The results are shown in Figure 7.

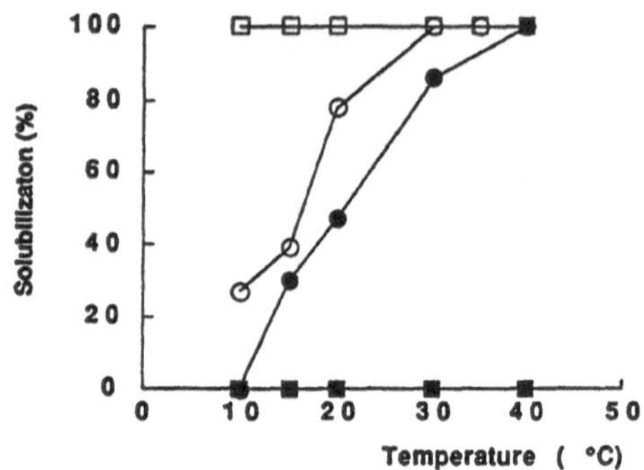

Figure 7- Temperature effect on the extraction of cytochrome b5 from an aqueous phase at pH 7.0 (□,■) or pH (○,●), with 0.1 (○,□) or 0.8 (●,■) M KCl to a reversed micellar phase of 100 mM CTAB/cyclohexane/10% (v/v) decanol.

As it can be seen, the extraction is only dependent on the temperature when the pH of the aqueous phase is close to the isoelectric point. As the temperature is lowered, the extraction of cytocrome b5 markedly decreases. At higher ionic strength (0.8 M KCl) the curve is shift to the left, i.e, the extraction seems to be more affected by the temperature, which is in agreement with the results shown in Figure 6.

The hydrophobic interaction between an acidic protein, cytochrome b5 and a cationic surfactant (CTAB) may be explained by the existence of an hydrophobic region at the cytochrome b5 surface that can interact with the micellar aggregate under conditions of pH close to the pI and high ionic strength. This region was identified in calf cytochrome b5, and was described as a hydrophobic patch which is surrounded by negatively charged residues with a total solvent exposed area of 350 Å^2 (22). The patch contains the hydrophobic groups Phe-35, Pro-40, Leu-70 and Phe-74. In the rat sequence Phe-74 is replaced by Tyr-74.

References

1. Hatton, T.A. (1989) "Reversed Micellar Extraction of Proteins" in J.F.Scamehorn and J.H.Harwell (Eds.), Surfactant-Based Separation, Marcel Dekker, N.Y., pp. 55-99.
2. Goklen, K.E. and Hatton, T.A. (1987) Sep. Sci. Technol., 22, 831-841.
3. Dekker, M., Baltussen, J.W.A., Van't Riet, K., Bijsterbosh, B.H., Laane, C. and Hilhorst, R. (1987) "Reversed Micellar Extraction of Enzymes: Effect of Non-Ionic Surfactants on the Distribution and Extraction Efficiency of α-amilase" in C. Laane, J. Tramper and M.D. Lilly (Eds.), Biocatalysis in Organic Media, Elsevier, Amsterdam, pp. 285-288.
4. Rahaman, R.S., Chee, J.Y., Cabral, J.M.S. and Hatton, T.A. (1988) Biotechnol. Progress, 4, 218-224.
5. Camarinha Vicente, M.L., Aires-Barros, M.R. and Cabral, J.M.S. (1990) Biotechnol. Tech., 4, 137-142.
6. Aires-Barros, M.R. and Cabral, J.M.S. (1991) Biotechnol. Bioeng., 38, 1302--1307.
7. Maitra, A. (1984) Phys. Chem., 88, 5122-5125.
8. Goklen, K.E. and Hatton, T.A. (1985) Biotech. Progress, 1, 69-74.
9.Imre, V.E. and Luisi, P.L. (1982) Biochem. Biophys. Res. Commun., 107, 538-545.
10. Gatfield, I. and Saud, T. (1981) European Pat., 0061023 AL (DE 3108927).
11. Haering, G., Meussdorffer, F. and Luisi, P. (1985) Biochem. Biophys. Res. Commun., 127, 911-915.
12. Bonner, F.Y., Wolf, R. and Luisi, P.L. (1980) J. Solid-Phase Biochem., 5, 255-268.
13. Dekker, M., Van't Riet, K., Weijers, S.R., Baltussen, J.W.A., Laane, C. and Bijsterbosch, B.H. (1986) Chem. Eng. J., 33, B27-B33.
14. Luisi, P.L., Bonner, F.J., Pellegrini, A., Wiget, P. and Wolf, R. (1979) Helvetica Chimica Acta, 62, 740-753.
15. Goklen, K.E. (1986) Doctoral Thesis, Massachussetts Institute of Technology, Cambridge, USA.
16. Leser, E.M., Wei, G., Luisi, P.L. and Maestro, M. (1986) Biochem. Biophys. Research Commun., 135, 629-635.
17. Misiorowsky, R.L. and Wells, M.A. (1974) Biochemistry, 13, 4921-4927.
18. Castro, M.J.M., Moura, J.J.G. and Cabral, J.M.S. (1988) "Liquid-liquid Extraction of Cytochrome c_3 from Desulfovibrio Using Reversed Micelles" in Advances in Gene Technology: Protein Engineering and Protein Production, Miami, USA.
19. Woll, J.M., Hatton, T.A. and Yarmush, M.L. (1989) Biotechnol. Progress, 5, 57-62.
20. Eremin, A.N. and Metelitsa, D.I. (1988), Appl. Biochem. Microbiol., 24, 35-42.
21. von Bodman, S., Schuler, M., Jolie, D., Sligar, S. (1986) Proc. Natl. Acad. Sci., 83, 9443-9447.
22. Mathews, F.S. and Czerwinski, E.W. (1984) in Martonosi, A.N. (Eds.) The Enzymes of Biological Membranes, Plenum Press, N.Y., 4, pp. 235-300.

PART 3:

ENVIRONMENTAL BIOTECHNOLOGY

ROLE OF CYANIDE ON POLYCYCLIC AROMATIC HYDROCARBON DEGRADATION IN MANUFACTURED GAS PLANT SOILS

LENORE S. CLESCERI, RAYMOND WATERBURY
AND YIFONG HO
Rensselaer Polytechnic Institute
Troy, NY 12180

1. Introduction

Polycyclic aromatic hydrocarbons (PAHs) are widespread environmental contaminants that are formed during the incomplete combustion of organic material. This paper reports on a study of soils from manufactured gas plant (MGP) sites that contain PAHs as well as other gas byproducts including ammonia, cyanide, coal tars and phenols. These methane production byproducts were removed from the gas prior to distribution. The byproducts were then disposed of on the land, and many sites still have high concentrations of these byproducts even though MGP plants in the United States have not been operating for the past 40 years.

2. The Study

The major PAHs found in the MGP soil studied are naphthalene, phenanthrene and anthracene as shown in Figure 1. PAHs with higher numbers of rings that are also found in MGP sites are shown in Figure 2. These compounds biodegrade aerobically according to the pathways illustrated in Figure 3 which shows the production of NADH in the bacterial pathway. The NADH then enters the electron transport scheme to generate ATP

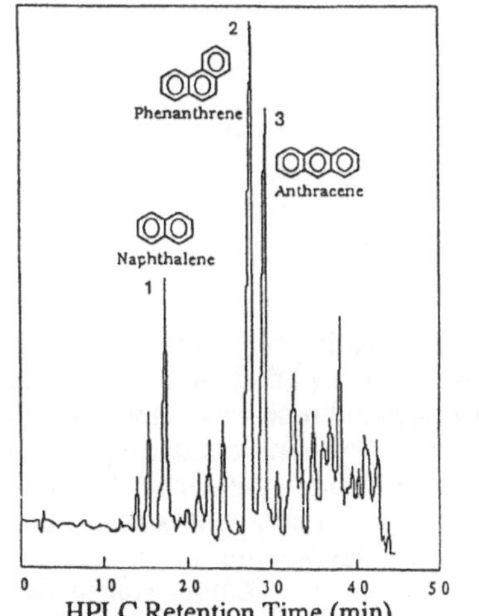

Figure 1. HPLC chromatograph of MGP soil extract.

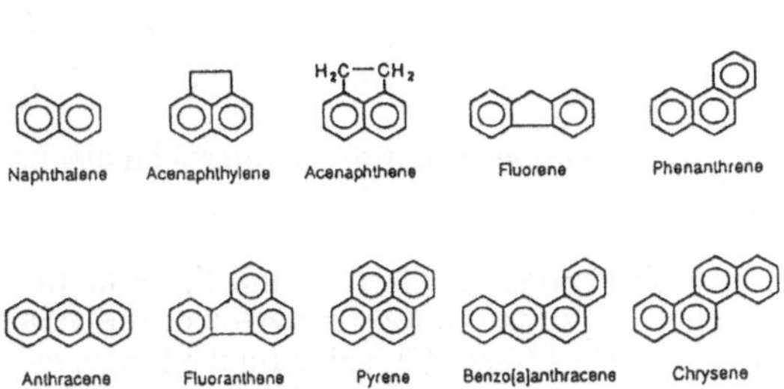

Figure 2. Structure of PAHs commonly present in MGP sites.

A.R. Moreira and K.K. Wallace (eds.), Computer and Information Science Applications in Bioprocess Engineering, 239-243.

for growth and metabolism and regenerate oxidized NAD. Figure 4 illustrates the blockage of this process by cyanide at cytochrome C oxidase

Figure 3. PAH degradation pathway for fungi and bacteria.

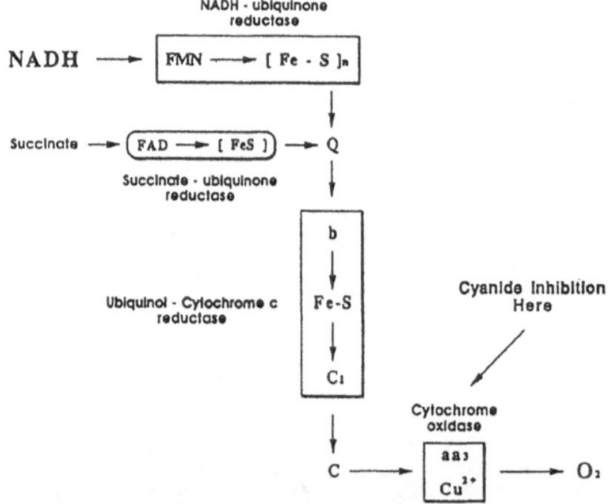

Figure 4. Cyanide inhibition on microbial electron carrier

The biodegradation of the three major PAHs is considerably enhanced when minerals are added to the rainwater of the area especially for naphthalene as shown in Figure 5. The effect of cyanide on the biodegradation of phenanthrene was determined in an experiment in which additions of cyanide were made to systems containing a soil isolate, mineral salts and phenanthrene. At concentrations of cyanide greater than 5 mg/l, the rate of phenanthrene biodegradation was affected. Figure 6 shows this effect and includes a control which is abiotic. The relative rates of biodegradation of anthracene, phenanthrene and naphthalene are depicted in Figure 7 which is a comparison of two systems (A and B) in which system B contains additional PAHs and an inoculum of an

MGP soil isolate capable of PAH degradation. System A as well as B contains an MGP soil slurry and extra mineral salts which have Been shown to enhance biodegradation (Figure 5)

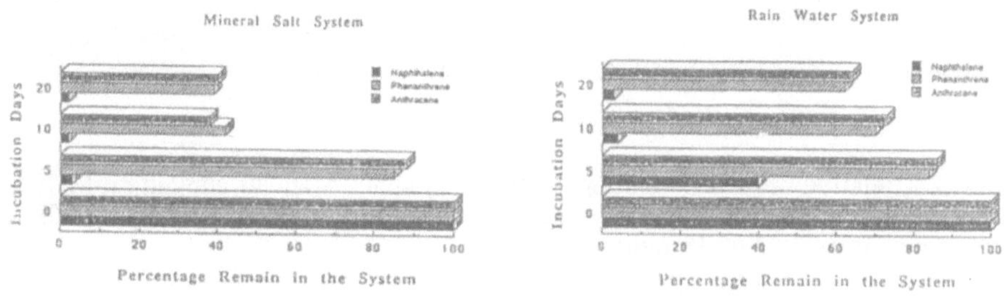

Figure 5. Effect of minerals on degradation of PAHs.

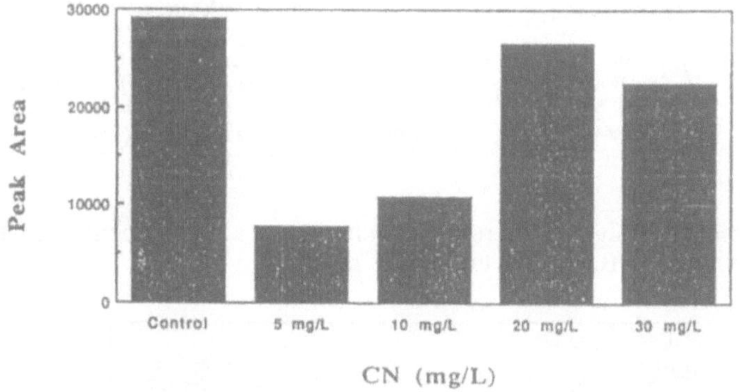

Figure 6. Effect of cyanide on PAH degradation.

Since cyanide has been shown to be inhibitory in these soils, an attempt was made to treat the soils with a microorganism reported to degrade cyanide to ammonia and carbon dioxide from concentrations as high as 0.1 M cyanide. The organism is a strain of *Bacillus pumilus* (ATCC 7061). The ability for the cyanide-oxidizing strain to function in cyanide removal in a system containing PAHs at a level inhibitory to PAH biodegradation was tested. Five systems were compared as illustrated in Figure 8. These systems contained increasing amounts of cyanide in flasks containing a PAH degrading isolate (*Pseudomonas aeruginosa*) with mineral salts and a mixture of PAHs. Two of the flasks, 3A and 3B, also contained the cyanide using *Bacillus pumilus*. The results from these experiments are shown in Figure 9. These results show that the presence of *Bacillus pumilus* allowed the PAH degradation to occur as if the cyanide were present at concentrations of 5 mg/l which is not inhibitory.

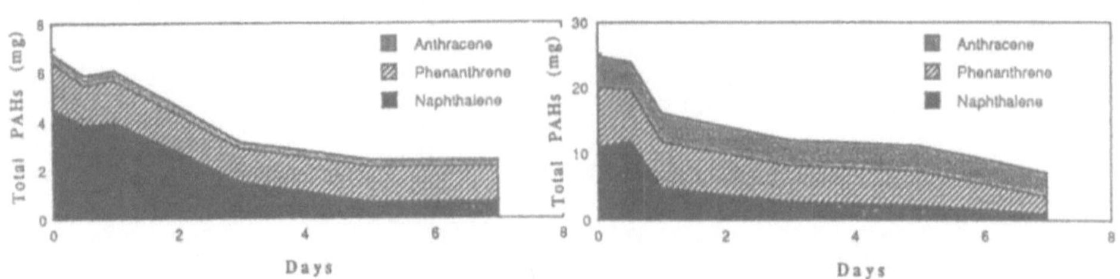

Figure 7. Relative rates of PAH degradation. Systems A and B contain an MGP soil slurry and mineral salts. System B also contains additional PAHs and an MGP soil isolate which is capable of degrading PAHs.

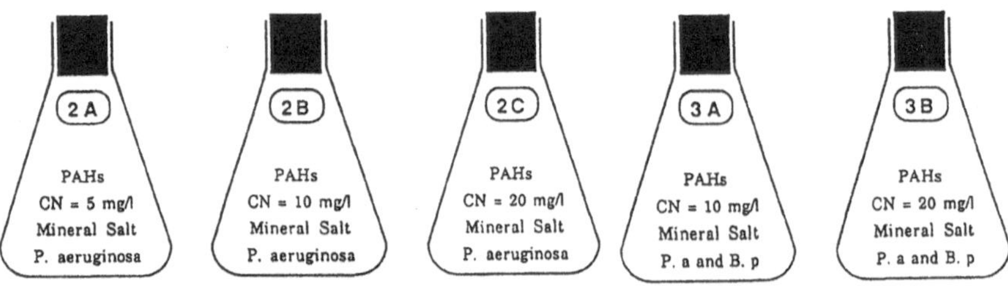

Figure 8. Experimental design of microbial cyanide removal experiment. P.a. refers to *Pseudomonas aeruginosa*. B.p. refers to *Bacillus pumilus*.

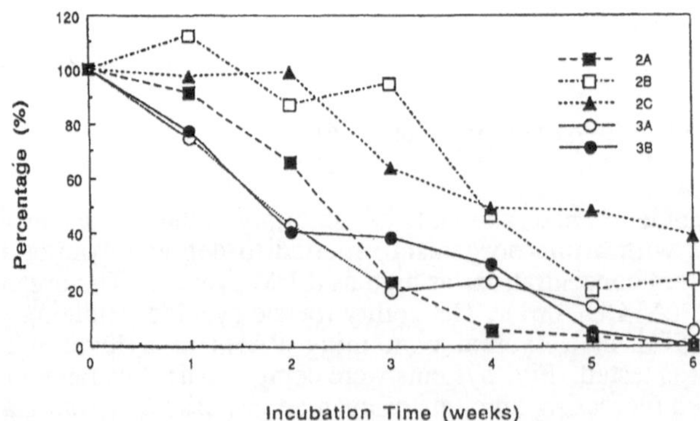

Figure 9. Percentage microbial cyanide removal. Numbers in figure legend are illustrated in Figure 8.

These results are encouraging and form the basis for further experimentation in the design of a two-step system for the bioremediation of MGP soils that are contaminated with PAHs as well as with cyanide. It is our contention that the presence of cyanide and absence of adequate minerals have been important factors in the recalcitrance of these soils to biodegradation.

3. References

Atkinson A. (1975) Bacterial Cyanide Detoxification. Biotechnology and Bioengineering, 27:457.

Castric, P.A., Stobel, G.A. (1969) Cyanide Metabolism by *B. megaterium.* J. Biol. Chem., 244, 15:4089-4094.

Knowles, C.J. (1976) Microorganisms and Cyanide. Bacteriological Reviews. 40, 3:652-680.

Meyers, P.R., Gokool, P., Rawlings, D.E., Woods, D.R. (1991) An Efficient Cyanide-degrading *Bacillus pumilus* Strain. J. Gen. Microbiol. 137:1397-1400.

Skowroski,B., Stroebel, G.A. (1968) Cyanide Resistance and t Cyanide Utilization by a Strain of *Bacillus pumilus.* Can. J. Microbiol., 15:93-98.

Wang, X., Yu, X., Bartha, R. (1990) Effect of Bioremediation of Polycyclic Aromatic Hydrocarbon Residues in Soil. Environ. Sci. Technol., 24:1086-1089.

BIOMASS REFINING

H. R. BUNGAY
Department of Chemical Engineering
Rensselaer Polytechnic Institute
Troy, NY 12180-3590, USA

ABSTRACT. Biomass refining is not yet ready to impact in a major way on the gluttonous demand for motor fuel, but a few factories that convert wood to a mix of several products should be highly profitable. There need to be credits from the sale of other chemicals or from excess plant energy in the form of steam or electricity. A biomass refinery will have three major products - fuel alcohol, crude syrup for feeding cattle, and lignin for such applications as adhesive in plywood or chipboard. With good recycle and careful management of plant streams, there will be almost no pollution of the environment. A computer spread sheet for a refining process is used as a teaching tool.

1. Introduction

1.1 GENERAL CONCEPTS

Chemicals from wood is not a new idea. Henry Ford built factories in the 1920's to roast wood to make a mixture of organic chemicals, but the advent of cheap oil as a starting material for these same chemicals ruined his economic prospects. The oil crisis of the early 70's redirected attention to such alternative feedstocks as wood. Alcohol can be the main product from wood when there is a fermentation step; plants in St. Petersburg and elsewhere have been doing this for many years.

The two biggest reasons that biomass refining has not become important in the US are inflated capital cost and poor attention to byproducts. The estimated capital cost is high because chemical engineers think first of stainless steel equipment and high technology. In contrast, fuel alcohol in Brazil is made from cane sugar using much lower technology; many of the fermenters are open tanks made of inexpensive materials. Fuel alcohol must compete with gasoline or must be inexpensive enough to justify blending with gasoline to improve combustion efficiency. Ethanol is but one of three oxygenated organic molecules that will satisfy new air pollution regulations demanding cleaner-burning motor fuel for population centers. If ethanol is too expensive, the new gasoline formulations will be based on methanol or MTBE (methyl-tert-butyl ether made petrochemically). Although wood can substitute for oil, natural gas, or coal in the high-temperature and high-pressure manufacture of many organic chemicals, the economics will be unattractive until the prices of fossil feedstocks climb much higher.

Meat refining and refining of petroleum address everything in the feedstock. The cuts of meat range from steak to hamburger, the hides are converted to leather, the horns and bones are used for gelatine or glue, and the scraps are fed to animals. Crude oil is converted to gas, lubricants, asphalt, and other fractions. Biomass refining to be profitable must also address its main components - cellulose, hemicellulose, and lignin (see Table 1). Anything that is not a product is a waste, and treatment and disposal of wastes add greatly to costs and help determine whether a factory is acceptable to the community.

A.R. Moreira and K.K. Wallace (eds.), Computer and Information Science Applications in Bioprocess Engineering, 245-252.

TABLE 1. Composition of Unprocessed Wood

Water	- 50
Cellulose	- 22
Hemicellulose	- 12
Lignin	- 11
Bark	- 3.5

Cellulose and hemicellulose are polymers that have to be broken down to their components. The building block of cellulose is glucose that is an excellent sugar for fermentation to make alcohol. Hemicellulose has several different sugars in its structure, but most of them are not very good in established processes for making alcohol. Lignin is an aromatic polymer with no sugars; it is used for making specialty chemicals.

Refining of lignocellulosic feedstocks such as trees and agricultural residues leads to lignin, sugars, and smaller amounts of chemicals derived from wood. A report prepared by Arthur D. Little, Inc. for the New York State Energy Research and Development Authority (NYSERDA) and the National Renewable Energy Laboratory (formerly the Solar Energy Research Institute, SERI) analyzed one of the promising methods for biomass refining and found that coproduct credits were crucial (Nystrom, et al., 1985). Fuel alcohol would have to sell for roughly $1.30 per liter when the only value derived from the other components is adding them to boiler fuel. On the other hand, refineries that sell lignin at prices that seem reasonable for applications such as manufacture of adhesives would be profitable even if the alcohol were given away.

Lignin extracted from wood using solvents is markedly different from the damaged and reacted lignin that is a waste product from conventional pulping to make paper. A thorough analysis of lignin processes and markets (Frank, et al., 1989) concluded that there would be a modest demand for specialty derivatives at high prices, but lignin at 22 cents/kg would have sizable markets for adhesives and the like. This low estimate for value of lignin is justified in the report, but others feel that the potential value of lignin should be high because it substitutes for phenol that sells for more than 30 cents/pound. Lignin can constitute about half of a type of phenol-formaldehyde adhesive that sells for more than 66 cents/kg.

1.2 FEEDSTOCK

The economics for biomass refining look good with commercial wood chips but only after markets develop for the byproducts. However, disposal of wastes that are rich in biomass is becoming terribly expensive. Profitability is assured when refineries get large credits for accepting wastes. In areas where most of the dumps have closed, waste wood from construction, demolition, broken pallets, crates, dimension lumber, and the like is sent to landfills at a cost of about $90 per ton. The current price for chips from forestry is $30 to $45 per ton on a dry basis. Cut areas can be replanted with species that grow rapidly and propagate from the roots after harvesting every 2 to 4 years with machines that are like big lawn mowers. Although the prices of chips from such plantations may be lower, there are disputes about the costs. Poplar and willow trees are suitable for plantations in North America and could be crops on farm lands now idle.

A biomass refinery will generate little waste and none of it will be toxic or hazardous. If located near farms or forest areas, the logistics of irrigation will be favorable. Refinery residues could be pumped to the growing area. Biomass refining liquid waste has some unreacted sugars, unchanged organic compounds found in wood, and small amounts of yeast cells and traces of unrecovered alcohol. The amount of wastewater will be insignificant for the thousands of acres of woodlands that supply the refinery, but many acres near the refinery can be irrigated. These will be by far the most productive areas, and proximity to the refinery will result in minimum transportation costs.

1.3 HYDROLYSIS OF CELLULOSE

Low cost sugars for the bioconversion steps are the key to producing most of the products that are considered opportunities for biomass refining to impact on the consumption of petroleum. Hemicellulose or xylan is so easily hydrolyzed that very little research or discussion is warranted. The principal polymer of wood, cellulose, is quite difficult to hydrolyze inexpensively.

Neither acid hydrolysis or enzymatic hydrolysis work well with intact biomass; subdivision is essential. Simple milling or grinding make the carbohydrate polymers in biomass accessible for hydrolysis while also enhancing yields. However, a very fine particle size that promotes enzymatic hydrolysis requires an uneconomical amount of energy for size reduction. Acid hydrolysis requires less pretreatment because the acid molecules are small and will diffuse fairly well into soaked biomass.

Disintegration of biomass structure by explosion is more cost effective than is milling or grinding. There is no need to waste steam pressure by venting from high pressure to atmospheric pressure, but some violent release of pressure to shatter the biomass seems advisable. In other words, the steam at roughly 40 atmospheres could explode to perhaps 10 atmospheres to disintegrate the biomass while still having a residual steam at a useful pressure for energy needs in the plant.

1.3.1 *Enzymatic Hydrolysis*. Major improvements have been made in the titres of cellulase enzymes. Numerous improved cultures have been reported. In fact, many groups tout their cultures. Nevertheless, Trichoderma cellulases still seem to be at the forefront.

1.3.2 *Acid Hydrolysis*. The first large-scale dilute acid hydrolysis project was a percolation system developed by Scholler in Tornesch, Germany in the 1920's. It used 0.4 percent sulfuric acid at 170°C and a pressure of 8 atmospheres. The acid solution moved in a semicontinous fashion. This produced a 4 percent or greater sugar solution with yields in the range of 50 to 55 percent. The Madison process was developed at the Forest Products Laboratory at Madison, Wisconsin towards the end of WWII. It represented an improvement on the Scholler-Tornesch process by being able to run the acid continuously and still maintain high yields.

1.3.3 *Concentrated Acid*. Hydrolysis of cellulose with concentrated acid has been practiced for over 100 years starting with the Rheinau process patented in Germany in 1880. Concentrated acid hydrolysis is based on the fact that concentrated mineral acids such as 72 percent sulfuric acid, 85 percent phosphoric acid, and 41 percent hydrochloric acid will dissolve crystalline cellulose at temperatures of 20°C or less. The original work at Rheinau led to the commercial operation of the Rheinau-Bergius process during WWII. It used 41 percent HCl, with the HCl recovered by distillation. Yields in excess of 90 percent and solutions in excess of 30 percent sugar were produced. This system suffered from poor economics due to inefficient recovery of the acid, which in part resulted from the formation of an acid/lignin complex.

2. Ethanol

2.1 PERSPECTIVE

Many experts consider ethanol as the prime opportunity for biomass refining to make inroads on the use of petroleum. Ethanol is a good motor fuel and can also improve gasoline by improving the blending of anti-knock ingredients. The major research areas for fuel ethanol are improved bioconversion through 1. better organisms (possibly Zymomonas), 2. more productive bioreactors such as beds packed with immobilized or flocculated cells, 3. relieving ethanol inhibition by removing the product during the course of the process, 4. better product recovery or better energy management, and 5. improving ways to ferment sugars (some hydrolysates formed with acids contain materials that inhibit the fermentation). There has been considerable interest in the bacterium *Zymomonas mobilis* that ferments some sugars to ethanol and high rates and achieves good titers. It should be easier to manipulate the genetics of this bacterium than to modify yeast, and bacteria grow faster than do yeast.

2.2 ORGANISMS THAT CONVERT PENTOSES TO ETHANOL

Most of the research on fermenting sugars from hemicellulose to ethanol has focussed on xylose because it is the predominant pentose. However, the yeasts *Candida shehatae* and *Pichia stipitis* that convert glucose, mannose, galactose, and xylose to ethanol fail to convert xylitol, arabinose, or rhamnose (du Preez, et al., 1986). Failure to convert all of the sugars from hemicellulose means that the fermentation residue will be high in organic matter and that the attempt to maximize ethanol production is not very successful.

Ethanol can be produced from xylose in at least two ways. In addition to a direct path, there is a path through the sugar xylulose. Xylulose is acceptable to *S. cerevisiae*, an excellent yeast in terms of both approaching theoretical yield and titer of ethanol. Moving the genes for isomerizing xylose to xylulose into *S. cerevisiae* has not met with success. Xylose bioconversion was reviewed recently by Skoog and Hahn-Hagerdahl (1988).

2.3 PRODUCT RECOVERY

An important advance in recovery of ethanol is the vapor-phase drying invented by Ladisch and Dyck (1979). This process has been in commercial operation for several years. Ground corn absorbs water from wet ethanol vapor to give the final dry product. Periodically the flow is stopped and the corn is dried with heat making it suitable for reuse. The liquid-vapor equilibrium curve for ethanol and water is highly unusual in having a wide separation between the liquid and vapor equilibrium lines at low concentrations of ethanol. Just 4 or 5 equilibrium contacts will take the ethanol entering at fermentation concentration to over 70 percent ethanol in the product stream. The curves almost converge at high concentrations of ethanol. When the curves are close, each step in distillation accomplishes very little increase in ethanol concentration. In fact, there is a constant boiling mixture (azeotrope) at about 95 percent ethanol that cannot be changed in composition by distillation. There are several ways to purify to concentrations higher than the azeotropic composition, but there must be a change in pressure or the addition of another organic liquid. In a few words, distillation works great for dilute ethanol as with fermentation broths but performs much much worse for high concentrations.

Alternatives to distillation such as various membrane processes have more or less failed for recovering ethanol from a fermentation. Even when results are good in the laboratory with pure test materials, real fermentation broths are high in solids, many of which are terrible in terms of fouling membranes. Solvent extraction of ethanol from fermentation broths has had poor results because of lack of a non-toxic solvent with a high selectivity for ethanol and because of emulsions that impair extraction. The most economical recovery at present is steam stripping followed by distillation and vapor phase drying. There is nothing to foul when steam flushes ethanol from the fermentation, but the concentration of ethanol in the residue is about 0.5 percent.

The bottom residue from the distillation step should be recycled. This recovers a small amount of ethanol, but more important are the reuse of the expensive nutrients (mainly those nitrogenous materials), saving on make up water, and greatly reducing the amount of liquid sent to waste. Waste treatment is expensive and doing a bad job can damage community relationships beyond repair.

2.4 ETHANOL FERMENTATION

The bioconversion of sugar to ethanol is not a major hurdle for commercialization of bioenergy. One prospectus for a factory to convert wood chips showed that the section for bioconversion of sugars to ethanol took about 20 percent of the capital investment. Major improvements in this fermentation are desirable, of course, but the cost impacts on the overall process would not be crucial. Energy savings depend on other features of the process. For example, a steam explosion process should have an excess spent steam that could be used for ethanol recovery as well as for evaporation of sugar solutions to syrups.

3. State-of-the-Art Process

3.1 STEAM EXPLOSION REFINERY

There are several other processes of interest, but one excellent process will illustrate what constitutes a biomass refinery. The harvesting, chipping, hauling, and handling of wood would be the same as the well-established steps for the pulp and paper industry. The IOGEN process, currently in the pilot plant stage at Iogen, Ltd. in Ottawa, Canada, impregnates wood chips with steam, and they disintegrate as the pressure is released suddenly. This is much like the old Masonite process that made a fluffy material for construction board. The cellulose is later hydrolyzed to the glucose that is converted by yeast into ethanol. Fermentation to produce ethanol is old technology, and the new modern twists do not make much difference in costs. Final products are ethanol, pentose syrup, carbon dioxide from the fermentation, and a lignin containing some unhydrolyzed cellulose. Figure 1 is a diagram of the process.

Steam explosion weakens cellulose because its molecular weight is lowered, destroys hemicellulose, and melts lignin. The chips are transformed to a wet brown powder that is nicely sized for further processing. Explosion cellulose is relatively easy to hydrolyze with enzymes. Hemicellulose is hydrolyzed to sugars during the steaming and explosion, but the reactions continue on to resinous and polymeric compounds. Yields of sugars from hemicellulose are only about 50 per cent of theoretical with severe conditions of steam explosion, but milder conditions with catalysts improve the yield.

Feeding of cattle is a proven option for using crude sytrup while better alternatives are explored such as fermenting the sugars to ethanol or to other organic chemicals. There is no doubt that the syrup can be sold at some price. Syrup is used to flavor and to bind animal feeds. The value is not based mainly on the sugar content but on properties such as thickness, tackiness, feel, binding strength, flavor, and odor. These properties can be manipulated to some extent, but prices depend on competition, customer service, and suitability.

Fuel alcohol and syrup for feeding cattle can be sold easily as soon as the refinery comes on line, but lignin poses major problems. Marketing lignin requires technical support to tailor properties to the customer's needs. It will take some time to develop customer confidence and to prove that the lignin works well and results in major savings. The fermentation evolves pure carbon dioxide that would find ready markets in the Northeast U.S.

3.2 ECONOMICS

A spreadsheet BIOREF.WK1 for mass balances and economics of a typical biomass refinery and an explanation file BIOREF.TXT are in the Public Domain. The best way to get copies is with the FTP system available at most computer centers. The address is FTP.RPI.EDU, directory PUB-READ/bungy. Both hydrolysis with dilute sulfuric acid or with enzymes are analyzed. The spreadsheet is in Lotus 1-2-3 format that is accepted by most programs. Table 2 shows typical results of the spreadsheet analysis.

Some conclusions from the spreadsheet analysis are:

- capital costs tend to dominate thus lower technology should be stressed.
- selling lignin at a reasonable price insures profitability.
- selling pentose syrup helps the economics but more valuable byproducts from the pentose fraction would be much more important.
- waste cellulosic materials as feedstocks could make profits soar.

3.3 SPECIAL FEATURES OF THE SPREADSHEET

3.3.1 *Extraction Of Lignin.* If the extraction of lignin aims for a high yield, far too much ethanol is required. Even with recovery yields of 99 percent of the ethanol used, the consumption of ethanol can be greater than the amount produced by the refinery. The following equation reflects our limited results:

Table 2 Typical Spreadsheet Results

Option		Required Ethanol Price	Profit if $1/ga ($ 0.26/L)
Baseline process with commercial chips			million
acid hydrolysis		$ 0.68 per liter	$ -4.67
enzymatic hydrolysis		0.70	-5.72
syrup bioconversion		0.52	2.12
Baseline process with scrap wood (U.S. ton and metric tonne assumed equal)			
chips at $	0.00 /t	$ 0.37 per liter	$ -0.61
	- 10.00	0.17	2.10
	- 20.00	-0.03	4.81
	- 30.00	-0.24	7.52

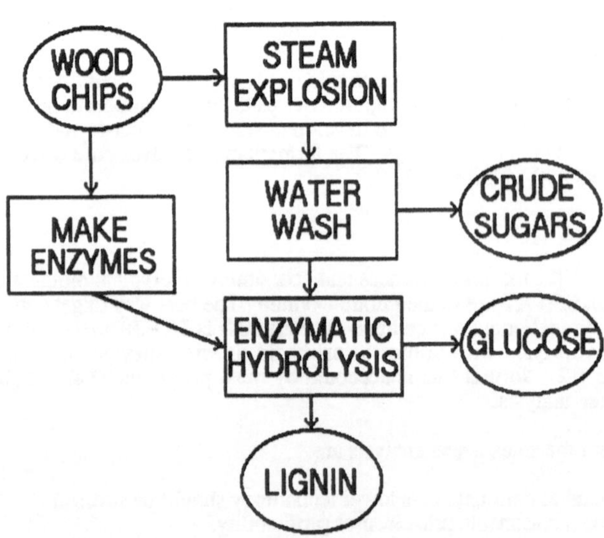

Figure 1. Biomass refining with steam explosion; fermentation of glucose not shown.

$$L = V/(1+V)$$

where L = fractional recovery of lignin

V = pounds of ethanol per pound of cellulose in the solids

The specification for V is entered at Cell H18 of the spreadsheet. Costs for ethanol recovery and for product lignin reflect this specification.

Poor removal of lignin may impair the step for acid hydrolysis of cellulose, but is not considered in the cost estimates. The most likely effect is additional filtration to remove residual solids during the hydrolysis, and this should be a minor expense.

A large supply of gasoline is needed for denaturing ethanol. The value of five per cent gasoline was found in several other economic analyses. The actual product alcohol in the spreadsheet is increased in volume because of the gasoline.

3.3.2 *Recovery Of Ethanol.* Recycle of ethanol from the extraction of lignin and recovery of ethanol from fermentation beer are handled separately because the final concentrations are different; however, the distillation columns will be located for efficient energy use. The ethanol for extraction of lignin will approach the azeotrope concentration of 95 per cent. The product ethanol will be 99.75 per cent, but fresh ethanol to replace losses for lignin extraction will be taken before the final dehydration step. As in the Arthur D. Little Report, vapor-phase drying will be used instead of distillation to achieve the final dryness.

3.3.3 *Hints For Using Spreadsheet.* Table 3 is a guide to important cells of the spreadsheet. When performing sensitivity studies or exploring permutations, find one or more blank cells near the relevant outputs and use them for input specifications. For example, if the parameter to change is percent sulfuric acid in Cell M20 and you are interested in capital cost for neutralization in Cell C64, a nearby blank cell is A64. In Cell M20, you enter "=A64". Now entering your specification in A64 will relay to M20.

TABLE 3. Index of Locations of Specifications

Acid hydrolysis assumptions	
final glucose concentration	M18
hydrolysis yield	M19
percent sulfuric acid	M20
Analysis of wood (labels in Column A)	S4-S11
Analysis of MSW (labels in Column A)	T4-T11
Byproduct credits	N27-N28
Days of operation per year	B94
Enzymatic hydrolysis assumptions	
enzyme per pound of cellulose, IU	M68
yield	M69
Ethanol extraction of lignin	
ethanol per pound of cellulose	F18
recovery of ethanol (fraction)	F20
Fraction of feedstock that is wood	S2
Labor	
salaries	H42-H46
numbers of employees	I42-I46
Material costs	
wood	H27
gasoline	H28
lime	H29

TABLE 3. Index of Locations of Specifications (continued)

Pentose fermentation assumptions	
Productivity	M78
yield of product from hexoses	M82
yield of product from pentoses	M83
Products	L26-L28
selling price (others assumed, EtOH to make profit)	M26-M28
revenue	N26-N28
Recycle assumptions	
Gr, ratio of glycerol to ethanol	P48
recycle ratio	P49
Solid waste tipping fee	H26
Water washing step	
water per pound of cellulose	D18
yield of hemicellulose after explosion	D20
hours for contacting	E28

4. Conclusion

Any biomass refinery built today could sell all of its fuel ethanol immediately, and the syrup of crude sugars would be a break-even or slighly profitable byproduct. Revenues from selling lignin could result in excellent profitability, but the markets may be slow to develop. The initial refineries should repay their investors handsomely because of the credits from accepting scrap wood. When lignin sales become significant, profits will soar.

The biomass industry can expand only a little before being swamped with unsold crude sugars and lignin. However, large reliable supplies of relatively inexpensive materials should motivate the search for additional uses. Converting pentose sugars to calcium magnesium acetate (CMA) for deicing highways and for cleaning coal and using lignin in paving materials would generate more income. At the scale required for fuel alcohol to have national significance, biomass refining will represent a large capital investment, many new jobs, and will have a major impact on national economies.

5. References

duPreez, J.C., Bosch, M. and Prior, B.A. (1986) 'Xylose fermentation by *Candida shehatae* and *Pichia stipitis*; effects of pH, temperature, and substrate concentration', Microbiol. Technol. 8, 360-364.

Frank, M.E., Mednick, R.L. and Stern, K.M. (1988) 'Use and value of reactive lignin', Report prepared for New York State Energy Research and Development Authority by CHEM SYSTEMS, INC. 944-ERER-ER-87.

Ladisch, M.R., and Dyck, K. (1979) 'Dehydration of ethanol: New approach gives positive energy balance', Science 205, 898-900.

Nystrom, J.M., Greenwald, C.G., Hagler, R.W., and Stahr, J.J., (June 1985) 'Technical and Economic Feasibility of Enzymatic Hydrolysis for Ethanol Production from Wood', Final Report for NYSERDA and SERI, Arthur D. Little, Inc.

Skoog, K. and Hahn-Hagerdahl, B. (1988) 'Xylose fermentation', Enzyme Microb. Technol. 10, 66-80.

ANAEROBIC METHANE FERMENTATION IN A PLUG-FLOW REACTOR TREATING ORGANIC WASTES

J. Mata-Alvarez and P. Llabrés-Luengo
Departament d'Enginyeria Química. Universitat de Barcelona.
C/ Martí i Franquès 1, 6p. 08028 Barcelona (Spain)

ABSTRACT. A simulation of the anaerobic biodegradation of organic solid wastes is carried out. As a basis, a plug-flow reactor with a fluid recycle is used. The effect of process parameters, solid retention time (SRT), pH, recycle ratio on process performance (degree of methanization and degradation) is analyzed by means of a program based on a Runge-Kutta subroutine and written in basic. Results show the optimal value of 7.2 for the pH together with the convenience of operating the reactor at a zero recycle ratio.

1. Introduction

Biomethanization of solid wastes is a mature, ready to use technology, as was recently pointed out in the international symposium on biomethanization of solid wastes held in Venice. Dry and semi-dry approaches definitively appears as successful processes that allow higher performances and easier management of solid organic wastes with respect to conventional processes (Mata-Alvarez and Cecchi, 1992). Semi-dry biomethanization is carried out in conventional stirred tank bioreactors. This type of flow pattern has been modelled and applied to the digestion of sewage sludge (Siegrist et al., 1992). Dry processes are carried out on digesters which approach more the plug flow pattern model. Thus, the aim of this paper is to study the behaviour of this type of bioreactors, by simulating a biomethanization process, based on the basic steps of anaerobic digestion.

2. The plug-flow reactor model

The anaerobic digestion process takes place in accordance with several steps (Gujer and Zehnder, 1983). In order to simulate the production of volatile fatty acids from solid organic wastes, a simplified model was built (Negri et al., 1992) based in a series of hypothesis which can be summarized in the following points: a) Insoluble solids are formed by spherical particles of uniform diameter. When they are metabolized the radius decreases in accordance with the hydrolysis rate; b) There is no diffusion control of the enzymes through the core. The controlling step is the hydrolysis, which is modeled in accordance with a first order kinetic model; c) A fraction of the soluble biodegradable solids cannot be metabolized readily and must be hydrolysed as well; d) Acidification and methanization are carried out in accordance

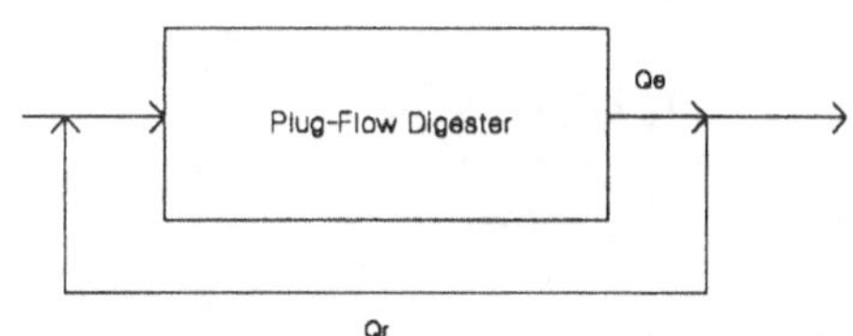

Figure 1. Schematic flow diagram of plug-flow digester with recycle.

A.R. Moreira and K.K. Wallace (eds.), Computer and Information Science Applications in Bioprocess Engineering, 253-263.

with a Monod model.

The model is based on mass balances for insoluble volatile solids, soluble volatile solids, volatile fatty acids (assumed acetic acid). These acids are converted into methane by the methanogenic bacteria. Balances for hydrolytic, acidogenic and methanogenic bacteria are also considered. These balances, set for a plug-flow reactor with liquid recycle (see Figure 1) are as follows (see at the end of the chapter for notation, more details can also be found in Negri et al, 1992):

Insoluble Volatile Solids

$$\frac{d\Phi}{dz} = - \beta X_h / v_s \tag{1}$$

Hydrolytic Biomass

$$\frac{dX_h}{dz} = (3\ Y_h\ Svi^*\ \beta\ \Phi^2 + Y_h\ k_{ho}\ Svs - km_h)\ X_h / v_{Bio} \tag{2}$$

Soluble Volatile Solids

$$\frac{dSvs}{dz} = \{[- k_{ho}\ Svs + km_h]\ X_h + km_a\ X_a + km_m\ X_m\} / v_l \tag{3}$$

Monomer Species

$$\frac{dMs}{dz} = \{- \frac{k_a\ pHIF_a\ Ms\ X_a}{K_a + Ms} + (1 - Y_h)(3\ Svi^*\ \beta\ \Phi^2 + k_{ho}\ Svs)\ X_h\} / v_l \tag{4}$$

Acidogenic Biomass

$$\frac{dX_a}{dz} = \{\frac{Y_a\ k_a\ pHIF_a\ Ms}{(K_a + Ms)} - km_a\}\ X_a / v_{Bio} \tag{5}$$

Volatile Acids

$$\frac{dC_a}{dz} = \{- \frac{k_m\ pHIF_m\ C_a\ X_m}{K_m + C_a} + \frac{(1-Y_a)\ k_a\ pHIF_a\ Ms\ X_a}{K_a + Ms}\} / v_l \tag{6}$$

Methanogenic Biomass

$$\frac{dX_m}{dz} = \left\{\frac{Y_m\, k_m\, pHIF_m\, C_a}{(K_m + C_a)} - km_m\right\} Xm / v_{Bio} \tag{7}$$

Boundary conditions for these equations are:

$$\Phi_{(z=0)} = 1 \tag{8}$$

$$X_{h(z=0)} = (X_h^0 + Rel\ f\ X_{h(z=L)}\ v_{Bio} / v_l) / (1 + Rel) \tag{9}$$

$$Svs_{(z=0)} = (Svs^0 + Rel\ Svs_{(z=L)}) / (1 + Rel) \tag{10}$$

$$Ms_{(z=0)} = (Ms^0 + Rel\ Ms_{(z=L)}) / (1 + Rel) \tag{11}$$

$$X_{a(z=0)} = (X_a^0 + Rel\ f\ X_{a(z=L)}\ v_{Bio} / v_l) / (1 + Rel) \tag{12}$$

$$C_{a(z=0)} = (C_a^0 + Rel\ C_{a(z=L)}) / (1 + Rel) \tag{13}$$

$$X_{m(z=0)} = (X_m^0 + Rel\ f\ X_{m(z=L)}\ v_{Bio} / v_l) / (1 + Rel) \tag{14}$$

where the following definitions have been used.

$$v_{Bio} = ((1 - f)\ v_s + f\ v_l) \tag{15}$$

$$v_s = Qe / (\pi\ D^2 / 4) \tag{16}$$

$$v_l = Qe\ (F + Rel) / (\pi\ D^2 / 4) \tag{17}$$

$$F = 1 - Svi^0 / 1000 \tag{18}$$

$$Svi^* = Svi^0 / (1 + Rel) \tag{19}$$

The model can handle pH dependence of the constants. This is carried out by means of a pH Inhibition Factor (Costello et al., 1991). This Factor is of the form:

$$(pHIF)^{1/m} = \frac{pH - pH_{LL}}{pH_{UL} - pH_{LL}} \tag{20}$$

This inhibition factor is used to model the influence of pH, both in the acidogenic and methanogenic step. The inhibition function is considered symmetric around the upper limit (pH_{UL}). Values used for pH_{UL} are 6.4 and 7.2 for acidogenic and methanogenic bacteria respectively. Similarly values for pH_{LL} are 4.0 and 5.6. Values for m are 3.0 and 2.2 for acidogenic and methanogenic steps respectively.

The program has been written in Turbo-Basic. A listing of it is presented in Table 1. Table 2, shows the values of the constants used in the simulation.

TABLE 1

Digester Simulation Basic Program

```
DEF FN FUNCION1(z,FI)=-beta*Xh/vs:DEF FN FUNCION2(z,Xh)=(3*Yhe*Svi*beta*FI^2+Yho*kho*Svs-kmh)*Xh/vbio
DEF FN FUNCION3(z,Svs)=((-kho*Svs+kmh)*Xh+kma*Xa+kmm*Xm)/vl
DEF FN FUNCION4(z,Ms):fun4=0  fun4=(-ka*pHIFa*Ms*Xa/(Ka+Ms)+(1-Yh)*(3*Svi*beta*FI^2+kho*Svs)*Xh)/vl
FN FUNCION4=fun4:END DEF:DEF FN FUNCION5(z,Xa)=(Ya*ka*pHIFa*Ms/(Ka+Ms)-kma)*Xa/vbio
DEF FN FUNCION6(z,Ca):fun6=0:fun6=(-km*pHIFm*Ca*Xm/(Km+Ca)+(1-Ya)*ka*pHIFa*Ms*Xa/(Ka+Ms))/vl
FN FUNCION6=fun6:END DEF:DEF FN FUNCION7(z,Xm)=(Ym*km*pHIFm*Ca/(Km+Ca)-kmm)*Xm/vbio
CLS:INPUT "ENTRAR EL NUMERO DE PUNTOS QUE SE DESEA =",NP:
INPUT "VALOR del pH =",pH:input "valor del SRT =",SRT
input "valor de REL =",REL:input "valor de beta =",beta
input "valor de kho =",kho:DIM Ca(NP),Xh(NP),Svs(NP),Xa(NP),Ms(NP),Xm(NP),FI(NP)
beta=0.01:rem kho=0.3:kmh=0.02:kmm=0.02:kma=0.2:km=5.46:
ka=35.7:Ka=5:Km=0.03:Yhe=0.05:Yho=0.05:Ym=0.055:Ya=0.22
f=0.1:D=0.4:L=6:pHhm=7.2:pHlm=5.6:pHha=6.4:pHla=4:ma=3:mm=2.2:Xho=0.025:Svso=17.5292:Mso=12:Xao=0.05:Cao=0.74
Xmo=0.1:Sio=16.99:Qe=L*3.1416*D^2/(SRT*4):vs=Qe*4/(3.1416*D^2):
vl=Qe*((1-(Sio/1000))+REL)*4/(3.1416*D^2):vbio=(1-f)*vs+f*vl
IF pH>pHha THEN Za=2*pHha-pH
        ELSE Za=pH
END IF
IF Za>pHla THEN pHIFa=((Za-pHla)/(pHha-pHla))^ma
        ELSE pHIFa=0
END IF
IF pH>pHhm THEN Zm=2*pHhm-pH
        ELSE Zm=pH
END IF
IF Zm>pHlm THEN pHIFm=((Zm-pHlm)/(pHhm-pHlm))^mm
        ELSE pHIFm=0
END IF
Xh=Xho:Svs=Svso:Ms=Mso:Xa=Xao:Ca=Cao:Xm=Xmo
1000 ITERACION=ITERACION+1:A=0:IF ABS((Ca-Caant)/Ca)<=0.001 then A=A+1
IF ABS((Xh-Xhant)/Xh)<=0.001 THEN A=A+1
IF ABS((Xa-Xaant)/Xa)<=0.001 THEN A=A+1
IF ABS((Xm-Xmant)/Xm)<=0.001 THEN A=A+1
IF ABS((Svs-Svsant)/Svs)<=0.001 THEN A=A+1
IF ABS((Ms-Msant)/Ms)<=0.001 THEN A=A+1
IF A=6 THEN GOTO 2000
Caant=Ca:Ca=(Cao+REL*Caant)/(1+REL):Xhant=Xh
Xh=((Xho+(REL*f*Xhant*vbio)/vl))/(1+REL):Svi=Sio/(1+REL):Xaant=Xa:
Xa=((Xao+(REL*f*Xaant*vbio)/vl))/(1+REL):Xmant=Xm
Xm=((Xmo+(REL*f*Xmant*vbio)/vl))/(1+REL):Svsant=Svs:Svs=(Svso+REL*Svsant)/(1+REL):Msant=Ms
Ms=(Mso+REL*Msant)/(1+REL):FI=1:z=0:VINT=L:NUMDIVI=10:INTERVALO=VINT/(NP*NUMDIVI)
FOR K=0 TO NP-1
        FOR J=1 TO NUMDIVI
                if FI<0 then FI=0
                R1=INTERVALO*FN FUNCION1(z,FI):R2=INTERVALO*FN FUNCION2(z,Xh)
                R3=INTERVALO*FN FUNCION3(z,Svs):R4=INTERVALO*FN FUNCION4(z,Ms)
                R5=INTERVALO*FN FUNCION5(z,Xa):R6=INTERVALO*FN FUNCION6(z,Ca)
                R7=INTERVALO*FN FUNCION7(z,Xm)
                S1=INTERVALO*FN FUNCION1(z+INTERVALO/2,FI+R1/2)
                S2=INTERVALO*FN FUNCION2(z+INTERVALO/2,Xh+R2/2)
                S3=INTERVALO*FN FUNCION3(z+INTERVALO/2,Svs+R3/2)
                S4=INTERVALO*FN FUNCION4(z+INTERVALO/2,Ms+R4/2)
                S5=INTERVALO*FN FUNCION5(z+INTERVALO/2,Xa+R5/2)
                S6=INTERVALO*FN FUNCION6(z+INTERVALO/2,Ca+R6/2)
                S7=INTERVALO*FN FUNCION7(z+INTERVALO/2,Xm+R7/2)
                T1=INTERVALO*FN FUNCION1(z+INTERVALO/2,FI+S1/2)
                T2=INTERVALO*FN FUNCION2(z+INTERVALO/2,Xh+S2/2)
                T3=INTERVALO*FN FUNCION3(z+INTERVALO/2,Svs+S3/2)
                T4=INTERVALO*FN FUNCION4(z+INTERVALO/2,Ms+S4/2)
                T5=INTERVALO*FN FUNCION5(z+INTERVALO/2,Xa+S5/2)
                T6=INTERVALO*FN FUNCION6(z+INTERVALO/2,Ca+S6/2)
                T7=INTERVALO*FN FUNCION7(z+INTERVALO/2,Xm+S7/2)
                U1=INTERVALO*FN FUNCION1(z+INTERVALO,FI+T1)
                U2=INTERVALO*FN FUNCION2(z+INTERVALO,Xh+T2)
                U3=INTERVALO*FN FUNCION3(z+INTERVALO,Svs+T3)
                U4=INTERVALO*FN FUNCION4(z+INTERVALO,Ms+T4)
                U5=INTERVALO*FN FUNCION5(z+INTERVALO,Xa+T5)
                U6=INTERVALO*FN FUNCION6(z+INTERVALO,Ca+T6)
                U7=INTERVALO*FN FUNCION7(z+INTERVALO,Xm+T7)
                FI=FI+(R1+2*S1+2*T1+U1)/6:Xh=Xh+(R2+2*S2+2*T2+U2)/6:
                Svs=Svs+(R3+2*S3+2*T3+U3)/6:Ms=Ms+(R4+2*S4+2*T4+U4)/6
                Xa=Xa+(R5+2*S5+2*T5+U5)/6:Ca=Ca+(R6+2*S6+2*T6+U6)/6:
                Xm=Xm+(R7+2*S7+2*T7+U7)/6:z=z+INTERVALO
        NEXT J
        Ca(K+1)=Ca:Xh(K+1)=Xh:FI(K+1)=FI:Svs(K+1)=Svs:Ms(K+1)=Ms:Xa(K+1)=Xa:Xm(K+1)=Xm
NEXT K
PRINT "z,  Ca   ;Xh   ;FI   ;Svs   ;Ms   ;Xa    ;Xm":PRINT z;Ca;Xh;FI;Svs;Ms;Xa;Xm
GOTO 1000
2000 PRINT:METANO=0
FOR I=1 TO (NP-1)
INTEGRAL=INTEGRAL+(km*pHIFm*VINT/2)*((Ca(I)*Xm(I)/(Km+Ca(I)))+(Ca(I+1)*Xm(I+1)/(Km+Ca(I+1))))
next I
METANO=3.1416*D^2*INTEGRAL/4:PRINT "METANO =",METANO
open "o", #1,"B:tolotus.asc":print #1,Sio:print #1,Svso:print #1,Mso:print #1,Cao
print #1,Xho:print #1,Xao:print #1,Xmo:print #1,FI:print #1,Svs:print #1,Ms:print #1,Ca:print #1,Xh
print #1,Xa:print #1,Xm:print #1,METANO:print #1,pH:print #1,SRT:print #1,REL:close #1:END
```

TABLE 2

Values of kinetic parameters used in digester simulation

NOMENCLATURE	DESCRIPTION	VALUE	UNITS
k_a	Acidogenic microorganism growth rate	35.7	1/d
K_a	Acidogenic saturation constant	5	kg/m^3
k_m	Methanogenic microorganism growth rate	5.46	1/d
K_m	Methanogenic saturation constant	0.03	kg/m^3
k_{ho}	Homogeneous hydrolysis reaction rate	0.3	m^3/kg.d
k_{mh}	Hydrolytic bacteria decay constant	0.02	1/d
k_{ma}	Acidogenic bacteria decay constant	0.2	1/d
k_{mm}	Methanogenic bacteria decay constant	0.02	1/d
ma	Coefficient pH acidogenic inhibition factor	3	dimensionless
mm	Coefficient pH methanogenic inhibition factor	2.2	dimensionless
pHha	pH upper limit for acidogenesis	6.5	dimensionless
pHla	pH lower limit for acidogenesis	4.0	dimensionless
pHhm	pH upper limit for methanogenesis	7.2	dimensionless
pHlm	pH lower limit for methanogenesis	5.6	dimensionless
Y_a	Acidogenic biomass yield	0.22	kg/kg
Y_m	Methanogenic biomass yield	0.055	kg/kg
Y_h	Hydrolytic biomass yield	0.05	kg/kg
β	Heterogeneous hydrolytic rate	0.025	m^3/kg.d

These constants are based on literature values and also on own experimental data. They are coincident with the simulation carried out for the production of volatile fatty acids (Negri et al. 1992). Table 3 presents the initial settings of the variables.

3. Results and discussion

The study has been addressed to analyze the influence of the following independent variables:

a) pH

b) Liquid recirculation ratio (Rel)

c) Solid retention time (SRT)

As dependent variables, the following are considered:

a) Percentage of insoluble volatile solids not degraded
b) Percentage of soluble volatile solids not degraded
c) Percentage of monomers not degraded and percentage of biomethanization achieved.

TABLE 3

Values of environmental (feed) parameters used in digester simulation

NOMENCLATURE	DESCRIPTION	VALUE	UNITS
pH	pH	7.0 and 7.5	dimensionless
D	Digester diameter	0.4	m
Svio	Insoluble volatile solids in the feed	16.99	kg/m^3
Rel	Recycle ratio (Qe/Qr)	0.5 to 5	dimensionless
Svso	Soluble volatile solids in the feed	17.5292	kg/m^3
Mso	Monomer fraction of soluble volatile solids	12	kg/m^3
SRT	Solid retention time	.1 to 50	day
Xao	Acidogenic microorganism concentration	0.05	kg/m^3
Xho	Hydrolytic microorganism concentration	0.025	kg/m^3
Xmo	Methanogenic microorganism concentration	0.025	kg/m^3

3.1. INFLUENCE OF pH

In accordance with the previous work, an optimal pH of around 6.4 was found to maximize volatile fatty acids production (Viturtia et al., 1991; Negri et al., 1992). At lower pH, the degree of biological activity decrease and at higher pH, methanization was favoured. Several runs with the program indicated that this step was optimal at pH 7.2. In fact, this result was expected because of the values chosen for the methanogenic constants (see equation 20). Results on the influence of the remaining variables are presented at two pH's 7 and 7.5, both bounding the optimum.

3.2. INFLUENCE OF SRT

The percentage of degradation of all the species studied, monomers, soluble and non-soluble polymers, increase with solid retention time in the reactor. Methanization also increase. Figures 2-5 show the influence of SRT in the variables mentioned at pH 7.5 and Figures 6-9 show the same influence at pH 7. Curves are represented at different values of the recirculation ratio (Rel). As a difference with VFA production, where an optimum value of 12 days was found for SRT, (Negri et al., 1992). In the present case, the

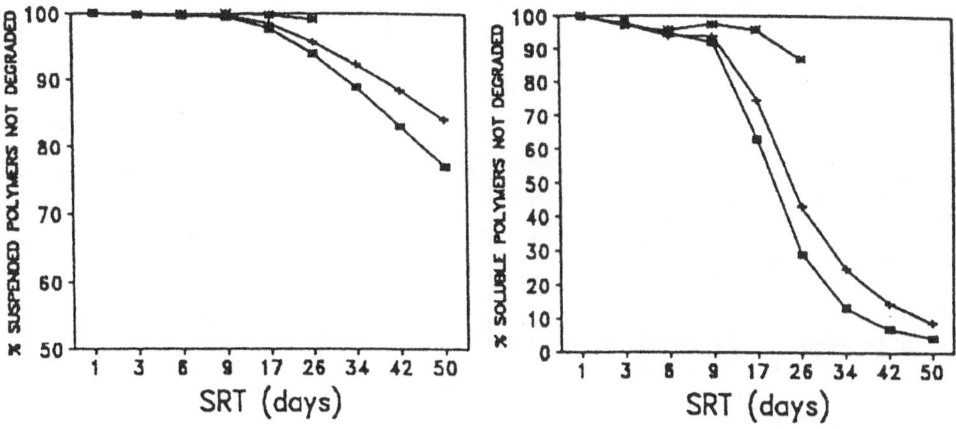

Figure 2. Simulation results at pH 7.5 and X_{mo}=0.025. % suspended polymers ungraded as SRT function •, Rel=0.5; +, Rel=1; *, Rel=5.

Figure 3. Simulation results at pH 7.5 and X_{mo}=0.025. % soluble polymers not degraded as function of SRT. •, Rel=0.5; +, Rel=1; *, Rel=5.

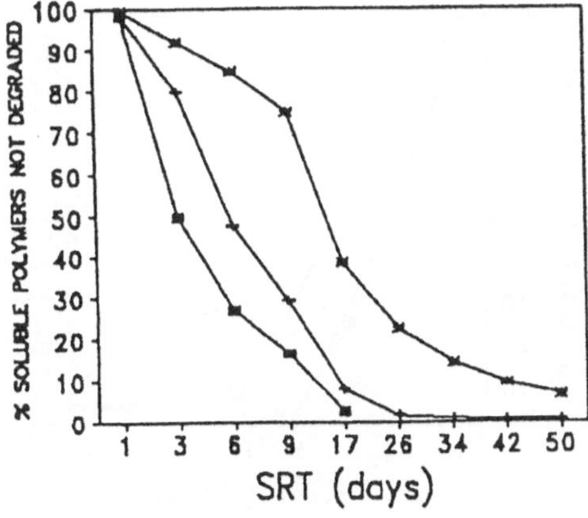

Figure 4. Simulation results at pH 7 and X_{mo}=0.025. % soluble polymers not degraded as function of SRT. •, Rel=0.5; +, Rel=1; *, Rel=5.

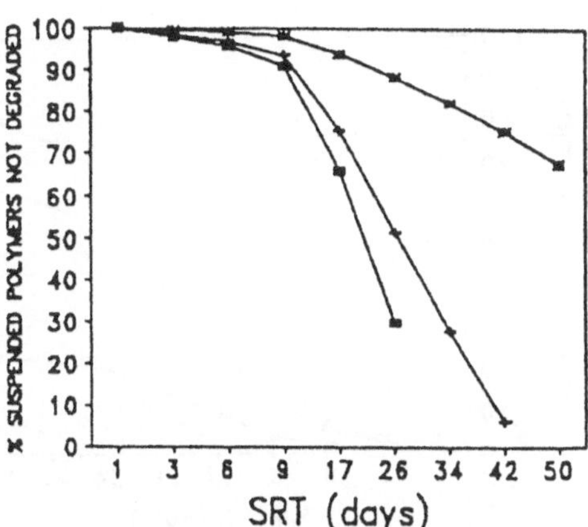

Figure 5. Simulation results at pH 7 and X_{mo}=0.025. %suspended polymers undegraded as SRT function. •, Rel=0.5; +, Rel=1; *, Rel=5.

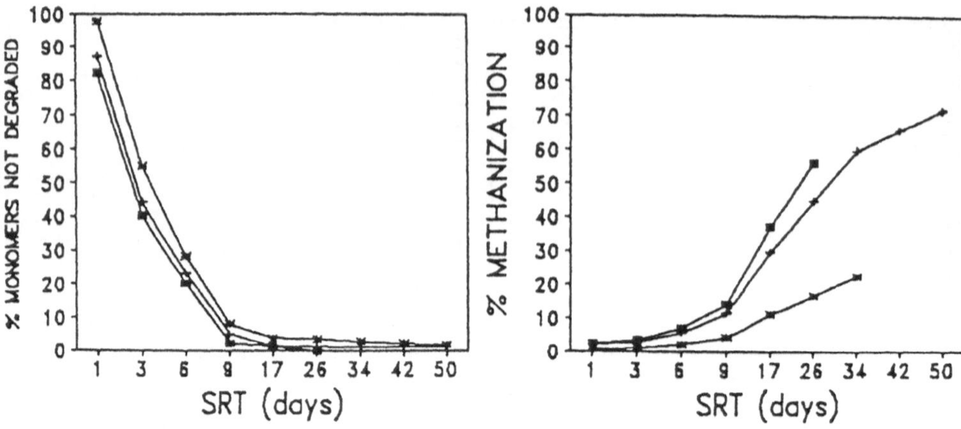

Figure 6. Simulation results at pH 7 and X_{mo}=0.025. % monomers undegraded as SRT function. •, Rel=0.5; +, Rel=1; *, Rel=5.

Figure 7. Simulation results at pH 7 and X_{mo}=0.025. % methanization as SRT function. •, Rel=0.5; +, Rel=1; *, Rel=5.

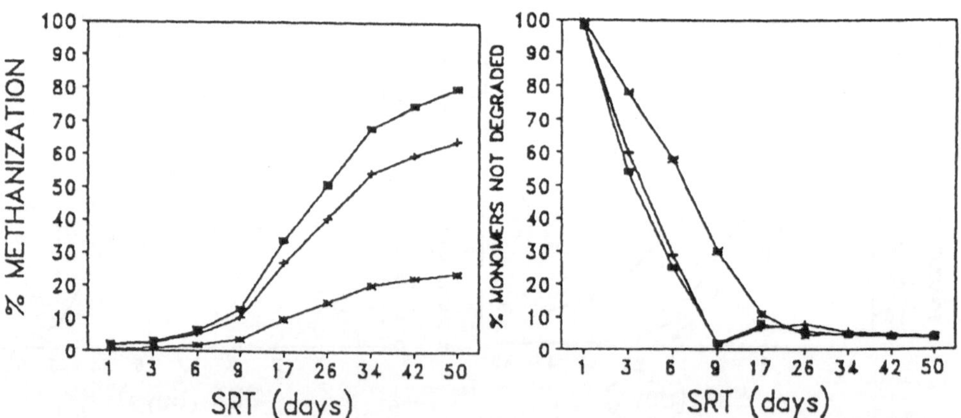

Figure 8. Simulation results at pH 7.5 and X_{mo}=0.025. % methanization as SRT function. •, Rel=0.5; +, Rel=1; *, Rel=5.

Figure 9. Simulation results at pH 7.5 and $X_{m\,o}$=0.025. % monomers undegraded as SRT function. •, Rel=0.5; +, Rel=1; *, Rel=5.

continuous increase of methanization precludes the existence of an optimum based on economic considerations. However, residence time (SRT) should be greater than ca. 25 days in order to have a substantial methanization. On the other hand, this SRT should be higher than 9 days if a significant conversion of non-soluble and soluble compounds is to be obtained.

3.3. INFLUENCE OF RECIRCULATION RATIO

Recirculation ratio (Rel) defined as the recirculation flow rate/reactor outlet flow rate influences negatively, that is, at large recirculation flow rates, methanization and percentage of degradation of the different compounds decrease. The explanation should rely on the type of overall kinetics, which are more favourable for a plug-flow reactor than for a stirred tank reactor. Inoculation by the recirculation stream seems to have no significative influence on the overall performance.

4. Conclusions

A mathematical model has been implemented in a TURBO-BASIC program to simulate the performance of a plug-flow reactor anaerobically degrading an organic waste to produce biogas. Results of the simulations show that optimal pH is situated around 7.2 and that larger solid retention times should be employed. Liquid recirculation flow-rate affects negatively the yields (percentages of degradations and methanization). For methanization purposes using a substrate with the characteristics given in Table 3, a residence time of at least 25 days has to be used.

5. Nomenclature

C_a	volatile fatty acids concentration, kg / m^3
f	fraction of microorganisms carried by the fluid phase, dimensionless
F	fraction of fluid in the reactor inlet, dimensionless
k_{ho}	homogeneous hydrolysis reaction rate, m^3 / kg d
k	microorganisms growth rate constant, 1 / d
km	bacteria decay coefficient, 1 / d
K	saturation constant, kg / m3
L	plug-flow reactor length, m
Ms	monomer fraction of soluble volatile solids, kg / m^3
pH_{LL}	lower limit of pH inhibition function, dimensionless
pH_{UL}	upper limit of pH inhibition function, dimensionless
pHIF	pH inhibition factor, dimensionless
Qe	plug-flow reactor input flow, m^3 / d
Qr	plug-flow reactor recycle flow, m^3 / d
r	insoluble volatile solid particle radius, m

r_A acidification reaction rate, kg / m^3 d
r_{HI} heterogeneous hydrolytic reaction rate, kg / m^3 d
r_{HS} homogeneous hydrolytic reaction rate, kg / m^3 d
r_M methanization reaction rate, kg / m^3 d
R_P initial insoluble volatile solid particle radius, m
Rel recycle relation, dimensionless
Svi insoluble volatile solid concentration, kg / m^3
Svi* insoluble volatile solid concentration at the reactor beginning ($Svi_{(z=0)}$), kg/m^3
Svs soluble volatile solid concentration, kg / m^3
t time, d
v phase velocity, m / d
X microorganisms concentration, kg / m^3
Y biomass yield, kg / kg
z axial coordinate (plug-flow reactor), m

Greek letters

ß heterogeneous hydrolysis rate, m^3 / d kg
Φ r/R_P, dimensionless insoluble volatile solid particle radius

Subscripts

$_a$ corresponding to the acidogenic step
$_{Bio}$ in the biomass phase
$_h$ corresponding to the hydrolytic step
$_l$ in the fluid phase
$_m$ corresponding to the methanogenic step
$_s$ in the solid phase
$_{SVI}$ corresponding to insoluble volatile solid

Superscripts

$_0$ inlet condition

6. References

Costello D.J., Greenfield P.F. and Lee P.L. (1991),"Dynamic Modelling of a Single-Stage High-Rate Anaerobic Reactor - I. Model Derivation". Wat. Res. 25, 847-858.
Gujer, W. and Zehnder, J.J. (1983).'Conversion processes in anaerobic digestion'. Wat. Science and Technology, 15, 127-133.
Siegrist, H., Renggli, D. and Gujer, W.. 'Mathematical modelling of single and two stage sewage sludge treatment'. Wat. Science and Technology (in press).

Mata-Alvarez, J. and Cecchi, F. (1992). "Conclusions of the symposium". International Symposium of Anaerobic Digestion of Solid Wastes". Venice, 14-17 April, 1992.

Negri, E.D., Mata-Alvarez, J., Sans, C. and Cecchi, F. 'A mathematical model of VFA production in a plug-flow reactor treating the organic fraction of municipal solid waste' Wat. Science and Technology (in press).

Viturtia A.Mtz. (1989), "Estudio de la Digestión Anaerobia en Dos Fases de Residuos de Frutas y Verduras". Ph.D. dissertation, Departament D'Enginyeria Química y Metal.lúrgia, Facultat de Química, Universitat de Barcelona, España.

ACTIVATED SLUDGE

H. R. BUNGAY
Department of Chemical Engineering
Rensselaer Polytechnic Institute
Troy, New York 12180-3590, USA

ABSTRACT. Teaching exercises are presented that increase understanding of the activated sludge process for biological waste treatment.

1. Introduction

The challenge for biological waste treatment is to convert dissolved and colloidal pollutants to cell mass, water, and gases. However, there is little advantage in changing pollutants into new cell growth that is too small to collect by any practical method, because when microorganisms are discharged from the treatment plant to a body of water, they create an increased demand for oxygen. The goal is to gather the cells into flocs or films large enough to collect by sedimentation or filtration.

A trickling filter has a solid matrix, with attached microorganisms, through which fluid and air percolate. In activated sludge the organisms are suspended in the liquid into which air is mixed. Lagoons or oxidation ponds are not much different in principle from activated sludge, except that algae usually predominate and provide additional oxygen by their photosynthesis. Land cost, construction costs, and operating costs may make one process the clear choice over another.

Activated sludge units constructed on the site are usually rectangular, while units fabricated by the supplier are most likely to be circular. In either design, the feed enters an aerated zone of air bubbles and liquid that are in good contact.

Some typical concentrations of materials that are important in domestic sewage are shown in Table 1. These values will provide a frame of reference for your simulation exercises.

A.R. Moreira and K.K. Wallace (eds.), Computer and Information Science Applications in Bioprocess Engineering, 265-270.

TABLE 1. Typical composition of sewage

Component			Concentration	
Solids	Total		700	mg/L
	Dissolved	Total	500	
		Volatile	200	
	Suspended	Total	200	
		Volatile	150	
	Settleable Solids		10	ml/L
B.O.D.			200	ppm
T.O.C.			200	
C.O.D.			500	
Nitrogen	Total as N		40	
	Organic		15	
	Ammonia (free)		25	
	Nitrate or nitrite		0	
Phosphorus	Total as P		10	
	Organic		3	
	Inorganic		7	
Chloride			50	
Alkalinity as CaCO			100	
Grease			100	

2. Teaching Programs

These computer programs are in the Public Domain. The best way to get copies is with the FTP system available at most computer centers. The address is FTP.RPI.EDU, directory PUB-READ/bungay.

2.1 SKETCHES OF ACTIVATED SLUDGE TANK

2.1.1. *Instructions.*

- If you are beginning a computer session, boot your system and get BASIC running.
- Type **RUN "ACTSLUD"** ENTER
- Respond to the prompts.
- Continue to the next exercise.

2.2 SIMULATION

One of the most important tools for systems analysis is simulation. As you gain understanding of a system, you can express relationships as equations. Simultaneous solution of the equations can be shown either as numbers or as graphs that can be compared to information from the real system. If the agreement is good, you have confirmation (but not proof) that your understanding has value. Poor agreement means that there are gaps in your knowledge. Computer simulation is a quick and relatively inexpensive way to handle your equations. In addition, formulating equations and constructing a model forces you to analyze your system in a way that leads to valuable insights and new ideas.

SIMBAS is a BASIC program for solving simultaneous, ordinary differential equations with a 4th-order Runge-Kutta integration scheme. The independent variable (usually time) is T. Names beginning with T and followed by an integer are reserved for the program and should not be used in other equations, e.g., T1, T2. Standard BASIC commands are available as are the usual math and trig functions.

The example built into the SIMBAS program comes from the reaction:

$$A \rightarrow B \rightarrow \text{Products}$$

$$\frac{dA}{dt} = - K_1 A$$

$$\frac{dB}{dt} = K_1 A - K_2 B$$

The concentration of A can only decline; B may peak as it is made from A, but then declines as its further reaction overshadows its formation because A is being used up. A simple rule will prevent omitting terms in differential equations that are based on kinetics. In the above reaction equation there is one arrow to or from A, so there is but one term in the differential equation. For B, there are two arrows and thus two terms.

BASIC notation is used with a special way of showing variables and derivatives. Derivatives (inputs to the integration operation) are denoted I(1), I(2), I(3), etc. (The letter I stands for IN). Dependent variables (outputs from integration) are denoted O(1), O(2), O(3), etc. (The letter O stands for OUT). In the example, I(1) stands for dA/dt and O(1) for A. Similarly, I(2) represents dB/dt and O(2) represents B. Thus, Command 1000 is the dA/dt equation:

```
1000 I(1) = K1 * O(1)
```

Command 1010 is the dB/dt equation:

```
1010 I(2) = K1 * O(1) - K2 * O(2)
```

2.2.1. Instructions.

- If you are beginning a computer session, boot your system and get BASIC running.
- Type **RUN "SIMBAS"** [ENTER].
  Observe the built-in example.
- When the run is over, Type **LIST - 30** [ENTER].
  If you have a printer, Type **LLIST.**
  This makes a listing of the entire program.
  If you have no printer, you can display parts of SIMBAS on your screen by listing the range of interest, e.g., **LIST 40 - 90** [ENTER].
- Inspect the commands mentioned in the following paragraphs:

In Command 40, K1 is set to 0.2. In Command 42, K2 is set to 0.1. The reactions being simulated have a chemical, A, added to water. The initial concentration of A is 100 millimolar and B is initially zero. This is specified in Commands 86 and 88.

Time control of the solution depends upon the particular set of equations and on the portion of the solution which is of interest. Specifying a smaller integration interval, T1 in Command 56, would give greater accuracy but with a longer wait for the answers.

Output is controlled by two portions of the program. The headings and axes for the graphs are printed early in the program because they are wanted only once. Commands in the 2000's cause time T, A, and B to be written out at each print interval. You may select numerical or graphical output or both. In the differential equations in the 1000's are commands for plotting. The listing of SIMBAS shows how to suppress numerical output if you want only a graph. There is no graphical output if the PSET commands are deleted.

- Type **40 K1 = 0.4** [ENTER].
  Then type **RUN** [ENTER].
- Experiment with specifications for K2, T1, T2, O(1), and O(2) by retyping the appropriate command as in Instruction 65. Be sure that you understand what happens.
- Continue to the next exercise.

2.3 TAPERED AERATION

Tapered aeration is a way to match supply and demand of dissolved oxygen better in an activated sludge tank. Conventional activated sludge with feed at the inlet and uniform aeration along the length of the tank has a high demand for dissolved oxygen at the inlet and almost no demand at the outlet. Aeration at the final portion of the tank is mostly wasted. With tapered aeration, there are more spargers in the initial region of the bioreactor and fewer near the end. This can be modeled in SIMBAS, and we will assume a linear variation in aeration. Of course, a real tank could have an arrangement of spargers that resulted in non-linear variation.

The model will use a modification of the Streeter-Phelps equations that are used for the dissolved oxygen sag curve that are widely used in environmental engineering. The change is to make the rate of metabolism a function of the dissolved oxygen concentration. The equations are:

$$\frac{dL}{dt} = -kL$$

$$\frac{dD}{dt} = kL - K_aD$$

$$k = \frac{k_{max}O_2}{K_o + O_2}$$

where L is concentration of pollution
D is the oxygen deficit
t is time
k and K_a are coefficients
k_{max} is the maximum uptake rate for oxygen
K_o is a half-saturation constant
O_2 is oxygen concentration

2.3.1. *Instructions.*

- If you are beginning a computer session, boot your system and get BASIC running.
- Type **RUN "TAPER"** [ENTER].
- Respond to the prompts with numbers from zero to six for reaeration coefficient. If you make the maximum and minimum the same, you are simulating standard activated sludge without tapered aeration.
  You can clear the screen by making the minimum greater than the maximum.

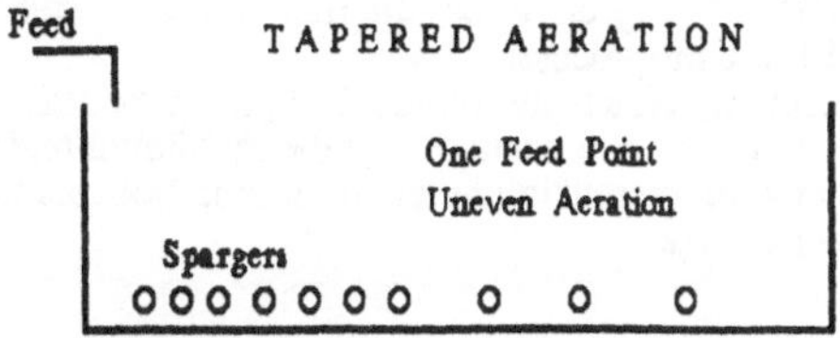

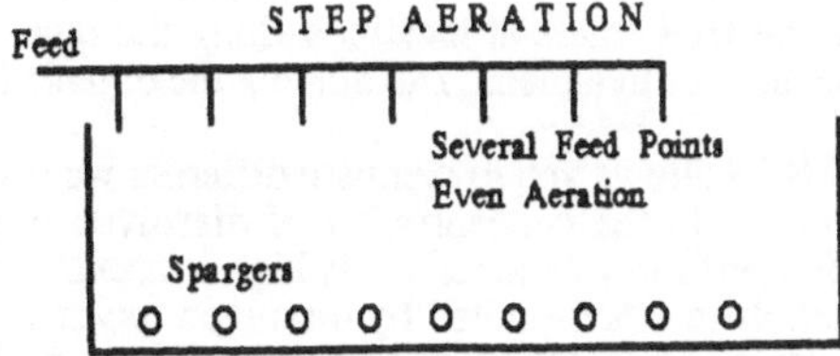

Figure 1 Options for Activated Sludge

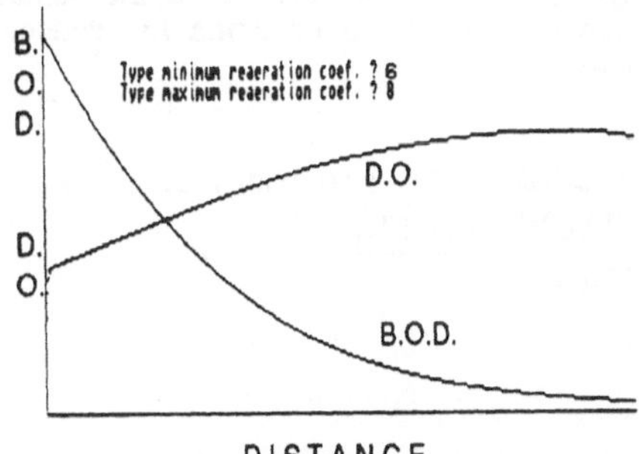

Figure 2 Typical Run for Taper Exercise

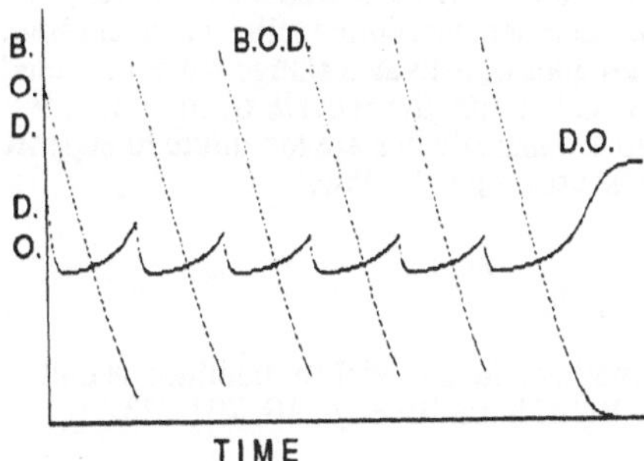

Figure 3 Typical Run for Stepaer Exercise

- After a few runs, LIST the program to be sure that you can associate the SIMBAS code with the equations for the bioprocess.
  Note that tapered aeration gives better removal of pollution than does standard aeration. You show this by a simple mental averaging of the maximum and minimum and comparing with a run with the maximum and minimum both set to this average.
- Continue to the next exercise.

2.4. STEP AERATION

Tapered aeration and step aeration are both methods for better matching of oxygen supply to demand. In step aeration, the feed enters at locations along the tank but not near the outlet because there must be some time for treatment. Distributing the organic load makes better use of the aeration.

The exercise STEPAER.BAS allows you to compare different ways of distributing the feed. The feature of greatest interest in the concentration of dissolved oxygen. Well-acclimated organisms in activated sludge perform well at relatively high concentrations of dissolved oxygen and slow their rate of metabolism when limited by dissolved oxygen. Metabolic rate can be limited either by concentration of organic nutrient or dissolved oxygen. When you respond with one for the number of feed points, the simulation corresponds to conventional activated sludge. Multiple feed points should result in plots of dissolved oxygen that are above the plot for conventional activated sludge most of the time thus limiting the metabolic rate less. At the same time, it makes no sense to have much of the bioreactor at low concentrations of organic matter that would also limit metabolic rate.

2.4.1. *Instructions*

- If you are beginning a computer session, boot your system and get BASIC running.
- Type **RUN "STEPAER"** [ENTER].
- Respond to the prompts.

3. Conclusion

These exercises are intended to provide practice in computer simulation directed towards continuous bioprocessing. The important messges are

- consumption of pollution and of oxygen are related stoichiometrically
- bioprocesses are rate-limited by the ingredient in lowest proportion
- activated sludge waste treatment is more efficient when aeration and feeding are matched

Other important configurations of activated sludge are not covered here. A crucial feature of most of these processes is collection and recycle of the microorganisms to maintain a high population. Otherwise most waste streams are too dilute to support a dense microbial culture, and treatment rates would be unacceptably slow.

4. Additional Reading

Andrews, J.F. (1968) 'A mathematical model for continuous culture of microorganisms using inhibitory substrates', Biotechnol. Bioengr. 10, 707-723.

Grady, C.P.L. and Lim, H.C. (1980) Biological Waste Treatment: Theory and Applications, Marcel Dekker, NY and Basel.

BIOTECHNOLOGY AND THE ENVIRONMENT

ANTON MOSER
Institut für Biotechnologie
TU Graz , Petersgasse 12
A-8010 Graz Austria

ABSTRACT . The role of biotechnologies for the environment will be elucidated and discussed in this paper. The main problems in the environment are the level of pollutions and the resulting loss of biodiversity Two different types of biotechnologies are to be distinguished:

1) Environmental Biotechnologies
2) Ecological Process Engineering ("Eco-technologies").

The first is understood as the application of biotechnologies in order to reduce the level of pollutions in air, water and soil in a holistic way. It is regarded as a short term but end-of-pipe activity (e.g. Verstraete , 1992). The second represents the new technology paradigm, where pollution is prevented with the aid of biotechnologies, which are able to replace polluting (chemical) productions (Moser, 1991).

Renewable raw materials and catalysts will be used and only (bio) degradable products formed. Waste must not necessarily be at zero level as the products are biodegraded by the assimilation capacity of natural cycles. This new area is a long-term activity, it means a full embeddedness of technosphere into the eco-sphere i.e. closing material cycles, obeying the principles of evolution: "ecoprinciples". This will lead to eco-restructuring and ecological sustainability representing an ecologically oriented economy in direction of a " biosociety" .

1. INTRODUCTION

In agreement with the preamble of the European Federation of Biotechnology (EFB) definition of biotechnology, that " biotechnology is directed towards the benefit of human mankind by obeying biological principles " the activities of biotechnology in the area of the environment are to be seen as a major contribution.

Over the last century, environmental technologies gradually developed sanitary engineering as an offshoot of civil engineering, where the "bio" component has largely been ignored. However, due to the great and increasing number of unsolved problems on earth (ozone layer, acid rain, recalcitrant pesticides, waste disposal, water, air, soil), a holistic approach is of utmost importance where biology and ecology are included in technical solutions. It is to be added on this place that one major environmental problem is the substantial loss of biodiversity (biological information in plants and animals), as man's activities result in the

A.R. Moreira and K.K. Wallace (eds.), Computer and Information Science Applications in Bioprocess Engineering, 271-278.

disappearance of about 10,000 species per year

Thereby, two different areas of (bio)technologies will be supplementary in the long run: "environmental biotechnology" and " ecological (bio) process engineering " .

2. Environmental Biotechnology

This branch of biotechnologies, which is defined as the application of biotechnologies to solve environmental problems and where the emphasis is on the "bio" as much as on " technology" is strongly needed at present. It is e.g. best represented in the working party on "environmental bio-technology" in the European Federation of Biotechnology (EFB).

This direction is best outlined in several papers from Verstraete (1992). It is to be seen as end-of-pipe technology looking for the reduction or recycling of existing pollutions in air, water and soil including living bodies.

A series of problem areas are to be mentioned, created by industrial production of pesticides, fertilizers and other "chemicals":

* potable water purification
* waste water treatment
* solid waste handling & disposal including sludge dewatering
* soil & sediment cleaning incl. bioremediation and food
* air & off-gas cleaning

No difference can be seen in these problems in the industrial and industrializing world, only a time shift may exist. Many scientific efforts are under way, as summarized by Verstraete (1992), who also proposes broad guidelines to link with agriculture, to integrate with down stream processing and to implement pollution legislation. Research has to focus on new aspects of natural processes e.g. microbial associations (mixed cultures with different forms of cooperation). Lots of finances are needed for R&D in this direction. Risk assessment is an important part in this "overpositivism", including the special risks of rDNA techniques . Who is going to pay for it? Is "environmental management" the best answer for the solution? Or do we need a deeper ecological movement?

3. Ecological (Bio)Process Engineering: "Eco-Technology"

EFB installed a task group in 1990 on this topic with the general aim to develop a mature concept until 1993, on how to integrate environmental protection directly in the production process. Thus the principle of prevention will be realized instead of the end-of-stream activities which dominate up to now.

The main aim of this innovation, thus, is to develop the concept of "ecological sustainable (bio) technology" as a contribution to "biosociety".

The following steps of activity are incorporated :

(1) *to analyse the background of technosphere*:

in environment (pollutions in air, water, soil and living bodies),
in society (gaps between 1st/3rd world, west/east)
in economy system (pure profit orientation dominates)

The roots of all these problems are in the existing world view i.e. the materialistic/ mechanistic paradigm.

Technology, acting according to this "old" view, can be defined as "end-of-pipe" activity e.g. "environmental management" looking for quick fixes (belief that everything can be made & manipulated by central governments /industries). This is in agreement with the dominant behaviour of industry: "only a fast growing industry can manage the problems".

This view, however, seems to have one weak point: the costs of the end-of-stream technologies are drastically high (see appendix 1). Based on the fact that all phenomena mentioned are interconnected, one has to consider the inequalities on earth. Extrapolating the western living standard to all people on earth (in the east and the south), one can easily understand that the old paradigm is unable to solve the problems. Thus, in order to find a solution for all the problems of all the people on earth, it is necessary to find another paradigm, which is already in development i.e. the holistic/ecologic paradigm.

The belief thereby is that all phenomena are interconnected and must be seen as an entirety: Thus, technosphere, which is in strong conflict with the biosphere at present, must be embedded into the ecosphere.

(2) *to analyse the laws of ecosphere:* ("learning by listening to nature")

Ecosphere evidently represents the intelligence of evolution. Thus, applying a system-analytical approach the "wisdom of nature" can be extracted resulting in a series of "biological/ecological principles" (see appendix 2). They can better be accepted by understanding the basic relationship between structure & function in general, which also plays a role in the field of restructuring the world.

(3) *to transfer the "eco-principles" to practice / technosphere:* "ecologically sustainable production processes"

One of the most essential principles to be transferred to practice is to close the cycles of all materials. The consequence is that e.g. (chemical) production processes in industry based on fossil raw materials will be replaced by (bio) processes based on renewable raw materials: **"large volume / low price" low cost bio-processes** in aqueous media!

On the basis of these principles a "Sustainable Process Index" (**SPI**) can be derived, which represents a guideline for ecological restructuring in industrial practice (see later on). It is used as a questionnaire to be sent out to industry. Technology pre-assessment will replace assessment !

A series of bioprocesses is compiled in appendix 3, representing the practical implementations of the eco-principles. These cases are regarded as case studies to show the production processes which already exist in practice.

(4) to disseminate the concept of "ecological sustainable technology "

An international network is established, shown in appendix 4.

The contribution of biotechnology to sustainable development was recently summarized in an UNIDO report (Swaminathan,1992) including the following essential points:

food security, third world concerns, socio-economic impact,
poverty, employment, biodiversity, biofuture, biosafety,
potential adverse impacts, sustainable agriculture,
proprietary and patent rights, pro-poor R&D,

4. Sustainable Process Index (SPI)

The SPI (Narosdoslawsky, Moser, 1992) includes:

a) <u>process oriented factors :</u>

* NEEDS: basic/industrial/agricultural/comfort/luxury/cultural/quality
* CONSUMPTION rate per year
* SIZE of production per year possible/ minimal/ maximal/ optimal
* TRANSPORT needs: raw materials/ energy/ products

b) <u>product related factors:</u>

* RAW MATERIAL CONSUMPTION: non-renewable/ renewable/ air/water/energy / energy distribution / human work
* PRODUCTS: main/ by-/ recyclable
*WASTES: (quantity & quality): solid/ liquid/ gas
* PRODUCT SPECIFICATION: degradable / life time/ recyclability/ survival time in environment ("if living product")

5. References

Verstraete W. (1990) in Proc. ECB5 vol. 1 , 75-84
Verstraete W. (1992) Holistic Environmental Biotechnology , in Microbial Control of Pollution (Fry J.G et al., eds.) Soc.Gen.Microb. Symp., vol.48, Cambridge Univ.Press
Moser A. (1991) Ecologic Process Engineering-The new technology paradigm , Biotech Forum Europe , vol.8 644-649
Narodoslwasky M. and Moser A. (1992) in Interim Report of EFB Task Group on Ecologic Bioprocessing (MoserA. and Narodoslawsky M., eds.) presented at EC-EFB meeting in Brussels , April, ÖGBPT (publ., Graz/A)
Moser A. and Narodoslawsky M. (eds.) 1992 Ecologic Bioprocessing-Chances in New Applications , Proc. EFB Task Group Intern.workshop in Graz, October 21-22 1991, ÖGBPT (publ.,Graz/A)
Swaminathan M.S. (1992) Contribution of Biotechnology to Sustainable Development within the Framework of the United Nations system , UNIDO report ,Vienna

APPENDIX 1: Situation in Environmental Management:

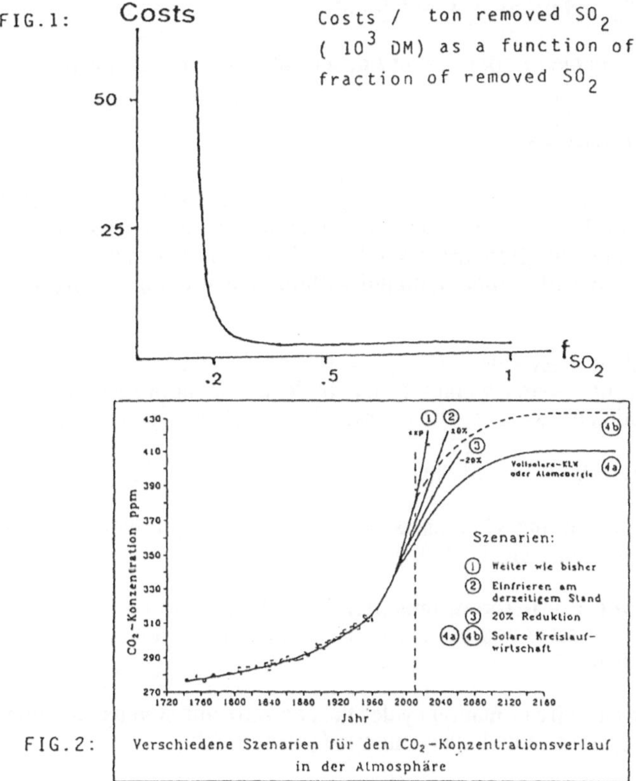

FIG.1: Costs / ton removed SO_2 (10^3 DM) as a function of fraction of removed SO_2

FIG.2: Verschiedene Szenarien für den CO_2-Konzentrationsverlauf in der Atmosphäre

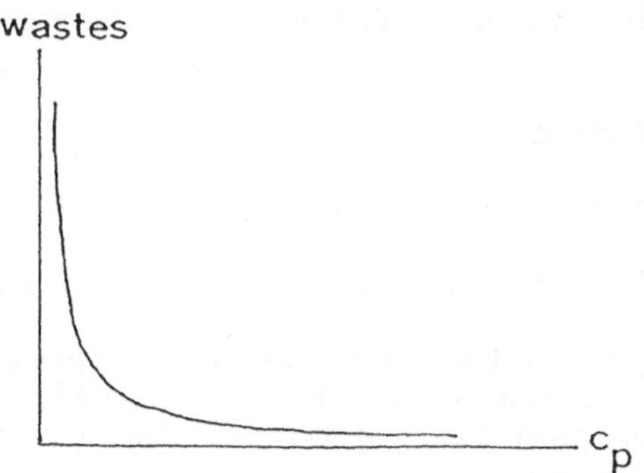

FIG.3: Tons of wastes produced per ton of used raw material as a function of the concentration of the raw material

APPENDIX 2: Bio/Eco Principles

(1) to be globally oriented but act locally

Establish niches of adapted / appropriate / decentralized technologies in order to satisfy local needs first

(2) think & act long-term

Ecology is long-term economy, benefit is long term profit, GNP will be replaced by Eco National Product or other measurements of economic accounting (Human Development Index, Index of Sustainable Economic Welfare). *Sustainability* is the capacity to satisfy the current natural needs of human mankind without jeopardizing the prospects of future generations:

1: rate of use of renewable resources > rate of their regeneration
2: rate of use of non-renewables > rate of development of sustainable substitutes
3: rate pollution emission > rate of assimilation capacity/ ecosphere

(3) use the potential of biosphere:

Genetic information in living systems as basis of industrial productions: renewable raw materials and catalysts (biological cells), degradable products;

coordination/ cooperation (commensalism!) is superior to competition (survival of the best coordinated system and not survival of the fittest), thus technosphere must be embedded in biosphere!

use the assimilation capacity of natural cycles thus no zero emission technologies needed! establish "technology mix" with high economic / social stability

(4) protect the potential of biosphere:

maintain the high diversified ecosphere as a resource for genetic information and assimilations; develop "failure friendly" and repairable technologies as pollution prevention pays!

(5) close the material cycles

is the essential technological principle to be realized on all levels.

(6) minimize fluxes of material & energy (efficiency), integrate:

known engineering principles, which have to be transferred also to biotechnologies in order to increase efficiency resulting in a coordinated supplement and better competitiveness to chemical processing, realize process integration at all levels (genetics,physiology,M&C, down & waste streams, ecology).

APPENDIX 3: Ecological Bioprocesses show cases 1991:

A series of bioprocesses can be mentioned here, falling in five categories depending on the type of replacement: raw materials (1), catalyst (2), products (3), byproducts (4), energy (5). The time of realization could also be used a criterion of systematization.

* biopolymers e.g. polyhydroxybuturic acid
* biocontrol agents (pesticides) e.g. biotoxins from *Bacillus thuringiensis*
* biofertilizers: e.g. algae, rhyzobium, Azotobacter for N2-fixation
* bioleaching of ores. e.g. Cu, U, Au,..
* biosorption of heavy metals from waste waters: e.g. Ag, Zn,Sn,Cr.
* enzyme technology as a whole is a clean technology e.g. paper industry, starch as raw material for bulk products.
* biodegradation/depolymerization of wastes & cellulosic materials: with enzymes (Trichoderma, cellulose, lignin & hemicellulose as raw materials, with bacteria (e.g. Aerobic thermophilic sludge)
* biodesulfurization : e.g. coal
* biodenitrification e.g. of drinking and ground water
* biodepestification e.g. ground and/or drinking water
* biofuel (biogas) e.g. esters from plant-oil water ("bio-energy")
* biodegreasing: with biosurfactants e.g. in electro-plating and metal-working industry, cleansing agents
* biodehairing e.g. leather-industry
* biodefatting e.g. recycle of Cr.
* bioremediation in general for the detoxification of water, soil
* biodrugs e.g. from the high diversity of exotic plants
* biocosmetics in general as high quality products
* bioflavours/bio-odors in general
* bioenrichment: soil by minerals in order to increase food quality
* eco-sustainable farming & pest control according to locally adapted old "technologies" in developing countries e.g. the "waru-waru", from Inca Indians 3.000 a.d.
* biomaterials e.g. for housing in third world
* biomembranes
* biosensors
* BIONICs
* "ecological bioreactor operation" based on multiple criteria

APPENDIX 4: International Network of EFB Task Group "Ecologic Bioprocessing"

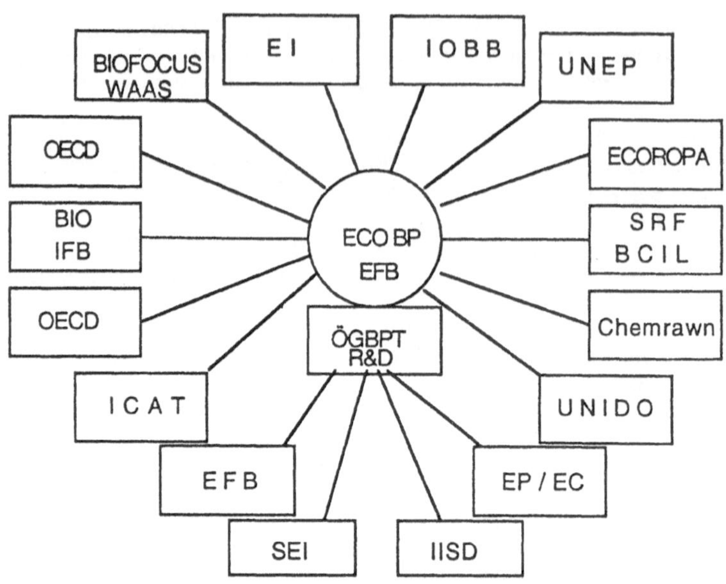

EFB European Federation of Biotechnology
ÖGBPT Austrian Association of BiorprocessTechnology (secretariat), Graz / A , A. Moser and Narodoslawsky
EC/EP European Commissions / Parlament , Brussels/Strasbourg
UNIDO Vienna (Vakataraman, Subrahmaniam)
UNEP group BT for cleaner production (Luyben, Kothuis)
IFB Inter. Forum Biophilosophy , Leuven/B (K. Simpson)
BIO Biopolitics Intern.Organisation , Athens /GR (A.Vlavianos-A.)
EI Elmwood Institute,Berkely/USA (F.Capra)
IOBB Intern.Org.Biotechnol & Bioengng, Guatemala, (C.Rolz)
R&D/A research project , Austria "sustainable economy"(Moser F.)
WAAS World Acad.Art & Scienc. ,BIOFOCUS Stockholm (C-G.Heden)
BCIL Biotech Consortium India Ltd ,New Delhi (Chandrasekar S.)
SRF Swaminathan Research Foundation,Madras/India (Swaminathan)
IACT Inter.Association of Clean Technology , Vienna
ECOROPA European Group for Ecological Action (F.Meissner-Blau)
ECO BP EFB task group Ecologic Bioprocessing (A.Moser)
SEI Stockholm Environmental Institute (M. Chadwick)
IISD Intern. Inst.for Sustainable Development (A.J.Hanson,Canada)
OECD Biotechnolog for clean environment (Dr.. S.Wald,Paris)
CHEMRAWN USSR Academy of Sciences, Moscow (V. Koptyug)

PART 4:

FOOD APPLICATIONS

AN EFFICIENT COMPUTER PROGRAMME FOR THE SIMULATION OF FOOD FREEZING OPERATIONS

A. M. FIGUEIREDO and A.M. SERENO*
University of Porto, Faculty of Engineering
Department of Chemical Engineering
Rua dos Bragas, 4099 PORTO - CODEX
PORTUGAL

ABSTRACT. An efficient and robust computer programme for the modelling and simulation of food freezing operation is described. Calculated results for the freezing times of slices of both potato and tylose as well cylinders and spheres of tylose compare with experimental times within 5.7%. The programme proved to be stable and fast when compared to others published algorithms.

1. Introduction

Freezing is one of the most common means to preserve foods. The combined effect of low temperature and partial crystallization of the water in the material, decreasing its water activity, prevents microbial development and slows down any enzymatic or chemical degradation process. Food freezing can be described as an heat conduction process with phase change. This phase change produces significant and steep changes on physical properties of the material requiring special care for its mathematical description.

Bonacina and Comini (1974) have used the description proposed by Lees (1966) to simulate freezing process. They used the concept of an apparent heat capacity which includes in a single quantity both sensible and latent heats of the product during the phase change. Later, Wood and Lewis (1975) and Cleland and Earle (1977a,b, 1979) used the same model with a convective boundary condition to describe the behaviour of cylindrical and spherical particles.

Freezing times calculated with mathematical models based on the concept of an apparent heat capacity generally agree with experimental times. However, the discontinuity on the first derivative of this function at the start of the freezing process may produce oscillations of the numerical integration; this may be avoided by very small time steps leading to large computation times. Dussinberre (1962) suggested a different approach. He used enthalpy as the main independent variable and temperature as an auxiliary variable used to estimate enthalpy along the calculation path; the method was used by Crowley (1978), Voller and Cross (1985).

* *Author to whom correspondence should be address.*

A.R. Moreira and K.K. Wallace (eds.), Computer and Information Science Applications in Bioprocess Engineering, 281-289.

Mannapperuma and Singh (1988) took this formulation and included corrective shape factors for the heat balance in the case of cylindrical and spherical geometries, leading to a better precision of the calculated predictions.

Pham (1985) presented an hybrid method based simultaneously on the apparent heat capacity and the enthalpic formulation. He used an explicit enthalpic algorithm to estimate the apparent heat capacity which is then used with the conventional formulation. This method shows the same advantages as the apparent heat capacity approach with the robustness typical of the enthalpy formulation, as stated by the author.

2. Mathematical Model

Most of the food materials being frozen may be adequately described by a simple unidimensional geometry. Heat conduction may then be described by the following equation based on the apparent heat capacity concept already mentioned :

$$C_a(T) = \frac{\partial H(T)}{\partial T} = C_p(T) + H_s \frac{\partial W(T)}{\partial T} \tag{1}$$

and

$$C_a(T) \frac{\partial T}{\partial t} = \frac{1}{r^m} \frac{\partial}{\partial r} (r^m K_r(T) \frac{\partial T}{\partial r}) \tag{2}$$

where C_a is the apparent volumetric heat capacity, W is the fraction of non-frozen water, K is the thermal conductivity and m is a geometric factor (m=0 infinite slab, m=1 infinite cylinder, m=2 sphere). The mathematical model used in this work was based in Equation 2 with the following initial condition:

$$t = 0 \quad : \qquad T = T_i \qquad ; \forall_r \tag{3}$$

Symmetrical heat flux at the geometric center and a convective type boundary condition at the surface leads to:

$$r = 0 \quad : \qquad K_r(T) \frac{\partial T}{\partial r} = 0 \qquad ; t > t_0 \tag{4}$$

$$r = R_0 \quad : \qquad K_r(T) \frac{\partial T}{\partial r} = h_c (T_a - T_s) \qquad ; t > t_0 \tag{5}$$

where T_a is the external ambient temperature, T_s is the temperature of the product at the surface and h_c is the convective heat transfer coefficient. Equation 5 assumes that heat transfer by radiation and vaporization is negligible. This proves to be a reasonable assumption at the low temperatures existing during freezing processes.

3. Numerical Algorithm

For the physical model described, simple unidimensional geometries and uniform boundary conditions, a finite difference algorithm is probably the best choice for the numerical integration of Equation 2. A grid of *np* node surfaces (plane, cylindrical or spherical) regularly spaced from the geometric center of symmetry (plane, axis or single point) to the external surface were considered leading to *np-2* elementary control volumes centered on each nodal surface plus two terminal half-thick elements. A similar procedure was followed for the time discretization. Figure 1 gives a simplified description of this time-space grid.

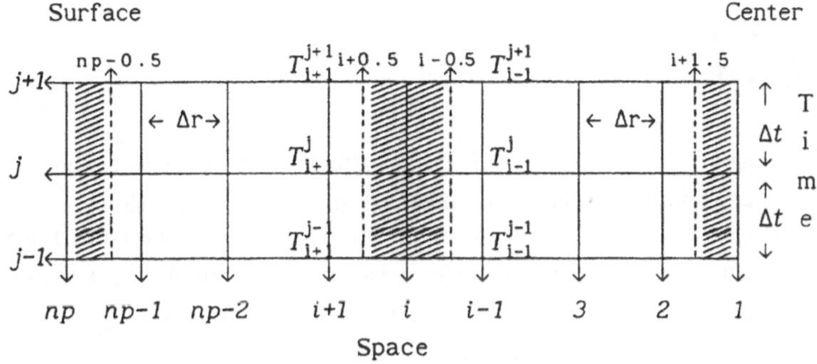

Figure 1 - Time-space nodal representation of product.

Making an heat balance on the elementary control volume *i* of unit cross section, between time nodes $t_{i,j-1}$ and $t_{i,j+1}$ the following expression is obtained based on a three time level average of centered finite differences with respect to space (Lees, 1966):

$$C_{ai,j} * (T_{i,j+1} - T_{i,j-1}) = \frac{2*\Delta t}{3*\Delta r^2*v_i} *\Big(a_{i+0.5}*k_{i+0.5,j}(T) * ((T_{i+1,j+1} + T_{i+1,j} + T_{i+1,j-1}) - (T_{i,j+1} + T_{i,j} + T_{i,j-1})) + a_{i-0.5}*k_{i-0.5,j}(T) * ((T_{i-1,j+1} + T_{i-1,j} + T_{i-1,j-1}) - (T_{i,j+1} + T_{i,j} + T_{i,j-1})) \Big) \quad \text{for } i=2,np-1 \text{ and } j=1,\ldots \tag{6}$$

In the above expression, a_i and v_i represent area and volume factors respectively as suggested by Mannapperuma and Singh (1988). A similar procedure was used for the terminal volume elements, taking into account initial and boundary conditions of Equations 3 to 5. The second term of Equation 6 is unconditionally stable (Bonacina and Comini, 1973) and represents the enthalpy gain by heat conduction of node *i* during the given time interval of amplitude 2*Δt. Equation 7 is an explicit approximation of this gain:

$$\Delta H_{i,j} = \frac{2*\Delta t}{\Delta r^2 * v_i} * \left(a_{i+0.5} * k_{i+0.5,j}(T) * (T_{i+1,j} - T_{i,j}) + \right.$$

$$\left. a_{i-0.5} * k_{i-0.5,j}(T) * (T_{i-1,j} - T_{i,j}) \right) \tag{7}$$

this value of $\Delta H_{i,j}$ may be used with the following equation to approximate temperature $T_{i,j+1}$:

$$T_{i,j+1} = f_T (f_H (T_{i,j-1}) + \Delta H_{i,j}) \tag{8}$$

and the apparent heat capacity:

$$C_{ai,j} = \frac{\Delta H_{i,j}}{T_{i,j+1} - T_{i,j-1}} \tag{9}$$

The auxiliary functions f_T and f_H are numerical approximations to calculate T from H and H from T values. These approximations were obtained by linear interpolation of pairs of experimental data for the heat capacity of the material. Values of $C_{ai,j}$, estimated by Equation 9 and based only on data calculated for time nodes $t_{i,j}$ and $t_{i,j-1}$, are used with Equation 6 to predict all the temperatures corresponding to time $t_{i,j+1}$. These temperature predictions are subsequently corrected by the following formula as suggested by Pham, (1985):

$$T_{i,j+1(\text{corrected})} = f_T \left(f_H (T_{i,j-1}) + C_{ai,j} * (T_{i,j+1} - T_{i,j-1}) \right) \tag{10}$$

4. Worked Examples

A FORTRAN 77 computer programme (CDA - Congelação/Descongelação de Alimentos) was written based on the algorithm previously described. Such programme was used to simulate the continuous freezing of 10 kg/hr of tylose slices. Tylose is a 23% methylcellulose gel, with thermophysical properties (Figure 2) very similar to lean beef (Cleland and Earle, 1984) and thus very popular in freezing experiments. Table 1 indicates the operational parameters considered. The calculated freezing time was 56.3 minutes and theoretical cooling requirements

Table 1 - Operational parameters considered in example simulation

Material	:	Tylose	Ambient temperature (T_a)	:	-25.0 °C
Geometry	:	Slice	Final center temperature (T_f)	:	-18.0 °C
Half thickness (R_o)	:	0.010 m	Heat transfer coefficient (h_c)	:	50.0 $Wm^{-2}{}^oC^{-1}$
Initial temperature (T_i)	:	20.0 °C	Mass flow rate (wg)	:	10.0 $Kg\ h^{-1}$

were 3260 KJ/hr (assuming no losses). Figure 3 shows the predicted temperature profiles in the tylose slices during the freezing process.

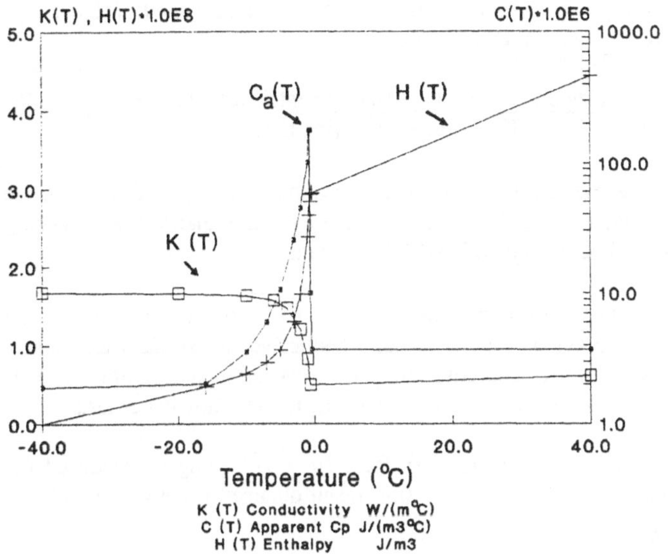

Figure 2 - Thermophysical properties of tylose with 77% of water (Cleland and Earle, 1984)

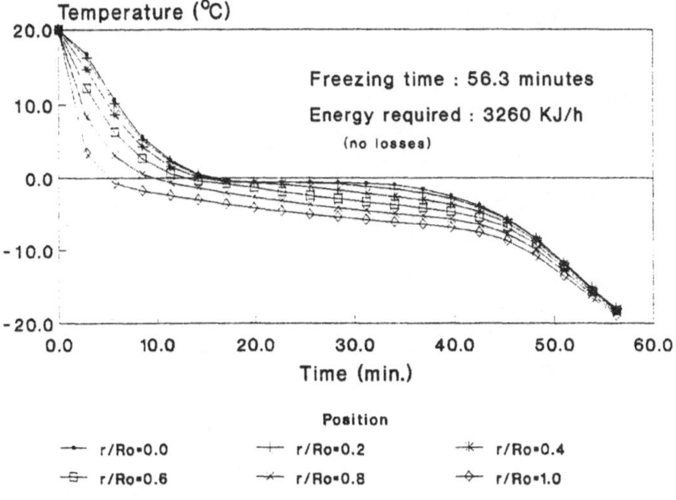

Figure 3 - Temperature profiles in tylose slice during the freezing process.

To assess the reliability of the model described, calculated freezing times for tylose and potato were compared with corresponding experimental data by Cleland and Earle (1977b,1979), and Hung and Thompson (1983). In Figures 4 and 5 experimental and calculated freezing times for slices, cylinders and spheres of both tylose and potato are compared. Table 2 summarize the results using a percent deviation D defined as :

$$D\ (\%) = \frac{\text{Calculated freezing time - Experimental freezing time}}{\text{Experimental freezing time}} * 100 \qquad (11)$$

All calculations were checked by a global heat balance technique, comparing the total heat transferred through the external boundary with the overall enthalpy change undergone by the product. Both calculations agreed with a 0.1 % tolerance.

A major objective of this work was to compare the efficiency and robustness of the algorithm described with others previously proposed and applicable to cylinders and spheres. Table 3 summarizes the results of a comparative study with two of such algorithms in the freezing of tylose. The three methods were compared according to the following criteria:

a) Time increment and number of nodal surfaces (*np*> *4*) leading to calculated freezing times agreeing within 1% with the "true" result (i.e. result obtained when very small time and space intervals are used).

b) Computer processing (cpu) time on a DecStation 3100 computer with processor RISC MIPS 2000 using ULTRIX 4.0

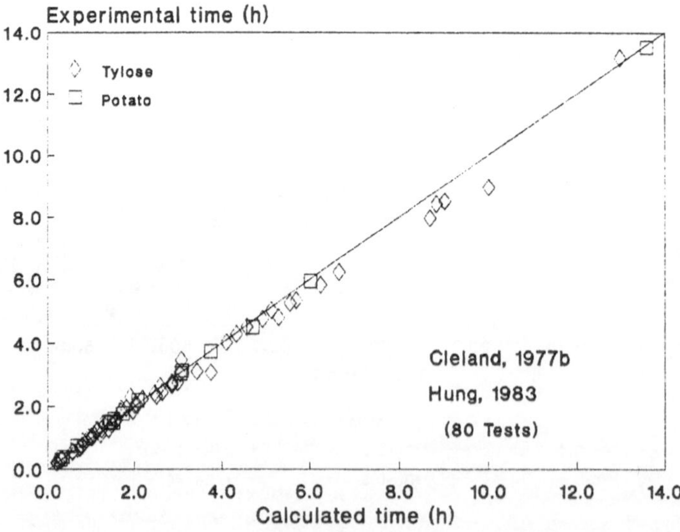

Figure 4 - Experimental and calculated freezing times for tylose and potato slices.

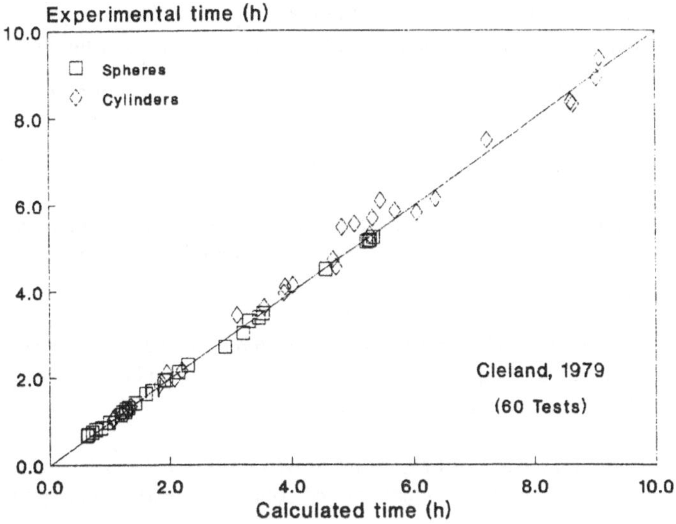

Figure 5 - Experimental and calculated freezing times for tylose cylinders and spheres.

Table 2 - Percent deviations in the prediction of freezing times

Geometry	Number of tests	Average	Mean	Standard	Maximum Deviation Negative	Maximum Deviation Positive
Slice	80	5.67	-1.14	6.92	-11.74	+11.80
Cylinder	30	3.80	-1.57	4.67	-11.96	+ 5.10
Sphere	30	2.54	0.15	3.46	-7.58	+ 7.41

Table 3 -Comparison of performance for different models.

Geometry	Parameters	Cleland and Earle, 1979	Mannapperuma and Singh, 1988	Present Model
Cylinder	Relative cpu time	27.2	1.7	1.0*
	np	40	5	5
	Δt	1.0	1.9	10.0
Sphere	Relative cpu time	38.0	1.7	1.2
	np	80	5	5
	Δt	1.0	1.25	5.5

* *Actual cpu time was 0.1 s on DecStation 3100 and 3.2 s on a 80386, 33MHZ based PC.*

As a closing note, it is important to stress that the same algorithm (and computer programme) works equally for thawing as well as for freezing, as the basic mechanisms involved are the same.

5. Conclusions

The results obtained have shown that mathematical modelling and computer simulation of freezing operations is possible and constitutes an important tool for the design and optimization of industrial freezing plants.

Calculated freezing times agree with experimental data within an average absolute deviation of 5.7 %. The programme can be used both for freezing and thawing and enables the prediction of several physical and operational parameters of such plants.

The programme based on the described algorithm proved very stable and fast allowing its easy integration in computer-based systems to monitor and control industrial freezing units.

Acknowledgements

Financial support by JNICT (Programa Ciência) and NATO (Programme SFS - Project PO-PORTOFOOD) are greatly acknowledged.

Nomenclature

- a_i - Area factor of node i
- C_a - Apparent volumetric heat capacity ($J\ m^{-3}\ {}^0C^{-1}$)
- C_p - Ordinary volumetric heat capacity ($J\ m^{-3}\ {}^0C^{-1}$)
- f_H - Function to calculate enthalpy from temperature ($J\ m^{-3}\ {}^0C^{-1}$)
- f_T - Function to calculate temperature from enthalpy ($m^3\ {}^0C\ J^{-1}$)
- h_c - Surface convective heat transfer coefficient ($J\ m^{-2}\ {}^0C^{-1}$)
- H - Volumetric enthalpy ($J\ m^{-3}$)
- H_s - Latent heat of solidification ($J\ m^{-3}$)
- K - Thermal Conductivity ($W\ m^{-2}\ {}^0C^{-1}$)
- m - Geometric factor
- np - Number of nodal surfaces
- r - Radial coordinate (m)
- Δr - Radial increment (m)
- R_0 - Radius/half thickness of product (m)
- t - Time (s)
- Δt - Time step (s)
- T - Temperature (${}^0C$)
- T_a - Ambient temperature (${}^0C$)
- T_i - Initial temperature of product (${}^0C$)
- T_s - Temperature at product surface (${}^0C$)
- v_i - Volume factor of node i
- W - Fraction of non-frozen water

Subscripts, Superscripts

- i - Spacial node
- j - Time node
- r - Radial direction
- 0 - Initial condition

REFERENCES

Bonacina, C., Comini, G., (1973) On the Solution of the Nonlinear Heat Conduction Equations by Numerical Methods', Int. J. Heat Mass Transf., 16, 581.

Cleland, A.C., Earle, R.L., (1977a) 'The Third Kind of Boundary Condition in Numerical Freezing Calculations', Int. J. Heat Mass Transfer, 20, 1029.

Cleland, A.C., Earle, R.L., (1977b) 'A Comparison of Analytical and Numerical Methods of Predicting the Freezing Times of Foods', J. Food Sci., 42(5), 1390

Cleland, A.C., Earle, R.L., (1979) 'A Comparison of Methods for Predicting Times of Cylindrical and Spherical Foodstuffs', J. Food Sci., 44(5), 958.

Cleland, A.C., Earle, R.L., (1984) 'Assessment of Freezing Time Prediction Methods', J. Food Sci., 49, 1034.

Crowley, A.B., (1978) 'Numerical Solution of Stefan Problems', Int. J. Heat Mass Transf., 24(3), 545.

Dussinberre, G.M., (1962) Heat Transfer Calculations by Finite Differences, International Book Company, Scranton, PA., pp.171.

Hung, Y.C., Thompson, D.R., (1983) 'Freezing Time Prediction for Slab Shape Foodstuffs by an Improved Analytical Method', J. Food Sci., 48., 555.

Lees, M., (1966) 'A Linear Three Level Difference Scheme for Quasilinear Parabolic Equations', Math. Comput., 20, 516.

Mannapperuma, J.D., Singh, R.P., (1988) 'Prediction of Freezing and Thawing Times of Foods Using a Numerical Method Based on Enthalpy Formulation', J. Food Sci., 53(2), 626.

Pham, Q.T., (1985) 'A Fast, Unconditionally Stable Finite-Difference Scheme for Heat Conduction with Phase Change', Int. J. Heat Mass Transf., 28(11), 2079.

Voller, V., Cross, M., (1985) 'Application of Control Volume Enthalpy Methods in the Solution of Stefan Problems', in R.W. Lewis, K. Morgan, J.A. Johnson, and R. Smith (eds.), Computacional Techniques in Heat Transfer, Pineridge Press, Swansea, U.K., pp.245.

Wood,W.C., Lewis, R.W., (1975) 'A Comparison of Time Marching Schemes for the Transient Heat Conduction Equation', Int. J. Num. Methods Engng., 9, 679.

MODELLING AND SIMULATION OF CONVECTIVE GREEN-PEA DRYING

G.L. MEDEIROS, A.M. SERENO*
University of Porto, Faculty of Engineering
Department of Chemical Engineering
Rua dos Bragas
4099 PORTO CODEX
PORTUGAL

ABSTRACT. Warm air drying of peas was studied to assess the effect of drying conditions on some relevant physical properties of the product. Experimental drying data was well correlated by a previously described model where an implicit variation of moisture diffusivity during the dehydration is considered. Using that model, the experimental data could be reproduced within 6.1%. Drying conditions have influenced shrinkage of the product, as expressed by apparent density, and moisture diffusivity but not sorption isotherm.

1. Introduction

Warm air drying is still an important mean of reducing moisture content of fruits and vegetables and achieve an enhanced resistance to their degradation due to the corresponding decrease in water activity.

In this study the behaviour of peas during drying by warm air was studied. The major objective was to test the ability of a simplified mathematical model recently proposed by Sereno and Medeiros (1990) to describe the drying behaviour of foods and to identify the effect of air temperature on relevant physical and transport properties of the product.

The model takes into account the effect of moisture-solid interaction at the drying surface by means of any sorption equation available and the change in solid density due to shrinking. Fourier's and Fick's laws describe the transfer of heat and mass in the solid. At the surface, mass and heat balances together with the chosen sorption equation are used to represent the vaporization of water.

Temperature and moisture-content profiles are obtained by integration of the resulting set of partial differential equations, an implicit-finite-differences algorithm being used.

2. Mathematical Model

The model used assumes that a temperature gradient and a concentration gradient are the sole driving forces responsible for the heat and mass transfer, respectively.

* *Author to whom correspondence should be addressed*

A.R. Moreira and K.K. Wallace (eds.), Computer and Information Science Applications in Bioprocess Engineering, 291-297.

Non-steady heat and mass conservation equations are used with Fourier's and Fick's laws of diffusion to model drying behaviour. Uniform boundary conditions expressing convective heat and mass balances at the surface are also used.

$$\rho \cdot c_p \frac{\partial T}{\partial t} = - \operatorname{div} q^T \tag{1}$$

$$\rho_s \frac{\partial X}{\partial t} = - \operatorname{div} q^X \tag{2}$$

$$q^T{}_r = - K \frac{\partial T}{\partial r} \tag{3}$$

$$q^X{}_r = - \rho_s \cdot D \frac{\partial X}{\partial r} \tag{4}$$

$$q^T{}_s = h\,(T_s - T_a) - \lambda\, k_p\,(p_s - p_a) \tag{5}$$

$$q^X{}_s = k_p\,(p_s - p_a) \tag{6}$$

As a major approximation the product $\rho_s \cdot D$ appearing in Equation 4 is assumed constant over the entire drying process and not D alone as considered by other models (Saravacos and Charm, 1962, Bimbenet et al, 1985):

$$\rho_s \cdot D = \rho_{s0} \cdot D_0 = \text{constant} \tag{7}$$

3. Material and Methods

Deep-frozen peas of "Agrilusa" brand were used. The peas had been stored at -18°C for about one month, and were thawed at 8°C for 12 hours. Peas were air dried in a pilot dryer, using constant temperature, humidity and velocity of the air during each run. Several sample sets of 20 peas each were individually positioned over a stainless steel net of 5 mm square mesh and were dried by a vertical upwards flow of air at 2.2 m/s.

At selected times along the drying operation each one of the 20 peas sample was withdrawn from the dryer. Each sample was analyzed for water activity, density and moisture content. Water activity was measured with an electric hygrometer (Thermoconstanter from DEFENSOR -Novasina AG, Switzerland) after a stable reading was reached. Density was measured by means of a 50 ml picnometer using n-heptane; after introducing the peas, the picnometer has been de-aerated under the vacuum of a common laboratory water jet. Two independent density values were obtained for each sample and averaged. Moisture content was evaluated from the weight loss of the sample after 24 hours in an electric oven at 50°C and atmospheric pressure, followed by another 24 hours period at 70°C and 60 mmHg absolute pressure.

For the numerical fitting of experimental data a non-linear least-squares regression programme based on the Levenberg-Marquardt algorithm and proposed by Press et al. (1988) was used.

4. Results and Discussion

Experimental apparent densities were correlated with the model proposed by Lozano et al. (1983):

$$\rho = C_1 + C_2 \cdot X + C_3 \cdot \exp(-C_4 \cdot X) \tag{8}$$

where: ρ - apparent density (kg/m^3)
X - moisture content (kg/kg db)

The calculated parameters C_1 to C_4 are presented in Table 1.

Table 1 - Parameters calculated for Lozano equation (Eq. 1).

	30°C	50°C	65°C
C_1	1148	1141	1167
C_2	-27.1	-24.2	-40.1
C_3	65.6	159	247
C_4	3.43	3.37	3.22
Average relative deviation (%)	1.2	1.3	0.75

Water activity was well correlated by the GAB model (Guggenheim, Anderson, de Boer):

$$X = \frac{Xm \cdot C \cdot k \cdot a_w}{(1 - k \cdot a_w)(1 - k \cdot a_w + C \cdot k \cdot a_w)} \tag{9}$$

where: a_w - water activity
X - moisture content (kg/kg db)
Xm = 0.0516 (kg/kg db)
C = 1.41
k = 0.954

Prediction of drying curves was made using the model proposed by Sereno and Medeiros (1990). This model assumes the combined modification of apparent density of dry solid matrix and water diffusivity, with moisture content of the solid during the entire operation.

The experimental data (Figure 1) was adequately described by that model using the operational parameters of Table 2. The best agreement of the predicted curves to the experimental data was obtained for the values of water diffusivity shown in Table 3. Average relative deviation of experimental and predicted values is 6.1%.

Table 2 - Values for the parameters used for the simulation of the drying curves.

R_0	0.0049 m
X_0	2.26 kg/kg dry solid
T_0	20°C
K	0.315 W/m/°C (Qashou et al., 1979)
c_p	3.3 kJ/kg/°C (Polley et al., 1980)
v_a	2.2 m/s
h	71.2 W/m²/°C

Table 3 - Absolute humidity and $\rho_s \cdot D$ values used for each temperature.

T_a (°C)	H_a (kg/kg dry air)	$\rho_s \cdot D$ (kg/m/s)	D_0 (m^2/s)
30	0.0065	$1.03 \cdot 10^{-10}$	$3.1 \cdot 10^{-10}$
50	0.0070	$1.53 \cdot 10^{-10}$	$4.6 \cdot 10^{-10}$
65	0.0075	$2.18 \cdot 10^{-10}$	$6.6 \cdot 10^{-10}$

ρ_s - ratio of dry matter to apparent total volume $[= \rho/(1+X)]$

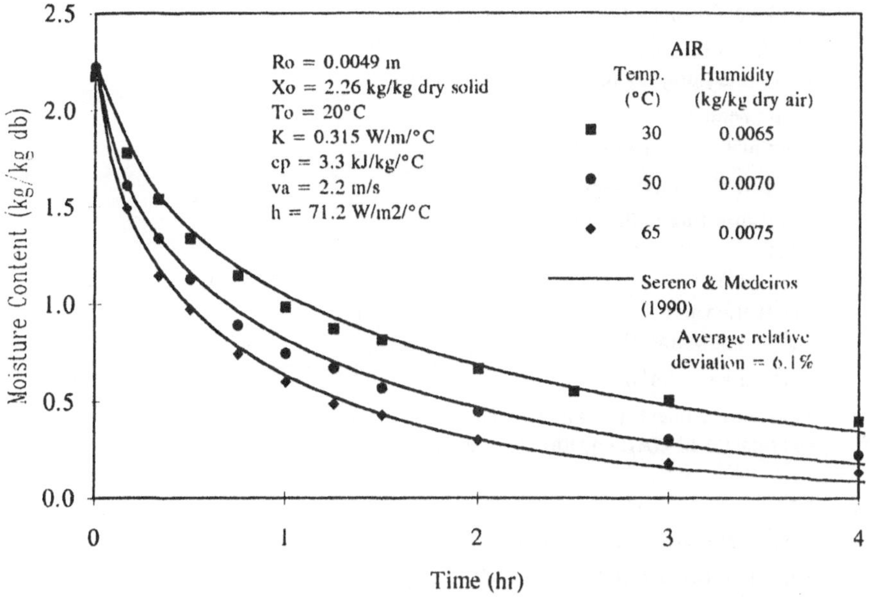

Figure 1 - Drying curves for peas. Experimental data and simulation results.

5. Conclusions

The results obtained indicate that at low moisture content, apparent density seems to increase with drying air temperature. Values of parameters for Lozano et al. (1983) model were calculated. Drying temperature has not shown any significant influence on the desorption isotherm at 25°C, and parameters for the GAB equation were calculated.

A good agreement was obtained between experimental drying data and the curves predicted by the model of Sereno and Medeiros (1990). The average relative deviation encountered was 6.1%.

6. Nomenclature

a_w - water activity

c_p - specific heat (kJ/kg/°C)

C - parameter for GAB equation (Eq. 2)

C_1 a C_4 - parameters for Lozano equation (Eq. 1)

D	- effective diffusivity of water in solids (m^2/s)
h	- surface heat transfer coefficient ($W/m^2/°C$)
H_a	- absolute humidity of air (kg/kg dry air)
k	- parameter for GAB equation (Eq. 2)
K	- thermal conductivity of solid (W/m/°C)
q^T	- energy flux in the solid ($J/m^2/s$)
q^X	- moisture flux in the solid (kg water/m^2/s)
R	- radius of solid particles (m)
t	- time (s)
T	- temperature (°C)
T_a	- air temperature (°C)
v_a	- air velocity (m/s)
X	- moisture content (kg/kg dry basis)
Xm	- parameter of GAB equation (Eq. 2) (kg/kg dry basis)
	Greek
ρ	- apparent density of solid (kg/m^3)
ρ_s	- apparent density of dry solid matrix (kg/m^3)
	Subscript
0	- initial state
e	- equilibrium
s	- air-solid interface
r	- radial coordinate

7. Acknowledgements

The authors acknowledge the financial support of JNICT - Junta Nacional de Investigação Científica e Tecnológica, and of NATO's Scientific Affairs Division through project NATO-SfS-PO-PORTOFOOD.

8. References

Bimbenet, J.J., Daudin, J.D., Wolff, E. (1985) 'Air Drying Kinetics of Biological Particles', in R. Toei and A.S. Mujumdar (eds.), DRYING'85, Hemisphere, Washington, pp. 178-185.

Lozano, J.E., Rotstein, E., Urbicain, M.J. (1983) 'Shrinkage, Porosity and Bulk Density of Foodstuffs at Changing Moisture Contents', J. Food Sci., **48**, pp. 1497-1502, 1553.

Polley, S.L., Snyder, O.P., Kotnour, P. (1980) 'A Compilation of Thermal Properties of Foods', Food Technol., **34** (11), pp. 76-94.

Press, W.H., Flannery, B.P., Teukolsky, S.A., Vetterling, W.T. (1988) Numerical Recipes, Cambridge University Press, Cambridge.

Qashou, M.S., Vachon, R.I., Touloukian, Y.S. (1972) 'Thermal Conductivity of Foods', Trans. ASHRAE, **78**, pp. 165-183.

Saravacos, G.D., Charm, S.E. (1962) 'A Study of the Mechanism of Fruit and Vegetable Dehydration', Food Technol., **16**, pp. 78-81.

Sereno, A.M., Medeiros, G.L. (1990) 'A Simplified Model for the Prediction of Drying Rates for Foods', J. Food Eng., **12**, pp. 1-11.

PART 5:

METABOLIC ENGINEERING

MODELLING OF THE BACULOVIRUS INFECTION PROCESS IN INSECT-CELL REACTOR CONFIGURATIONS.

CORNELIS D. DE GOOIJER, RICK H.M. KOKEN, FRANK L.J. VAN LIER, MARCEL KOOL[+], JUST M. VLAK[+], and JOHANNES TRAMPER
Wageningen Agricultural University
Food and Bioprocess Engineering group
P.O. Box 8129
6700 EV Wageningen
The Netherlands
[+] Department of Virology.

ABSTRACT. In this paper two mathematical models of the infection of insect cells with baculovirus in a continuously-operated reactor configuration are presented. The reactor configuration consists of one bioreactor in which insect cells (*Spodoptera frugiperda*) are grown, followed by one or two bioreactors in which cells are infected by a baculovirus (*Autographa californica* nuclear polyhedrosis virus). The first model shows that the steady-state infection level can be described by a first-order reaction-rate constant. The second model is based on the hypothesis that the limited runtime of series of continuously operated bioreactors is associated with the occurrence of a virion which is defective and has interfering properties. Combined with the assumption that not all non-occluded virions are capable of establishing a correct infection leading to new virus production, the infection process in continuously operated reactor configurations can be described well with the model. The difference in run time between a configuration with one or with two infection vessels with the same overall residence time can be attributed to the occurence of defective interfering virions.

1 Introduction.

Baculoviruses are attractive biological agents for control of insect pests in agriculture [13]. In addition, these viruses can be genetically engineered for the production of recombinant proteins [12,18,22]. To obtain commercially attractive levels of productivity and at the same time meet most of the registration requirements, the production of baculoviruses or recombinant proteins can best be achieved in insect-cell bioreactors [5,19], preferably operated in a continuous [16] or semi-continuous [7] fashion.

Two phenotypically different forms of a baculovirus exist. The occluded form, where the rod-shaped virus particles are present in proteinatious bodies (polyhedra), is infectious for insect larvae, whereas the non-

A.R. Moreira and K.K. Wallace (eds.), Computer and Information Science Applications in Bioprocess Engineering, 301-316.

occluded virus form (NOV), circulating in the insect blood, is infectious for insect cells. Polyhedra of *Autographa californica* nuclear polyhedrosis virus (AcNPV), often used as a model virus, can be produced over long periods of time in a system consisting of an upstream bioreactor in which insect cells are cultured, followed by one or two

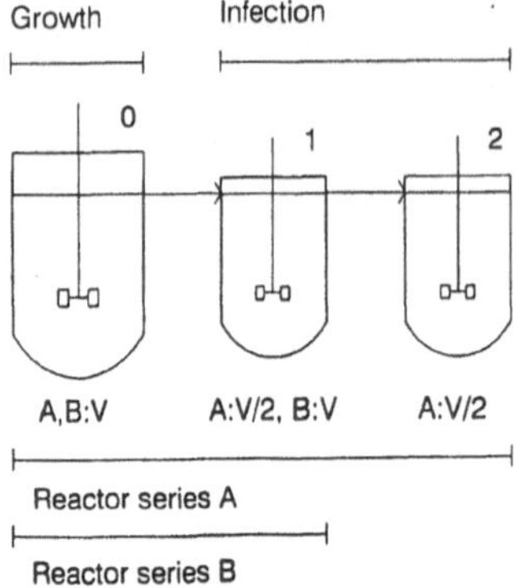

Figure 1. Schematic representation of the experimental continuously operated reactor configurations as used by Van Lier et al. [17] (series A) and Kompier et al. [8] (series B) (adapted from de Gooijer et al. [1]).

bioreactors where infection with the non-occluded form of AcNPV takes place (Fig. 1). The continuous production can be maintained up to several weeks, after which production declines, as shown by Kompier et al. [8] for reactor configuration B.

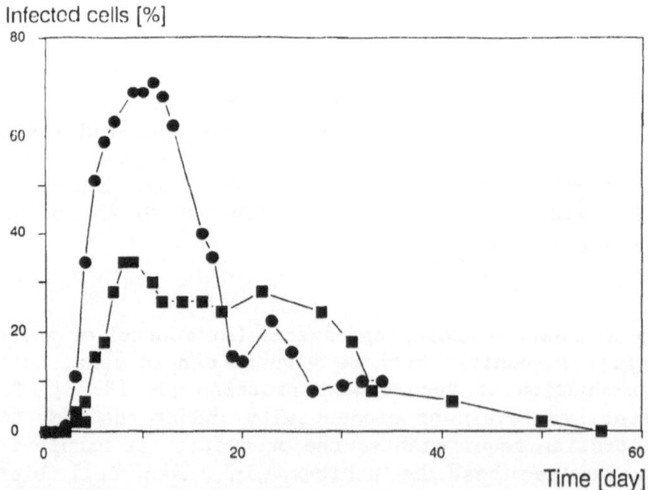

Figure 2. Percentage of visibly infected cells present in reactor 2 of configuration A (●) or in reactor 1 of configuration B (■) versus time (adapted from de Gooijer et al. [1]).

The level of infection can be increased by increasing the number of infection reactors as shown by Van Lier et al. [17] for reactor configuration A. However, the time this system can be operated at this enhanced level of infection decreases significantly to less than three weeks (Fig. 2).
Kool et al.[9,10] showed that the reduction in the production of infectious virus and/or recombinant proteins is due to the occurrence of mutant virions present in the infection reactor(s) which interfere with the replication of intact virus.

2 Theory.

2.1 First-order infection rate.

Two products are generated in the viral infection process: NOVs and polyhedra. As shown earlier [1], about 200 NOV particles are produced per insect cell. Regarding the viable, non-infected, insect cells and NOVs as substrates, and both the NOVs and the polyhedra (P) as products resulting from the infection 'reaction', the following reaction equation can be derived:

$$\text{Insect Cell} + 20\ \text{NOV} \rightarrow 200\ \text{NOV} + n\ \text{P} \tag{1}$$

In eq. 1, the denominator 20 denotes the Multiplicity of Infection (MOI), the number of infectious NOV (I-NOV) per insect cell. In our laboratory we found for n, the number of polyhedra released per insect cell, a value of 25, but this may vary with reactor operating parameters [8]. Eq. 1 shows the large amount of NOVs produced in the infection process, relatively to the number of insect cells. This value, determined from anchored-cell cultures [1], may differ under varying reaction conditions.
Under the next assumptions: i) the growth rate of the cells is constant both in time and place, ii) cell growth is a first-order process, iii) the NOVs are, in steady state, available in excess and therefore eq. 1 can be described by a first-order reaction rate with respect to the cell concentration, the following equations can be derived :

$$r_s = k_r C_m \tag{2}$$

$$r_g = k_g C_m \tag{3}$$

where r_s is the first-order infection rate, $[dm^{-3}.h^{-1}]$
r_g is the first-order growth rate, $[dm^{-3}.h^{-1}]$
k_r is the first-order reaction-rate constant, $[h^{-1}]$
k_g is the first-order growth-rate constant, $[h^{-1}]$
C is the cell concentration, and subscript $[dm^{-3}]$
m denotes the reactor number in the series. [-]

With a mass balance for viable, non-infected cells for the mth reactor in steady state it can be deduced that :

$$k_r = \frac{C_{m-1} + (k_g \tau_m - 1) C_m}{C_m \tau_m} \tag{4}$$

A special case arises if the series of reactors contains one reactor for growth followed by one reactor for infection, with both volumes equal, so that $\tau_0 = \tau_1$ (configuration B). Then, since the growth rate of cells adapts to the mean residence time in the growth vessel, *i.e.* $k_g = 1 / \tau_0$, equation 4 reduces to :

$$k_r = \frac{C_0}{C_1 \tau_1} \tag{5}$$

All values for the parameters in equations 4 and 5 can be measured easily, except for the reaction rate constant k_r.

2.2 Passage effect.

The passage effect, as described by Faulkner [3], manifests itself as a decrease in the production of polyhedra at higher undiluted passages of baculovirus NOV preparations. This effect is also accompanied by a decrease in the number of I-NOVs that is produced per insect cell. One passage is defined [3,6] as the process of an I-NOV entering an insect cell, transport of the genetic information into the cell nucleus, production of new I-NOVs, transport to and budding through the cell membrane, and secretion of the virions into the extracellular fluid. After completion of this process the passage number is increased by one. The secreted virions will maintain the infection process in the continuously operated infection vessels. Schematically this is illustrated in Figure 3.

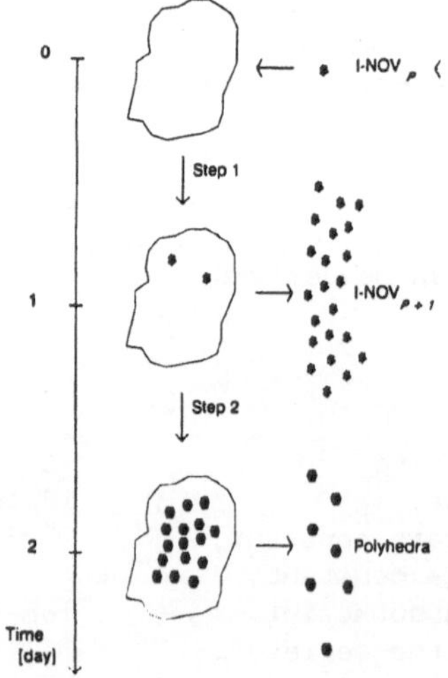

Figure 3. Schematic representation of the replication cycle of baculoviruses in insect cells (adapted from de Gooijer et al. [2]). *p* and *p+1* denote passage numbers. See text for more details.

Apart from this infectious I-NOV, two additional types of virions are introduced here. The first additional type of virion was described by Kool et al. [9,10]. They showed the existence and established some of the properties of a defective interfering virion, present in the medium of the reactor during continuous runs. This D-NOV lacked about 44% of the viral genome, among which the polyhedrin gene and the gene coding for DNA-polymerase. Due to this deletion the D-NOV needs the intracellular presence of an intact I-NOV as a helper for multiplication. Since their genome is smaller, D-NOVs are likely to be reproduced faster than I-NOVs. For our experiments, with a total residence time of 60 hours, it could be calculated that 68% of the cells had a residence time of more than 24 hours and therefore should have been visibly infected in the case of one infection reactor, whereas this is 82% in the second reactor for two infection vessels with the same total residence time [11] (Configurations B and A in figure 1, respectively). This theoretical value, however, could never be reached in our laboratory (Fig. 2). As we also did not reach a 100% infection in batch cultures [2], we assume that a second additional virion type is present which is characterized by its inability to complete a replication cycle as discussed above. These virions have the same physical properties as I-NOVs and for example may attach to the cell surface but fail to form an endosome, or may fail to leave the endosome [14]. In these cases no reproduction takes place, whilst cell receptors are being occupied. To express this assumption mathematically, a NOV leading to an abortive infection (A-NOV) is included in our model. An A-NOV does occupy an entry site, but will not lead to any reproduction. Volkman et al. [23] reported a ratio I-NOV/A-NOV of less than 0.01 for AcNPV produced in *Trichoplusia ni* cells. In our laboratory, a ratio of 1 I-NOV to 60 A-NOVs could be calculated [2].

2.3 Reactor model.

A general mass balance for I-NOVs becomes hard to solve for higher passages [2]. Therefore, a discrete mass balance is introduced :

$$x_{t+\Delta t} = x_t + ((r_{x,in} - r_{x,out}) + r_{x,prod} - r_{x,use})\Delta t \tag{6}$$

where t is the time, [h]
Δ t is one time step, [h]
x is a compound (number of cells or virions), [-]
$r_{x,in}$ and $r_{x,out}$ are the in and outgoing flows of x, [h^{-1}]
$r_{x,prod}$ is the production rate of the compound, and [h^{-1}]
$r_{x,use}$ is the consumption rate of x. [h^{-1}]

Equation (6) can be applied to all types of NOV as well as insect cells. For the I-NOV this can be done for each virion passage. This leads to:

$$v_{i,m,p,t+\Delta t} = (1 - D_m \Delta t).(v_{i,m,p,t} + v_{i,in,m,p,t}) + (R_{i,m,p} - I_{i,m,p})\Delta t \tag{7}$$

where D is the dilution rate in a reactor, $[h^{-1}]$
I is the virion consumption rate used for infection, $[h^{-1}]$
R is the rate of virion release by the cells, $[h^{-1}]$
v is the amount of virions, subscript [-]
i denotes the infectious I-NOV,
p denotes the passage number.

For D-NOVs and A-NOVs it is not necessary to calculate with separate passages, so equation (7) is simplified to :

$$v_{(d \vee a),m,t+\Delta t} = (1 - D_m \Delta t).(v_{(d \vee a),m,t} + v_{(d \vee a),in,m,t}) + (R_{(d \vee a),m} - I_{(d \vee a),m})\Delta t \quad (8)$$

where d denotes the defective interfering virion (D-NOV), [-]
and [-]
a denotes the virion leading to an abortive infection (A-NOV).

Due to the fact that in the cell-growth vessel in the series no virion production takes place, v_0 is equal to zero for all NOV types.
The total amount of I-NOV in vessel m can be calculated from:

$$v_{i,m,tot,t+\Delta t} = \sum_{p=0}^{p=p_{max}} v_{i,m,p,t+\Delta t} \quad (9)$$

where p_{max} is the maximum number of passages at time $t+\Delta t$, [-]
and subscript [-]
tot denotes the total number of I-NOVs.

The amount of virion release by the insect cells is controlled by the number of infected cells some time before. As shown previously [1], the virions are not released at once. Therefore, a time distribution of the I-NOV and A-NOV release is introduced in the model, as described in [1]. Kool et al. [9,10] showed that D-NOVs have a 44% smaller genome than I-NOVs. If D-NOV DNA and I-NOV DNA are replicated at the same speed, the replication cycle of a D-NOV will be finished sooner. Since the process of transport to and from the cell nucleus may also be of importance, the D-NOVs are arbitrarily assumed to have the same time distribution but moved forward in time by two hours.
Equation (6) can also be applied to viable insect cells:

$$CV_{m,t+\Delta t} = (1 - D_m \Delta t).(CV_{m,t} + D_{m-1} \Delta t CV_{m-1,t}) + k_g(CV_{m,t} - G_m)\Delta t \quad (10)$$

where G is the number of cells that is infected in one time step. [-]

Note that $CV_{0,t}$ is the number of cells in the growth vessel. Cells that are not infected can grow in infection vessels with the same speed as in the cell-growth vessel. Infection of insect cells with baculovirus will annihilate cell division [3].

2.4 Infection model.

With the three virion types (I-NOV, A-NOV, and D-NOV), three important modes of infection can be distinguished. The first mode is formed by correct infections, being an insect cell infected with at least one I-NOV, and not infected by any D-NOV. Such cells will produce I-NOV, A-NOV, and a small amount of D-NOVs. The second mode consists of defective infections, where an insect cell is infected with at least one I-NOV and at least one D-NOV. Such an insect cell will produce I-NOV, A-NOV, and a large quantity of D-NOVs. The third mode is formed by an abortive infection, defined by an insect cell that is infected with an A-NOV and/or D-NOV, but without an infection with infectious I-NOV. Insect cells infected this way will not produce virions. For both the first and second mode of infection the ratio of the numbers of virions produced is constant with time.

To be able to calculate the fractions of each mode of infection, the following assumptions were made:

a) Each insect cell has the same amount of 'entry sites' for NOVs. This number is constant during one run and is much less than the number of receptors [25].
b) All entry sites are equal.
c) All NOVs have identical binding sites. Hence, the probability that a NOV attaches is the same for all types of NOV.
d) Attachment of a NOV to a cell is irreversible.
e) The three NOV types cannot change into alternate types spontaneously.
f) Binding of NOVs to cells will take place within one time step.

With these assumptions the fractions of the different modes of infection can be calculated for each time step if the probabilities for the different modes of infection are known.

This problem, where a known number of three types of virions can occupy a known number of entry sites, is the same as the problem where red, white and blue balls are to be dispended over a number of small boxes. It is known from the latter that the probability of a box remaining empty has a Poisson distribution. For our case, the probability (P_o) of an entry site remaining empty is given by :

$$P_o = e^{-\frac{v_{tot}}{BCV}} \tag{11}$$

where B is the number of entry sites per cell, and subscript [-]
tot denotes the total number of all virions. [-]

From that, the probabilities ($P_{i,d,a}$) of an entry site being infected with an I-NOV, D-NOV, or A-NOV are described by:

$$P_{i,d,a} = \frac{v_{i,d,a}(1 - P_o)}{v_{tot}} \tag{12}$$

The sum of these probabilities equals unity.

With these probabilities the fractions of the various modes of infection of the cells can be calculated [2]. The fraction of the cells that are not infected (F_o) then is:

$$F_o = F(B_a = 0 \wedge B_i = 0 \wedge B_d = 0) = P_o^B \tag{13}$$

The fraction of the cells that are abortively infected (F_a) becomes:

$$F_a = F(B_i = 0 \wedge (B_a + B_d > 0)) = \sum_{\alpha \geq 0, (\beta+\lambda)>0}^{\alpha+\beta+\lambda=B} \frac{B!}{\alpha!\beta!\lambda!} P_o^\alpha P_a^\beta P_d^\lambda \tag{14}$$

where α is the number of entry sites per cell that are not occupied by any virion, [-]
β is the number of entry sites per cell that is occupied by A-NOVs, and [-]
λ is the number of entry sites per cell that is occupied by D-NOVs. [-]

The fraction of cells that are correctly infected (F_i) is calculated from:

$$F_i = F(B_i > 0 \wedge B_d = 0) = \sum_{\kappa>0, \alpha, \beta \geq 0}^{\alpha+\beta+\kappa=B} \frac{B!}{\alpha!\beta!\kappa!} P_o^\alpha P_a^\beta P_i^\kappa \tag{15}$$

where κ is the number of entry sites per cell that is occupied by I-NOVs. [-]

The fraction of cells with a defective infection (F_d) is can be found from:

$$F_d = F(B_i > 0 \wedge B_d > 0) = \sum_{\kappa, \lambda>0, \alpha, \beta \geq 0}^{\alpha+\beta+\kappa+\lambda=B} \frac{B!}{\alpha!\beta!\kappa!\lambda!} P_o^\alpha P_a^\beta P_i^\kappa P_d^\lambda \tag{16}$$

The sum of all fractions is equal to one.
The actual amounts of cells and virions, as required in the equations (7), (8), and (10) can easily be calculated from these fractions. Note that the implementation of equations (14), (15), and (16) also limits the theoretical number of entry sites that can be handled in the model to 1546 for numerical reasons. With the above equations the concentration in time for each of the virion types can be calculated. With these concentrations the decline in infection level in continuously operated bioreactors can be described.

3 Materials and Methods.

Cells and viruses

Spodoptera frugiperda cells [20] were maintained in TNM-FH medium [4] without egg ultra-filtrate, but supplemented with 10% fetal-calf serum. For suspension cultures 0.1% (w/v) methylcellulose was added. The E2-

strain of *Autographa californica* nuclear polyhedrosis virus (AcNPV)[15] was used. The stock solution of the virus contained 10^8 $TCID_{50}$ units [1] per cm^3.

Reactor configuration

The continuously-operated reactor configuration with one infection vessel has been described by Kompier et al.[8], and the configuration with two infection vessels by van Lier et al.[17].

Assays

In order to determine the infectivity of the virus, 4-cm^3 aliquots of the samples were centrifuged (1600 *g* for 15 min). The supernatant was filtered (0.45 μm), and the infectivity was measured using the end-point dilution method, described by Vlak [21]. The presence of polyhedra in the cell nucleus was determined using an inverted microscope (magnification 400x). A Neubauer hemocytometer was used to measure the cell number.

Serial passages

The procedure for the serial passages was described by de Gooijer et al.[2].

4 Results and Discussion.

First-order infection rate

For the reactor series with two infection vessels, the viable, non-infected cell concentration in the first infection vessel (table I) was used (eq. 4) to calculate the reaction rate constant in the first infection vessel.

Parameters		A (2)	B (1)
C_0	Cell concentration in reactor 0	8.1 10^5	7.8 10^5
C_1	Viable cell concentration in reactor 1	4.9 10^5	4.2 10^5
τ_1	Residence time in reactor 1	30	60
I_1	Infected fraction in reactor 1	31	55
k_r	Reaction rate constant	1.07 10^{-5}	0.86 10^{-5}
C_{2p}	Predicted cell conc. in reactor 2	3.0 10^5	-
C_{2m}	Measured cell conc. in reactor 2	2.1 10^5	-
I_2	Infected fraction in reactor 2	59	-

Table I. Results with two reactor configurations. A and B are referred to in the text. Numbers between parenthesis denote the number of continuous operated infection vessels.

If the presented first-order reaction-kinetics model is valid, the same value for the reaction-rate constant should apply for the second infection vessel, and therefore the viable cell concentration in the second infection vessel can be predicted. Moreover, using the same cell line and virus in the other reactor configuration, the observed reaction-rate constant (eq. 2) has to be in the same range. Experimental results, taken from Kompier et al.[8] and from van Lier et al.[17], and calculated reaction rates are listed in table I. Due to the increase in residence time of the first

infection vessel going from reactor series A to reactor series B, an increase in the infected fraction in the first reactor can be expected, as is clearly shown by our results in table 1. Moreover, as stated by Kompier et al.[8], the infected cell fraction in the second vessel of reactor series A should be higher than in the infection vessel of series B. We indeed could observe this effect in practice (Table I), however not as strongly as theoretically could be expected.
In reactor series A, one growth vessel followed by two infection vessels, the reaction-rate constant in the first infection vessel is $1.07\ 10^{-5}\ s^{-1}$. With this reaction rate, the concentration of viable, non-infected cells in the second infection vessel can be calculated to be $3.0\ 10^{5}$ $cells.cm^{-3}$.

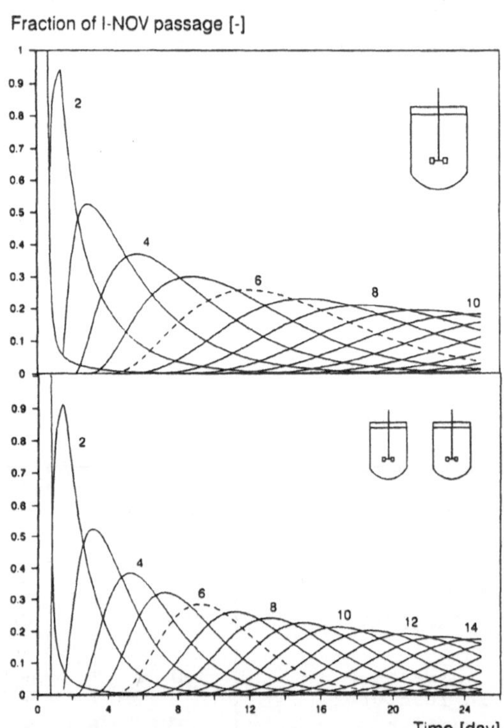

Figure 4. Passage distribution of I-NOVs, depicted as a fraction ($I\text{-}NOV_{p}/I\text{-}NOV_{tot}$) occurring in time in the last vessel for two different continuously-operated reactor configurations with one and two infection vessels (adapted from de Gooijer et al. [2]).

The actual concentration observed was $2.1\ 10^{5}$ $cells.cm^{-3}$. For reactor series B, one growth vessel followed by one infection vessel, a value for the reaction-rate constant of $0.86\ 10^{-5}\ s^{-1}$ can be calculated. The results thus match our theory rather well.

Passage-time distribution of I-NOVs

With the model discussed above, the effects of different reactor configurations were studied. For one set of parameters, the total residence time in the series was kept constant by varying the reactor volumes for each reactor at a constant flow. The fractions of infectious NOV of each passage in time were calculated for reactor configurations with one and two infection reactors, with the infection reactor volume being 1.2×10^{-3} and 0.6×10^{-3} m^3, respectively. Results are shown in figure 4, where the line representing the sixth passage is dotted. Figure 4 clearly shows that higher passages occur sooner if the number of vessels in the series of infection reactors is increased, i.e. if plug flow is more closely approximated.

To assess the passage effect in more detail, virions were serially passaged in two independent series of batch experiments and the amount of NOV produced in each passage was determined. The results are shown in figure 5. At the ninth passage a sharp drop in the number of I-NOVs produced per cell was observed. At higher passages, a further decrease in the number of polyhedra per cell was found. Furthermore, the polyhedra showed morphological aberrations, which is a known consequence of the passage effect[3].

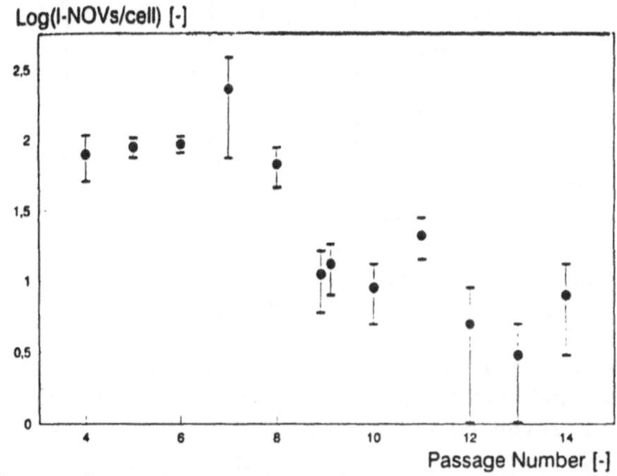

Figure 5. Prolonged passages of I-NOV : the logarithm of the averaged number of produced I-NOVs per cell as a function of the passage number (adapted from de Gooijer et al. [2]).

Infection process

Matching the model on the data points was done on the basis of visibly infected cells. We assumed that of the synchronously-infected cells, the first ones become visibly infected after 20 hours, and the last after 44 hours. Visibly infected cells remain visible for 60 hours, then they lyse. After including the three NOV types into the program all our experimental continuous runs could easily be described with the model. Input variables are given in table II. Fitted parameters were the ratio between I-NOVs and A-NOVs at start-up, the virion production, the number of entry sites, the D-NOV production rate at correct infections, and the number and ratio of virions produced at a defective infection. The ratio of I-NOV to A-NOV was

in the same range as discussed earlier.
Two examples of fits on the data of van Lier et al.[17] and Kompier et al.[8] are given in figure 7 and 6.

Variable	A	B
Reactor working volume	2 x 600	800
Reactor start-up volume	2 x 300	400
Flow	$2x10^{-2}$	$1.33x10^{-2}$
Number of I-NOVs at start-up	$7.1x10^{6}$	$1.6x10^{3}$
Number of A-NOVs at start-up	$1.1x10^{8}$	$6.2x10^{4}$
Number of D-NOVs at start-up	0	0
Cell concentration in growth vessel *(Co)*	$9.5x10^{5}$	$9.0x10^{5}$
Time step (Δ *t*)	2	2
Number of entry sites *B*	40	33
Number of virions produced at a correct infection	3340	4000
of which A-NOVs (%)	94	97.5
of which D-NOVs (%)	10^{-6}	10^{-6}
Number of virions produced at a defective infection	1600	1600
of which A-NOVs (%)	71	77
of which D-NOVs (%)	25.5	22.9

Table II. Input variables for the infection model. Values under A and B are used in figures 7 and 6, respectively.

From these figures it can be seen that at the time at which the decrease in I-NOV concentration occurred, the number of D-NOVs increased. The model behaved this way with all experimental runs analyzed so far. Note that in figure 7, the concentration of virions does not decrease to zero. This was experimentally observed, and can be predicted by the model assuming that the cells that are either infected with I-NOV and D-NOV and/or A-NOV will still produce some I-NOVs.
Kool et al.[9,10] showed with a recombinant virus that D-NOVs selectively accumulate during continuous runs at each point in time, with a clear increase at the end of the run. Apparently, after a period of low occurrence, D-NOVs are produced in large amounts due to their faster rate of synthesis. After the exponential increase in the amount of D-NOVs, there are too few correct infections to produce enough I-NOVs to support D-NOV reproduction (helper function), and the infection process ceases. This seems a valid explanation for the passage effect.
As reported by Von Magnus [24], prolongation of the serial passages will eventually result again in an increase in the number of I-NOVs. Since D-NOVs need I-NOV to be reproduced, the number of D-NOVs will drop when there is too little I-NOV. After this phase, at low multiplicities of infection of the three virion types the chance increases that correct infections will occur, thereby starting the I-NOV production again.

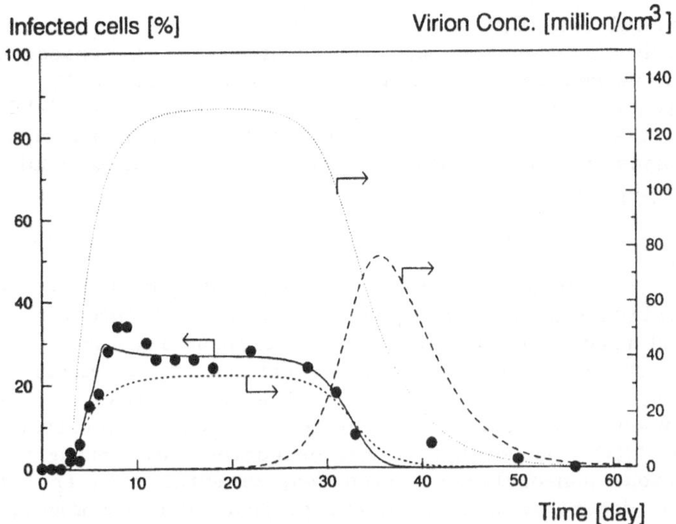

Figure 6. Fit of the model on the percentage of visibly infected cells in one continuously operated infection reactor (adapted from de Gooijer et al. [2]). Parameters are in table II under A. ● = experimental data; —— is the model description of the percentage of visibly infected cells.---- and —— are model descriptions of the I-NOV and D-NOV concentration, respectively, and .... is the A-NOV concentration divided by 10.

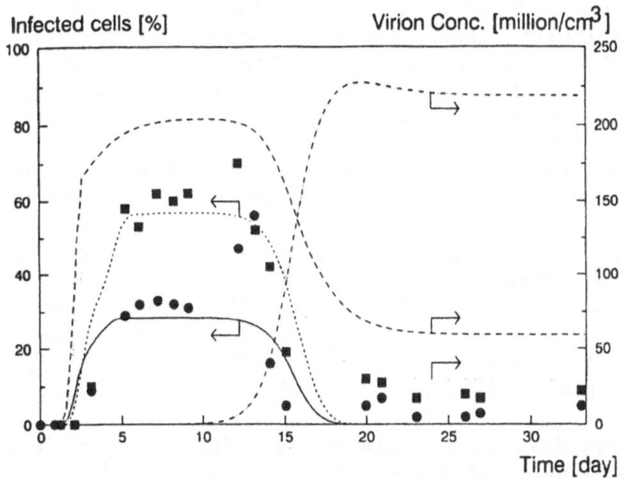

Figure 7. Fit of the model on the experimental (●, ■) percentage of visibly infected cells in the first (——) and second (---) reactor of a series of two continuously operated infection reactors with the model (adapted from de Gooijer et al. [2]). ...., —··— and — — — are the concentrations in the second vessel of I-NOV, D-NOV, and the A-NOV concentration divided by 10. Parameters are in table II under B.

In a continuous culture high passages of I-NOVs occur soon, and eventually D-NOV production will decrease and subsequently, as a consequence, the D-NOVs will be washed out of the reactor. Then, a low multiplicity of infection is reached and if not all I-NOVs are washed out, I-NOV reproduction starts again since the chance that insect cells become correctly infected increases in that situation. This reactor behaviour could also be simulated with the model.

5 Conclusions.

In the case of a continuously operated reactor configuration, an increased number of infection reactors with the same total residence time resulted in an increased pseudo-steady-state level of infection, but a decreased run time. The steady-state levels of infection can be described by a first-order reaction-rate constant. The observed decrease in infection level in continuously operated series of reactors can be attributed to the passage effect. With the concept of the defective interfering virion and the non-infectious non-occluded virion, experimental data from the bioreactors can be described well with the proposed dynamic model, thereby giving an accurate description of the passage effect. Our current research is focused on the determination of the actual numbers of the three types of NOVs occurring in the continuously operated reactor series, in order to validate the model presented here.

6 References.

1. De Gooijer, C.D., van Lier, F.L.J., Van den End, E.J., Vlak, J.M., Tramper, J. (1989). A model for baculovirus production with continuous insect cell cultures. Appl. Microb. Biotechnol. **30**:497-501.
2. De Gooijer, C.D., Koken, R.H.M., van Lier, F.L.J., Kool, M., Vlak, J.M., Tramper, J. (1992). A structured dynamic model for the baculovirus infection process in insect-cell reactor configurations. Biotechnol. Bioengin. in press.
3. Faulkner, P. (1981). Baculoviruses. In: *Pathogenesis of invertebrate microbial diseases*. E.A. Davidson, Ed. Osmun & Co., Allanheld, NJ, pp 3-37.
4. Hink, W.F. (1970). Established cell line from the cabbage looper, *Trichoplusia ni*. Nature **226**:466-467.
5. Katinger H, Scheirer W. (1985). Mass cultivation and production of animal cells. In: Spier R.E, Griffiths J.B. (eds) Animal Cell Biotechnology. vol. I, pp 167-193. Academic Press.
6. Kelly D.C. (1982). Baculovirus replication. J General Virology 63:1-13.
7. Klöppinger, M., Fertig, G., Fraune, E., Miltenburger, H.G. (1990). Multistage production of *Autographa californica* nuclear polyhedrosis virus in insect cell cultures. Cytotechnol. **4**:271-278.
8. Kompier, R., Tramper, J., Vlak, J.M. (1988). A continuous process for the production of baculovirus using insect-cell cultures. Biotechnol. Lett. **10**:849-854.

9. Kool, M., Voncken, J.W., van Lier, F.L.J., Tramper, J., Vlak, J.M. (1991). Detection and analysis of a mutant with defective interfering properties during the continuous production of *Autographa californica* nuclear polyhedrosis virus in bioreactors. Virology **183**:739-746.
10. Kool, M., van Lier, F.L.J., Vlak, J.M., Tramper, J. (1990). Production of (recombinant) baculoviruses in a two stage bioreactor system using insect-cell cultures. In: Agricultural Biotechnology in focus in the Netherlands. J.J. Dekkers, H.C. van der Plas and D.H. Vuijk, Eds. Pudoc, Wageningen, the Netherlands, pp. 64-69.
11. Levenspiel, O. (1972). Chemical reaction engineering. J. Wiley & Sons, New York, pp. 253-296.
12. Luckow, V.A., Summers, M.D. (1988). Trends in the development of baculovirus expression vectors. Bio/Technology **6**:47-55.
13. Martignoni M.E. (1984). Baculovirus : An attractive biological alternative. In: Garner W.Y, Harvey J. (eds) Chemical and Biological Controls in Forestry. pp 55-67. American Chemical Society. Washington.
14. Seth, P., Fitzgerald, D., Willingham, M., Pastan, I. (1986). Pathway of adenovirus entry into cells. In: Virus attachment and entry into cells. Proceedings of an ASM conference held in Philadelphia, PA, 10-13 April 1985, Crowell, R.L., Lonberg-Holm, K., Ed. pp 191-195.
15. Smith, G.E., Summers, M.D. (1978). Analysis of baculovirus genomes with restriction endonucleases. Virology **89**:517-527.
16. Tramper, J., Vlak, J.M. (1988). Some engineering and economic aspects of continuous cultivation of insect cells for the production of baculoviruses. Ann. N.Y. Acad. Sci. **469**:279-288
17. Van Lier, F.L.J., Van den End, E.J., de Gooijer, C.D., Vlak, J.M., Tramper, J. (1990). Continuous production of baculovirus in a cascade of insect-cell reactors. Appl. Microb. Biotechnol. **33**:43-47.
18. Van Lier, F.L.J., van der Meijs, W.C.J., Grobben, N.G., Olie, R.A., Vlak, J.M., Tramper, J. (1992). Continuous β-galactosidase production with a recombinant baculovirus insect-cell system in bioreactors. J. of Biotechnol. **22**:291-298.
19. Vaughn J.L. (1976). The production of viruses for insect control in large scale cultures of insect cells. In: Maramosch K. (ed) Invertebrate Tissue Culture. pp 295-303. Acad. Press.
20. Vaughn, J.L., Goodwin, R.H., Tompkins, G.J., McCawley, P. (1977). The establishment of two cell lines of the insect *Spodoptera frugiperda* (*Lepidoptera:Noctuidae*). In Vitro **13**:213-217.
21. Vlak, J.M. (1979). The proteins of nonoccluded *Autographa californica* nuclear polyhedrosis virus produced in an established cell line of *Spodoptera frugiperda*. J. Invertebr. Pathol. **34**:110-118.
22. Vlak, J.M., Keus, R.J.A. (1990). The baculovirus expression vector system for production of viral vaccines. In: Advances in Biotechnological Processes. A. Mizrahi, Ed. Alan R. Liss Inc., New York, **14**:91-128.
23. Volkman, L.E., Summers, M.D., Hsieh, C-H. (1976). Occluded and nonoccluded nuclear polyhedrosis virus grown in *Trichoplusia ni*: comparative neutralization, comparative infectivity, and in vitro growth studies. J. Virol. **19**:820-832.

24. Von Magnus, P. (1959). Incomplete forms of influenza virus. Adv. Virus Res. **2**:59-79.
25. Wickham, T.J., Granados, R.R., Wood, H.A., Hammer, D.A., Shuler, M.L. (1990). General analysis of receptor-mediated viral attachment to cell surfaces. Biophys. J. **58**:1501-1516.

USE OF FED-BATCH OPERATION TO IMPROVE SECONDARY METABOLITE PRODUCTION

KIMBERLEE K. WALLACE[1], MARILYN K. SPEEDIE[3] AND GREGORY F. PAYNE[1,2]
[1]*Department of Chemical and Biochemical Engineering*
[2]*Center for Agricultural Biotechnology*
University of Maryland Baltimore County
Baltimore, MD., 21228, USA
[3]*Department of Biomedicinal Chemistry*
School of Pharmacy, University of Maryland at Baltimore
Baltimore, MD., 21201, USA

ABSTRACT. Increases in secondary metabolite biosynthesis are traditionally obtained by empirical strategies. However, the extent of these methods is limited due to the lack of knowledge as to why these strategies are successful. By combining traditional engineering strategies with knowledge of the intracellular environment, rational strategies can be developed to overcome specific biosynthetic rate limitations and to increase titers.

1. Introduction

Traditionally, empirical strategies were employed to obtain production increases for secondary metabolic compounds. For example, in the development of the penicillin fermentation, empirical mutation strategies initially resulted in exponential increases in penicillin production (Demain, 1973). However, successive mutations of these high producing strains resulted in increased titers of only a few percent each year. To obtain further increases in secondary metabolite production, it will be necessary to develop rational approaches which combine manipulations of the extracellular environment with knowledge of the intracellular limitations to biosynthesis.

The objective of this paper is to demonstrate how traditional engineering strategies to increase secondary metabolite production can be combined with knowledge of the microbial physiology in order to extend the production phase. To explore this objective, we have chosen as our model system production of the antitumor, antibiotic streptonigrin by *Streptomyces flocculus*. General nutritional requirements, fed-batch operation and biochemical regulation (i.e. enzyme and substrate availability) of streptonigrin biosynthesis were examined to illustrate this interdisciplinary approach.

2. Nutritional Requirements of Biosynthesis

Initially, we examined streptonigrin production from a nutritional standpoint to determine how this system is regulated by carbon, nitrogen and phosphate availability. Although

A.R. Moreira and K.K. Wallace (eds.), Computer and Information Science Applications in Bioprocess Engineering, 317-323.

carbon and phosphate levels did not adversely affect streptonigrin production, we observed that biosynthesis could be correlated to the availability of nitrogen when ammonium chloride was the sole nitrogen source in the medium. Figure 1 shows that streptonigrin is actively produced during the period of ammonium consumption. In contrast to other nitrogen regulated systems (for reviews see Wallace et al., 1992; Shapiro, 1990), streptonigrin production ceased upon depletion of ammonium from the medium.

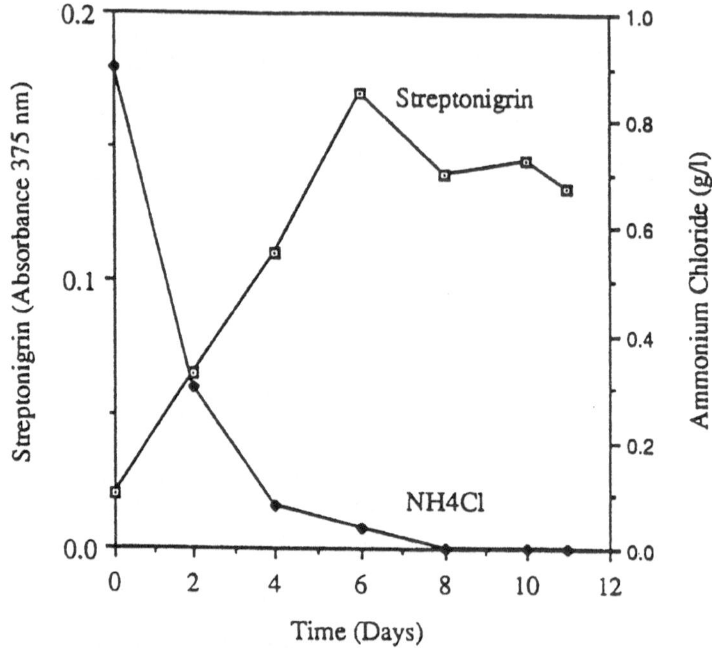

Figure 1. Fermentation profiles for streptonigrin production and ammonium consumption by *Streptomyces flocculus.* Cultures were grown in a defined medium as described in Wallace et al., 1990.

This correlation between ammonium availability and streptonigrin biosynthesis suggested to us that in order to increase production, the initial ammonium concentration available to the cells needed to be increased. However, Figure 2 shows that increasing the initial nitrogen concentration available to the culture adversely affected streptonigrin biosynthesis. Although it is commonly observed in secondary metabolic systems that increased concentrations of required nutrients often negatively impact production (for reviews see Wallace et al., 1992; Shapiro, 1989; Martin; 1989, Demain, 1989), we could not strictly correlate the negative effects with the extracellular ammonium concentration (Wallace et al., 1990).

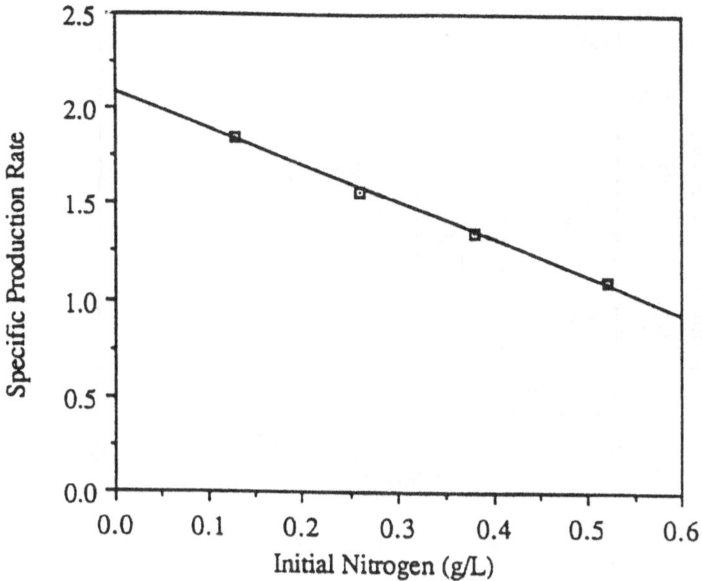

Figure 2. Specific production rate of streptonigrin as a function of initial nitrogen concentration. Units of specific production rate are (absorbance Units)/(g cells/L)/day. Adapted from Wallace et al., 1990.

3. Fed-Batch Operation

To overcome this negative nutrient regulation while still meeting the nutritional demands of the culture, fed-batch operating strategies are typically employed. Ryu and Hospodka (1980) demonstrated that the specific penicillin production rate could be improved by supplying nutrients at a rate which coincided with the uptake rate of the culture. Subsequently, Gray and Vu-Trong (1987) showed that tylosin productivity could be enhanced by cyclic feeding of glucose and glutamate to minimize catabolite repression/inhibition.

In our system, we maintained a low but finite concentration of ammonium in the medium by periodically supplementing the culture with ammonium. As seen in Figure 3a, streptonigrin production in the control culture, which had received an initial ammonium concentration of 0.5 g/l, ceased upon depletion of ammonium from the medium. By limiting the extracellular concentration of ammonium using an ammonium feeding strategy, we were able to extend the production phase of the culture. However, Figure 3b shows that despite the presence of ammonium, streptonigrin production ceased after 8 days. Thus, based upon the use of traditional engineering feeding strategies, it is not clear how to further extend the streptonigrin production period. To rationally design further operating strategies to enhance production, we feel that the biosynthetic rate limitation must be determined. In order to determine this limitation, we have focused our efforts on understanding the intracellular environment at the time that production ceases.

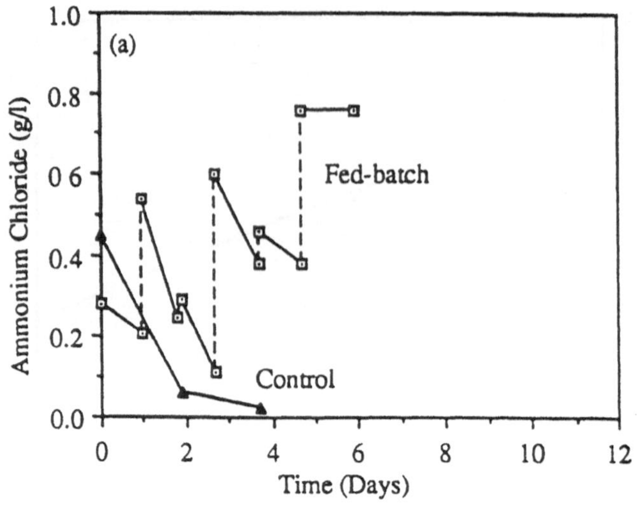

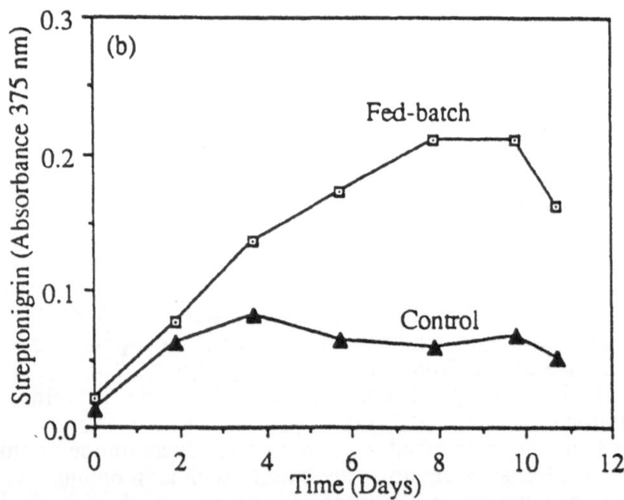

Figure 3. a) Ammonium chloride concentration profiles for control and nitrogen fed-batch cultures in a defined medium. b) Extension of streptonigrin production using an ammonium fed-batch operating strategy. (Adapted from Wallace et al., 1990)

4. Biosynthetic Rate Limitations

Secondary metabolites are biosynthesized from complex biosynthetic pathways, many of which have not been completely elucidated. By analogy to amino acid biosynthesis, pathway entry points are likely sites of metabolic regulation. Thus, we have chosen to concentrate our studies on the first step of the streptonigrin biosynthetic pathway. In this first step, the primary metabolite tryptophan is committed to secondary metabolite biosynthesis by an S-adenosylmethionine requiring C-methyltransferase. Analysis of this step reveals two possible biosynthetic rate limitations: lack of availability of the necessary enzyme or lack of availability of the required substrates.

Initially, we focused on the area of substrate availability. The first step in streptonigrin biosynthesis requires the availability of the co-substrates tryptophan and S-adenosylmethionine. If either of these compounds was found to be limiting streptonigrin biosynthesis, operating strategies involving tryptophan or methionine feeding could be developed to increase intracellular precursor supplies and overcome this specific limitation.

To determine whether pools of these compounds could be inadequate for supporting continued production, we initially determined the affinity constants for the C-methyltransferase substrates, using partially purified enzyme. Secondly, we measured the intracellular pools of tryptophan and S-adenosylmethionine over the course of cultivation to determine whether the intracellular levels met the enzymatic requirements. Table 1 shows that the minimum intracellular concentrations of both substrates exceeded the K_m values for the C-methyltransferase. Thus, lack of substrate availability does not appear to be responsible for the cessation of streptonigrin biosynthesis.

TABLE 1. Affinity constants and intracellular concentrations of the co-substrates S-adenosylmethionine and tryptophan.

	K_m	Minimum Intracellular Concentration
S-adenosylmethionine	10 uM	530 uM
Tryptophan	1 mM	9.3 mM

Secondly, we focused on whether availability of the C-methyltransferase limited streptonigrin biosynthesis. Previous studies have shown that secondary metabolite production may be limited by the need for continued synthesis of biosynthetic enzymes (Grootwassink and Gaucher, 1980) and by the instability of many of these enzymes (Agathos and Demain, 1988). As seen in Figure 4, we observed a qualitative relationship between C-methyltransferase activity and streptonigrin biosynthesis. Production commences following enzyme induction and ceases upon a drop in enzyme activity. This apparent correlation suggests that enzyme availability may represent a biosynthetic limitation to continued production. Thus, to further extend the production phase, it will be necessary to extend the period of enzyme availability. By undertaking a similar study using an ammonium fed-batch operation, we hope to determine whether the extension in production was specifically due to an extension of the period of enzyme availability.

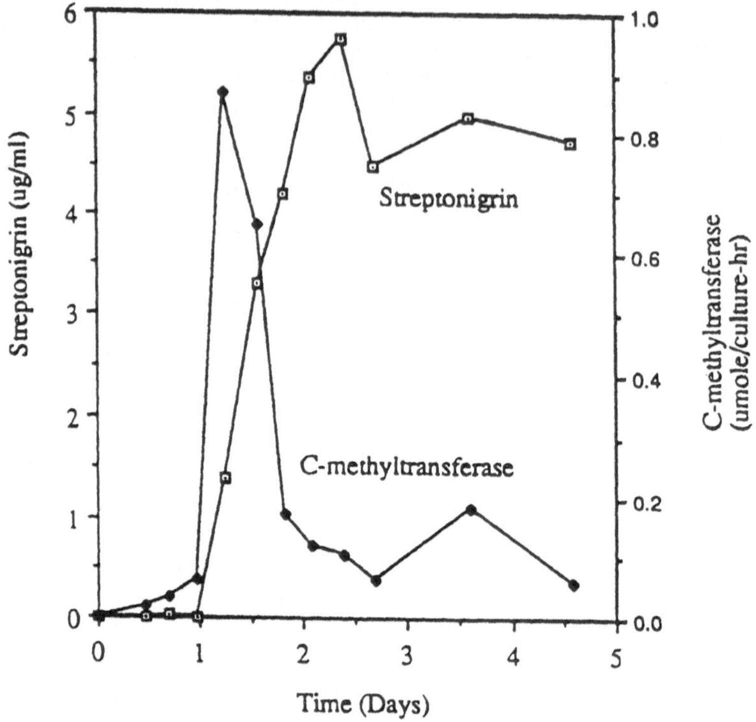

Figure 4. Time course profiles for C-methyltransferase activity and streptonigrin biosynthesis in a complex medium consisting of 4 g/l yeast extract and 10 g/l malt extract.

5. Summary

Due to the complexity and lack of knowledge concerning secondary metabolic pathways, traditional strategies to obtain production increases involved empirical mutational or nutritional manipulations. Although recent advances in genetic and metabolic engineering provide the technology to overcome specific biosynthetic limitations, the problem of identifying these limitations remains. By using an interdisciplinary approach to gain an understanding of intracellular factors associated with secondary metabolite production, we feel that rational engineering strategies can be developed to overcome specific biosynthetic bottlenecks.

Acknowledgments: This work was supported by NSF grant #EID-9021048.

6. References

Agathos, S.N. and Demain, A.L. (1988) 'The *in vivo* longevity of antibiotic synthetases' in S. Aiba (ed.), Horizons in Biochemical Engineering, Oxford Press, New York, 147-161.

Demain, A.L. (1973) 'Mutation and production of secondary metabolites', Adv. Appl. Microbiol. *16*, 177-202.

Demain, A.L. (1989) 'Carbon source regulation of ideolite biosynthesis in actinomycetes' in S. Shapiro (ed), Regulation of Secondary Metabolism in Actinomycetes, CRC Press Inc., Boca Raton, Florida, pp. 128-134.

Gray, P.P. and Vu-Trong, K. (1987) 'Production of the macrolide antibiotic tylosin in cyclic fed-batch culture', Biotechnol. Bioengin., *29*, 33-40.

Grootwasssink, J.W.D. and Gaucher, G.M. (1980) 'De novo biosynthesis of secondary metabolism enzymes in homogeneous cultures of *Penicillium urticae*', J. Bacteriol., *141*, 443-455.

Martin, J.F. (1989) 'Molecular mechanisms for the control by phosphate of the biosynthesis of antibiotics and other secondary metabolites' in S. Shapiro (ed), Regulation of Secondary Metabolism in Actinomycetes, CRC Press Inc., Boca Raton, Florida, pp. 214-237.

Ryu, D.D.Y. and Hospodka, J. (1980) 'Quantitative physiology of *Penicillium chrysogenum* in penicillin fermentations', Biotechnol. Bioengin., *22*, 289-298.

Shapiro, S (1989) 'Nitrogen assimilation in actinomycetes and the influence of nitrogen nutrition on actinomycete secondary metabolism' in S. Shapiro (ed), Regulation of Secondary Metabolism in Actinomycetes, CRC Press Inc., Boca Raton, Florida, pp. 136-211.

Wallace, K.K., Payne, G.F. and Speedie, M.K. (1990) ' Ammonium effects on streptonigrin biosynthesis by *Streptomyces flocculus*' J. Industrial Microbiol., *6*, 43-48.

Wallace, K.K., Payne, G.F. and Speedie, M.K. (1992) 'Streptomyces Bioprocessing: From secondary metabolites to heterologous proteins' in P. Todd, S. Sikar and M. Bier (eds.), Frontiers in Bioprocessing II, American Chemical Society, Washington D.C., pp. 168-180.

BIOREACTOR DESIGN FOR ANIMAL CELL CULTURES: WITH SPECIAL REFERENCE TO MASS TRANSFER

R.E. SPIER
School of Biological Sciences
University of Surrey
Guilford, Surrey, U.K.

ABSTRACT. This paper examines the design of bioreactors for animal cells from the points of view of individuals who wish to choose an "off-the-shelf" design and those who wish to start from first principles. In the former case, it is necessary to be fully aware both of the environment within the establishment where the reaction is to function as well as the biological and physicochemical parameters of the system that is under consideration. In the design from first principles, the emphasis is on the way the limiting nutrient, oxygen, may be supplied to either homogeneous suspensions of cells or to high local concentrations of cells in inhomogeneous cell masses. It is concluded that designs based upon the provision of oxygen result in workable and efficient systems for the generation of products from animal cells in culture and that this is commercially opposite a way to the manufacture of prophylactic and therapeutic materials.

1. Introduction

The increasing demand for products made from animal cells in culture (Spier, 1991a) has resulted in a situation where it is necessary to consider further the design of bioreactors for animal cells. There are two principle cases that require this examination; the one is when a laboratory or company seeks to make animal cells in culture for the first time while the second is when an existing animal cell based activity seeks to expand its production or change its mode of operation. The subsequent sections of this paper will address both these issues and will not only provide a check-list of questions that have to be answered but will also seek to establish the way boundary conditions may be envisaged and possibly circumvented. To help answer some of these questions programmes written in simple (devoid of screen commands) BASIC have been used. These programmes have been appended to this paper. A design begins with the Design Brief, the formulation of which will be dealt with next.

2. The Design Brief

This stipulates that the instigator of the project defines what is required of the bioreactor to be designed. These requirements are held in the answers to a set of questions. Such a set of questions is portrayed in Table 1.

A.R. Moreira and K.K. Wallace (eds.), Computer and Information Science Applications in Bioprocess Engineering, 325-347.

Table 1. The basic components of the Design Brief for the bioreactor type for animal cell cultures.

Definition of the Bioprocess

1. Type of Cell
   - Surface dependent
   - Surface independent
   - Hardy
   - Fragile

2. Definition of the Scale of Operation
   - Smallest scale
   - Largest scale

3. Existing In-House Capability/ Capacity
   - Present
   - Non-existent

4. Process Type
   - Batch
   - Continuous

5. Degree of Control
   - Homogeneous
   - Heterogeneous
   - Basic system
   - Expanded system

6. Data Requirements
   - Real-time
   - On-line
   - Derived parameters
   - Record
   - Archive
   - Relay/Network

7. Local Cell Concentration
   - 10^6 cells/ml
   - 10^7 cells/ml
   - 10^8 cells/ml

8. Bioreactor Format
   - Simple (reliable/Robust)
   - Complex

It may be that some of the answers to the questions present some difficulty. For example, it is not always clear that once one has set the overall size of the bioreactor that it would not be more appropriate to have two bioreactors half the size or four a quarter of the size, etc. The programme in Appendix 1 attempts to cast some light on this problem by envisaging a production operation making 10 kg of animal cell derived product per annum with different levels of loss due to contamination or other failures and with the addition of costs of quality control added on a per bioreactor basis, as one bioreactor might be construed to be a batch by the regulatory authorities.

Having answered the questions in Table 1, it may be possible to define the type of bioreactor in general terms using the bioreactor characteristics as set out in Table 2.

Once the choice of bioreactor type has been made then it is necessary to consider the supplementary questions in Table 3.

Table 2. Basic Bioreactor Characteristics.

BR-Type	Cell Type	Hardy ?	Scale (L)	Batch Cont.	Homo-Hetero-	Contl B/E	Data On-ln	Locl Cell Conc *E?
STR*	I	H	.01/20K	B	HO	Y	Y	6
STR/P	I	H	2/500	C	HO	Y	Y	7
AL	I	H/F?	10/2K	B	HO	Y	Y	6
PB	D	F	.02/1K	B/C	HE	Y	Y	7
MC	D	F	.01/20K	B	HO/HE	Y	Y	6
MC/P	D	F	5/70	C	HO/HE	Y	Y	7
PPB	I/D	F	.01/1?	C	HE	Y	Y	8
PPS	I/D	F	.02/20	C	HE	Y	Y	8
HF	I/D?	F	.002/2	C	HE	Y	Y	8
SF	D	F	.001/1	B	HO	N	N	6
RB	D	F	.25/6	B	HO	N	N	6

* STR, Stirred Tank Reactor;STR/P Stirred Tank Perfused; AL, Airlift Reactor;MC, Microcarrier;MC/P Porous Microcarier;PPB, Porous Packed Bed;HF, Hollow Fibre;SF, Static Flask; RB, Roller bottle.

Table 3. Design Brief for a chosen bioreactor type

Height/Diameter Ratio	System for Oxygenation
Materials of Construction	Without Gas/Liquid Interface Bubbles
Quality of Finishes	The Patent Situation
Connections for Inputs and Outputs of solids, liquids and gases	Availability of System/Parts
Method for Affecting Fluid Flows Within the bioreactor Outside the bioreactor	Servicing Back-up
Method of sterilization	Cost
Method of Sampling	Delivery
Sensor designation and nstallation	Ease of Modifications to Hardware Software
Environmental Protection Issues Off-gas Fluid discharge	Installation Costs
Emergency pressure release	Existing in-house Skills
	Existing in-house Documentation
	Availability of Literature Data
	Regulatory Agency Familiarity

Having considered the issues depicted in the above tables, it is necessary to be clear as to the criteria that are going to be used to assess the productivity of the system. This is not a trivial issue as there are many parameters that can be chosen to establish this system attribute. Table 4 delineates such variables.

At this juncture, there is a fundamental dichotomy of approach. It may be possible with the answers to the questions of Tables 1-4 to be able to write a design brief that can be given to a manufacturer of animal cell bioreactors to purchase an off-the-shelf item without further ado. However, it is also possible to dispense with the "off-the-shelf" approach and begin the design of the bioreactor from first principles.

Table 4. Value for money issues (Efficiencies)

Cells/ml in Bioreactor	Product/Cell
Cells/ml in Medium Used	Product/Cell/Unit of Time
Cells/Bioreactor	Product/Bioreactor
Cells/Unit Surface Area	Product/Bioreactor/Unit of Time
Cells/Unit Volume of Carrier	Product/ml Medium Used
Cost/Unit of Product	Time to Market

3. The Approach to Bioreactor Design from First Principles

In this section, the two extreme cases of a homogeneous system operating at 10^{-6} to 10^{-8} cells/ml and the heterogeneous case of an operation with 10^{-8} cells/ml will be considered. The strategy of the first principles approach is outlined in Table 5.

Table 5. Design from first principles.

Knowledge of Cell Physiology

- Nature of Intermediary Metabolism
- Cellular Inputs and Outputs (Medium Design)
  - For growth
  - For product formation
- Intracellular Energy Relationships

Knowledge of Heat/Mass Transfer

- Particularly Oxygen and Glucose
- Mixing

Knowledge of Scale-up Relationships

3.1 CELLULAR ANATOMY AND PHYSIOLOGY

While there has been much work effected on modeling the production of monoclonal antibodies from hybridoma cells (A. Moreira, lecture at conference), this author is unaware of the practical use of this data in the design of bioreactors. (It would seem that the main thrust of modeling of such systems is the definition of process conditions (medium addition rate and composition), that could be used to improve the productivity of a particular hybridoma culture, rather than the determination of the parameters of the bioreactor.) However, the observations of the way the cell system can respond to changes in its chemical environment can influence productivity and hence the size of the bioreactor required to affect a desired amount of product.

Other models that are based on population dynamics (Faraday et al) or the measurable components of intermediary metabolism including the levels of "energy" compounds (ATP, ADP, ATP/ADP ratios, proton gradients etc.) can be construed to describe systems on which the measurements have been taken to obtain the relevant proportionalities (Barford and Harbour, 1991). These models have not been used in circumstances beyond those for which the model was constructed in the first instance.

Much thought has been given to the way the designer of an animal cell bioreactor needs to provide an environment for the allegedly shear sensitive animal cells that would not cause the delicate cells to be disrupted. As it is unlikely that the cells are particularly fragile, it is worth examining the issues that pertain to this concept in more detail.

3.2 ARE ANIMAL CELLS SHEAR SENSITIVE?

When an animal cell culture fails to thrive, it is often held that the problem lies with the shear sensitive nature of the cells; the lack of a "protective cell" wall provides a rational if erroneous basis for such conjectures. However, most workers have had difficulty from time to time growing cells in stationary flasks or bottles and there had been little systematic work demonstrating that animal cells are particularly shear sensitive. More recent work by this author amongst others has shown that cells require shear forces in excess of 3-5 N/m^2 in order for them to be detached from a surface that they are growing on and that they will attach to such surfaces when shear forces are below a critical level of 0.01-0.09 N/m^2 (Crouch et al., 1985). Using concentric cylinder shear stress generating devices, the cells have been shown to withstand shear forces of 1-2 N/m^2 (calculated from Wudtke and Schugerl, 1987).

3.2.1. *Do We Need to Generate Hydrodynamic Environments that Cause Animal Cells to Experience Shear Stresses that Could Be Damaging?* Although the previous section (3.2) indicated that animal cells can withstand a degree of shear stress, need such stresses be engendered in homogeneous cultures of animal cells? This is a more difficult question to answer as the hydrodynamic environment deep within a culture is not as well defined as the laminar shear field in a concentric, cylinder viscometer. Not withstanding this difficulty, Kolmogorov has defined a length which relates to the size of the smallest eddies in a turbulent flow system and when eddy diameters are generated that compare with the size of the particle in the suspension culture, deleterious hydrodynamic effects may be expected. This has been demonstrated for cultures of cells grown on the surface of 200 um microcarriers by Papoutsakis and others (Cherry and Papoutsakis, 1990), but has not been fully tested for cells growing as a monodisperse cell suspension. To determine the limits to which it is possible to stir a cell suspension, a programme is provided in **Appendices 2 and 3** that uses the formula for the determination of the Kolmogorov length and calculates that dimension for different sizes of bioreactors with a variety of impeller diameters rotating at a number of selected speeds. From the table of data generated, it is possible to define the boundary conditions for the design of a vessel and that way it operates such that the Kolmogorov length does not approach the size of the cell, if monodisperse, or the diameter of the aggregate of cells, if such are required or the diameter of the microcarrier when that it is used.

A conclusion from these considerations is that in animal cell culture systems, it is a useful design concept **that the aeration of the culture should be achieved independently of the mixing of the culture.** In this way, it is not necessary to supply energy (that will dissipate in hydrodynamic friction) to decrease the size of the bubbles used for aeration rather it is practical to achieve small bubble diameters at the time of bubble generation. Thus mixing and aeration can be separated and treated independently.

The remainder of this paper will discuss the problems of mass transfer in animal cell cultures for it can be reasonably argued that such parameters dominate what can be achieved in such cultures. In particular attention will be focused on oxygen transfer in two contrasting and extreme system types, namely, the homogeneous culture of animal cells at concentrations of 10^{-6}-10^{-8} cells/ml. This will be followed by an examination of the rate of glucose diffusion in the heterogeneous case to show that this is less likely to be a limiting parameter than the rate of oxygen diffusion.

3.3 PROVISION OF OXYGEN TO HOMOGENEOUS CULTURES OF ANIMAL CELLS HELD AT CONCENTRATIONS OF 10^{-6} -10^{-8} CELLS/ml

3.3.1. *Bubble-free Aeration Systems.* There are a number of alternative methods for the provision of oxygen to cultures of animal cells held in homogeneous suspensions. It is a commonly held view that as animal cells in culture express a heightened fragility that it is wise to exclude from the culture the possible damaging effects of the gas-liquid interface as introduced in a system that is aerated by bubbling. This has led to much ingenuity in the use of coils of silicone tubing, or other oxygen permeable membranes. Such systems work well for bench scale reactors but at the larger commercial scales (over 2000 l volumes), the length of the tubing that would be required would be about 4000m. An alternative system based on recycled silicone oil emulsions has also been used to transfer oxygen to animal cell cultures. This method suffers from the need to separate, recycle and reoxygenate the oil that necessarily involves an external loop and pumps. This additional hardware adds to the complexities of the system and the procedures required during the equipment sterilisation process; it thereby adds to the possibilities for failure. Other ingenious solutions have been proposed such as *in situ* generation of oxygen using illuminated chlorella but such systems have not reached the stage of publication as yet. As bubble aeration is the common means of providing oxygen to other aerobic microbial bioreactors and such reactors are commonly run at the 200 m^3 scale, it would be rational were this situation used to provide an analogous model for the aeration of animal cell based microbial cultures.

It should be noted that the cells experience the oxygen as dissolved in the culture medium. On the one hand with homogeneous suspensions of cells, it is possible to generate a mixture of medium, bubbles and cells. In bioreactors where the cells are held at high concentrations locally, it is not possible to bring the bubbles, cells and medium together in one vessel so that it is more usual to bubble aerate the medium in a second chamber and pass the aerated bubble-free medium past the cells. In some cases, hollow fibre aerators are used in place of the bubble aeration systems in such applications. In the later case, the chief question is whether the cells can be provided with sufficient oxygen to maintain their productive status. This matter will be dealt with after a consideration of the issues of the bubble aeration of homogeneous suspension cultures of animal cells.

3.3.2. *Bubble Aerated Cultures of Animal Cells.* It is not doubted that it is possible to so formulate a culture medium such that when the culture of animal cells is aerated using bubbles of air the cells will not thrive as well as control cultures that aerated from the top of the surface culture (Handa-Corrigan, 1989).

The mechanism of action of the Pluronic F68 (MW c. 7000D), (a block copolymer of polyoxypropylene as the central element and polymers of polyoxyethylene on either end), has not yet been ascertained in detail. Hypotheses vary from an influence on the physiological state of the cells, (as a result of their increased permeability to key nutrients), to the separation of the cells and the bubbles during the disengagement of the bubbles at the surface of the culture. Others workers stress the slipperiness of the bubbles so that there is a decrease in the adhesive interactions between the cells and the bubbles with consequential decrements in the shear stress that the cells experience. Nevertheless, it is clear that the greater the size of the bubble disengagement zone to the volume of the culture the more likely it is that the bubble damage will be observed in an appropriately formulated culture medium.

Table 6. The influence of medium composition on bubble damage of cells

Medium Type (Aerated)	Maximum Viable BHK Suspension Cells $ml^{-1}(*10^5)$ Bubble Column	Control Surface
Serum Free	2.0	6.4
RPMI + 2% FCS	8.0	12.6
RPMI + 5% FCS	8.8	14.4
RPMI + 5% FCS+ 0.1% Pluronic F68	16.2	16.3
RPMI + 5% FCS + 4 g/l TPB + 4.5 g/l Glucose	32.0	32.0
RPMI + 10%FCS	14.2	14.4

Can bubbles supply sufficient oxygen for a concentrated cell culture containing more than 108 cells ml^{-1}? The answer to this question hinges upon three variables; the rate at which oxygen will transfer across the air liquid interface in a culture medium replete with surface active agents, the amount of surface area of bubbles that can be held within the culture volume which is dependent on the two parameters of bubble diameter and number coupled with the residence time of the bubble in the culture (see Diagram 1).

Oxygen transfer rates. Much of the traditional oxygen transfer rate determinations have been effected with reference to non-animal cell microbial cultures. As such, the medium has generally been a simple salts solution or more generally still, plain water. The question that this provokes is the determination of the effect that the surface complex surface active materials of an animal cell culture could have on that transfer rate. (Even in a serum-free medium there could be antifoams, Pluronic F68 and albumins). Experiments to determine such effects were undertaken by this author and J.P. Whiteside (Spier and Whiteside, 1990). It was found that a fully formulated medium expressed an oxygen transfer rate of about half of that which could be obtained with the least conductive water it was possible to obtain readily by distillation followed by passage through the MilliQ ion exchange/organic removal column marketed by Millipore. In addition, it was clear from this experiment that under the least vigorous conditions the lowest figures obtained for the oxygen transfer rate were 10 ug oxygen cm^{-2} hr^{-1} while figures in excess of 50 ug oxygen cm^{-2} hr^{-1} were not uncommon and often exceeded. From this data, it is possible to generate a theoretical air flow rate requirement for a culture respiring at a given rate that is aerated by bubbles with a defined diameter and residence time in the reactor. (The respiration rate of cells can be inferred from data in Spier and Griffiths (1984)). To examine the feasibility of such bubble aeration systems, the programme in **Appendix 4** was written. This demonstrated that with bubbles of less than 1 mm in diameter it was possible to aerate cultures respiring at their maximal rates (10 pg cell^{-1} hr^{-1}), at cell concentrations of greater than 10^8 cells ml^{-1}. These theoretical considerations were tested using bubbles whose diameters were measured at less than 0.07 mm (70 um).

Oxygen transfer coefficients achieved with bubbles of 0.04-0.05 mm diameter. The objectives of the experiments initiated by this author and his colleague, A. Handa-Corrigan, were to determine the effect of the surface active components of common animal cell culture medium on oxygen transfer coefficients and to elucidate whether it would be possible to aerate cultures of over 10^8 cells ml^{-1} respiring at maximal rates. We found that we could achieve a k_la of 80 hr^{-1} using sintered glass dispersion devices, and the inclusion of pluronic in the serum-containing medium improved the k_la probably as a result of a decrease in the bubble diameter (Zhang et al., 1992). (Such a k_la would support the respiration of 10^8 cells ml^{-1} respiring at about half the maximum rate).

3.3.3. *Supplying Oxygen to Cell Cultures Where the Cells Are Held at High Concentrations Locally.* **The evidence from *in vivo* tissues** (McCullough and Spier, 1990). From the existing physiology literature, it is possible to obtain figures for the respiration rate of cells and for the average distance of a cell from its supply of oxygen. With regard to the former parameter, it is heuristic to note that the tissues of the body may be supplied with blood to between 500 ml of blood per 100 gm of tissue min^{-1} and 1 ml of blood per 100 gm of tissue min^{-1}. Secondly, it is possible using the difference in

Diagram 1

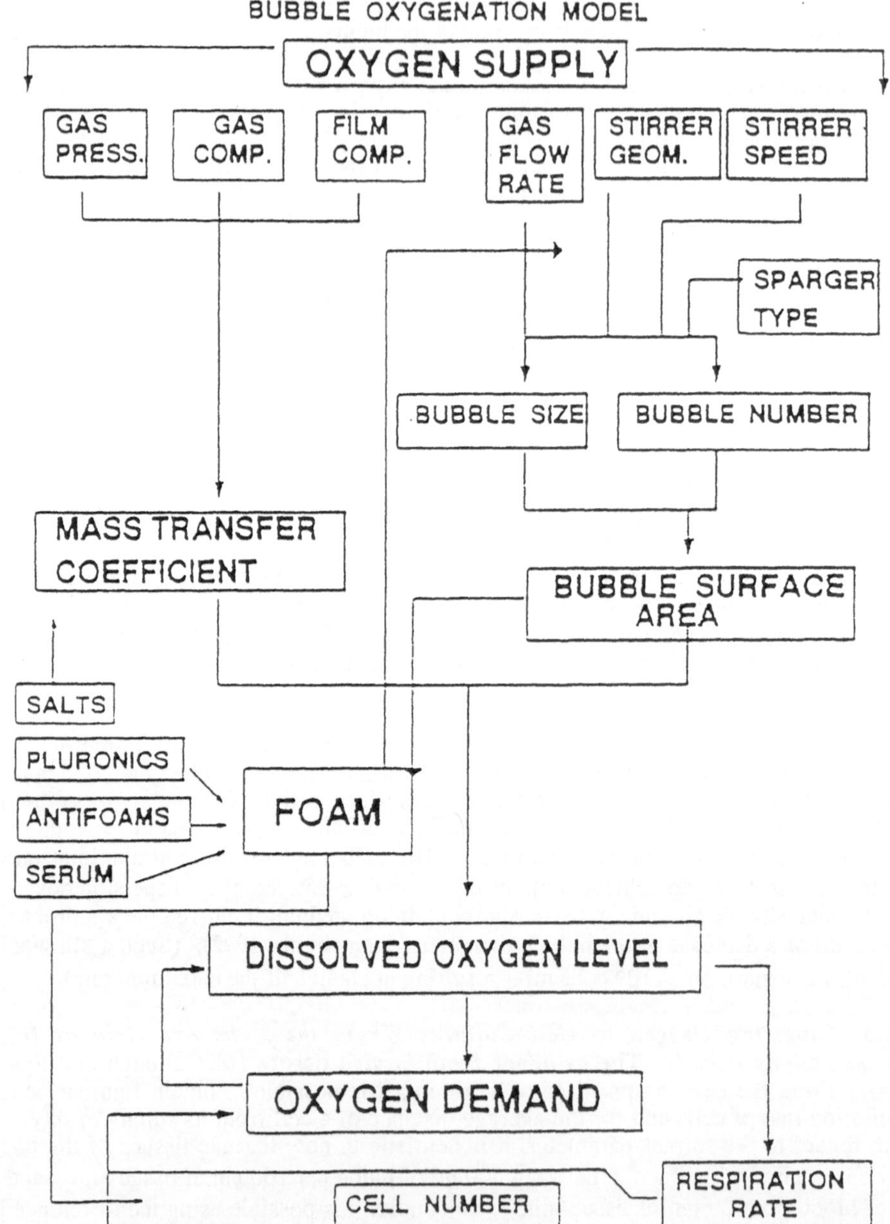
BUBBLE OXYGENATION MODEL
OXYGEN SUPPLY
GAS PRESS.
GAS COMP.
FILM COMP.
GAS FLOW RATE
STIRRER GEOM.
STIRRER SPEED
SPARGER TYPE
BUBBLE SIZE
BUBBLE NUMBER
MASS TRANSFER COEFFICIENT
BUBBLE SURFACE AREA
SALTS
PLURONICS
ANTIFOAMS
SERUM
FOAM
DISSOLVED OXYGEN LEVEL
OXYGEN DEMAND
RESPIRATION RATE
CELL NUMBER

the oxygen content of arterial blood versus that in venous blood to calculate the amount of oxygen used by a cell in a defined organ. A sample calculation given in the above reference results in a value of 3.6 pg $cell^{-1}$ hr^{-1} which compares well with the data compiled by Spier and Griffiths (1984). A third more ambitious calculation can be effected to determine the distance of an average cell from the capillary supplying it with oxygenated blood. Thus if all the cells of the body were smeared over the surface area of all the capillaries of the body, there would result a layer of cells some 256 um thick or **21 cell diameters** of 12 um diameter cells. (It should be noted that the amount of oxygen carried by the blood stream can be 6.3 mg ml^{-1} (cf. water at approx 7 mg ml^{-1}) and that the process of the liberation of that oxygen to the solution is determined by the pH of the solution). Armed with this information one can proceed to the *in vitro* situation and determine both theoretically and practically the thickness of the cell mass that can be supported by the dissolved oxygen of a nearby nutrient culture medium.

The essential question is **what is the distance that oxygen can travel through a concentrated mass of cells in culture when it is supplied to such cells as an aqueous solution?** Clearly this distance will depend on the cell concentration, the rate of respiration of the cells, the dissolved oxygen concentration and the diffusivity of oxygen in the cell mass/culture medium system. Using figures that result in a cell concentration of $5*10^8$ cells ml^{-1}, a respiration rate of 2.2 mMol of oxygen per 10^9 cells day^{-1} with a solubility of oxygen of 7 mg ml^{-1}, (while it is a commonly assumed figure, refinements of it based on the composition of the fluid, with about 0.1 M NaCl, do not change it radically), the distance oxygen can diffuse through such a cell mass may be calculated using Fick's diffusion relationship. This is based on

> The amount of material transferred = the constant of diffusivity * the concentration gradient

There is a complication when considering how far oxygen will penetrate into a cell mass in that the cells are using the oxygen for respiration while it is in the process of diffusing; thus we have diffusion with consumption (or reaction). This can be effectively modeled in two ways. In **Appendix 5**, a short basic program calculates the time it takes the oxygen to pass across a single layer of cells and in so doing decreases its concentration by one second worth of respiratory activity. The time is then calculated to cross the second layer of cells, etc. The program concludes when a full second has been used up when it returns with the number of layers of cells the oxygen has been able to travel through. (The data given result in a cell mass some **29 layers** thick; this compares well with the *in vivo* situation described above). The second approach to determine this parameter is based on calculus (Murdin et al., 1987). When the oxygen concentration/cell layer relationship is so determined it is indistinguishable from that obtained from Appendix 5's program.

The practical data that have been generated to date are in accord with the dimensions which have been derived from the above theoretical considerations. In a set of experiments to determine the thickness of the cell layer that could be supported by oxygenated medium on the surface of an agarose immobilised cell mass, it was found that when the gel contained 10^7 cells ml^{-1} that the thickness of the cell layer that could be maintained was 200-300 um (Wilson and Spier, 1988). Also the growth of cells within the hollow spheres seemed to result in cell masses within the sphere with diameters of up to 500 um. This again concurs with unpublished observations by H. Katinger who grows cell aggregates which express an active surface layer of cells of 200 um even though the diameter of the particle is over 500 um. Additionally, aggregates of cells between 90-360 um were found by Hu (Goetghebeur and Hu, 1991) to be formed when CHO cells were used in aggregate promoting culture medium. Thus, the observed phenomena concur with the concept that cell layer thicknesses of about 200 um (20 cells of 10 um diameter), are practicable and that this can lead to spherical particles of up to 500 um, but not beyond. Yet the question remains as to whether oxygen is the dominant parameter controlling the dimensions critical for the design of the bioreactor.

4. A Theoretical Analysis of the Diffusion of Glucose in Animal Cell Culture Systems

It is possible from the literature to cull a reasonable value for the diffusivity of glucose in cell (yeast) loaded agarose particles (Estape et al., 1992). This value can be substituted into the programme which is logically identical to the one used for the determination of the distance oxygen will travel in a cell mass. Such a programme is presented in **Appendix 6**. Using the relevant values for glucose, that is, a normal concentration in the culture medium and a usual rate of glucose uptake (McCullough and Spier, 1990), it is possible to show that under these circumstances the glucose will travel twice the distance that oxygen will. This means that glucose is unlikely to be one of the limiting nutrients and that from the point of view of bioreactor design considerations based on this material can be omitted.

5. Conclusions

There are two basic approaches to the design of a bioreactor for animal cells. One route takes advantage of existing expertise and, having made basic decisions about the type of cell line and process that is required, surveys the "on-the-shelf" bioreactors and effects a selection that is most appropriate for the task as defined in the design brief. A second route begins with the design brief and uses the fundamental properties of the cells, the culture medium and the physicochemical (i.e. mass transfer) and hydrodynamic environments of the bioreactor to evoke a design. It may be that a combination of approaches can also provide a design solution in that an "off-the-shelf" bioreactor may be modified according to some concept based on the fundamental properties of the system.

While there exists the view that animal cells in culture are fragile there will be a market for bioreactor designs that seek to cosset the cells against the putatively dangerous hydrodynamic environment. However, the arguments presented in this paper rely on more objective criteria for the determination of the damaging environment and conclude that such environments are not necessarily engendered in the provision of animal cell culture systems. Secondly, it is held that bubbles can damage animal cells in culture. Although this damage can be achieved when using culture media that are inappropriately formulated, the addition of Pluronic F68 normally eliminates such deleterious effects. Most importantly, the design of animal cell bioreactors is largely dominated by the need to provide oxygen to the respiring cells. As the animal body distributes the oxygen via cells and capillaries so do our humanly designed bioreactors distribute oxygen using bubbles and/or hollow fibres. Bubble aeration can provide oxygen to homogeneously suspended cell masses held at concentrations of up to 10^8 cells ml^{-1} providing the bubble diameters are small (typically below 1 mm in diameter), the residence times of the bubbles approach several seconds and the oxygen transfer rates do not fall below 10 ug cm^2 hr^{-1}. With inhomogeneous systems of cells held at high concentrations locally there is a limit to the distance oxygen will diffuse into a respiring cell mass and this may be determined theoretically and confirmed practically. (It is about 20-30 cell diameters with cells respiring at less than half the maximum rate of 10 pg $cell^{-1}$ hr^{-1}). On the basis of such understandings it is possible to design bioreactors for animal cells that are efficient, practical and provide products at costs that promote the continuing advance of this important area of biotechnology.

6. References

Barford,J., Harbour, C. 1991. A computer simulation of the kinetics and energetics of hybridoma cell growth and antibody production. In Production of Biologicals from Animal Cells in Culture. Eds R.E. Spier, J.B. Griffiths, B. Meignier. Butterworth-Heinemann, Oxford. pp 631-633.

Cherry, R.S., Papoutsakis, T. (1990). Understanding and controlling fluid-mechanical injury of animal cells in bioreactors. In Animal Cell Biotechnology, Vol. 4. Eds R.E. Spier and J.B. Griffiths. Academic Press, London, pp 72-122.

Crouch, C.F., Fowler, H.W., Spier, R.E., (1985). The adhesion of animal cells to surfaces, the measurement of critical shear stress permitting attachment or causing detachment. J. Chem Technol. Biotechnol. vol 35B, pp 273-281.

Estape, D., Godia, F. and Sola, C. (1992). Determination of glucose and ethanol effective diffusion coefficients in calginate gel. Enzyme and Microbial Technology, vol 14, pp. 396-401.

Goetghebeur, S. and Hu, W-S., (1991). Microsphere induced aggregate culture of animal cells. In Production of Biologicals from Animal Cells in Culture. Eds R.E. Spier, J.B. Griffiths, B. Meignier. Butterworth-Heinemann, Oxford. pp 423-428.

Handa-Corrigan, A. (1986). Gas-liquid interfacial effects on the growth of hybridomas and other suspended mammalian cells. Thesis submitted for the Ph.D. at the University of Birmingham. U.K.

Handa-Corrigan, A., Emery, A.N. and Spier, R.E. (1989). Effect of gas-liquid interfaces on the growth of suspended mammalian cells; mechanisms of cell damage by bubbles. Enz. Mic. Technol. vol 11, pp. 230-235.

McCullough, K. and Spier, R.E. (1990). Chapter 5 In Monoclonal Antibodies in Biology and Biotechnology: Theoretical and Practical Aspects. pp. 265-315.

Murdin, A.D., Wilson, R., Kirkby, N.F. and Spier, R.E. (1987). Examination of a simple model for the diffusion of oxygen into dense masses of animal cells. In Modern Approaches to Animal Cell Technology. Eds R.E. Spier and J.B. Griffiths, Butterworth-Heinemann. pp. 353-364.

Spier, R.E. and Griffiths, J.B. (1984). An examination of the data and concepts germane to the oxygenation of cultured animal cells. Developments in Biological Standardisation. vol 55, pp. 81-92.

Spier, R.E. and Whiteside, J.P. (1990). The oxygenation of animal cell cultures by bubbles. In Animal Cell Biotechnology. vol 4, Eds R.E. Spier and J.B. Griffiths, Academic Press, London, pp. 133-147.

Spier, R.E. (1990a). Recent advances in animal cell biotechnology. In K. Sasaki and K. Ikura, Eds., Animal Cell Culture and Production of Biologicals, Kluwer Academic Publishers, Netherlands, pp 41-46.

Wilson, R. and Spier, R.E. (1988). Cytometabolism diffusion test (CMDT) for the determination of animal cell bioreactor parameters. Enzyme and Microbial Technology, vol 10, pp. 161-164.

Wudtke, M., Schugerl, K., (1987). Investigations of physical environment on the cultivation of animal cells. In Modern Approaches to Animal Cell Technology. Eds R.E. Spier and J.B. Griffiths, Butterworth-Heinemann. pp. 297-315.

Zhang, A., Handa-Corrigan, A. and Spier, R.E. (1992). Oxygen transfer properties of bubbles in animal cell culture media. Biotech. Bioeng. vol 40, in press.

Appendix 1

```
 10REM > BR.NO
 20PRINT"ASSUMPTIONS"
 30PRINT" "
 40PRINT"24 HR OPERATION"
 50PRINT"BATCH SYSTEM, BATCH LENGTH 6 DAYS, TURNAROUND TIME 1 DAY"
 60PRINT"A 100L VESSEL (INSTRUMENTED SIMPLY) COSTS $100,000"
 70PRINT"VESSEL CAPITAL COST INCREASES BY 3 FOR A 10 FOLD SCALE UP"
 80PRINT"VESSELS DEPRECIATE TO $0 AFTER 10 YEARS (STRAIGHT LINE)"
 90PRINT"LABOUR COST BASED ON SMALLEST TEAM OF 3 FOR UP TO 4 VESSELS"
100PRINT"THEN 1 ADDITONAL FOR EACH TWO VESSELS"
110PRINT"LABOUR WORKS A 1750 HOUR YEAR"
120PRINT"LABOUR COSTS $20/HR WITH 100% OVERHEAD = $40/HR"
130PRINT""
140PRINT"PLANT TO PRODUCE 10 KG PRODUCT PER ANNUM"
150PRINT"CELLULAR PRODUCTIVITY 30PG/CELL/DAY"
160PRINT"OPERATES 280 DAYS/ANNUM"
170PRINT"2E9 CELLS/LITRE OF MEDIUM FOR GROWTH; MEDIUM COST $2/LITRE."
180PRINT"INITIAL CELL CONCN 2E5; FINAL CELL CONCN 2E6"
190PRINT"SEED DEVELOPMENT RATIO 1:10"
200PRINT"INITIAL SEED VOL (FROM FREEZER) ENOUGH FOR A 1 LITRE CULTURE"
210PRINT"MODEL SYSTEM BASED ON AN STIRRED TANK REACTOR"
220PRINT"CONTAMINATION/FAILURE RATES EXAMINED; 2%, 10%, 30%"
230PRINT"COSTS OF QA/QC PER FINAL SCALE BIORACTOR; 0.5*, 1*, 2*, 5* BATCH"
240DIM TBCQAC(5)
250DIM QAQC(4)
260DIM TBC(5)
270DIM MC(3)
280DIM TVC(5)
290DIM VC(5)
300DIM NPV(5)
310DIM PVV(5)
320DIM SVN(5)
330DIM TVN(5)
340DIM LC(5)
350DIM MEDVOL(3)
360REM CACLULATION OF VOLUME OF MEDIUM REQUIRED
370DIM FR(3)
380MEDVOLL=10000/(2E6*30E-12*1000)
390FOR N=1 TO 3
400DATA 1.02,1.1,1.3
410READ FR(N)
420NEXT
430FOR N=1 TO 3
440MEDVOL(N)=MEDVOLL*FR(N)
450NEXT
460FOR Z = 1 TO 3
470REM CALCULATION OF MEDIUM VOLUME PER BATCH
480MEDVOLB=MEDVOL(Z)/40
490REM VOLUME OF PRODUCTION VESSELS
500FOR N=1 TO 5
510PVV(N)=MEDVOLB/N
520NEXT
530REM CALCULATION OF NUMBER OF SEED VESSELS FOR THE LARGEST FINAL VESSEL
540 N=1
550X=1
560Y=1
570IF PVV(N)/(10*Y) < 1  THEN GOTO 610
580X=X+1
590Y=Y*10
600GOTO 570
610SVN=X-1
620REM TOTAL NUMBER OF VESSELS
630FOR N=1 TO 5
640TVN(N)=N+SVN
650NEXT
```

```
660REM LABOUR COST
670FOR N=1 TO 5
680LC(N)=(((3*40)*1750)+((TVN(N)-4)*40)*1750/2)
690NEXT
700FOR N=1 TO 5
710NEXT
720NEXT
730FOR Z=1 TO 3
740NEXT
750REM CALCULATION OF COST OF PRODUCTION VESSELS
760FOR N=1 TO 5
770VC(N)=(((LOG(PVV(N))+8)/2))
780VC(N)=EXP(2.303*VC(N))
790NEXT
800REM COST OF SEED VESSELS
810SVC=EXP(2.303*(((LOG(PVV(1))/10)+8)/2))+EXP(2.303*(((LOG(PVV(1))/100)+8)/2
820REM TOTAL VESSEL COSTS
830FOR N=1 TO 5
840TVC(N)=(VC(N)*N)+SVC
850TVC(N)=TVC(N)/10
860NEXT
870REM TOTAL MEDIUM COSTS
880FOR Z = 1 TO 3
890MC(Z) = MEDVOL(Z)*2
900NEXT
910REM TOTAL BATCH COSTS (MEDIUM+LABOUR+VESSEL) FOR DIFFERENT FAILURE RATES
920FOR Z=1 TO 3
930FOR N= 1 TO 5
940TBC(N)=MC(Z)+TVC(N)+LC(N)
950NEXT
1000NEXT
1010REM ADDING IN THE COST OF QA/QC
1020FOR A= 1 TO 4
1030DATA .5,1,2,5
1040READQAQC(A)
1050NEXT
1060REM TOTAL COST OF BATCH WITH QA/QC COSTS ADDED
1070FOR A=1 TO 4
1080FOR Z =1 TO 3
1090FOR N=1 TO 5
1100TBCQAC(N)=INT(MC(Z)+TVC(N)+LC(N))+((MC(Z)+TVC(N)+LC(N))*(QAQC(A)))
1110NEXT
1120NEXT
1130NEXT
1140REM PRINT OUT
1150PRINT
1160PRINT
1170PRINT
1190PRINT"  TOTAL COST/ANNUM  ($) FOR 10 KG ANIMAL CELL PROTEIN"
1200PRINT
1210PRINT"      QA/QC   N.PRODVESS  2%FAIL    10%FAIL    30%FAIL"
1220PRINT
1230FOR A=1 TO 4
1240FOR N=1 TO 5
1250PRINT QAQC(A);
1260PRINT N;"    ";
1270FOR Z=1 TO 3
1280PRINT INT(MC(Z)+TVC(N)+LC(N))+INT(((MC(Z)+TVC(N)+LC(N))*(QAQC(A))));" ";
1290NEXT
1300PRINT
1310NEXT
1320PRINT
1330NEXT
```

ASSUMPTIONS

24 HR OPERATION
BATCH SYSTEM, BATCH LENGTH 6 DAYS, TURNAROUND TIME 1 DAY
A 100L VESSEL (INSTRUMENTED SIMPLY) COSTS $100,000
VESSEL CAPITAL COST INCREASES BY 3 FOR A 10 FOLD SCALE UP
VESSELS DEPRECIATE TO $0 AFTER 10 YEARS (STRAIGHT LINE)
LABOUR COST BASED ON SMALLEST TEAM OF 3 FOR UP TO 4 VESSELS
THEN 1 ADDITONAL FOR EACH TWO VESSELS
LABOUR WORKS A 1750 HOUR YEAR
LABOUR COSTS $20/HR WITH 100% OVERHEAD = $40/HR

PLANT TO PRODUCE 10 KG PRODUCT PER ANNUM
CELLULAR PRODUCTIVITY 30PG/CELL/DAY
OPERATES 280 DAYS/ANNUM
2E9 CELLS/LITRE OF MEDIUM FOR GROWTH; MEDIUM COST $2/LITRE.
INITIAL CELL CONCN 2E5; FINAL CELL CONCN 2E6
SEED DEVELOPMENT RATIO 1:10
INITIAL SEED VOL (FROM FREEZER) ENOUGH FOR A 1 LITRE CULTURE
MODEL SYSTEM BASED ON AN STIRRED TANK REACTOR
CONTAMINATION/FAILURE RATES EXAMINED; 2%, 10%, 30%
COSTS OF QA/QC PER FINAL SCALE BIORACTOR; 0.5*, 1*, 2*, 5* BATCH

TOTAL COST/ANNUM ($) FOR 10 KG ANIMAL CELL PROTEIN

QA/QC	N.PRODVESS	2%FAIL	10%FAIL	30%FAIL
0.5	1	941052	981052	1081053
0.5	2	1039381	1079382	1179382
0.5	3	1127046	1167046	1267047
0.5	4	1209190	1249191	1349191
0.5	5	1287807	1327807	1427808
1	1	1254736	1308070	1441404
1	2	1385842	1439176	1572510
1	3	1502728	1556062	1689396
1	4	1612254	1665588	1798922
1	5	1717076	1770410	1903744
2	1	1882105	1962105	2162106
2	2	2078764	2158764	2358765
2	3	2254093	2334094	2534094
2	4	2418382	2498382	2698383
2	5	2575615	2655615	2855616
5	1	3764211	3924212	4324212
5	2	4157530	4317530	4717530
5	3	4508188	4668188	5068189
5	4	4836765	4996765	5396766
5	5	5151231	5311231	5711232

Appendix 2

```
L.
  100CLS
  110PRINT"BIORVOL    RPS     IMPDIAM(M) KOLRAD(MU)"
  120REM THIS PROGRAMME ASSUMES THE VALUE OF 6.9 FOR THE IMPELLER NUMBER (WEBB)
  130REM AND THE FLUID BEING LIKE WATER WITH A DENSITY OF 1000 KG/M^3
  140REM THE RESULTING POWER TERM IS IN WATTS EQUIVALENT TO J/S AND NM/S
  150 REM THE DYNAMIC VISCOSITY OF WATER IS TAKEN AS 0.7 CENTIPOISE
  160 REM WHICH IS EQUIVALENT TO .0007 KG/M.SEC
  170 REM THIS MAKES THE KINEMATIC VISCOSITY (=DYNAMIC/DENSITY) =.0007/1000=7E-7
  180BIORVOL=1
  183RPS=.1
  186IMPDIA=.01
  190FOR X=1 TO 4
  200FOR Y=1 TO 8
  210FOR Z=1 TO 7
  220 POWER=6.9*1000*RPS^3*IMPDIA^5
  230 PPVOL=POWER/BIORVOL
  240 KOLRAD=((7E-7^3)/PPVOL)^.25
  250 KOLRAD%=KOLRAD*1E6
  260PRINT""; BIORVOL;"          ";RPS;"          ";IMPDIA;"        ";KOLRAD%
  270IMPDIA=IMPDIA*2
  280NEXT
  290 IMPDIA= 0.01
  300PRINT
  310PRINT"BIORVOL    RPS     IMPDIAM(M) KOLRAD(MU)"
  320PRINT
  330RPS=RPS*2
  340NEXT
  350RPS=.1
  360PRINT
  370PRINT"BIORVOL    RPS     IMPDIAM(M) KOLRAD(MU)"
  380PRINT
  385BIORVOL=BIORVOL*10
  390NEXT
>RUN
```

Appendix 3

BIORVOL	RPS	IMPDIAM(M)	KOLRAD(MU)
1	0.2	0.01	2807
1	0.2	0.02	1180
1	0.2	0.04	496
1	0.2	0.08	208
1	0.2	0.16	87
1	0.2	0.32	36
1	0.2	0.64	15

BIORVOL	RPS	IMPDIAM(M)	KOLRAD(MU)
1	0.4	0.01	1669
1	0.4	0.02	701
1	0.4	0.04	295
1	0.4	0.08	124
1	0.4	0.16	52
1	0.4	0.32	21
1	0.4	0.64	9

BIORVOL	RPS	IMPDIAM(M)	KOLRAD(MU)
1	0.8	0.01	992
1	0.8	0.02	417
1	0.8	0.04	175
1	0.8	0.08	73
1	0.8	0.16	31
1	0.8	0.32	13
1	0.8	0.64	5

BIORVOL	RPS	IMPDIAM(M)	KOLRAD(MU)
1	1.6	0.01	590
1	1.6	0.02	248
1	1.6	0.04	104
1	1.6	0.08	43
1	1.6	0.16	18
1	1.6	0.32	7
1	1.6	0.64	3

BIORVOL	RPS	IMPDIAM(M)	KOLRAD(MU)
1	3.2	0.01	350
1	3.2	0.02	147
1	3.2	0.04	62
1	3.2	0.08	26
1	3.2	0.16	10
1	3.2	0.32	4
1	3.2	0.64	1

BIORVOL	RPS	IMPDIAM(M)	KOLRAD(MU)
1	6.4	0.01	208
1	6.4	0.02	87
1	6.4	0.04	36
1	6.4	0.08	15
1	6.4	0.16	6
1	6.4	0.32	2
1	6.4	0.64	1

BIORVOL	RPS	IMPDIAM(M)	KOLRAD(MU)
1000	1.6	0.01	3319
1000	1.6	0.02	1395
1000	1.6	0.04	586
1000	1.6	0.08	246
1000	1.6	0.16	103
1000	1.6	0.32	43
1000	1.6	0.64	18

BIORVOL	RPS	IMPDIAM(M)	KOLRAD(MU)
1000	3.2	0.01	1973
1000	3.2	0.02	829
1000	3.2	0.04	348
1000	3.2	0.08	146
1000	3.2	0.16	61
1000	3.2	0.32	25
1000	3.2	0.64	10

BIORVOL	RPS	IMPDIAM(M)	KOLRAD(MU)
1000	6.4	0.01	1173
1000	6.4	0.02	493
1000	6.4	0.04	207
1000	6.4	0.08	87
1000	6.4	0.16	36
1000	6.4	0.32	15
1000	6.4	0.64	6

BIORVOL	RPS	IMPDIAM(M)	KOLRAD(MU)
1000	12.8	0.01	697
1000	12.8	0.02	293
1000	12.8	0.04	123
1000	12.8	0.08	51
1000	12.8	0.16	21
1000	12.8	0.32	9
1000	12.8	0.64	3

Appendix 4

```
 10  REM
 20 PRINT
 30PRINT
100DATA .1,.3,1,3
110 DIM BUBDIAM(10)
120DIM OTR(10)
130DIM RESTM(10)
140 DIM OXDEM(10)
150FOR X=1 TO 4
160 READ BUBDIAM(X)
170NEXT
180DATA 10,25,50
190FOR X=1 TO 3
200 READ OTR(X)
210NEXT
220DATA .1,.3,1,3
230FOR X=1 TO 4
240READ RESTM(X)
250NEXT
260 DATA .01,.1,1,10
270 FOR X=1 TO 4
280 READ OXDEM(X)
290 NEXT
300 FOR B=1 TO 4
310PRINT "OXYGEN DEMAND = ";OXDEM(B);"GM/L/HR"
315PRINT"BD(MM) OTR(UG/SCM) REST(S) AFR(L/HR)"
320 FOR A=1 TO 4
330 FOR C=1 TO 3
340 FOR D=1 TO 4
350PROCCALC
360 PRINT"";BUBDIAM(A);"    ";OTR(C);"    ";RESTM(D);"    ";AFR
370NEXT
371NEXT
372 NEXT
373NEXT
500DEF PROCCALC
510 BUBSA=4*PI*(BUBDIAM(A)*.05)^2
520 BUBNO=OXDEM(B)/((OTR(C)*BUBSA)*1E-6)
530 BUBVOL=4*PI*((BUBDIAM(A)*.05)^3)*BUBNO*1E-6
540 AFR=BUBVOL*3600/RESTM(D)
550ENDPROC
700 FOR B=1 TO 4
>
```

OXYGEN DEMAND = 1E-2GM/L/HR

BD(MM)	OTR(UG/SCM)	REST(S)	AFR(L/HR)
.1	10	0.1	0.18
.1	10	0.3	6E-2
.1	10	1	1.8E-2
.1	10	3	6E-3
.1	25	0.1	7.2E-2
.1	25	0.3	2.4E-2
.1	25	1	7.2E-3
.1	25	3	2.4E-3
.1	50	0.1	3.6E-2
.1	50	0.3	1.2E-2
.1	50	1	3.6E-3
.1	50	3	1.2E-3
.3	10	0.1	0.54
.3	10	0.3	0.18
.3	10	1	5.4E-2
.3	10	3	1.8E-2
.3	25	0.1	0.216
.3	25	0.3	7.2E-2
.3	25	1	2.16E-2
.3	25	3	7.2E-3
.3	50	0.1	0.108
.3	50	0.3	3.6E-2
.3	50	1	1.08E-2
.3	50	3	3.6E-3
	10	0.1	1.8
	10	0.3	0.6
	10	1	0.18
	10	3	6E-2
	25	0.1	0.72
	25	0.3	0.24
	25	1	7.2E-2
	25	3	2.4E-2
	50	0.1	0.36
	50	0.3	0.12
	50	1	3.6E-2
	50	3	1.2E-2
	10	0.1	5.4
	10	0.3	1.8
	10	1	0.54
	10	3	0.18
	25	0.1	2.16
	25	0.3	0.72
	25	1	0.216
	25	3	7.2E-2
	50	0.1	1.08
	50	0.3	0.36
	50	1	0.108
	50	3	3.6E-2

OXYGEN DEMAND = 10GM/L/HR

BD(MM)	OTR(UG/SCM)	REST(S)	AFR(L/HR)
0.1	10	0.1	180
0.1	10	0.3	60
0.1	10	1	18
0.1	10	3	6
0.1	25	0.1	72
0.1	25	0.3	24
0.1	25	1	7.2
0.1	25	3	2.4
0.1	50	0.1	36
0.1	50	0.3	12
0.1	50	1	3.6
0.1	50	3	1.2
0.3	10	0.1	540.
0.3	10	0.3	180
0.3	10	1	54
0.3	10	3	18
0.3	25	0.1	216
0.3	25	0.3	72
0.3	25	1	21.6
0.3	25	3	7.2
0.3	50	0.1	108
0.3	50	0.3	36
0.3	50	1	10.8
0.3	50	3	3.6
1	10	0.1	1800
1	10	0.3	600
1	10	1	180
1	10	3	60
1	25	0.1	720
1	25	0.3	240
1	25	1	72
1	25	3	24
1	50	0.1	360
1	50	0.3	120
1	50	1	36
1	50	3	12
3	10	0.1	5400
3	10	0.3	1800
3	10	1	540
3	10	3	180
3	25	0.1	2160
3	25	0.3	720
3	25	1	216
3	25	3	72
3	50	0.1	1080
3	50	0.3	360
3	50	1	108
3	50	3	36

Appendix 5

Computation of the number of layers of cells supportable by 1 ml of culture fluid when the diffusivity of oxygen is taken into account and the oxygen consumption of the cells is included

---

Additional assumptions

1 Diffusivity of oxygen at 37°C in water is 2.5×10^{-5} cm^2/s.

Program

```
100  OX DEM = 4.1 E-10               oxygen demand (5 × 10^-5
                                     × 8.3 × 10^-16)
110  DIFCO = 2.5 E-5                 (diffusion coefficient of
                                     oxygen in water)
120  LAYTH = 1 E-3                   (thickness of cell sheet)
130  DISOX = 7 E-6                   (dissolved oxygen concentration
                                     in water)
150  T1 = (OXDEM × LAYTH)/(DIFCO * DISOX)
                 (T1 = time taken for oxygen to move to
                       the top of the first layer)
160  T=T+T1
170  DISOX = DISOX - OXDEM
175  N = 2                           (for second layer of cells)
180  TA2 = OXDEM * N * LAYTH
185  TB2 = DIFCO * DISOX
190  T2 = TA2/TB2
195  T=T+T2
200  PRINT"";N;"";T
205  IF T>1 THEN END
210  DISOX = DISOX - OXDEM
220  N = N + 1
230  GOTO 180
```

Running this program gives N = 29 layers for T = 1 s.

Appendix 6

```
 10REM GLUCOSE MASS TRANSFER
 20REM SAVE UNDER "GLUMAST"
 30PRINT"ASSUMPTIONS"
 40PRINT
 50PRINT"THE GLUCOSE DIFFUSIVITY IS 5E-6 CM2 SEC-1"
 60PRINT"GLUCOSE DEMAND IS 6 MILLIMOLES OR 1.08GM/1E9CELLS/DAY"
 70PRINT"ONE LAYER OF CELLS CONTAINS 5E5 CELLS/CM2"
 80PRINT"THE THICKNESS OF A LAYER OF CELLS IS 1E-3CM (10MICRONS)"
 90PRINT"GLUCOSE CONCENTRATION IN THE MEDIUM IS 2GM/L OR 11 MILLIMOLAR)"
100GLUDEM = 5E5*(1.08/(1E9*24*3600))
110GLUDIFCO = 5E-6
120LAYTH = 1E-3
130GLUCONC = 2E-3
140T1 = (GLUDEM*LAYTH)/(GLUDIFCO*GLUCONC)
150T = T + T1
160GLUCONC = GLUCONC - GLUDEM
170N = 2
180TA2 = GLUDEM * N* LAYTH
190TB2 = GLUDIFCO * GLUCONC
200T2 = TA2/TB2
210T = T+T2
230IF T>1 THEN 300
240GLUCONC =GLUCONC - GLUDEM
250N = N+1
260GOTO 180
300PRINT:PRINT"THE NUMBER OF CELL LAYERS THAT CAN BE SUPPORTED IS ";N :END
>
>
>RUN
ASSUMPTIONS

THE GLUCOSE DIFFUSIVITY IS 5E-6 CM2 SEC-1
GLUCOSE DEMAND IS 6 MILLIMOLES OR 1.08GM/1E9CELLS/DAY
ONE LAYER OF CELLS CONTAINS 5E5 CELLS/CM2
THE THICKNESS OF A LAYER OF CELLS IS 1E-3CM (10MICRONS)
GLUCOSE CONCENTRATION IN THE MEDIUM IS 2GM/L OR 11 MILLIMOLAR)

THE NUMBER OF CELL LAYERS THAT CAN BE SUPPORTED IS 57
```

GENERAL METHODOLOGY IN BIOPROCESS ENGINEERING

ANTON MOSER
Institut für Biotechnologie, TU Graz,
A-8010 Graz, Petersgasse 12
AUSTRIA

ABSTRACT. **Commercialization** of biotechnologies not only depends on intelligent products for the market but also on a rapid and secure methodology for bioprocess design & development. The "old" empirical way of trial & error is to be replaced by a more systematic procedure according to a recent methodology in bioprocess technology, called the "formal macroapproach", based on a holistic mode of thinking, operating with interaction-networks (integration of biological and physical phenomena). It is based on macroscopic process variables and on formal analogies adapted to experiments, which are used for mathematical modeling. A better scale-up is achieved by a scale-down approach, where regime analysis is used to determine the process bottlenecks, which then are experimentally simulated in the scale-down "model-bioreactors", representing the new integrating strategy. Mathematical modeling plays a central role representing a simplified but adequate approximation to real process behaviour.

1. Introduction

Practice in biotechnologies dominates engineering sciences and research activities so that a major part of investigations follows the purely empirical approach on a pragmatic level. Thereby, exploratory research in lab-scale units is the first step, followed directly by pilot plant work for pragmatic evaluation of the bioprocess for industrial realization (Fig. 1). Here, the overall behavior is quantified mainly in terms of economics. This still is a powerful approach although only singular situations are treated under special conditions with restricted chances for generalization and without deep scientific ambitions.

Supplementary to this straightforward approach, some research groups are engaged in developing a more systematic strategy in bioprocess analysis & design (e.g. Moser, 1988), which increasingly becomes competitive to the purely pragmatic approach. The systematics, thereby, include a step of systematic process analysis and design (Figure 2), which is carried out on bench-scale bioreactors with special characteristics and special research tools. In the long run, time and money can be saved by using this modern engineering mentality & methodology according to a holistic approach.

A.R. Moreira and K.K. Wallace (eds.), Computer and Information Science Applications in Bioprocess Engineering, 349-364.

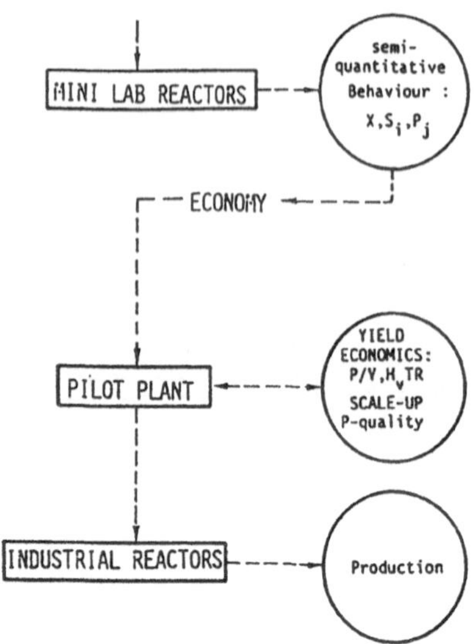

Figure 1. The purely empiric-pragmatic approach in bioprocess design.

2. Mode of Thinking and Working

In order to show the main differences between the purely empiric approach (Fig. 1) and the more systematic approach (Fig. 2), two facts are important :

* The supplementary step 2 in Figure 2 using bench-scale bioreactors, called "perfect" bioreactors and "model bioreactors" (Moser, 1989)

* The meaning of the pilot-plant-units.
While pilot reactors following the pragmatic approach are mainly used for quantifying economics for the process (i.e. power consumption, heat transfer, productivity and product quality), the pilot-plants according to the systematic approach are supplementary useful for the elucidation of the process model as a first working hypothesis. Pilot plants verify or falsify the process model before it can be used as a network-basis for further process considerations (process behaviour and scale-up). Thus, pilot plant work reintegrates the aspects of biology and physics from bench-scale work with the special interest of the interactions between them, thus, leading to generalizations.

However, the entire concept of this biotechnical methodology does not only include the separation and re-integration of biological and physical phenomena in the step of systematic process analysis (see Fig. 2) in order to elucidate these significant parts with the aid of mathematical models, but includes also transitional work between industrial scale, pilot-scale and lab-scale work as summarized in the scheme of Figure 3 (Moser, 1991).

As a prerequisite, the fully developed methodology is expected to build the basis of thinking & working. It should be mentioned that this methodology serves also as a tool for adequate comparison of different bioreactors in their performance using so-called " biological test systems" (Meyer ,1987).

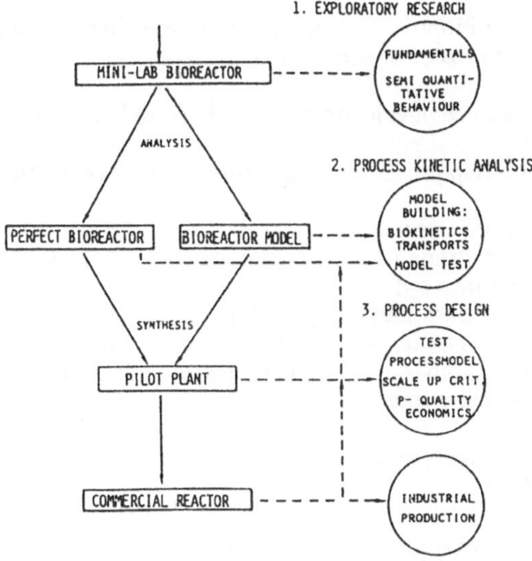

Figure 2. The systematic approach in bioprocess design using carefully designed experiments together with mathematical models. Step 2 represents the systematic process analysis as a basis of later process design.

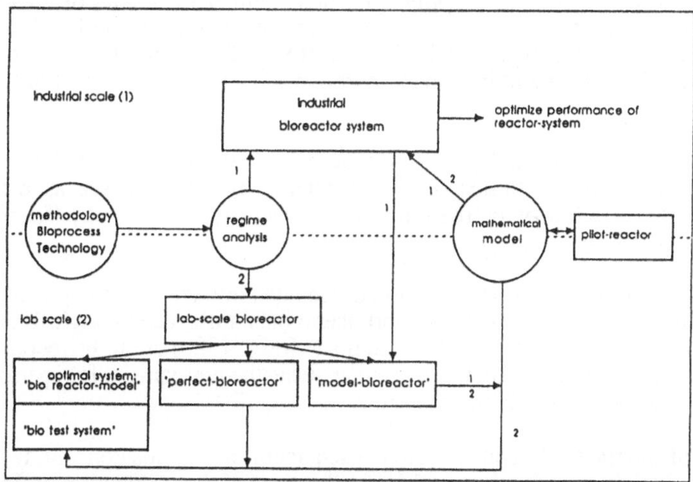

Figure 3. Integrating / holistic strategy of bioprocess technology for the systematic design of bioprocesses and comparison of bioreactor system: Application of regime analysis to industrial (1) or lab scale processes using scale-down bioreactors(lab-scale bioreactors) including "bioreactor models", "perfect bioreactors" and "model bioreactors" leading to mathematical models which then can be tested in pilot scale for further use in process design & optimization .

Bioprocess design, as another case of application of this methodology, has to start with systematic analyses including "regime analysis" of

* The industrial scale bioprocesses in the case of existing & running processes, which need to be optimized for better competition

* A new bioprocess to be developed completely from lab-scale.

Regime analysis is a four-step procedure (Oosterhuis and Kossen, 1985 ; Moser, 1988):

1) defining significant process variables
2) quantifying them, resulting in characteristic values for process rates (rate constants k_{ch} or characteristic times t_{ch})
3) comparing all individual values of k_{ch} resp. t_{ch}
4) the result will be that one rate of a process variable is the slowest, this is then the process bottleneck. All other non-limiting rates, are not neglected but are taken into account as a quasi-steady-state .

Examples of such a regime analysis are demonstrated in the literature (Sweere et al., 1986; 1989).

The most significant value of regime analysis is that it enables one to focus on the process bottleneck, which may be changing during the process.

Very often, the behaviour of the bioprocess in dependence of process bottlenecks (e.g. gradients of concentrations, mixing, viscosity, shear etc.) is not known. Thus, research should concentrate on this point. Only a limited number of results on this subject are described in literature, and have recently been summarized by the author (Moser, 1990). For this purpose a special type of lab-scale reactor, called a "model-bioreactors" which realizes a defined environment in respect to one of the above mentioned bottlenecks, is to be used.

Table. 1 compiles so-called "model-bioreactors" realizing strictly defined physical environment of lab-scale-studies of the transients of bioprocesses as a function of interfacial mass transfer, mixing, foaming, shear, viscosity, morphology (for references in detail see Moser, 1990).

At the same time (Fig. 3) the quantification of "ideal" biokinetics of the process is needed, where ideal means that no physical transport phenomena influence biological reaction rates. For this purpose so-called "perfect" bioreactors (Moser, 1988) are to be recommended for use. Applying the working concept of regime analysis the prerequisite for "perfect" reactors can be defined: the test of pseudohomogeneity

A third type of bioreactor in bench-scale is used according to this concept, representing a small-scale model (geometrical and other similarities) of the production-scale bioreactor. Here physical transport phenomena are to be quantified under pure physical conditions in the first run, but also under biological conditions as already mentioned ("biological test systems").

Tab.1 : List of model-bioreactors

bottleneck	model-bioreactor	characteristics	case studies
G/L transport	thin layer tubular	a = constant k_L = calculable no velocity profile sterile	bakers yeast Trichosporon cutaneum
L/S mass transport	sloping plane	a = constant biofilm thickness d_F= constant non-sterile	waste water
	fluidized bed	d_F = constant sterile	waste water
	tubular biofilm	a = constant d_F = constant sterile	Aspergillus niger
L mixing	"2-compartment" 2 stirred tanks(ST)	maximum mixed (m.m.)	Gluconobacter E.coli(rDNA) bakers yeast Bacillus. subtilis
	ST + plug flow (PF)	m.m. + t.s. (total segregated)	bakers yeast E.coli Penicillium
	loop (PF+PF)	t.s. + t.s. aerated+nonaerated	Asterophera lycoperdoides Aspergillus niger bakers yeast
Foam	ST + PF	m.m. + t.s. aerated + nonaerated	bakers yeast
Shear	L-jet in ST	shear stress = calculable	Chlorella Methanococcus Chlamydomonas Tetrahymena
	PF rheometer (coaxial rotating drums)	shear = calculable	Penicillium chrysog.
Viscosity	ST + helical ribbon impeller	shear = uniform = calculable	Aspergillus niger Penicillium chrysog.

The results of all these lab-scale activities represent the starting point for the set-up of mathematical models, which serve as a basis for all further engineering considerations of predicting process behaviour and process evaluation & economics.

3. Mathematical Model Approach

3.1. GENERAL

The quantification of bioprocess kinetics has had a long tradition since Monod carried out his fundamental work. In accordance with the mechanistic paradigm which still dominates the scientific world - although some strong indications exist that we are at a turning point where we are beginning to switch to a holistic view - much knowledge has accumulated for the clarification of the mechanisms, i.e. to "know why" . On the other hand, industrial practice and process engineers handle bioprocesses, under the stress of commercialization, quite differently and on a purely empirical basis. Somehow - on a basis of their feelings - the engineers use the mechanistic facts as a background for their "know how", which is very often treated as a matter of top secret. With this situation the exchange of knowledge between the pragmatic and the scientific levels is hindered, leading to an under-developed methodology in process design.

It is the ambition of this paper to build a bridge over this gap by first demonstrating the validity of the (formal) macroapproach as the scientific justification of this engineering approach, and then to show the supplementary enrichment of the formal macroscopic view to the microscopic mechanistic approach.

The holistic view in bioprocessing can be summarized as follows:
A living organism is an integral entity and must be understood as a network of coordinating activities non-linearly interacting with the physical & chemical environment and not as a sum of separate parts of metabolism, genes, chromosomes, cell tissues and environment in the reactor.

The consequences of this holistic view are taken into account when setting-up mathematical models of a bioprocess, which is graphically shown in Figure 4, representing the modeling cycle in general (Moser, 1988).

From the problem the formulation of a verbal model leads to the set-up of a mathematical model incorporating variables x, model function f with parameters k . Verification follows as indicated.

It is to be stated here that five different levels of engineering activities exists, which must be seen as supplements in a holistic view (Figure 5).

3.2. APPROACHES IN BIOPROCESS KINETIC MODELING

It is well known from the literature (Moser, 1988) that three main kinds of mathematical models exist:

- the (pure) numerical fit models,
- the (mechanistic) microkinetic models,
- the (macroscopic) formal kinetic models.

However, their evaluation is rarely clearly explained.

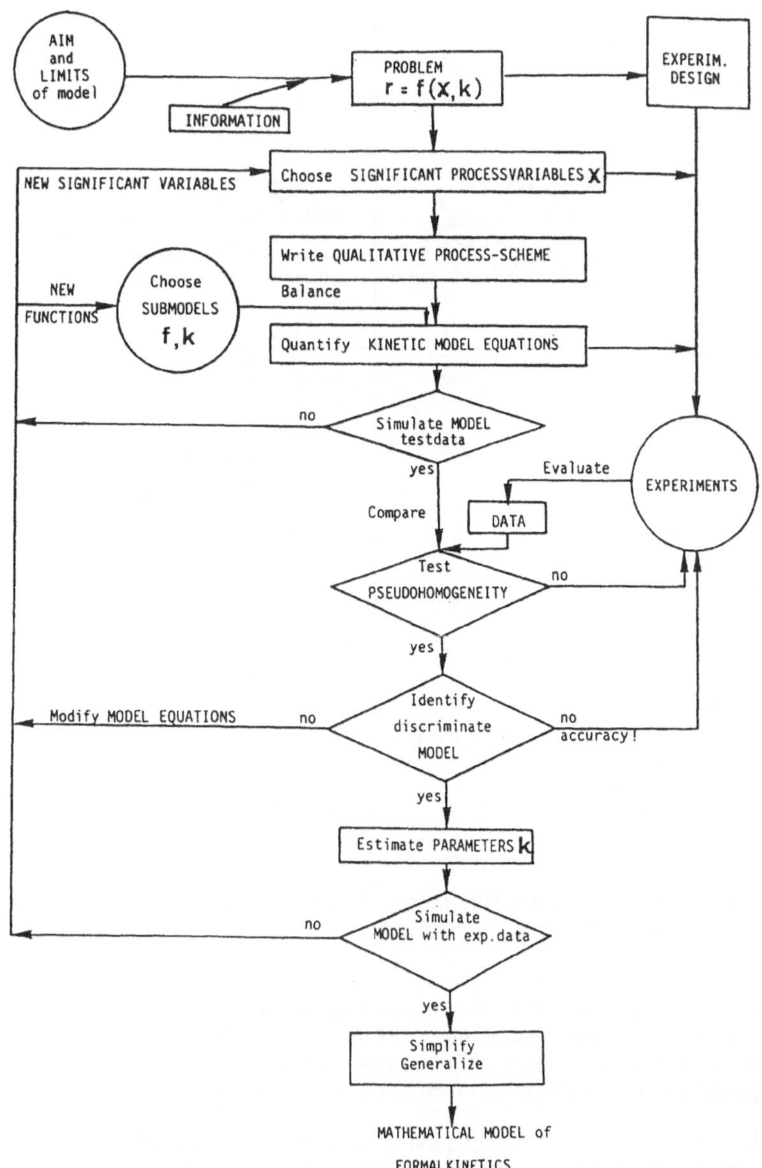

Figure 4. The modeling cycle in setting up mathematical models of bioprocesses based on verbal models using : process variables (x) and mathematical functions (f) including model parameters (k) .

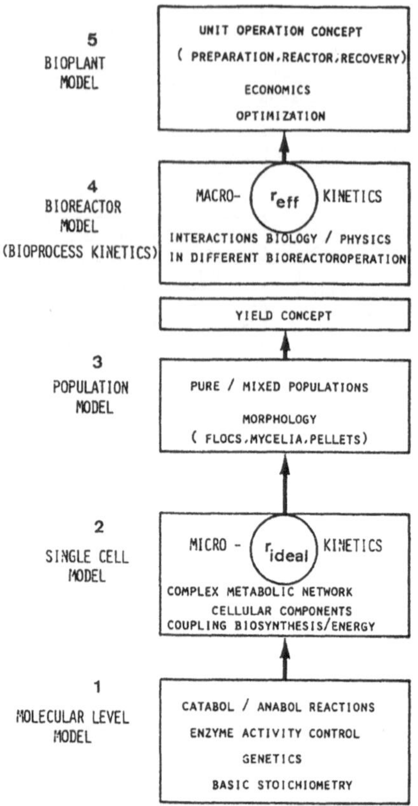

Figure 5. Overview of the five levels of process research and mathematical modeling of bioprocessing and bioreactor operation.

Pure numerical fitting methods are without biological relevance, as they use "blind" parameters in arbitrary mathematical functions e.g. polynoms. Owing to their simplicity and usefulness, they are used when poor knowledge is given and preferably by control engineers, as they represent an overall picture of individual cases.

Microkinetics, based on mechanisms derived from fundamental work, are widespread in biochemistry as they represent a "better" understanding. These types of models include parameters which can clearly be interpreted in detail on the basis of mechanisms (e.g. enzyme kinetics : Michaelis-Menten coefficients, and other descriptions of mechanisms, including modifications concerning inhibitors etc.). Microkinetics are thus quite complex; all parameters are not measurable and any satisfactory interpretation depends on the significance of mechanisms, which is a matter for speculation.

Formal kinetics, on the other hand, take into account that no mechanism can be regarded as true, that mechanisms cannot be clarified in "economic" time for engineering tools, and that mechanisms are far too complex to be handled for process design purposes. Even in the century of computers, the main problem is the experimental verification of model parameters.

In this situation, it can be recommended to use the concept of formal kinetics, as they do not pretend to be the truth. However, they are based on carefully-designed experiments, physical transport phenomena must be clearly excluded, so that they do not influence the bioprocess. Therefore special bench-scale bioreactors are needed as mentioned before: "perfect" bioreactors or "high-performance" bioreactors. These laboratory-scale reactors achieve pseudohomogeneity, e.g. by applying the highest possible power input for the sake of transport processes for heat, impulse and mass).

It should be mentioned here that e.g. Monod-type kinetics are formal kinetics . The Monod equation has the same form as the Michaelis-Menten equation. It should be clear that the difference lies in the meaning of the parameters (K_S vs. K_m and $q_{s,max}$ vs. v_{max}). It is typical for the formal approach, that the parameters stem from adapting them to experiments, therefore they are called adaptation parameters, with apparent character but without mechanistic interpretability (e.g. K_S).

Formal kinetics must not necessarily be of the unstructured type of model. Biomass (X), often represented by rather unsatisfactory measurements of cell dry weight, can easily be understood and handled by multicompartment models .

In the case of a two-compartment biomass model (Figure 6), the active, structural compartment includes the protein synthesizing system and RNA (K) , while the genetic compartment contains mainly DNA in the genome (G).

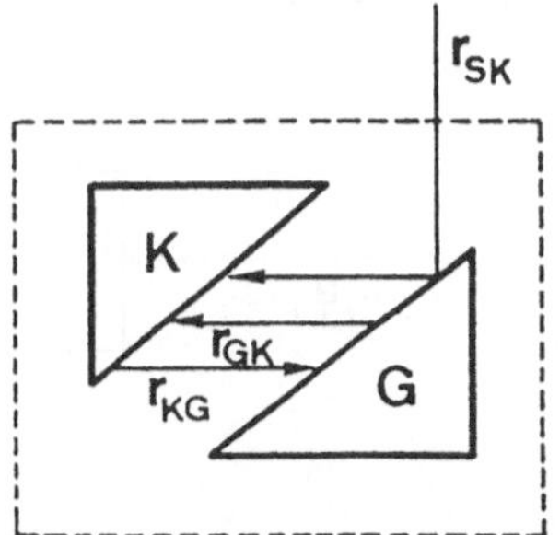

Figure 6 Block diagram of a two-compartment model (K and G) of cells following the concept of Williams. The rates are r_{SK} = rate of conversion of substrate to K compartment, r_{KG} = rate of conversion of K to G compartment, and r_{GK} = rate of depolymerization of G to K compartment

Most important to the understanding of such structured biomass models is the fact that an exact experimental verification does not exist. Thus, both compartments again represent a formal approach with some increase in plausibility resulting from the more detailed consideration.

Furthermore, **structuring of models** does not only involve structuring the biomass, but also includes structuring the metabolic pathways (Figure 7, the cell function e.g. cell age, activity and synchronization and structuring also the reactor itself due to different states of mixing. Figure 8 illustrates the situation of combining a well mixed L-phase (L_1) with a segregated zone (L_2) and a well mixed G-phase including an exchange stream from the pumping capacity F_P of the impeller.

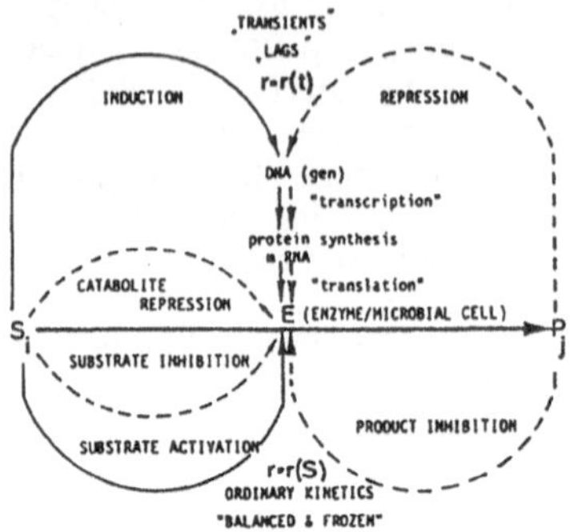

Figure 7. Schematic representation of enzyme regulation principles as a basic for biokinetic model building. While ordinary kinetics are analogous to chemical kinetics (heterogeneous catalysis), enzyme induction and repression are typical biological phenomena creating transients.

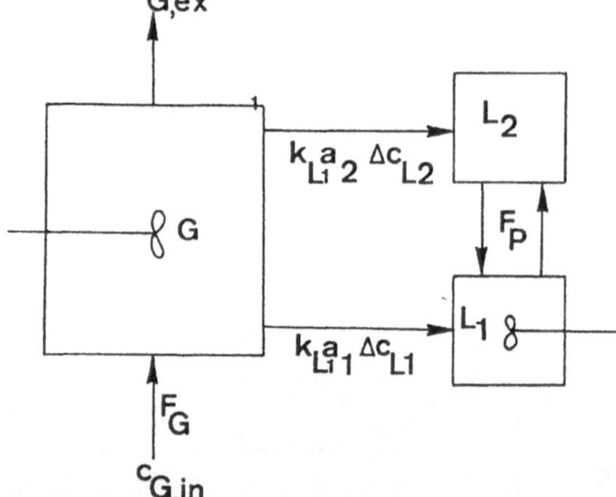

Figure 8. Schematic representation of structuring bioreactors including a well-mixed region and a region of bypass liquid flow.

The formal macroapproach is in full agreement with the holistic view in science (see Varela F., Maturana H.R., Uribe R., 1974:"Autopoiesis: the Organization of Living Systems" Biosystems 5, 187), stating that " One can hope to find a *much simpler* description of systems, when looking for *dynamic rules* of the procedure of process and for criteria of the overall behavior in a higher semantic level. Thereby a new "*macroscopic simplicity*" will be discovered, which is not on the microscopic/mechanistic level " (see also Jantsch E., 1982: The Selforganization of the Universe, p. 76).

3.3 THE FORMAL MACRO-APPROACH

The formal macro-approach is based on the quantification of the significant macroscopic process variables ("key variables") i.e. biomass X (unstructured), substrates S_i, O_2, CO_2, products P_j, heat H_v. According to the macroscopic principle, it is sufficient to verify process behaviour at the 95% level of all balances (Roels, 1983). All intermediary metabolites do not influence the macrobalances of a bioconversion system in a reactor. This is in agreement with Einstein's dictum that "everything should be made as simple as possible, but not simpler". Microscopic aspects of bioprocesses can also be handled on the same level, e.g. microscopic stoichiometry can be reduced by handling only Carbon, Hydrogen, Oxygen, Nitrogen and sometimes Sulfur and Phosphorus. Macroscopic stoichiometry, e.g. yield coefficients Y, can be explained by this approach. It must be stated here that stoichiometry (balancing methods) plays a powerful role in process design and must be regarded as an essential supplement to kinetics (Roels, 1983; Moser, 1988).

The formal macroapproach is graphically illustrated in the scheme of a bioprocess in Figure 9. The macroscopic input/output variables X, S, P, O, C, H_v, are indicated together with the areas of biological phenomena (kinetics, stoichiometry, thermodynamics), physical phenomena (mass, heat and impulse transfers) and interactions between them (mainly viscosity, shear, morphology of biomass , see lecture 4 and 23). The message of utmost importance in connection with the formal macro-approach, assuring its powerful status, is the fact that *all* (macroscopic) phenomena are to be handled together, and not as a sum but as an integral entirety, representing the bioprocess by means of a simplified but adequate network.

4. Process Modeling

Due to the high complexity of the interaction network in bioprocessing much research effort is made towards a systematic elucidation. Thereby, the use of mathematical models is of essential importance, as models - representing an abstract picture of the real process - serve as a spiritual approach to complex and unseparable phenomena.

Process modeling includes both activities of modeling the reactor with its physical transports as well as modeling the biokinetics including all significant interactions of them e.g. viscosity, morphology. Figure 10 illustrates graphically all the areas of problems in bioprocessing indicating at the same time the set-up of mathematical models.

It should be mentioned, that models - being a human invention - are made for a special aim and on basis of given process data and variables. Not necessarily one model can serve all purposes in the field of biotechnology at the same time. At least one has to distinguish the different tools for mechanistic studies (in physiology, genetics), process control and process engineering calculations, of which last aim lies in the center of this article.

In addition, a big variety of different types of mathematical models can be observed in the literature, like unstructured/structured models unsegregated/segregated (taking into account cell cycle with different types of cells), quasi-steady state/dynamic modeling and homogeneous/heterogeneous models according to bioreactor operation.

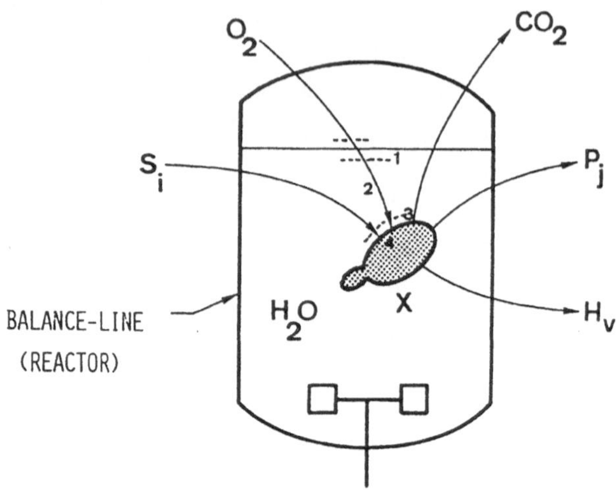

Figure 9. The macroscopic principle applied to bioprocessing expressed with pseudohomogeneous observable process variables in the liquid phase (L): biomass X, substrates Si, oxygen O, products Pj, carbon dioxide C, and volumetric heat Hv. Pseudohomogeneitiy is checked by considering a series of mass transfer steps: L film at the gas phase-L interface (1), L bulk (2), L film at the L-Solid phase (S) interface (3), cell wall and membranes (4), resp. S-phase cell mass with cytoplasm.

Stoichiometry (and thermodynamics), as can be seen from Figure 10 are included in these bioprocess engineering considerations. A procedure for the set-up of kinetic models using joint activities between kinetic & stoichiometric work is given in Figure 11, representing an integrating approach of greatest practical value. Thereby, the equation of conservation of mass is used extensively consisting of two parts, the exchange flows n and the reaction rates r :

$$\left(\frac{\partial C_i}{\partial t}\right) = -\nabla n_i^{'} \pm r_i \tag{1}$$

which takes the following form under steady state conditions .

$$\sum_i \nabla n_i^{'} = -\sum_i v_i r \tag{2}$$

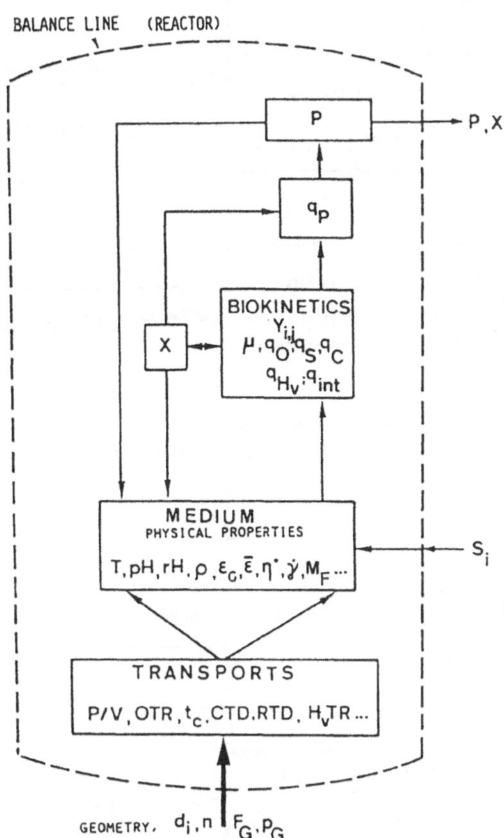

Figure 10. Schematic representation of the concept of interactions between physical transport phenomena (oxygen transfer rate OTR; heat transfer rate H_VTR, power consumption P/V; circulation time t_C, and distribution CTD ; residence time distribution RTD) and biokinetics (specific rates of growth; of consumption of substrates q_S, resp. consumption of oxygen, q_O; of production of heat, q_{Hv}, of CO2 q_C, of product q_P; and of all internal cell compartments q_{int}). The physical properties of the medium (temperature T, pH value, redox potential rH, density, gas hold-up G, mean energy dissipation, shear rate, respectively "specific viscosity" as a quantitative estimation of "engineering morphology", and M_F = morphology factor) represent the missing link for modeling the interaction concept .

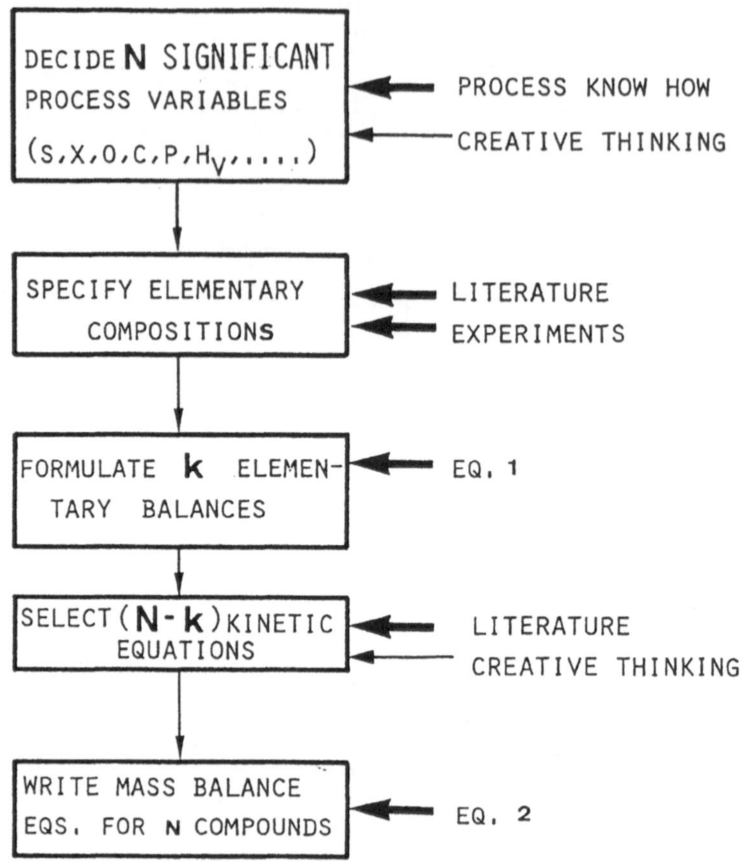

Figure 11. Flowchart of the construction of unstructured biokinetic models as a consequence of joint activity between kinetic work and stoichiometry.

The advantage is that based on a number n of process variables only a number of N-k kinetic equations must be measured in case that k elementary balanced can be verified! The application of such elemental balances based on C,H,O,N is graphically shown in Figure 12.

Macroscopic methods are fairly straightforward and easy to use, thus should be taken as supplements to the microscopic methods dealing with metabolic pathways in detail (Stephanopoulos and San, 1984; Papoutsakis, 1984). Metabolic pathway methods require more information and a more complicated mathematical manipulation of data. However, sometimes this microscopic method results in fewer degrees of freedom than the macroscopic method. Thus, a criterion for selecting stoichiometry methods is to be applied as recently proposed (Tsai and Lee,1989; Verhoff and Spradling,1976).

Concentrating now on a typical example of process modeling the case of antibiotic production is chosen, as the strategy can clearly be described on the basis of interesting papers (see lecture 4 and 23).Thereby the interactions between morphology of mycelia, viscosity and oxygen transfer are essential.

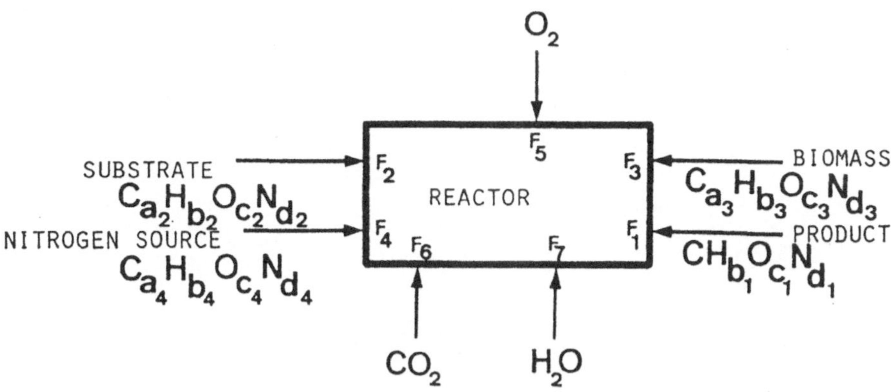

Figure 12. Graphical representation of the set-up of elemental balances of bioprocesses including the seven significant process variables from Fig. 8 concentrating on the four main elements (C,H,O,N) according to the macroscopic principle.

5. References

Meyer, H.P. (1987) in Physical Aspects of Bioreactor Performance, (W. Crueger et al., eds.) EFB working party Bioreactor Perfor- mance, Dechema/ Frankfurt (publ.) , chap. 9.

Moser, A. (1988) Bioprocess Technology-Kinetics and Reactors,Springer Inc. New York, Vienna.
Moser, A. (1989) in Proc. 8th Internat. Biotechnol. Symp. (Paris) , G.Durand et al.,(eds.) vol. 1, 386.
Moser, A. (1990) Biotech Forum Europe, vol. 7, no.4, 321-327
Moser A. (1991) Bioprocess Engng. 6, 205-211
Oosterhuis, N. and N.Kossen (1985) in Biotechnology (Rehm, H.J. and Reed, G., eds.) Verlag Chemie, Weinheim, vol. 2, chap. 24.
Papoutsakis, E.T. (1984) Biotechnol. Bioengng. 26, 174
Roels, J.A. (1983) Energetics and Kinetics in Biotechnology ,Elsevier Biomedical.
Stephanopoulos, G. and San, K.Y. (1984) Biotechnol. Bioengn. 26, 1174.
Sweere, A. et al. (1986) in Proc.Int.Conf.Bioreactor Fluid Dynamics, BHRA Cambridge, 217.
Sweere, A. et al. (1989) Bioprocess Engng. 4, 11.
Tsai, S.P. and Lee, T.H. (1989) Biotechnol.Bioengng. 33, 1347.
Verhoff, F.H. and Spradling, I.E. (1976) Biotechnol. Bioengng. 18, 425

INDUSTRIAL SCALE ANTIBIOTIC PRODUCTION: PROCESS KINETIC ANALYSIS AND PROCESS DESIGN

ANTON MOSER
Institut fur Biotechnologie, TU Graz
A-8010 Graz, Petergasse 12
Austria

ABSTRACT. The main aim of this paper is to illustrate the systematic approach in bioprocess analysis and design in industrial scale production by applying the strategy described in lecture 1 (i.e. the "formal macroapproach" as a holistic mode of thinking & working). Thereby, it is essential that mathematical modeling, computer simulation and experimental work is seen as an entirety of joint efforts where accuracy is reduced to an engineering level in accordance with models as approximations . Batch processes in 150 l bioreactors are kinetically analyzed resulting in characteristics of repression of product formation and inhibition of growth due to concentrations of substrate and product. Optimization of fed-batch processes using simulations based on this data result in feeding strategies with optimum values for productivity yield and product concentration. Finally, a generalization seems possible: the flux of substrate (q_S), internal in the cell, is the process limiting step as a consequence of respiration as the true bottleneck in cell metabolism .

1. Introduction

Industrial production of antibiotics is a typical example of biotechnology, where development & success are achieved mainly with the aid of the empiric pragmatic approach which primarily focus on the fundamental level of microbiological and genetic work while neglecting process engineering. Bioprocess engineering science, in its methodology as outlined before (Moser,1992a), will however play an eminent role in the future. As time and money for the design of new bioprocesses must be saved, the development must occur with more security and reproducibility and less trial & error. Thereby a joint strategy with a combination of modeling and experimental work is to be recommended, an essential element is intuition and belief in analogies.

This quite unconventional statement will better be accepted when considering the situation of bioengineers in practice, where any scientific effort must prove its practicability and powerful value.

2. Materials and Methods

All details including analytical methods are reported in the literature (Schneider and Moser, 1987).

A.R. Moreira and K.K. Wallace (eds.), Computer and Information Science Applications in Bioprocess Engineering, 365-376.

2.1 STRAIN

Inocula were taken from deep freezed sampüles of a mycelia suspension of the strain *Clitopilus pesseckerianus (Pleurotus)*, a mutant from Biochemie Kundl / Tirol in Austria, where also the experiments have been carried out. The strain exhibited a typical growth behaviour of fungi, where a mixture between mycelium and pellets appeared during the process.

2.2. MEDIA

The liquid contained a series of necessary compounds, essentially consisting of two carbon sources (glucose S_1 and several oils S_2).

2.3. FERMENTATION

Basically the process of pleuromulin production is carried out as fed-batch process with glucose feed, as schematically shown in Figure 1.

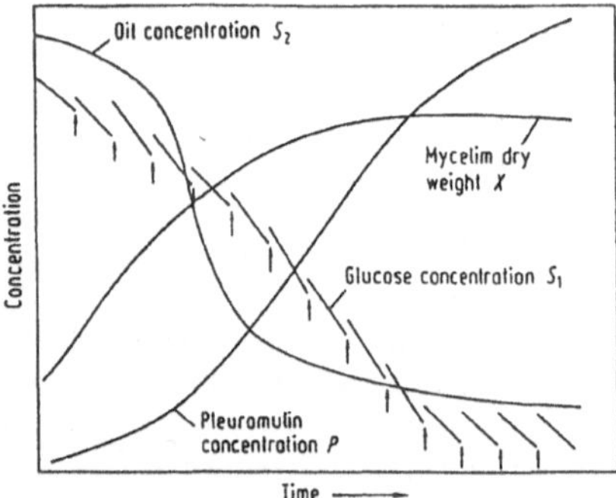

Figure 1. Graphical presentation of a typical concentration vs. time plot of the fed-batch process of the pleuromulin fermentation with arbitrary units.

Due to difficulties in evaluating the process kinetics and stoichiometry under the conditions of the technical scale (10-40 m^3) incl. stoichiometry , further data were collected from batch fermentations in bench scale (150 l) and later applied to the design of the fed-batch industrial process.

3. Experimental Results

3.1. STOICHIOMETRY

Balancing is the first part of scientific elucidation, clarifying the quantitative relations between consumed and produced compounds. In this work macroscopic stoichiometry was executed according to the macroscopic approach dealing with yield coefficients $Y_{i/j}$: As mentioned before, fed-batch data was difficult to evaluate as true fed-batch rates could not be defined since the addition of substrate was quite stochastic. Therefore, batch data are shown here.

Figures 2 and 3 give an idea of the behaviour in plotting substrate consumption rates r_{S1} and r_{S2} versus growth rate r_X. Batch data thus correlate quite well resulting in the chance to evaluate yield coefficients $Y_{X/S1}$, $Y_{X/S2}$ and also the maintenance coefficient in case of S_2 (m_{S2}).

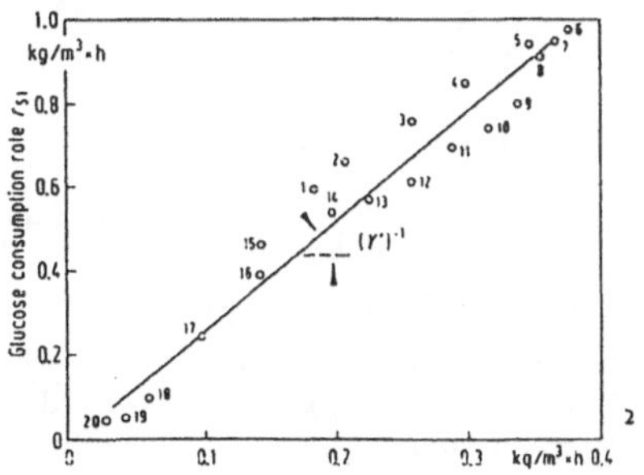

Figure 2. A plot of experiments on substrate consumption rate with glucose as substrate vs. growth rate from a batch process showing a possible linear relationship for the estimation of an overall yields coefficient $Y_{X/S1}$ over a certain range of time.

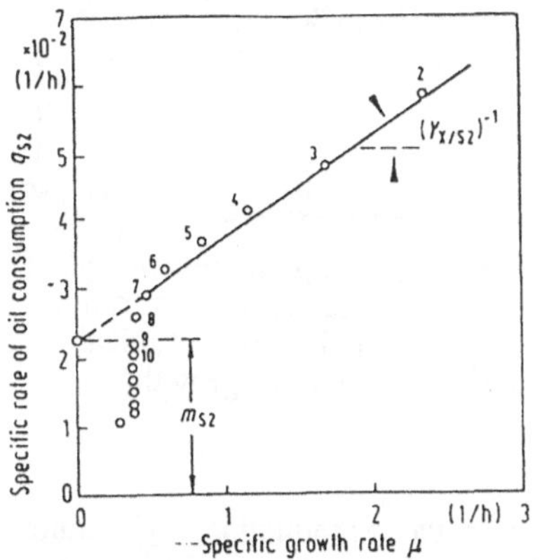

Figure 3. A plot of experiments on rates of substrate consumption S_2 (oil) vs. growth rate from a batch run leading again to the estimation of a yield coefficient $Y_{X/S2}$ from the slope of the linearization but also of a coefficient for maintenance m_{S2} from the intercept.

3.2. GROWTH

Although it was expected that complex behaviour would occur, the typical growth pattern shown in Figure 4 was reproducible; some form of a saturation curve appeared at lower substrate concentrations (less then 50 g/l), while υ increased later and decreased finally at the highest concentration, indicating substrate inhibition. Another typical property of this type of bioprocess was that product concentration had a strong influence on the growth rate as illustrated in Figure 5.

At the same time apparent viscosity, which was measured using a Brookfield type rheometer, increased drastically (450 mPas) and fell down again at a later phase (300 mPas), when water was added to the process in order to decrease viscosity and to facilitate transfer processes (O_2).

3.3 PRODUCT FORMATION

The pattern of product formation was also quite complicated. Repeated experiments are summarized in a plot of specific product formation rate q_P versus specific growth rate u both terms in dimensionless form, as shown in Figure 6.

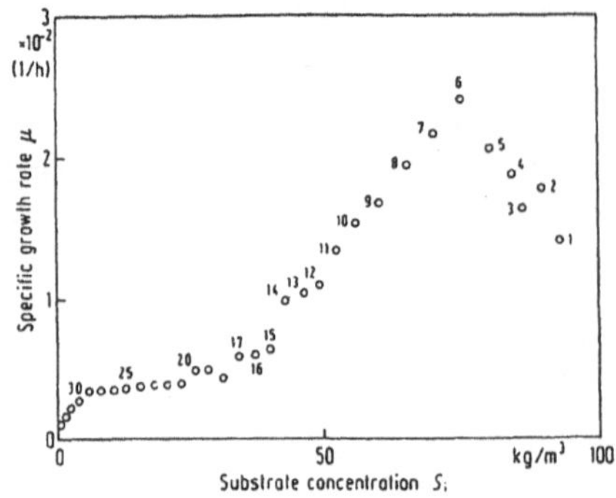

Figure 4. Plot of experiments on specific growth rate vs. glucose concentration of a batch process exhibiting a typical, reproducible pattern of growth.

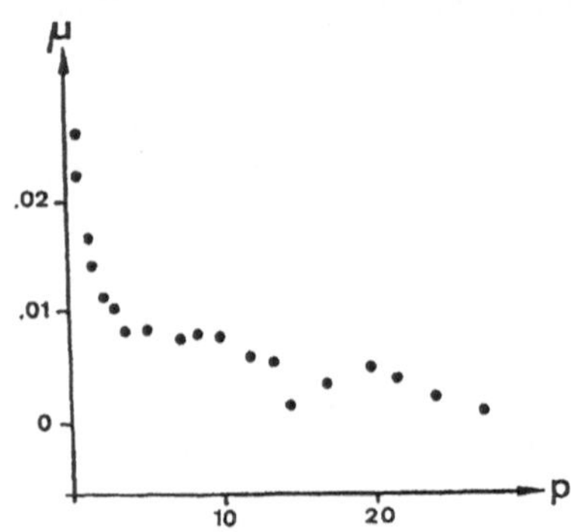

Figure 5. Plot of experiments on growth rate in formal dependence of product concentration suggesting P-inhibition type kinetics.

The pattern was reproducible, high values for qP are achieved at low growth rates, a typical behaviour of secondary production processes.

Further elucidation led to the graph given in Figure 7, where qP is dependent on the substrate concentration . Typical is the maximum at medium values of S indicating inhibition and repression type kinetics .

Trials were executed at the same time in case of rough data from the industrial scale resulting in similar trends of growth and production pattern.

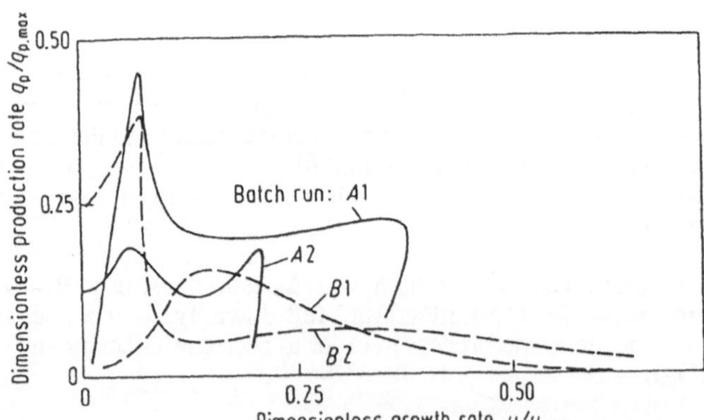

Figure 6. Plots of experimental results on specific product formation rate qP versus specific growth rate u in dimensionless form exhibit a variety of shapes with different substrates (curves A for S1 and curves B for S2) and different concentrations (A1 or B1 : 100 g/l; A2 or B2 200 g/l) .

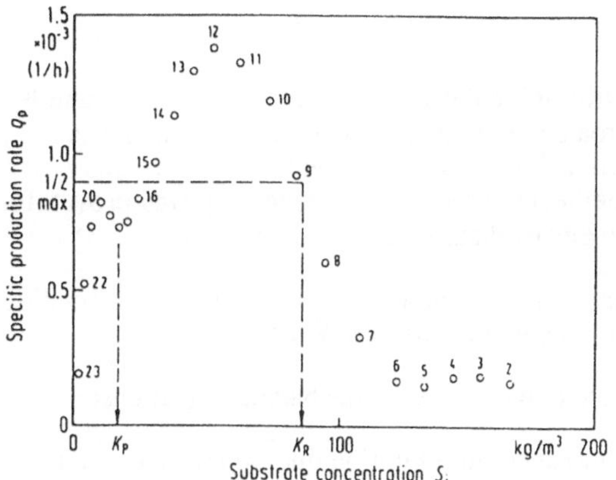

Figure 7. Plot of specific product formation rate q_P as a function of substrate concentration showing a saturation type kinetic behaviour with substrate inhibition analogy (repression type formal kinetics with a repression constant K_R).

4. The Set-up of Mathematical Model

4.1. KINETIC MODEL of BATCH PROCESS

As a preamble it is to be repeated that the procedure followed here gives only hints and no mechanistic explanations. The basis of such formal kinetics are experiments which are carefully planned but without mechanistic speculation. Thus, formal relationships are obvious from plotting different rates, and it remains unclear if a causal nexus is really given (e.g. plots between q_p vs. u or u vs. p but also the effect of repression or inhibition in P-formation).

The set of equations, based on the formal macroapproach which by experiments shown in section 3 is adequate for this type of bioprocess is summarized below for X, S, P, in the case of batch operation (Moser and Schneider ,1988).

Process scheme :

$$S + 0 \longrightarrow X + P \tag{1}$$

Balance equations for batch processing :

$$r_x = \mu(s,p)x - k_d x \tag{2}$$

$$-r_s = \frac{1}{Y_{x/s}}\mu(s,p)x + \frac{1}{Y_{p/s}}q_p(s) + m_s x \tag{3}$$

$$r_p = q_p(s)x \tag{4}$$

For a complete description the balance equations for O_2, CO_2 and heat should also be included. They are not treated here, as the main interest in this paper is in elucidating the basic behaviour and relationship between s, x and p. The missing variables oxygen ,CO_2 and heat h_V have been neglected as pseudohomogeneity was proven for the bench scale process before starting kinetic evaluations as a precondition.

Heat transfer may become a process bottleneck under industrial conditions like OTR and mixing in L-phase, as outlined before (Moser,1992a,b)

Two kinetic expressions are to be inserted in the balance equations:

Growth rate has to take into account inhibition due to p (K_{IPX}, see Fig. 5) but also due to s (K_{ISX}, see Figure 4):

$$\mu(s,p) = \mu_{\max} \frac{s}{s(1 + s/K_{ISX}) + \tilde{K}_S/(1 + p/K_{IPX})} \tag{5}$$

Product formation has to include the repression/inhibition type (K_R , Fig. 7):

$$q_p(s) = q_{p,\max} \frac{s}{\tilde{K}_p + s(1 + s/K_R)} \tag{6}$$

These model equations represent a first working hypothesis and will be used for quantification of process behaviour in batch mode in bench scale as well as for process design in fed-batch mode of operation in industrial scale.

Figure 8 shows the simulation of eq. 5 in order to mimic the characteristic behaviour of growth in the pleuromulin fermentation illustrated in Figure 4 . As can be seen it is possible to quantify the pattern using K_{ISX} and K_{IPX}.

Figure 9 then represents the simulation of the characteristic pattern of product formation in a fromal plot of qP vs. u as in Figure 6 . With eqs. 5 and 6 it is again possible to describe the process behaviour using the value of K_R together with both inhibition constants of growth . This result is in agreement with previous work assuming only K_R
(Bajpaj and Reuss ,1981).

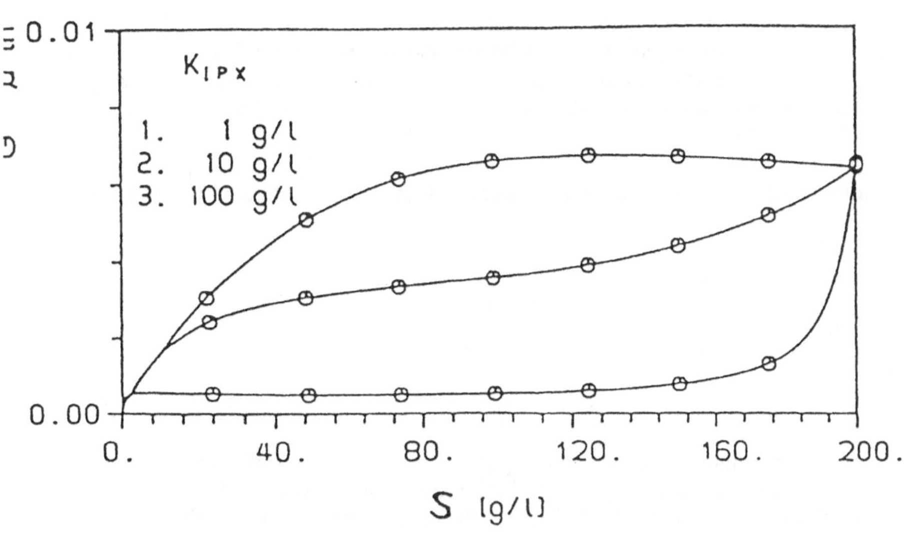

Figure 8. Computer simulation plot of u vs S according. to eq. 5 for varied values of the inhibition constant K_{IPX}. Compare to figure 4!

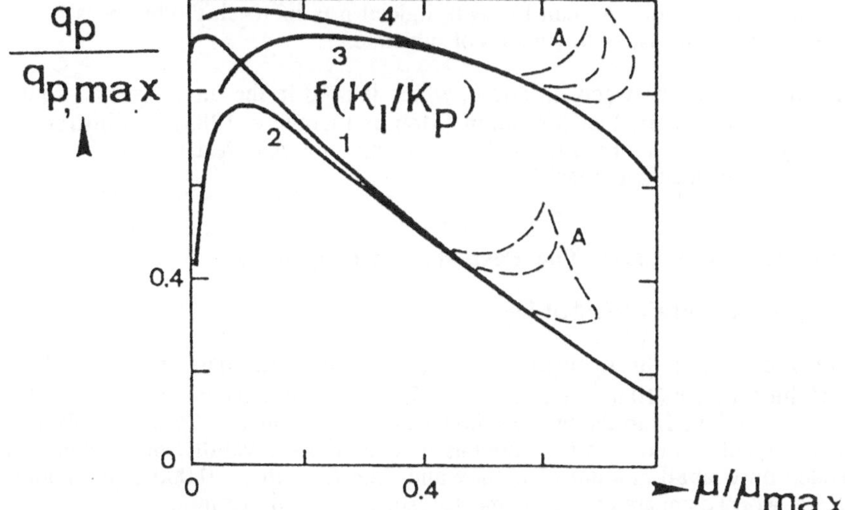

Figure 9. Normalized computer simulation plot of q_p vs. u for different values of the repression constant KR as well as with varied inhibition constants K_{ISX} and K_{IPX} according to eqs. 5 and 6.

4.2. PROCESS KINETIC MODEL of INDUSTRIAL FED-BATCH

The application of the before mentioned batch model & data from bench scale to industrial size fed-batch processes is based on the belief that the set of batch model data can be applied when adapting the balance equations to the fed-batch mode of operation by including terms for the volumetric substrate feed rate F_S with a substrate feed concentration S_F.

While in all equations the dilution effect through S-feed is to be corrected by the term

$$-\frac{C_i}{V}\cdot\frac{dV}{dt} \tag{7}$$

it is necessary to add the term in eq. 3 for S balance :

$$+ F_S\,(t) \tag{8}$$

With the rewritten balance equations incorporating eqs. 7 and 8 and inserting again eqs. 5 and 6 for the kinetics, it is now possible to look with the aid of computer simulations for an optimum feeding strategy in order to increase the economics (yields, productivity, product concentration) of the technical production, saving time and money for experimental search. More work on this subject is reported in the literature (Heijnen et al., 1979; Bajpaj and Reuß, 1982 ; Guthke and Knorre, 1981).

One typical result of feed rate dependence is represented in Figure 10 in the form of a concentration vs. time plot. As can be seen, indeed p is increasing when x is kept at moderate growth rates in case of low values of substrate.

The interconnectedness between q_p and u, which is used in the literature as a control parameter for optimum product formation (Mou and Cooney, 1983), is illustrated in Figure 11, in the case of the runs from Figure 10, giving a manifestation of the formal network pattern of the macroapproach.

5. Generalizations Using The Formal Macroapproach

5.1. CASE STUDY PLEUROMULIN

As a practical result from this evaluation work based on the batch data, it was possible to realize on technical scale (10 m^3) at Biochemie Kundl an improvement of more then 10% although the predictions from the process model were in the range of 25% improvement. This final statement can be seen somehow as judgment of the validity of the approach, which needed only small amount of money and time (less then 20.000 $ and about 6 months of real work) compared to a fully mechanistic research programme.

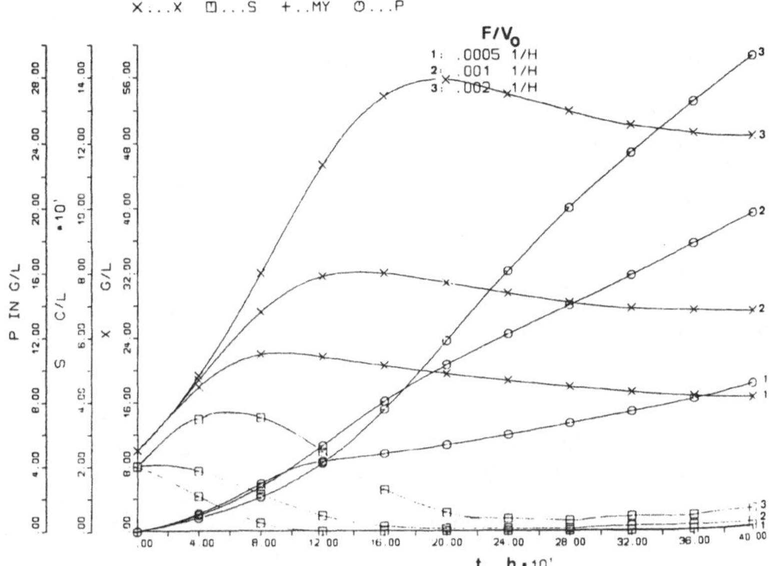

Figure 10. Computer simulation plot of concentrations of x, s and p vs. time in a fed-batch operation with varied values for the substrate feeding rate F_S showing the drastic influence of appropriate substrate concentration or flux on product formation.

5.2. GENERALIZATIONS

Beyond the case study considered, the approach used exhibits another property which seems to be of much more general interest.

As stated, many relations are regarded as formal interconnections. When looking for a deeper explanation of the experimental findings with higher plausibility, the use of the mathematical model shows its usefulness.

It is easy to perform a plot as shown in Figure 12, where q_P is related to q_S.

The result seems complex on first view, but reflects more essential consequences in the long run, including the concept of process bottleneck presented before, see Figure 4 (Moser, 1992 b). The central magnitude in these considerations is the substrate flux occurring internal in the cells but being in equilibrium with the external concentration change. Sub-, and supra-critical flux is to be distinguished, the first leading to loss of substrate due to maintenance requirements, the latter resulting in a loss of substrate for product formation due to increasing production of biomass.

This concept of a value of $q_{S,critical}$ directly can elucidate the experimental findings of optimum feeding strategies . However, not the concentration itself is governing the process but the flux of substrate. There exist indications that the final process bottleneck is respiration and, thus, $q_{s,crit}$ is just the consequence of a respiration bottleneck. Similar results were extracted from a study of yeast metabolism as another case of over-flow metabolism.

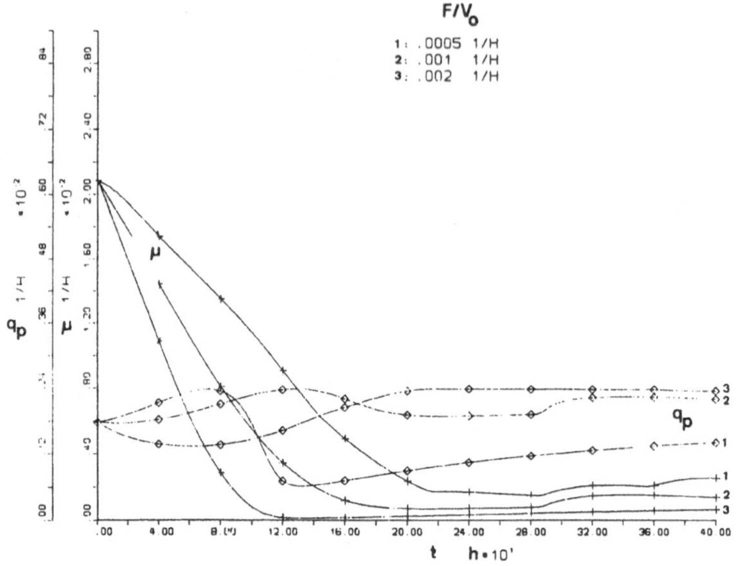

Figure 11. Computer simulation plot of the time dependence of q_p and u for a fed-batch process with different substrate feed rates according to figure 10

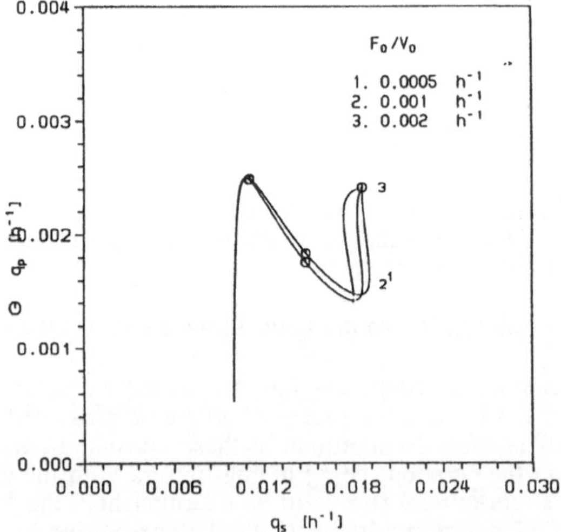

Figure 12. Computer simulation plot of qP vs. qS in direct connection to fig. 11 stressing the importance of the concept of a "critical substrate flux", which seems to govern process rates internally in the cells ($q_{s,crit}$) .

Concerning bioprocess engineering methodology two final statements are to be made for completeness :

* scale-up strategies will increasingly be used which are based on structured mixing modeling (e.g. , Reuss and Bajpaj, 1991; Mayr et al., 1992)

* scale-down approach thereby plays a central role, where e.g. the influence of mixing and oxygen gradients according to the structured model has to be incorporated in the process model (e.g. Larsson and Enfors, 1985), the equation for irreversible inhibition (I_{OUR}) of respiration (resp. product formation ?) is with $t_{c,2}$ being the residence time in an anaerated tubular reactor as second compartment :

$$\ln \frac{1}{1 - I_{OUR}/100} = 0.048 \cdot \bar{t}_{c,2} \qquad (9)$$

6. Typical Values Of Process Kinetic Parameters And Process Variables case: pleuromulin production

u_{max} = 0,05 (h^{-1})
K_{S1} = 2 (g/l)
K_{ISX} = 50 (g/l)
K_{IPX} = 20 (g/l)
$q_{P,max}$ = 0,003 (h-1)
K_P = 0,5 (g/l)
K_R = 30 (g/l)
$Y_{X/S1}$= 0,75
$Y_{P/S}$ = 0,71
$Y_{X/O}$ = 1,25
$Y_{P/O}$ = 6,25
K_O = 0,001 (g/l)
k_d = 0,001 (h^{-1})
k_La = 500 (h^{-1})
o* = 0,006 (g/l)
m_{s2} = 0,01 (h^{-1})
x_o = 0,05 (g/l)
s_o = 20 (g/l)
s_F = 500 (g/l)
V_o = 35 (m^3)
F_S = 0,002 (g/l.h) with $F_S(t) = F_L/V_o \cdot s_F$

7. References

Bajpaj R.K. and Reuss M. (1980) Biotechnol.Bioengng. 23 , 717
Bajpaj R.K. and Reuss M. (1981) J.Chem.technol.Biotechnol., 30, 332
Guthke R. and Knorre W. (1987) Bioprocess Engng., 2, 169
Heijnen J.J. et al., (1979) Biotechnol. Bioengng., 21, 2175
Larsson G. and Enfors S.O. (1985) Appl.Microbiol.Biotechnol.,21,228
Mayr B., E.Nagy, P.Horvat, A.Moser(1992) Bioprocess Engng. in press
Moser A. (1988) Bioprocess Technology, Springer Inc. N.Y., Vienna
Moser A. and Schneider H. (1988) in Proc. Intern.Conf.Computer Appl. in Fermentation Technol., (Fish N. et al., eds.) SCI London,93-103
Mou D.G. and Cooney Ch. (1983) Biotechnol.Bioengng.23, 225 and 257
Reuss M. and Bajpaj R.K. (1991) in Biotechnology 2nd ed., Rehm H.J. and Reed G. (eds.) vol. 4 , 299-348
Schneider H. and Moser A. (1987) Bioprocess Engng.,2 , 129 - 135

OXYGEN TRANSFER AND MICROBIAL GROWTH

ANTON MOSER
Institut für Biotechnologie, TU Graz
A-8010 Graz, Petersgasse12
Austria

ABSTRACT. Oxygen, as an essential nutrient for aerobic bioprocesses, needs special attention on both sides, the consumption kinetics and the supply capacity of bioreactors. This paper will clarify the background on measurement techniques for respiration and transfer by showing engineering correlations for the value of saturation concentration in broths and k_La as well as kinetic equations. The interactions between physiology and physics will be illustrated and the significant effect of biomass concentration and morphology on viscosity and finally on the O_2-transfer capacity .The structuring of bioreactors based on their mixing behaviour especially in the case of industrial scale will be mentioned for completeness.

1 Introduction

Real bioprocesses have their bottleneck quite often in the limited respiration capability of microbial cells as a result of process kinetic analyses. It is clear that maximum oxygen transfer is often seen as the key for success and optimum process output in practise. The main problem with oxygen as an essential nutrient comes from the fact that it is a gas with a quite low solubility in aqueous phases like fermentation broths.

As shown before (Moser, 1992a), the process engineering strategy is based on the interaction network between metabolic reaction kinetics of the living cells and physical transport phenomena of the reactors (see Figure 1). It is thus of primary interest to understand both parts (i.e. the consumption and the supply side) and to handle the quantification techniques properly for the sake of adequate data for better process design. The advantage of applying the more systematic strategy is the ability of predicting the process behaviour for design purposes.

This paper describes the measuring techniques for respiration and oxygen transfer (OTR), which are based on mass balances including problems in evaluation like electrode response and gas dynamics, the rate of oxygen consumption as a type of double - substrate kinetics, the transport coefficient characteristics including engineering correlations and the interactions between physiology and physics in case of viscosity effects.

A.R. Moreira and K.K. Wallace (eds.), Computer and Information Science Applications in Bioprocess Engineering, 377-394.

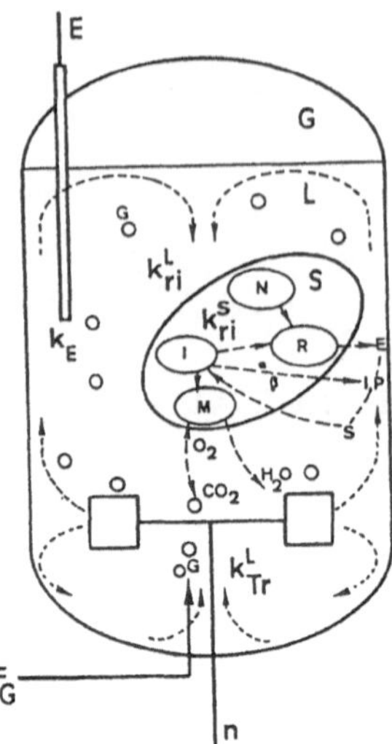

Figure 1. Schematic representation of the basic experimental situation in bioprocess/bioreactor analyses, where the interactions between physical transports (k_{Tr}) and biokinetic rates (k_r^L) in the liquid phase are thought to be representative for the process rates in the solid phase of cell mass (k_r^S). At the same time, response lags of measuring electrodes (k_E) have to be taken into account. G, gas phase; L. liquid phase; S. solid phase or substrate; E, enzyme or electrode; I, intermediary metabolites or products; P, end product; N, nucleus; R, ribosomes; M, mitochondria, anabolism, catabolism; F_G= gas flow rate; n= agitators rotational speed.

2. Measuring Techniques

The mass balance equations for oxygen (o) have the following form for well-mixed tanks and represent the basis for all engineering considerations:

For the L-phase ,where transfer (k_La) and respiration is occuring (q_o) :

$$\frac{d\sigma_L}{dt}=\frac{F_{L,in}}{V_L}\sigma_{G,in}-\frac{F_{L,ex}}{V_L}\sigma_{L,ex}+k_La\left(\sigma_L^*-\sigma_L\right)-q_0x \tag{1}$$

For the G-phase

$$\frac{d\sigma_G}{dt}=\frac{F_{G,in}}{V_G}\sigma_{G,in}-\frac{F_{G,ex}}{V_G}\sigma_{G,ex}+k_La\left(\sigma_L^*-\sigma_L\right)\frac{V_L}{V_G} \tag{2}$$

2.1 SOLUBILITY OF OXYGEN IN BROTHS

The above equations are coupled by the term of solubility of oxygen in the L-phase, which is in equilibrium with the partial pressure in the G-phase according to Henry's law:

$$o^*_L = p_G / He \cdot RT \tag{3}$$

with He = dimensionless Henry distribution coefficient (H=He.RT).
This saturation value of oxygen concentration in the L-phase, which is in the range of 5-10 mg/l, is furthermore directly dependent on temperature (Mihaltz and Hollo 1980)

$$\sigma_L^* = p_{tot} \cdot i / 0.1353 \cdot 10^6 T^2 - 31.73 \cdot 10^6 T \tag{4}$$

and it varies also significantly with the presense of salts and nutrients (Popovic et al., 1979)

$$\log\left(\sigma^* / \sigma_{eff}^*\right) = \log\left(\sigma^* / \sigma_{salt}^*\right) + \log\left(\sigma^* / \sigma_s^*\right) \tag{5}$$

with a term of ionic strength (i) measured with conductivity (l)

$$\log\left(\frac{\sigma^*}{\sigma_{salt}^*}\right) = \alpha_0 + \alpha_i \lambda + \alpha_2 \lambda^2 \tag{6}$$

and another factor for the influence of glucose (s)

$$\sigma_s^* = \sigma_s \left(1 - 0.0012 \cdot C_s\right) \tag{7}$$

Schumpe et al. (1985) reviewed the quantification methods for o*, which are quite difficult to evalute during real bioprocessing as the direct reading of elecrodes are in terms of partial pressure and not concentration.

2.2. THE DRIVING FORCE

OTR is acheived on the basis of two factors, the transfer coefficient (k_La) and the driving force (o^*_L-o_L), which need further elucidation due to problems in quantification. In larger volume reactors, the situation of balancing is more complex as the inlet stream (in) has a different concentration than the outlet stream (ex). Thus, the logarithmic mean value is to be formed for quantification purposes when the L-phase is well mixed and the G-phase is in plug flow region

$$\overline{\Delta\sigma}=\frac{(p_{\sigma,in}-p_L)-(p_{\sigma,ex}-p_L)}{\ln(p_{\sigma,in}-p_L)/(p_{\sigma,ex}-p_L)} \tag{8}$$

2.3.RATE COEFFICIENTS FOR RESPIRATION (q_o) AND TRANSPORT (k_La)

Generally, a number of quantification methods exists for OTR including
1) gas phase (G) analysis ,
2) dynamic method in liquid phase (L) and
3) biological method using "reference bioprocesses".

The use of an oxygen consuming chemical reaction like sulphite oxidation in the presence of Co^{++} or Cu^{++} - ions as catalyst is not discussed here, although is was widely spread in use in the past, as this method suffers from the fact that transfer enhancement can easily occur and that the conditions are quite different from real bioprocesses.

2.3.1 Gas Analysis . This method is based upon oxygen balance in inlet and outlet gas streams, where y = mole fraction of oxygen in air. The parameters in eq.9 can be determined using flow meters , pressure gauges, temperature sensors and a gaseous oxygen analyzer (van Meyenburg and Fiechter,1968) and including corrections for the influence of CO_2 and N_2 .

Gas analysis is superior to pO_2-electrodes due to better stability in practise(sterility, membrane, exchange of sensors during run) and is preferably used in industry for the estimation of q_o versus q_c as well as for k_La -values (Lehmann et al., 1982) , e.g. for q_o:

$$q_0x=\frac{32}{22.4}\cdot 10^6\left(\sigma_{in}-\sigma_{ex}\frac{C_{N_2in}}{1-\sigma_{ex}-C_{CO_2,ex}}\right)\frac{F_{G,in}}{V_L} \tag{9}$$

2.3.2 Dynamic Method Using pO_2 -Electrodes in the L-Phase. This well known method illustrated in Figure 2 in its measurement set-up (a), the curve of the concentration of O_2 as a response to the dis-turbance from a (step change to zero at t_0 and to saturation at t_1) in part b and their evaluation plot in part c . From the balance equation before (eq.1) the following equation becomes valid in this case

$$\frac{d\sigma}{dt}=k_La(\sigma_L^*-\sigma_L)-q_0x \tag{10}$$

which can be rearranged to give the basis for the plot c, resulting in the evaluation of k_La:

$$\sigma = \frac{1}{k_L a}\left(\frac{d\sigma}{dt} + q_0 x\right) + \sigma^* \tag{11}$$

This simple method, often used in research labs, however, suffers from the influence of the sensor itself exhibiting a response time behaviour. In order to achieve an undisturbed value it is necessary to eliminate the electrode characteristics, quantified with a response value k_e versus t_e. At the same time, the dynamics of the gas phase, changing in accordance with the disturbance shown in Fig.2a, can be eliminated in similar way according to the following approach using a characteristic value k_G resp. t_G (Ruchti et al.,1981):

$$t_{1/e} = \frac{1}{k_L a} + \frac{1}{k_e} + \frac{1}{k_{V,G}} \tag{12}$$

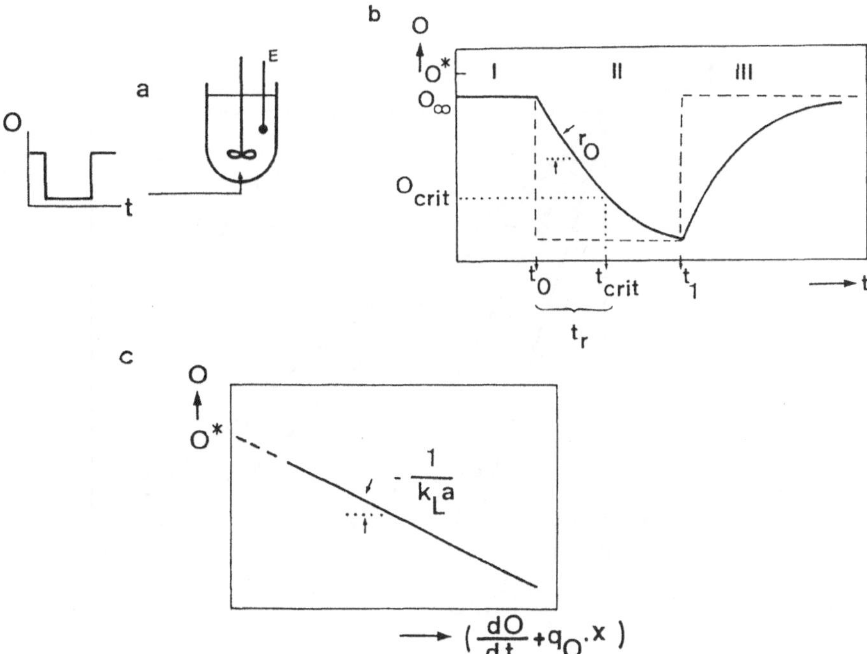

Figure 2. Experimental set-up and evaluation of the O_2 transfer rate characteristics of bioreactors using the dynamic mehod for measuring $k_{L1}.a$. (a) Measuring set-up using a step change in inflowing O_2 concentration (b) Response of the dynamic method (c) Evaluation plot

The plot shown in Figure 3 represents the evaluation procedure according to eq. 12. Using regime analysis (see methodology paper), it can be shown that the electrode reponse does not influence the k_La-value in the case of quick responding sensors.

2.3.3 Biological Method for OTR Measurement. This relatively new method has the advantage that OTR values are given under real process conditions. The main idea is to directly apply a bioprocess with known and reproducible, stable kinetics (" reference fermentations" or " biolog. test systems" according to Adler and Fiechter, 1982 and Lehmann et al., 1985) in order to quantify the aeration capacity of an unknown bioreactor. Under steady-state conditions the balance equation reduces to

$$OTR_{\max} = k_L a\left(\sigma_L^* - \sigma_L\right)_{\sigma_L \to 0} \equiv q_0 x \tag{13}$$

so that the OTR versus k_La can be calculated from the biomass obtained. As model strain an obligators yeast strain is recommended (Trichosporon cutaneum).

Thereby different bioreactors can be compared in their performance, as summarized in Table 1.

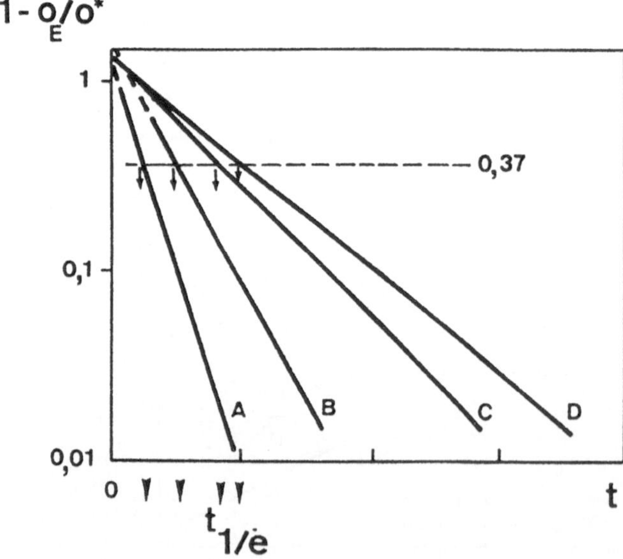

Figure 3. Semilogarithmic plots of (1 - o_E/o^*) versus time t of aeration curves at different conditions (curves B-D) and a step response experiment to determine k_E (curve A). Shifting the curve to intersect the ordinate at unit value yields the time constants $t_{1/e}$ at the 0.37 line (Ruchti et al., 1981).

Table 1 : Quantification of OTR using Trichosporon cutaneum in different bioreactors (typical range, ST = stirred tank) :

reactor type	V_L m^3	P/V W/m^3	F_G m^3/h	$r_{x,max}$ $kg/m^3.h$	OTR_{biol} $kg/m^3.h$	E_{OTR} kg O_2/kWh
Jet Loop	0.35	4.2	0.07	8.5	4.6	1.1
CFST	0.39	12	0.1	8.9	5.2	0.43
CR	0.07	3.7	0.1	6-12	4.3	1.2
VST	0.04	2.5	0.04	3.8	1.5	0.6
"	0.03	5	0.05	9.8	3.6	0.7
HST	0.04	6.2	0.05	8	3.5	0.56
HE	0.01	0.6	0.46	4.6	2.3	3.8
"	0.01	2.4	1.07	4.6	6.2	2.6

CFST=completely filled stirred tank, CR=cycle ring, VST=vertical ST, HST=horizontal ST, HE = hydroejector

3 Respiration Kinetics

The consumption of oxygen in the L-phase obeys the same laws of biokinetics as known with other nutrients i.e. the saturation type kinetics e.g. Monod type. The model parameters then are K_o and $q_{o,max}$. It is worthwhile to note that the value for K_o is quite low and is a function of cell diameter (in the range of 0.002 - 2 umol/l), while maximum respiration rate is 0.1-3 h^{-1}).

It is to be added that the value of critical O_2-concentration is of importance (o_{crit}), beyond which value the cells become OTR limited. The estimation of o_{crit} is indicated in Figure 2b, showing a dependence on growth rate of the strain.

Oxygen is to be regarded as an essential nutrient for aerobic cells together with the others (e.g. glucose). Thus, an adequate kinetic equation for such double-substrate kinetics must be available, which has the following form :

$$\mu(s,o) = \mu_{max} \cdot f(s) \cdot f(o) \qquad (14)$$

where f(s) and f(o) are Monod functions with K values for s and o.

This type of kinetics was created as a formal analogy basically and was explained afterwards with the aid of an *interactive* model (Bader,1982), where both essential substrates interact in metabolic pathways directly.

Another case of multi-substrate consumption exhibits a kinetic type, where both terms are additive (case of growth enhancing substrates e.g. homologues according to a *non-interactive* model).

The role of oxygen in metabolism of living cells is quite well understood, its significance is also well recognized in practice. Respiration is recently regarded as a central process bottleneck, demonstrated with the aid of applied regime analyses to metabolism and kinetics (see Moser, 1991).

The situation is illustrated in Figure 4, showing that the amount of substrate flux is governed by the respiration bottleneck, leading to overflow metabolism (e.g. ethanol in case of bakers yeast, loss of productivity in case of antibiotic productions exhibiting repression type kinetics) and thus drastic influences from substrate feeding strategies (fed batch processing).

4. Oxygen Transfer Characteristics

A series of terms are to be considered for a full characterizaton of OTR performance of bioreactors : OTR (kg O_2 / m^3.h), $k_{L1}a$ (h-1), k_{L1} and k_{L2} (m/s), $a_{G/L}$ (m^2/m^3) and $a_{L/S}$, P/V (kW/m^3), E_{OTR} (kg/kWh) the oxygen-efficiency, the degree of oxygen utilization in gas, the degree of "hinterland" HL=1/a. d, superficial gas velocity v_{sG} (m/s) etc.(Moser, 1988). Combining these factors leads to typical engineering correlations and plots, where dimensionless numbers are used, e.g sorption versus aeration number

$$N_{OTR} = k_La / (F_G.V) \quad \text{versus} \quad N_A = F_G / N.d_i^3 \tag{15}$$

where N = rotational speed and d_i = diameter of impeller .

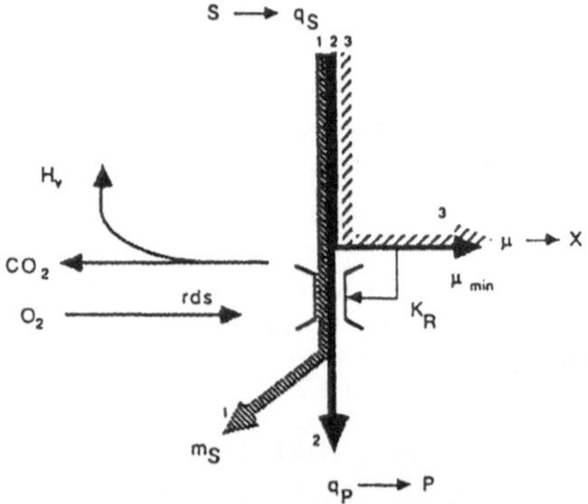

Figure 4. Concept of $q_{s,crit}$ in a dynamic flux diagram indicating a process bottleneck in case of antibiotic production as a result of a formal macro-kinetic analysis . The case of subcritical flux (1, flux for maintenance m_s), critical flux (2) for m_s and derepressed production with a minimum growth rate corresponding to the process bottleneck respiration (?) and supracritical flux (3) with higher growth rate and with repressed r_p.

Figure 5 represents a correlating plot showing the influence of power supply and gas velocity for pure water and salt solutions based on the following empirical equation with four coefficients (α,β,γ,ω):

$$k_{L1}a = \alpha(P/V)^{\beta} \cdot (v_{sG})^{\gamma} \cdot (\eta)^{\omega} \tag{16}$$

It is interesting to note that power consumption mainly is needed for increasing interfacial area a, while k_L is more or less in the same order of magnitude (1-4 $.10^{-2}$ cm/s) . The dependence of the surface area on P/V is illustrated in Figure 6 in case of different aeration systems.

Similar correlations are manyfold in use e.g. generally

$$k_{L1}a = (4.78 \cdot 10^{-13}) \cdot N^3 \cdot F_G^{\ 0.5} \tag{17}$$

or especially in case of coalescing systems in stirred tanks (STs)

$$k_{L1}a = 0.026 \cdot (P_G/V)^{\ 0.4} \cdot (v_{sG})^{\ 0.5} \tag{18}$$

versus for non-coalescing systems in STs

$$k_{L1}a = 0.002 \cdot (P_G/V)^{\ 0.7} \cdot (v_{sG})^{\ 0.2} \tag{19}$$

while the following correlation is to be preferred in case of a bubble column versus a bubble region in a ST :

$$k_{L1}a = 0.32 \cdot (v_{sG})^{\ 0.7} \tag{20}$$

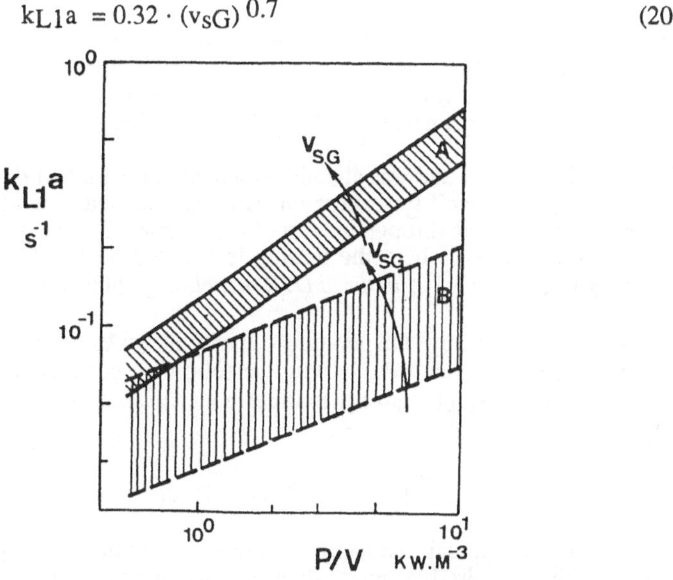

Figure 5. Typical plot of empirical engineering correlation between G/L transfer coefficient $k_{L1}a$ and specific power consummption P/V as a function of superficial gas velocity $v_{S,G}$. The shaded areas (A for pure water and B for ionic solutions) represent the range of validity of correlations.

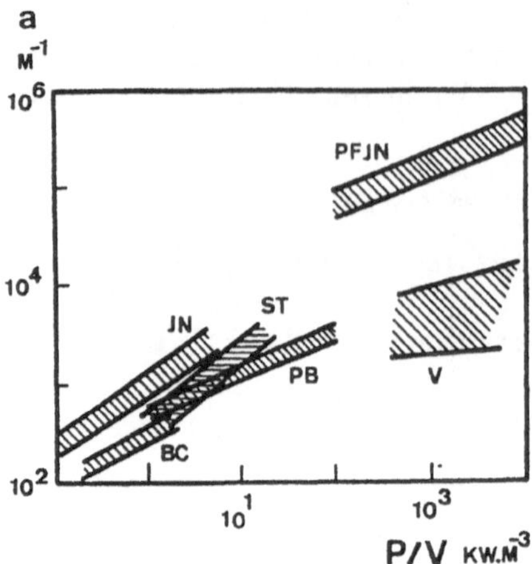

Figure 6. Comparison of G/L reactors of different constructions in a plot of specific interfacial area a versus specific power consumption P/V: stirred tanks (ST), bubble columns (BC), jet nozzles (JN), packes beds (PB), venturi washers (V) and plug flow-jet nozzles (PFJN).

This last equation is of special interest when setting up so-called *"structured reactor models"*, where the whole reactor liquid volume is treated in more then one compartment. This case is of importance when considering large scale industrial reactors, where the state of maximum mixedness can never be achieved with the aid of stirring.

It is indicated in Figure 1 and was already presented in Figure 8 of the methodology paper (Moser, 1992 a), that a well mixed region is to be assumed around the impeller, while liquid bypass streams cycle through the rest of the volume. For the quantification of such a structured reactor model, where the mixing is handled in detailed structure, eq. 19 is needed. The set-up of such a *structured OTR model* is graphically shown in Figure 7.

One significant value concerns the pumping capacity F_P (m^3/s) of the stirrer resulting in the circulating flow, which is calculable with the following correlation valid in the case of aerated reactors with $N_{P,G}$ = power number under gassed conditions ($N_{P,G} = P / \rho \cdot N^3 \cdot d_i^5$):

$$F_P = 0.2 \cdot N_{P,G} \cdot N \cdot d_i^3 \qquad (21)$$

It is to be added that a typical characteristic of reactors and their mixing and aeration device is connected with the hydrodynamics, which are quantified using a correlation between Np and the Reynolds-number N_{Re} ($Re = v \cdot L / \upsilon$) shown in Figure 8 for different types of stirrers.

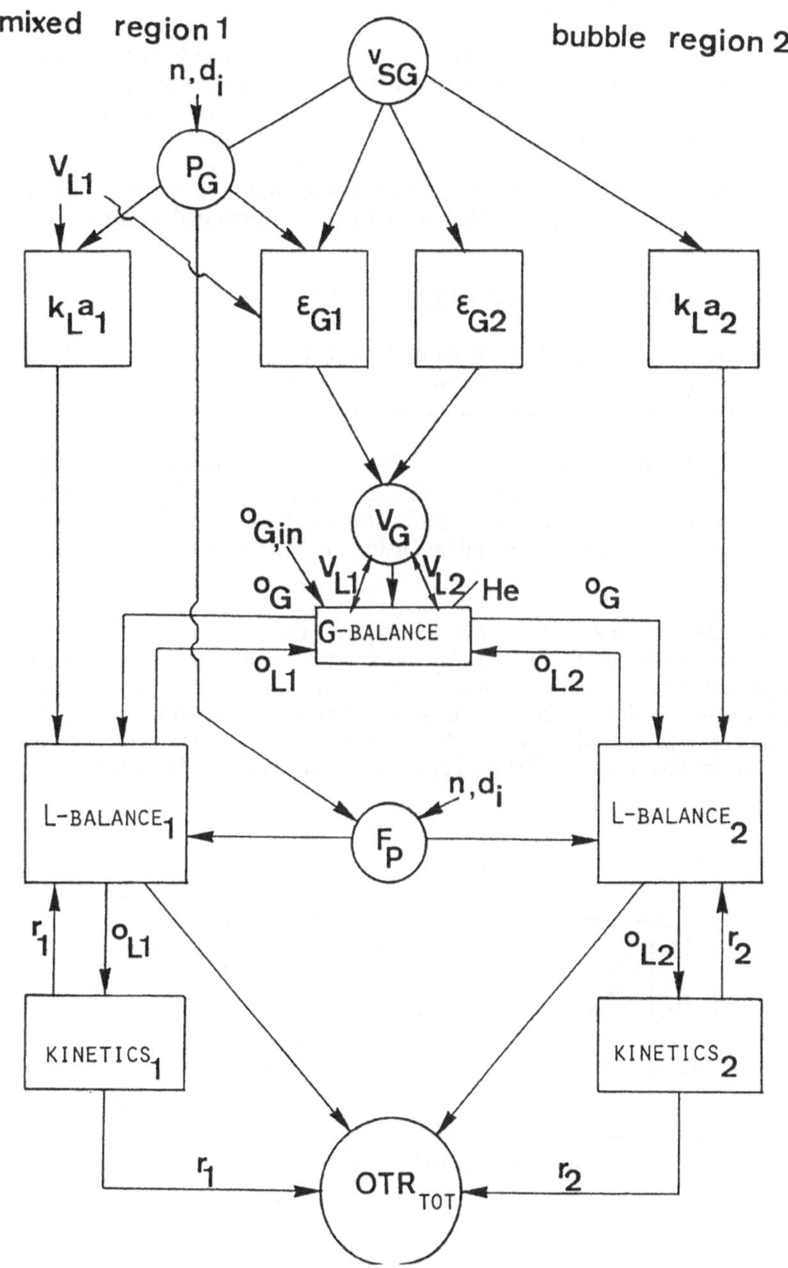

Figure 7. Block diagram of the two-compartment model for oxygen transfer in a production scale bioreactor including parallel work for a mixed zone (1) and a bubble zone (2) according to Oosterhuis (1984) in agreement with Figure 8 from lecture 1.

5. Interactions Between OTR and Physiology

As known from the methodology presented before, the interactions between physics and reactions play a central role for the understanding or real bioprocess performance in bioreactor systems. Thereby three cases can be distinguished: 1) external transport limitations 2) internal transport limitations 3) transport enhancement.

For a detailed study of this interaction pattern a text book is to be referred (e.g. Moser,1988). Here, only some typical phenomena will be illustrated in connection with this pattern.

5.1. BIOPROCESS RATES IN DEPENDENCE OF OTR

Here the phenomena of linear growth caused by OTR limitation (case of external limitation) will be presented and the appearance of apparent dependence of respiration rate on power input of the reactor (case of internal limitation) .

5.1.1. Growth Rate in Dependence of $k_{L1}a$. Based on the equations described before (eqs.1 and 13) for oxygen balance coupled with respiration and growth using Monod-type kinetics it is easy to understand that linear specific growth can occur under OTR limitations. This is shown in Figure 9 and will be simulated using the ISIM software OXENZ (Dunn et al., 1992) .

5.1.2 Respiration Rate as a Function of Power Input.. On the first view an estonishing phenomenon was described in literature (Reuss et a., 1982) and is quantified in Figure 10. A clarification can only be found when following the theory of mass transfer with simultaneous reaction inside biofilms. In this case, transport limitation is caused by the increased diameter of bioparticles (flocs or film), which are suspended in the liquid phase. A similar case is simulated in the ISIM package BIOFILM (Dunn et al., 1992).

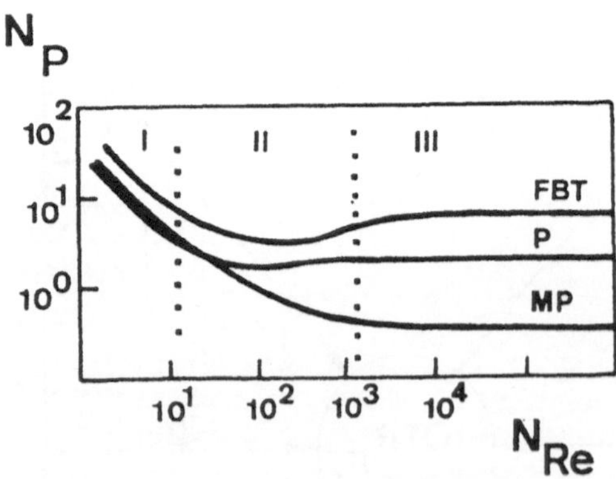

Figure 8. Typical graphical plot of dimensionless numbers of power consumption N_P versus N_{Re} in case of flat-blade turbine (FBT), paddles (P), and marine propellers (MP). The laminar transient and turbulent regions are indicated (I,II,III).

5.2 OTR AS A FUNCTION OF VISCOSITY AND AS A FUNCTION OF MORPHOLOGY

Concentrating now on a a typical example of process modeling the case of antibiotic production is chosen, as the strategy can clearly be described on the basis of literature data (Schneider and Moser, 1987).

The set of equations based on the formal macroapproach which are thought to be adequate for this type bioprocess for X, S, P, O_2 and heat H_v in case of batch operation is described in lecture 23 (Moser and Schneider ,1989)

The interactions between physiology and physics contain the following equations :

* OTR as a function of viscosity (see Figure 11):

$$k_{L1}a\frac{V_L}{F_G} = f\left(\frac{P_G}{F_G \cdot \rho_L\left(\upsilon \cdot g^{2/3}\right)}\right) \qquad (22)$$

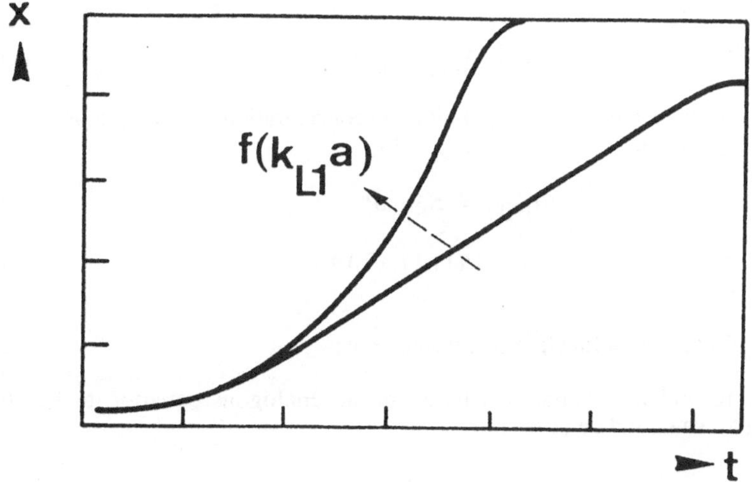

Figure 9.Linear growth as a consequence of transport limitation in case of O_2 quantified with $k_{L1}a$ in a concentration / time graph.

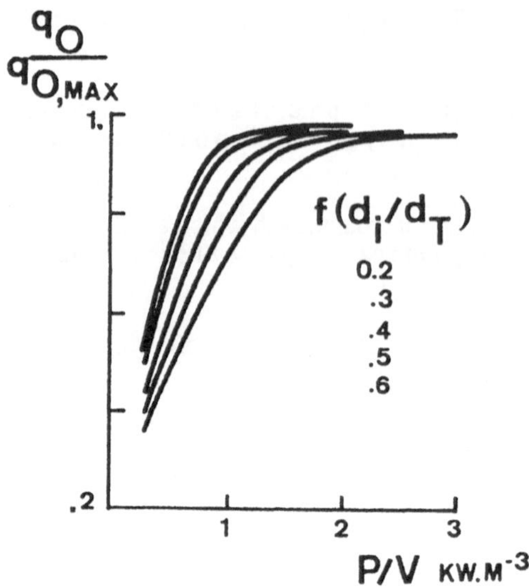

Figure 10. Simulations of oxygen consumption kinetics $q_o/q_{o,max}$ at different impeller/tanks diameter ratios d_i/d_T as a function of power consumption P/V, as a typical example of macrokinetics in a 20 l vessel (Reuss et al.,1982).

Kinematic viscosity is to be replaced by apparent viscosity in case of Non-Newtonian fluids (see Figure 12), in which case the simple power law is thought to be adequate:

$$\tau = \eta_{app} \cdot \gamma^{m} \quad . \tag{23}$$

where all constants are dependent on biomass concentration according to Reub et al., 1979 (see also Figure 13):

$$\eta_{app} = K_c \cdot x^p \tag{24}$$

and

$$m = 1/(1+k_i \cdot x^q) \tag{25}$$

* viscosity as a function of morphology :

using the formal macroapproach based on an analogous equation to Einstein`s law according to van Suijdam (1987)

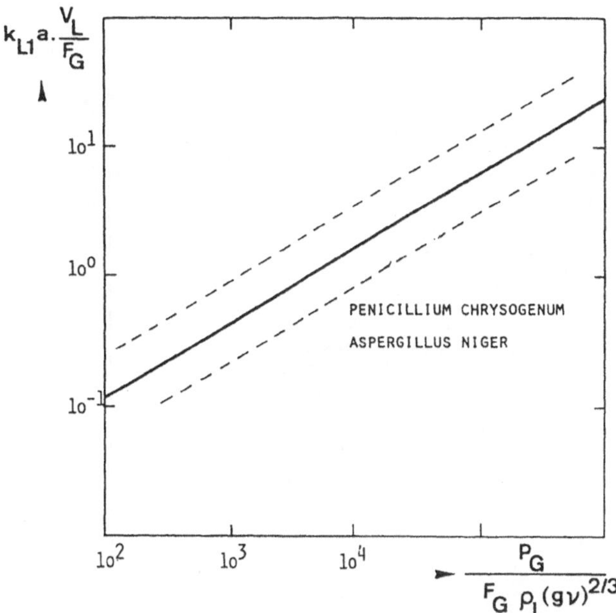

Figure 11. Application of chemical engineering correlation for oxygen mass transfer according to eq. 22 to technical bioprocessing in the case of fermentations with *Penicillium chrysogenum* and *Aspergillus niger* under various conditions. The dotted lines indicate observed deviations in the bioprocess.(Reuss et al., 1980)

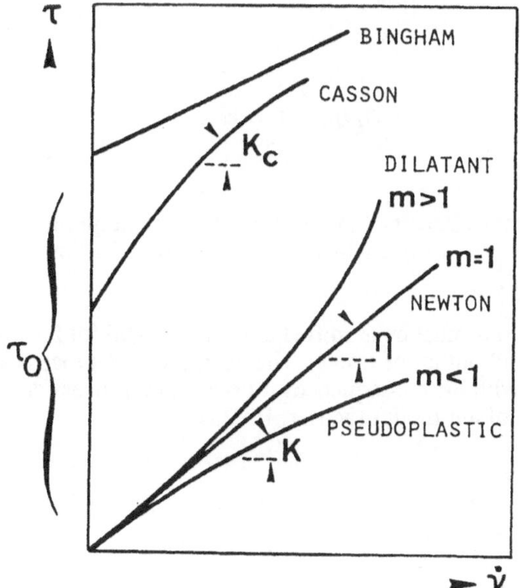

Figure 12. Rheogram of fermentation fluids in the case of Newtonian and Non-Newtonian fluids (cf. eq. 23).

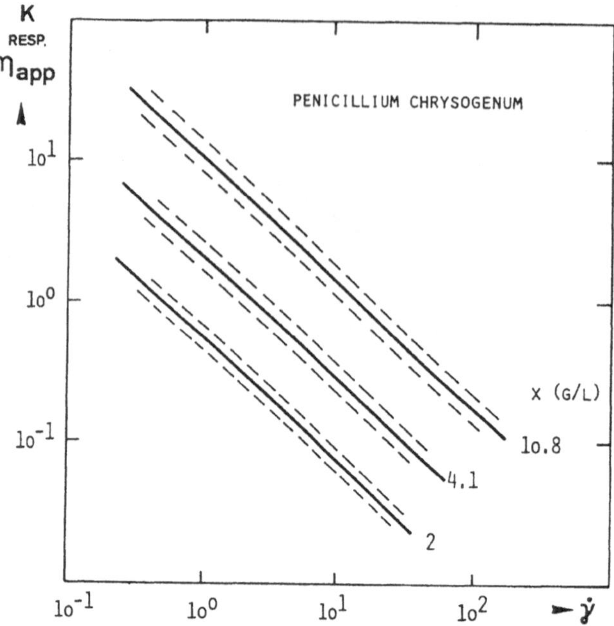

Figure 13. The correlation according to Fig. 11 can be verified by using a concept of the dependence of apparent viscosity on shear rate as a function of biomass concentration according to eq. 23 (Reuss et al. 1980)

$$\eta/\eta_0 = 1 + M_F \cdot x \tag{26}$$

where the relative apparent viscosity can be shown to depend on biomass concentration x with a proportionality constant M_F, which is the (macroscopic) morphology factor. Eq. 26 is plotted in Figure 14.

The set of balance equations must be modified to become valid for fed-batch operations by adding terms of input and output of flows . The estimation of process kinetic parameters will be shown together with the consequences for optimum process design in the paper on the industrial production of antibiotics (Moser, 1992b).

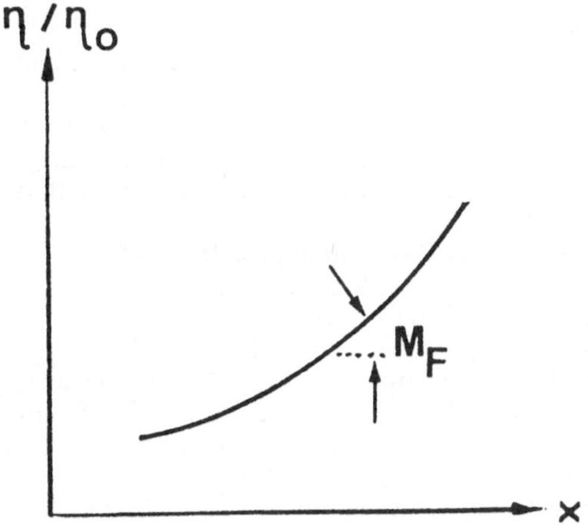

Figure 14. Graphical plot of eq. 26 for the estimation of the morphology factor M_F based an viscosity measurements (app)

6. List of Typical Values of Process Kinetics Parameters and Process Variables .

Case : OTR interacting with microbial growth

u_{max} = 0,104 (h^{-1})
K_{S1} = 0,1 (g/l)
$Y_{X/S1}$ = 0,2
$Y_{X/O}$ = 0,88
K_O = 0,001 (g/l)
k_La = 10- 500 (h^{-1})
o^* = 0,008 (g/l)
x_o = 0,05 (g/l)
s_o = 10 (g/l)

7. References:

Adler I. and Fiechter A. (1982), Chem. Ing. Techn. 55, 322-323

Bader F.G. (1982) in Bazin M.J. (ed.) Microbial Population Dynamics, Boca Raton Fla., CRC Press, chap. 1

Dunn I.J., E.Heinzle, J. Ingham , J.E. Prenosil (1992) Biological Relation Engineering , Verlag Chemie, Weinheim
Lehmann J. et al. (1982) in Arbeitsmethoden der Biotechnologie Frankfurt/D, Dechema (publ.)
Lehmann J. et al. (1985) in Rehm H. J. and Reed G. (eds),Biotechnology, vol. 2, Deerfield Beach, Fla. and Basel, Verlag Chemie, Weinheim, chap. 25
Meyer H.P. (1987) in Physical Aspects of Bioreactor Performance (Crueger W. et al., eds) Dechema; Frankfurt/D (publ.) chap.5
Mihaltz P. and Hollo J. (1980), Acta Biotechnologica, 0, 39-45
Moser A. (1988) Bioprocess Technology, Springer Inc., New York, Vienna
Moser A. (1990) Biotech Forum Europe 7, 321-327
Moser A. (1991) Dynamic Flux Approach, in Proc. Mathem. &Intelligent models in systems simulation IMACS Symp., Brussels, 747-751Moser A. (1992a) General Methodology in Bioprocess Engineering, lecture 1 , (NATO-course), this volume
Moser A. (1992b) Industrial Scale Antibiotic Production: Process Kinetic Analysis and Process Design, lecture 23 (NATO-course), ibid
Moser A., Schneider H. (1989) in Computer appl. in Ferm. Technology,(Fish N. M. et al., eds.) SCI London, Elservier Publ., 93-105
Oosterhuis N. and Kossen N.W.F. (1985) in Biotechnology ,Rehm H.J. and Reed G.(eds.), Verlag Chemie,Weinheim chap.24
Popovic M. et al. (1979) Europ. J. appl. microbiol. Biotechnol., 8, 1-17
Reuß M., et al. (1979) Europ. J. appl. microbiol. Biotechnology, 8,169-178
Reuß M. et al. (1982) Chem. Eng., June, 233-245
Ruchti G. et al. (1981) Biotechnology Bioengng., 23, 277
Schneider H., Moser A. (1987) Bioprocess Engng., 2, 129-135
Schumpe A. er al. (1985) Biotechnology, vol 2, chap. 10, VerlagChemie, Weinheim
van Suijdam H., Duisseljee P. (1987) in Crueger W. et al (eds.) Physical Aspects of Bioreactor Performance, Dechema , chap. 6
van Meyenburg M., Fiechter A. (1968) Biotechnol. Bioengng. ,10, 535

CLASSROOM ADAPTATION OF *ESCHERICHIA COLI* SINGLE CELL MODEL

GARY J. STANLAKE*
Biology Department,
Hardin-Simmons University
Abilene, Texas 79698

MICHAEL L. SHULER
School of Chemical Engineering,
Cornell University
Ithaca, New York 14853

ABSTRACT. Adaptation of a structured single cell model to use on MS DOS computer systems made this learning tool widely available. Adaptation involved creating screen displays, user interfaces and run time graphics in addition to the FORTRAN to BASIC translation. Some classroom uses for the simulation are suggested.

1. Introduction

Computer models are often mathematical/logical descriptions of real life systems. As such, they lend themselves to being manipulated as simulations. Model construction is a learning process which frequently enhances one's perception of the system being depicted. The educational value of the model, however, can extend beyond its construction when it is used as a simulation to study the behavior of a system. Such is the case for this adaptation for the Cornell Single Cell model for *Escherichia coli*. For example as a research tool, it has been used by Domach et al (1984) to test proposed mechanisms for control of chromosome replication and cell growth. In the General Microbiology classroom, the simulation functions to acquaint students with the intracellular dynamics of a bacterial cell cycle.

In actuality, the Cornell Single Cell model is a family of related models simulating the growth of an *E. coli* single cell under aerobic conditions (Domach et al., 1984), anaerobic conditions (Ataai and Shuler, 1985), and with plasmids (Ataai and Shuler, 1986; Kim and Shuler, 1990a). Additionally, Kim and Shuler (1991) have used the model to analyze the effects of plasmid multimerization on segregational instability. In addition, Domach and Shuler (1984) created an asynchronous population model for *E. coli* using a finite number of single cell models as subsystems. Kim and Shuler (1990b) have extended this population modeling approach to plasmid-containing cell populations. Shu and Shuler (1991) have extended the model to recognize the effects of supplementing minimal medium with amino acids. The classroom model is based on Domach et al.'s (1983) glucose-limited aerobic single cell model, but the model could be readily extended to include effects of anaerobes, amino acid supplementation or plasmid insertion and plasmid-encoded gene expression.

*Corresponding Author

A.R. Moreira and K.K. Wallace (eds.), Computer and Information Science Applications in Bioprocess Engineering, 395-400.

2. Model Formulation

The aerobic single cell model as described by Shuler and Domach (1983) is a structured model with all major metabolic functions grouped into 20 components for which masses are calculated repeatedly during a growth simulation, see Figure 1. Component concentrations are the factors which govern physiological and morphological changes in the model cell during its growth cycle. Most model parameters are free of prior constraints and are based on data collected from cells in exponential growth or cell-free systems. Exceptions to the above are: 1) values associated with the ratio of cell envelope used for cross wall and extension formation must be positive, 2) rate equations for the formation of cell envelope and cell envelope precursors are adjusted within the range of 0 to 0.5 gm/cc of cell volume, and 3) a parameter for unidentified energy consumption is used to bring predicted growth yields in line with experimental measurements. At the end of each simulated generation, values for cell factors are prepared showing values for the 20 metabolic components, cell length, cell width, cell volume, and generation time.

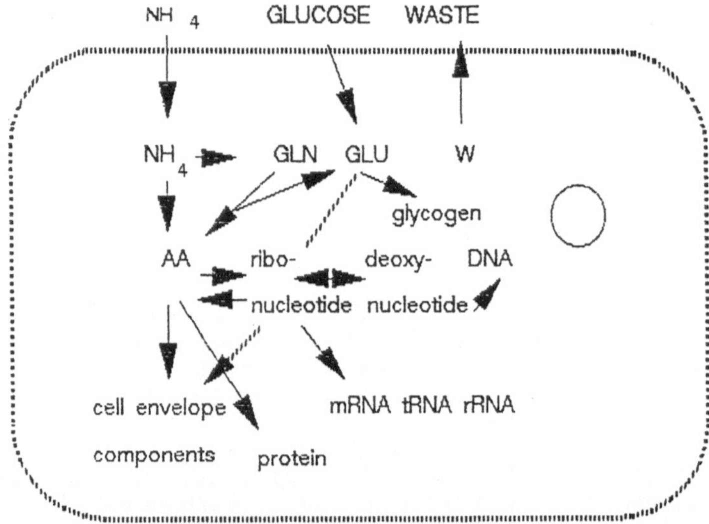

Figure 1. Model *E. coli* B/r cell.

The original computer program of the Cornell Single Cell model as created by Shuler, Leung and Dick (1979) has been modified by each subsequent investigator to meet his/her particular objective. A flow chart of the Classroom Adaptation program is shown in Figure 2. The heart of the program is a Runge Kutta routine for mass balance calculations. Chromosome related events, including the scheme for governing the initiation of replication comes next followed by a section coding for envelope elongation and cross septum formation. Near the end of the program is a decision fork which tests for completion of septum formation. If septum formation is complete, cell division ensues, if it is not complete, the program cycles back to the Runge Kutta routine for another iteration.

```
INPUT CONDITIONS AND NO.GENERATIONS

READ INTRACELLULAR PARAMETERS FROM DATA

CONVERT CONCENTRATION VALUES TO MASS VALUES

EULER START-UP FOR RUNGE KUTTA
with predictor corrector

CELL CHROMOSOME RELATED EVENTS
replication fork position
initiation of replication
repressor protein amount
gene dosage amount

CELL SEPTUM FORMATION AND ENVELOPE ELONGATION

CELL WIDTH DETERMINATION

DOES CELL DIVIDE?

YES                              NO

print end of run data            display time
divide all cell masses           display cell status graphics
end or cycle to next gen.        recalculate cell masses
                                 return to euler start-up
```

Figure 2. Program Listing.

3. Education Model Adaptation

Conversion of Shuler and Domach's (1983) version of the Cornell Single Cell model to a teaching tool suitable for widespread use at the undergraduate level began with the translation of the program from FORTRAN to interpreted BASIC. Except for difficulties caused by BASIC's lack of local variables, the translation was straight forward. Upon the translation of the model into BASIC, the problem became that of simulation run time. The FORTRAN version ran ten generations in ten minutes on a mainframe computer, while a personal computer (IBM XT clone) took over an hour to run one generation. Personal computers with faster microprocessors would, and did, shorten run times but software adaptations also helped. The Runge Kutta was shortened from a fourth order to a second order and the subroutine which housed the equations for calculating new mass values was split so some mass values were recalculated only every tenth iteration. Finally, the interpreted version of the program was translated into a complied version. This opened the door to the use of a math coprocessor which greatly reduced run time. Additional advantages of the compiled version were double precision mathematics and the creation of stand alone programs which ran in the MS DOS environment free of compiler support. The run time for the compiled version is now approximately two minutes on an XT or 386 computer with math coprocessors and substantially less on computers with the 486 microprocessor.

As a classroom tool, the simulation needed to be interactive to allow the operator to choose the number of generations per run and the glucose and ammonium concentrations in the medium (Figure 3). Graphics were added to show the status of cellular growth in length, width, volume and septum formation during the execution of the simulation (Figure 4). Finally, those pieces of end of run data deemed most useful to classroom use of the simulation were presented as a screen display (Figure 5), and all other end of cell cycle data were stored for an optional printer dump at the discretion of the operator.

```
                 THE CORNELL SINGLE CELL MODEL


ENTER VALUE FOR NUMBER OF GENERATIONS; 1 - 10
?
NUMBER OF GENERATIONS EQUALS 1

ENTER VALUE FOR EXTRACELLULAR AMMONIUM CONCENTRATION
SUGGESTED DEFAULT VALUE IS 0.001; RANGE IS 1E-9 TO .1
?
EXTRACELLULAR NH4 = .001

ENTER VALUE FOR EXTRACELLULAR GLUCOSE CONCENTRATION
SUGGESTED DEFAULT VALUE IS 0.001; RANGE IS 1E-9 TO .1
?
EXTRACELLULAR GLUCOSE CONCENTRATION IS .001
```

Figure 3. Opening Screen Display.

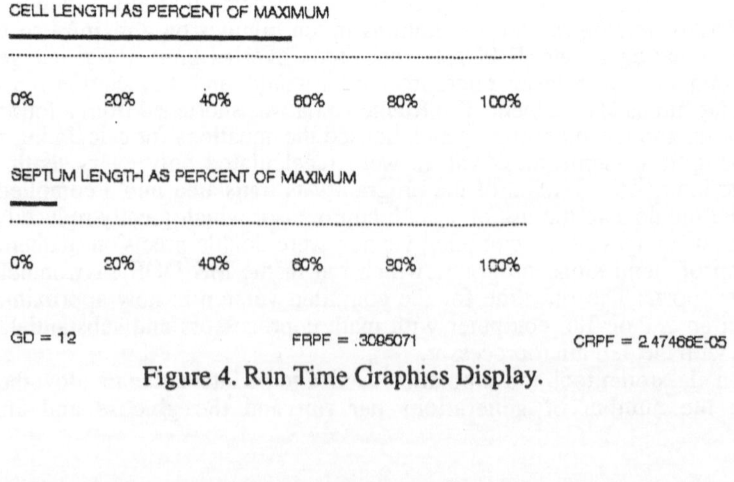

```
TIME = 2.403 MIN.            DT = .0016018            CYCLE 30

CELL VOLUME AS PERCENT OF MAXIMUM

0%      20%      40%      60%      80%      100%

CELL LENGTH AS PERCENT OF MAXIMUM

0%      20%      40%      60%      80%      100%

SEPTUM LENGTH AS PERCENT OF MAXIMUM

0%      20%      40%      60%      80%      100%

GD = 12                 FRPF = .3095071          CRPF = 2.47466E-05
```

Figure 4. Run Time Graphics Display.

```
                END OF GENERATION CELL DATA

GENERATION TIME = .7455604076385498
CELL VOLUME =  9.347415419816249E-007 MICROGRAMS
CELL LENGTH =  3.14334702417603 MICROMETERS
INTERNAL CONCENTRATIONS
        AMMONIA CONCENTRATION     = 1.572083856444806E-004
        GLUCOSE CONCENTRATION     = 1.661417307332158E-003
        AMINO ACID CONCENTRATION = 9.450917132198811E-004
        RNA CONCENTRATION         = 6.249810010194778E-002
        DNA CONCENTRATION         = 1.244362629950047E-002

GROWTH RATE FOR GENERATION NO. 1  = 1.341272940132845

DO YOU DESIRE PRINTED OUTPUT OF MACRO USE VALUES? ANS. Y OR N
```

Figure 5. End of Generation Data Screen.

4. Classroom Use

The model is presently used in a General Microbiology classroom when the topics of bacterial cell growth and population growth are introduced. Use of the simulation helps the students to better understand the key steps in the cellular growth cycle and the relationship of cellular growth to population growth.

By running a series of cell cycle simulations with different nutrient concentrations, students may view the relationship between concentration of a limiting nutrient and such cellular growth characteristics as growth rate, or changes in cell morphology (length, width, volume) at the time of cell division. At the end of the run, if the student wishes information on the status of intracellular macromolecular constituents, these values can be directed to the printer. While it is not presently an option, intermediate values for these macromolecules could be printed for the study of intracellular flux during growth. This could lead to a comparison of DNA, RNA and protein concentrations during periods of unbalanced growth. Such future applications are possible because the Cornell Single Cell model is chemically structured and based on real parameters of cellular growth.

Further adaptations could include aspects of plasmid replication (e.g. effect of multimerization) or growth rate depression due to high-level expression of plasmid-encoded protein synthesis. Effects of anaerobic culture or amino acid supplementation could be readily added.

Acknowledgments. The authors acknowledge the financial support of the Cullen Fund for Faculty Enrichment.

5. References

Ataai, M., and Shuler, M. (1985) 'Simulation of the Growth Pattern of a Single Cell of *Escherichia coli* under Anaerobic Conditions', Biotechnology and Bioengineering, 27, 1027-1035.

Ataai, M., and Shuler, M. (1986) 'Mathematical Model for the Control of ColE1 Type Plasmid Replication', Plasmid, 16, 204-212.

Domach, M. and Shuler, M. (1984) 'A Finite Representation Model for an Asynchronous Culture of *E. coli*', Biotechnology and Bioengineering, 26, 877-884.

Domach, J., Leung, S., Cahn, R., Cocks, G., and Shuler, M. (1984) 'Computer Model for Glucose-Limited Growth of a Single Cell of *Escherichia coli* B/r-A', Biotechnology and Bioengineering, 26, 203-216.

Kim, B.-G. and Shuler, M.L. (1991) 'Kinetic Analysis of the Effects of Plasmid Multimerization on Segregational Instability of ColE1 Type Plasmids in *Escherichia coli* B/r', Biotechnology and Bioengineering, 37, 1076-1086 (Special issue in honor of Elmer Gaden).

Kim, B.-G. and Shuler, M.L. (1990) 'A Structured Segregated Model for Genetically Modified *E. coli* cells and its use for Prediction of Plasmid Stability', Biotechnology and Bioengineering, 36, 581-592.

Kim, B.-G. and Shuler, M.L. (1990) 'Analysis of pB322 Replication Kinetics and its Dependency on Growth Rate', Biotechnology and Bioengineering, 36, 233-242.

Shu, J. and Shuler, M.L. (1991) 'Prediction Effects of Amino Acid Supplementation on Growth of *E. coli* B/r' Biotechnology and Bioengineering, 37, 708-715.

Shuler, M.L. and Domach, M.M. (1983) "Mathematical Models of the Growth of Individual Cells' in H.W. Blanch, E. Papoutskis and G. Stephanopoulos (eds) Foundations in Biochemical Engineering, ACS Symposium Series, no 207 ch.5, 9 3-133, American Chemical Society, Washington, D.C.

Shuler, M.L., Leung, S and Dick, C.C. (1979) 'A Mathematical Model for the Growth of a Single Bacterial Cell', Annals New York Academy of Science, 326, 35-55.

KINETICS OF CONTINUOUS MICROBIAL CULTURE : PLUG-FLOW, CHEMOSTAT, TURBIDOSTAT AND FED-BATCH PRINCIPLES

VANDAMME, E.J. and VAN SPEYBROECK, M.

Laboratory of General and Industrial Microbiology,
Faculty of Agricultural Sciences,
University of Gent, Gent, Belgium

ABSTRACT. The study of continuous flow culture has contributed much to our fundamental knowledge of the growth kinetics, structure and physiology of microorganisms. The continuous culture principle represents an "open" culture system, with plug flow culture and chemostat culture as two basic types; turbidostat culture and fed batch culture can be considered as elaborations on the chemostat principle. This chapter deals with the theory of these four continuous culture types and their basic equations are described. Applications of continuous flow culture in research and industry are indicated.

1.FUNDAMENTALS

The principle of continuous culture of microorganisms consist in continuous addition of fresh nutrient(s) and continuous harvesting of products and/or cells; it represents an "open" growth system. There are two basic types: plug flow culture and chemostat culture, both with possible elaborations.

In plug flow culture, ideally the culture moves along a tubular bioreactor without mixing.

A chemostat culture should consist of a perfectly mixed culture to which (fresh) medium is fed at a constant rate and the culture is harvested at that same rate, such that the culture volume remains constant. The liquid effluent has the same composition as the bioreactor contents. Turbidostat culture and fed batch culture can be considered as elaborations on the chemostat principle.

Relevant parameters and symbols used are defined as follows:

D = dilution rate(F/V)
F = medium flow rate
K_S = saturation constant, Monod constant
q = metabolic quotient = μ/Y
S = limiting substrate concentration
t_r = replacement time (t_r) = V/F

A.R. Moreira and K.K. Wallace (eds.), Computer and Information Science Applications in Bioprocess Engineering, 401-417.

V = volume of culture
x_0 = initial biomass concentration
x = biomass concentration
X = total biomass
Y = yield coefficient = dx/dS
μ = specific growth rate = 1/x . dx/dt
μ_m = maximal specific growth rate

2.PLUG FLOW CULTURE

Plug flow culture simulates a batch culture in an open system, but offers not directly improved culture control over batch culture. An interesting feature of plug flow culture is that the growth phases of batch culture (lag, log, stationary and death phase), which are separated in time, are separated spatially in plug flow culture. In ideal conditions, the inoculum and the medium are mixed upon entry into the tubular bioreactor and then the culture flows at constant velocity through this fermentor without mixing. All elements of the culture take the same residence time to pass through the fermentor (fig. 1).

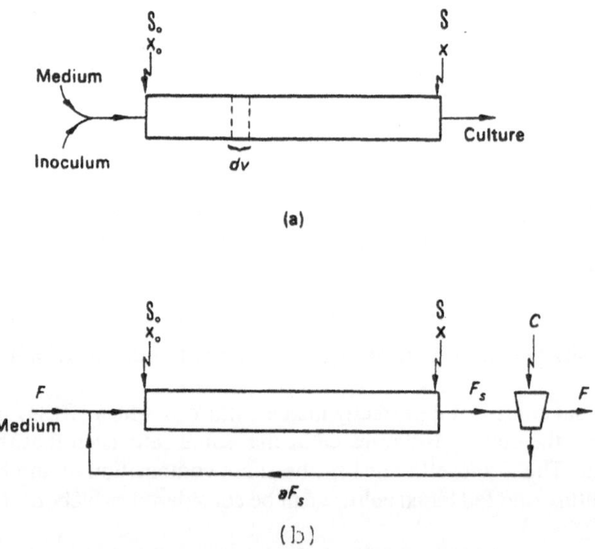

Figure 1:Plug flow culture: (a) without biomass feedback; (b) with biomass feedback. F, F_s, and aF_s are flow rates at various points. C is a device to concentrate biomass.

If V is the volume of the fermentor, and F the rate of culture flow through the fermentor, then the culture residence time or "replacement" time t_r is given by:

$$t_r = \frac{V}{F}$$

In practice there is inevitably always some mixing and a velocity gradient is formed across a cross section of the tubular fermentor with fastest flow in its centre, such that there will be a Gaussian distribution of residence times around the me   value t_r.

When the initial concentration of growth limiting substrate $S_0 >> K_S$, growth in an elementary compartment of the culture dv can be given by the batch growth equation (with $\mu = \mu_m$ until S_0 is exhausted):

$$x = x_0 \, e^{\mu t} \qquad (1)$$

If v is the culture volume displaced from the fermentor in time t, then
t = v/F or t = v/VD, with dilution rate D = F/V. Substitution for in equation (1) gives:

$$\ln x = \ln x_0 + \mu_m \frac{v}{VD}$$

The plug flow fermentor can be made independent of an external inoculum if part of the biomass is fed back to the inlet of the bioreactor.

The realisation of true plug flow culture has so far proved difficult. Problems are the occurrence of laminar flow in tubular vessels and the slow flow rate at the reactor walls which promote adhesion of the biomass. If aeration is required, mixing is almost inevitable, thereby disturbing the ideal plug flow mode. A good approximation to the ideal plug flow principle is obtained by a number of chemostats in series.

On the large scale, the plug flow culture principle is used in the activated sludge process, in the anaerobic fermentations for dairy products and in the digestive tract (PIRT, 1975).

3.CHEMOSTAT CULTURE

3.1.General

The fundamental and practical impact of continuous culture, according to the chemostat principle, became apparent only after the formulation of the basic theory by MONOD (1942) and NOVICK and SZLIZARD (1950). This theory indicates that it is possible to fix the specific growth rate μ of the biomass x at any given value from zero to its maximum μ_m.

Before it was implicitly assumed that the only stable growth rate was the μ_m, corresponding to the doubling time t_D in a simple, exponentially growing, batch culture.

Chemostat culture has opened new horizons in the study of microbial cell composition, their physiology and genetics and has also impact on several industrial fermentation processes (PIRT, 1975; RIGHELATO and ELSWORTH, 1970).

In a chemostat fermentor or continuous stirred tank reactor (CSTR), the addition of nutrients and removal of an equal fraction of the culture are maintained continuously; if the flow rates, the working volume in the fermentor and all other reaction parameters are kept constant, and if the mixing is ideal, a steady state of population characteristics will be reached.

The simplest setup consists of a single stage chemostat; this principle can be modified by the number and size of stages of feedback of cells and by applying multistage systems, whereby the overflow of one bioreactor serves as the feed for the next one (fig. 2).

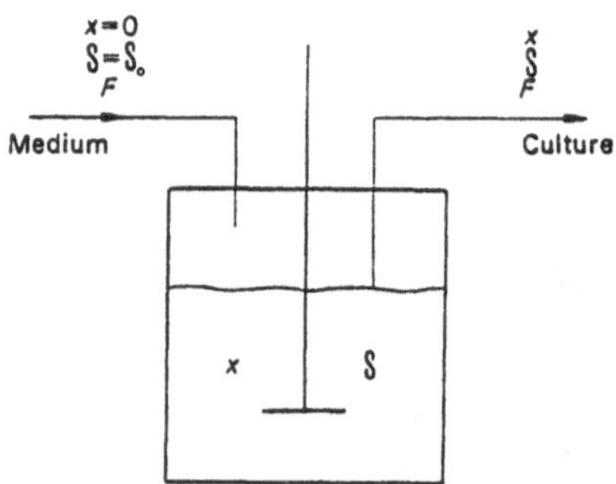

Figure 2: Chemostat

3.2.Principle

In a perfectly mixed CSTR, as described above, a drop of medium entering the fermentor is instantly distributed uniformly through the culture; the time required to achieve this is small compared with the replacement time t_r given by V/F. The inverse of the t_r value, F/V is called the dilution rate, D.

If one considers a culture, in which biomass growth is limited by the amount of a single nutrient S_0, all other nutrients being in excess, and where growth initially is allowed to proceed batchwise with no addition of fresh nutrients, when the flow of medium is started, the outcome can be one of the following three possibilities (fig. 3):

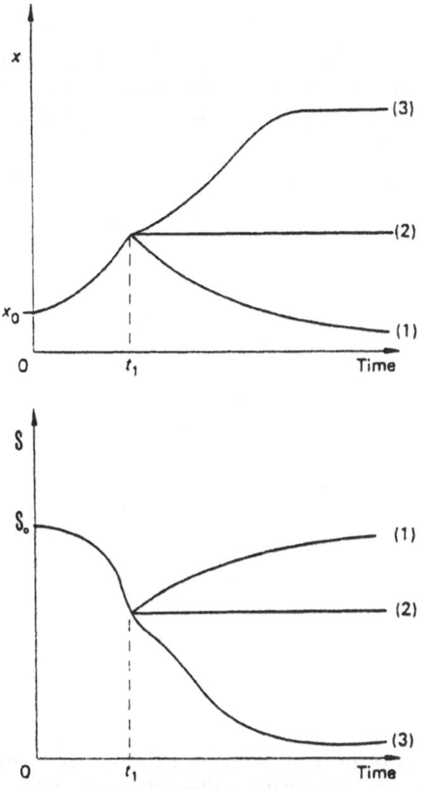

Figure 3: Possibilities in the biomass (x) and limiting substrate (S) concentrations in a chemostat, in function of time. The flow of medium is started at t_1. Numbers refer to the cases, explained in the text.

1)The rate of wash out of cells will exceed the maximal growth rate, such that the biomass concentration will decrease and the residual concentration of the growth limiting substrate will tend towards S_0.

2)The rate of wash out will exactly balance the growth rate; the cells will grow at μ_m. A steady state is obtained here in which the biomass concentration x and the growth limiting substrate concentration remain constant. However, it will not be a stable equilibrium since any temporary change in culture conditions affecting biomass and substrate concentration , will permanently change these concentrations.

3)The wash out rate of the cells is initially less than μ_m. In this case, x will continue to increase; however, eventually the concomitant decrease in growth limiting substrate concentration must decrease until the biomass growth rate equals the wash out rate; the there can be no further changes in the biomass and growth limiting substrate concentrations. Here, the specific growth rate of the biomass is less than μ_m and is governed by the medium flow rate F, or if V is constant, by $F/V = D$. This steady state is a self regulating one. A drop in biomass concentration x, will be associated with a rise in the residual substrate concentration S. This will increase the growth rate μ and act so as to restore the steady state conditions. A rise in biomass concentration will have the opposite effect (fig. 4).

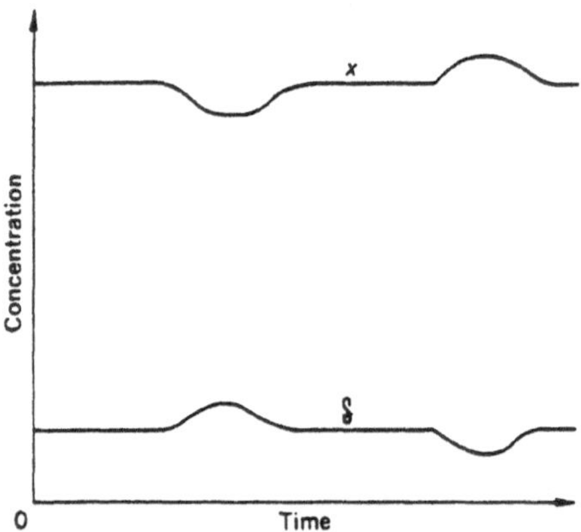

Figure 4:Self regulating mechanism in a chemostat.

The object of the quantitative chemostat theory is to predict or impose the values of the growth rate and the concentrations of biomass and substrate under different cultural conditions. Variables, which can be controlled at will are the limiting substrate concentration S_0 and the medium flow rate
F or with V constant, the dilution rate D, which is the flow rate per unit volume.

3.3.Biomass balance

The increase in biomass concentration is given by the biomass balance:

[inflow] + [growth] - [overflow] = [net increase in biomass]

or:

If the entering medium is sterile, $x_0 = 0$

$$\frac{x_0F}{V} + \mu x - \frac{\bar{x}F}{V} = \frac{dx}{dt}$$

For the time interval dt, the balance is:

$$Vdx = V\mu x dt - xFdt$$

Dividing by V.dt gives

$$\frac{dx}{dt} = \mu x - \frac{xF}{V}$$

OR $$\frac{dx}{dt} = \mu x - xD$$

OR $$\frac{dx}{dt} = (\mu - D)x \qquad [2]$$

In the steady state when dx/dt = 0, it follows that

$$\mu = D \qquad [3]$$

It follows that in a steady state condition in a chemostat, μ equals D. By imposing a fixed value of D (or F) to the chemostat culture, a fixed corresponding μ will be established.

3.4.Growth limiting substrate balance

The balance for the growth limiting substrate is:

[substrate input] - [substrate used for growth] - [substrate output] =
[net increase in substrate]

or:

$$\frac{S_0.F}{V} - \frac{\mu x}{Y} - \frac{S.F}{V} = \frac{dS}{dt}$$

It follows from Y= dx/dS and dx/dt = μ.x

that substrate consumption for growth is given by:

$$\frac{dS}{dt} = \mu.x.\frac{dS}{dx} = \frac{\mu.x}{Y} = q.x$$

In this equation the amount of substrate which is used for maintenance, is neglected.

For the time interval dt, the balance is:

$$V.dS = F.S_0.dt - V.\frac{\mu.x}{Y}.dt - F.S.dt$$

dividing by V.dt gives:

$$\frac{dS}{dt} = \frac{F}{V}.S_0 - \frac{\mu.x}{Y} - \frac{F}{V}.S$$

with F/V = D;

$$\frac{dS}{dt} = D.S_0 - \frac{\mu.x}{Y} - D.S = D.(S_0 - S) - \frac{\mu.x}{Y}$$

In the steady state dS/dt = 0, such that

$$\frac{\mu.\bar{x}}{Y} = D.(S_0 - \bar{S})$$

where $\bar{x}$ and $\bar{S}$ denotes a steady state value

With μ=D, it follows that:

$$\bar{x} = Y.(S_0 - \bar{S}) \qquad [4]$$

To obtain $\overline{x}$ and $\overline{s}$, it can be derived from the Monod equation (fig. 5):

$$\mu = \mu_m \cdot \frac{S}{S + K_S}$$

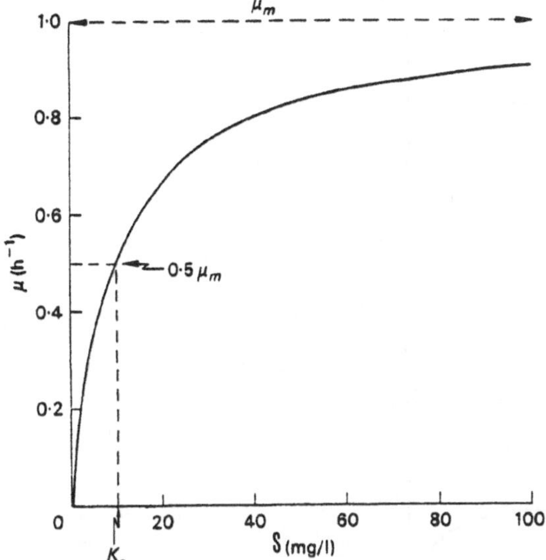

Figure 5: Specific growth rate in function of the substrate concentration

and substituting μ = D and S = $\overline{S}$, that

$$\overline{S} = K_S \cdot \frac{D}{\mu_m - D} \qquad [5]$$

Substituting this formula (5) for S in equation (4), gives

$$\overline{x} = Y.[S_0 - K_S \cdot (\frac{D}{\mu_m - D})] \qquad [6]$$

These important equations allow to calculate the steady state values of $\bar{x}$ and $\bar{S}$ for any given value of D (or F) and S_0, provided that the values of the constants μ_m, K_S and Y are known.

Plots of $\bar{x}$ and $\bar{S}$ against D are given in figure 6.

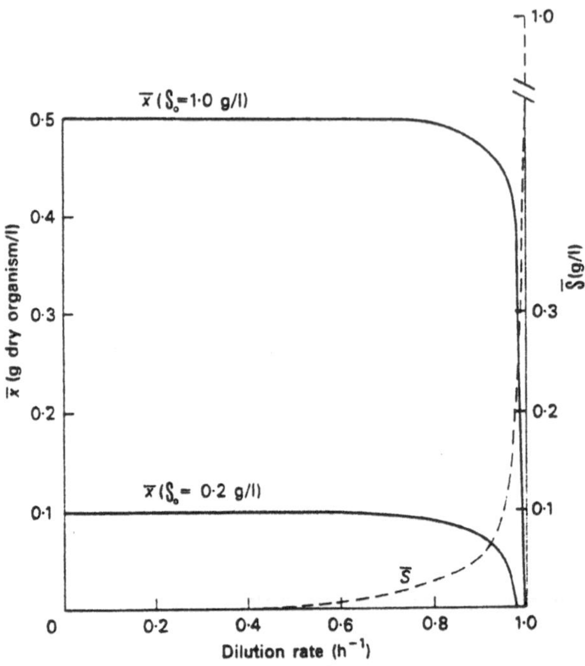

Figure 6: Steady-state values of $\bar{x}$ and $\bar{S}$ in a chemostat.

3.5.Critical dilution rate D_C

The maximal growth rate is obtained when $S \approx S_0$
Substitution of this value in the Monod equation, gives

$$\mu = D_c = \mu_m \cdot \frac{S_0}{S_0 + K_S} \qquad (7)$$

Normally $S_0 >>> K_S$, such that $D_c \approx \mu_m$

3.6.Determination of μ_m

The value μ_m can be obtained from a batch culture as the slope of the curve of ln x versus t.

Also $\mu_m \approx \mu = \ln 2/t_D$

Alternatively, when $S >>> K_S$, we can put $\mu = \mu_m$ in equation (2):

$$\frac{dx}{dt} = (\mu - D)x$$

and upon integration:

$$\ln x = (\mu_m - D)t + \ln x_0 \qquad (8)$$

If we run a chemostat at $D > D_c$, the biomass decreases or wash out occurs, according to (8) and the slope of the log plot is μ_m - D, which gives the value of μ_m.

3.7.Determination of K_S

K_S can be derived from equation (5). It follows from the Monod equation that:

$$K_s = \bar{S}.\frac{\mu_m - D}{D}$$

If D is set at $\mu_m/2$, then $K_S = \bar{S}$. It follows that the determination of $\bar{S}$ under chemostat condition of $D = \mu_m/2$, gives automatically the K_S value.

Based on the Monod equation, K_S and μ_m can also be derived graphically, by plotting $1/\mu$ versus 1/S; the intercept on the abscissa is $-1/K_S$ and the intercept on the ordinate is $1/\mu_m$ (fig. 7).

3.8.Biomass output

In a chemostat culture, the rate of output of biomass per unit culture volume is given by D.x.

It follows from equation (6) that

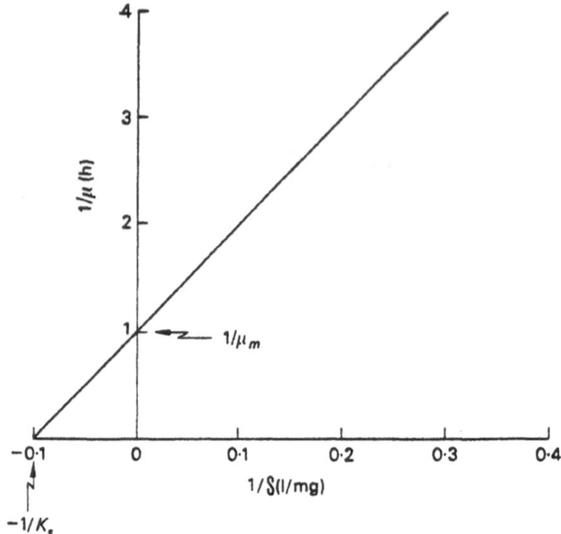

Figure 7: Graphical derivation of K_s and μ_m

$$D.x = D.Y.[S_0 - K_s.\frac{D}{\mu_m - D}] \qquad (9)$$

The biomass output reaches a maximum at dilution rate D_m, which is obtained by differentiating equation (9) with respect to D and equating to zero.

$$D_m = \mu_m.[1 - \sqrt{\frac{K_s}{S_0 + K_S}}]$$

The steady state biomass concentration x_m at D_m is obtained by substituting D_m in equation (6):

$$\overline{x_m} = Y.[S_0 \dot{+} K_s - \sqrt{K_s.(S_0 - K_s)}]$$

The maximal output rate is

$$D_m \overline{x}_m = \mu_m . Y . S_0 . [\sqrt{\frac{K_S + S_0}{S_0}} - \sqrt{\frac{K_S}{S_0}}]^2$$

If $S_0 >>> K_S$, it follows that

$$D_m \overline{x}_m \approx \mu_m . Y . S_0 \qquad [10]$$

3.9.Unique features of chemostat culture

-The chemostat permits biomass growth rate to be varied without changes in the environment other than the concentration of the growth limiting substrate.
-The chemostat permits the growth rate to be fixed, while the environment is altered.
-The chemostat permits to maintain substrate limited growth with a constant growth rate.
-The chemostat can be used for rational mutant selection
-The chemostat allows to obtain balanced growth (time independent state)
-The chemostat allows to maintain maximum output rate of biomass or product.
(PIRT, 1975; FIECHTER, 1981)

4.TURBIDOSTAT CULTURE

A turbidostat originally is a chemostat provided with a photoelectric cell for measuring the turbidity of a culture and adding more medium when the biomass density rises above a fixed level (fig. 8). Turbidity measurements are limited to cultures of unicellular organisms. Other methods of biomass measurement (CO_2-release, acid production, etc.) are now used to extend the applicability of the method, but here too the term turbidostat remains in use (PIRT, 1975).
A photocell activates the medium pump, when the culture turbidity exceeds a preset value. The culture volume is also kept constant. Eventually a steady state is reached in accordance with equation (4). By means of turbidostat control, the biomass density ($\approx$x) is set and the dilution rate adjusts itself to the steady state value; this is in contrast to the single chemostat, where the dilution rate D is fixed and x adjusts itself. The turbidostat provides a means of maintaining cultures in a constant environment with excess of substrates. The system will select for faster growing strains. Turbidostatic control should make the chemostat culture more stable at dilution rates close to $D_C = \mu_m$.

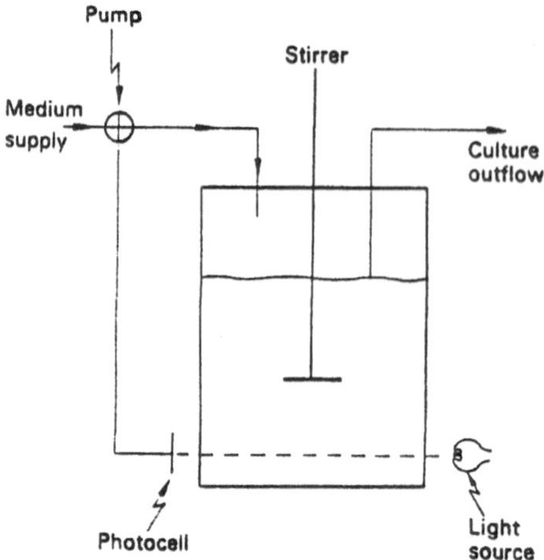

Figure 8: Diagrammatic representation of a turbidostat

5.FED BATCH CULTURE

5.1.General

A batch culture of microorganisms fed continuously with fresh medium is described as a "fed batch culture". When a portion of the "fed batch culture" is withdrawn at intervals, the system is described as a "repeated fed batch culture" (fig. 9) (PIRT, 1974; DUNN and MOR, 1975).

A fed batch culture may reach a "quasi-steady state" in which the specific growth rate μ virtually equals the dilution rate D.

In a quasi steady state, the specific growth rate μ gradually decreases. A unique feature of a fed batch culture is that it allows continuous reproduction of the transient conditions between two specific growth rates, which can be chosen at will. The volume variation distinguishes it from a chemostat culture, where it is essential to maintain a constant volume V.

5.2.The quasi steady state

In a homogenous batch culture with a growth limiting substrate, we have

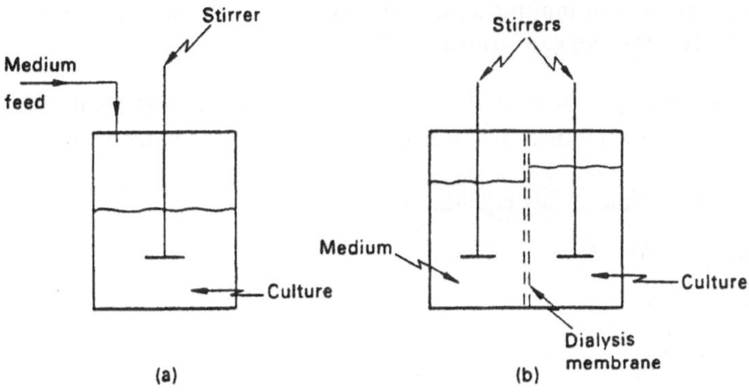

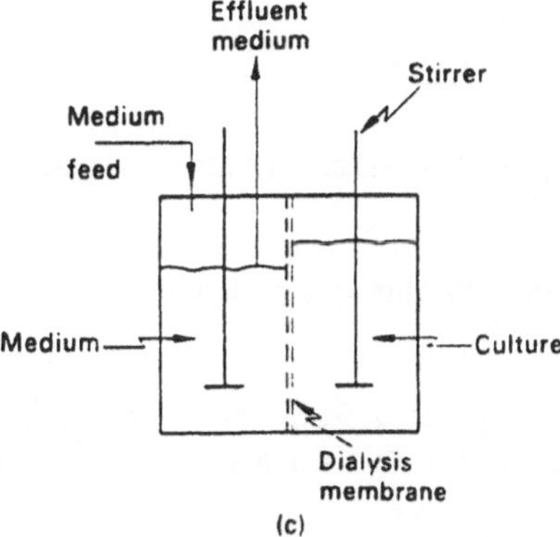

Figure 9 (a)Fed batch culture. (b) Batch culture dialysed against medium . (c) Batch culture dialysed against a stream of medium.

$$x = x_0 + Y.(S_0 - S)$$

with x_0 *= inoculum concentration*

When x reaches x_m, the growth limiting substrate is practically exhausted so that S <<< S_0. Assuming that x_0 is small, we can write $x_m \approx Y.S_0$

When $x_m \approx Y.S_0$, a medium feed is started at flow rate F with the growth limiting substrate at S_0. Total biomass X in the culture is X = x.V, with V, culture volume at time t.

Since x= X/V, differentiation of this equation gives

$$\frac{dX}{dt} = \frac{V.\frac{dX}{dt} - X.\frac{dV}{dt}}{V^2}$$

Substitution of dX/dt by μ.x and dV/dt byF leads to

$$\frac{dX}{dt} = (\mu - D).x \qquad (11)$$

This is the characteristic equation for growth in fed batch culture, provided that Monod-type growth is reached.

Consequently, the specific growth rate in this quasi steady state culture is equal to continuous cultivation e.g. $\mu = \mu_m.S/(S+K_S)$

When $S_0 >>> K_S$, over most of the range of μ from zero to $\approx\mu_m$, the growth limiting substrate will be almost completely utilized so that when x = $x_m \approx Y.S_0$, then dx/dt = 0

Then it follows from equation (11) that μ = D

If S_T = total growth limiting substrate, its balance can be given by:

[rate of increase] = [rate of input] - [rate of consumption]

$$\frac{dS_T}{dt} = F.S_0 - \frac{\mu.x}{Y}$$

When $X = V.x_m$, it means that the substrate is virtually all consumed as fast as it enters the culture so that $F.S_0 \approx \mu.X/Y$, hence dS_T/dt and dS/dt are approximately zero.

The state when $dX/dt \approx 0$, $dS/dt \approx 0$ and $\mu \approx D$ is called quasi steady state where the concentration of growth limiting substrate is a function of D and is given by

$$S = D.\frac{K_s}{\mu_m - D}$$

The rate of increase of total biomass is given by

$$\frac{dX}{dt} = F.Y.S_0 \qquad thus \qquad X_m = X_0 + F.Y.S_0.t$$

In a chemostat culture and in a fed batch culture, we have in both cases $\mu = D$, but whereas D is constant in a chemostat, it is decreasing in a fed batch culture and μ is assumed to decrease at the same rate. In the quasi steady state in a fed batch culture, the biomass is in a transient state with the growth rate under control. Several microbial (secondary) metabolites are often synthesised at maximal rates during such transient state (PIRT, 1975; WHITAKER, 1980).

6.REFERENCES

Dunn, I.J. and Mor, J.R. (1975) 'Variable volume continuous cultivation', Biotechnol. Bioeng., 17, 1805-1822.

Fiechter, A. (1981) 'Batch and continuous culture of microbial, plant and animal cells' in Rehm, H.J. and Reed, G. (eds.), Biotechnology, volume I, Verlag Chemie, Weinheim, pp. 453-505.

Monod, J. (1942) 'Recherches sur la croissance des cultures bactériennes' (thèse), Hermann, Paris.

Novick, A. and Szlizard, L. (1950), 'Description of the chemostat', Science, 112, 715-716.

Pirt, S.J. (1974), 'The theory of fed batch culture with reference to Penicillium fermentation', J. Appl. Chem. Biotechnol., 24, 415-424.

Pirt, S.J. (1975), 'Principles of microbial cell cultivation', Blackwell Scientific Publications, Oxford.

Righelato, R.C. and Elsworth, R. (1970), 'Industrial applications of continuous culture', Adv. Appl. Microbilogy, 13, 399-417.

Whitaker, A. (1980), 'Fed batch culture', Process Biochemistry, 15 , 10-15.

A SPREADSHEET SIMULATION OF A DIFFERENTIAL REACTOR FOR ESTIMATING KINETIC CONSTANTS

J. MATA-ALVAREZ* and P. CLAPES**
*Department d'Enginyeria Quimica
Universitat de Barcelona
Marti i Franques 1, 6p
08028 Barcelona (Spain)
** Unitat de Quimica i Bioquimica de Peptids C.I.D. (CSIC)
Jordi Girona Salgado 18-26
08034 Barcelona (Spain)

ABSTRACT. A spreadsheet simulation of a recirculation reactor is carried out. The resolution includes a Runge-Kutta integration procedure and the Newton method for solving implicit equations. The use of the spreadsheets as a tool for fitting constants is also emphasized. The importance of considering the complete mathematical model is pointed out. A brief discussion of diffusional effects is also included.

1. Introduction

The use of enzymatic methods in chemical reactions present several advantages such as smooth reaction conditions, high optical and chemical yields and with minimum protection of the other reactive sites in the molecule (1-3). In a given step, e.g. during scale-up procedures, the study of enzyme catalyzed reactions requires the consideration of the involved kinetics.

The kinetics of enzyme-catalyzed reactions are usually studied in batch stirred tank reactors. However, when an immobilized system is being considered, the latter type of reactor may not be the best choice. A more suitable system could be depicted that depicted in Figure 1 (4,5). This experimental device can be used for different purposes in enzyme technology: from immobilizing the enzyme to activating the support, to either determine the activity of an immobilized enzyme or to fit a kinetic model to an enzymatic reaction which takes place in the system (6-8).

The experimental system consists of two different devices: a tubular reactor of a small volume V_1 (differential reactor) and a stirred tank with a volume V_2. The flow pattern of the first approaches a plug flow, whereas the second is assumed to be completely mixed. The stirred tank acts as a substrate source and as a product sink for the tubular reactor. The tubular reactor is packed with a solid support, with the enzyme conveniently attached. A pump recirculates continuously the contents of one reactor to the reservoir. The flow rate, q, of this stream dictates the retention time in both reactors. The solid packing in a tubular reactor can have significant effects on the flow conditions, including departure from plug-flow. Diffusion can take place in the radial and in the axial directions. The last one would tend to invalidate the plug-flow assumption, whereas the first one would bring the reactor performance closer to it.

A.R. Moreira and K.K. Wallace (eds.), Computer and Information Science Applications in Bioprocess Engineering, 419-433.

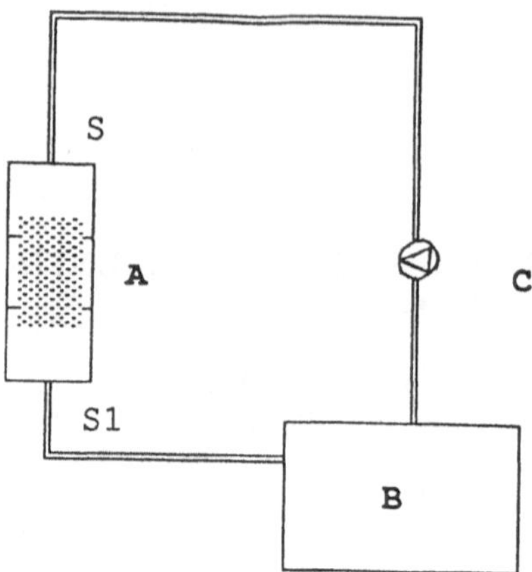

Figure 1. Recirculation reactor system: (A) Immobilized reactor; (B) Reservoir; (C)Pump

The use of a system like that presented in Figure 1 for activity determinations presents several advantages. Among them is that it allows testing of mass transfer limitations of a given support at a given flow-rate. Besides, the sampling procedures are much easier because the liquid is always free of support. A point favoring the reliability of data is that no support is withdrawn out of the system, as occurs in the operation of a Stirred Tank Reactor (STR). It can also be a useful device when on-line measurements are to be made. It just requires the installation of the appropriate device in the line coming from or going to the tank.

However, some precautions should be taken when dealing with this system. In some specific cases, this system behaves identically as if all the active support contained in the differential reactor were at the stirred tank. This is not so and in general, the profiles obtained by monitoring the concentration leaving the tank are not quite the same as if they were obtained by taking samples of a stirred batch reactor.

2. System Model

When enzymatic reactions are involved, isothermal conditions can be assumed. As a consequence, for the biocatalyzed reaction:

$$Substrate(S) \longrightarrow \Pr oduct(P)$$

Mass balances in both reactor and tank will suffice to establish the model. They will be developed in the following sections.

2.1. PLUG FLOW REACTOR

For the plug flow reactor (PFR), the integral balance of substrate S over the reactor length yields:

$$\frac{V_1}{q} = \Theta_1 \int_S^{S_1} \frac{dS}{-r} \tag{1}$$

where $\Theta_1 = V_1/q$ is the residence time for the reactor at a given flow rate q. Plug flow assumption is made, given the small length of the reactor.

2.2. RESERVOIR

The substrate balance for the completely mixed tank is as follows:

$$\frac{dS}{dt} = -\frac{1}{\Theta_2}(S - S_1) \tag{2}$$

where $\Theta_2 = V_2/q$ is the retention time in the tank. No generation term is included, as there is no enzyme in the liquid phase.

3. Zero Order Model. Activity Determinations

For activity determinations and when the conversion is low, it can be assumed that the variation of substrate concentration vs. reaction time is linear. Therefore, and at the beginning of the reaction, a zero order kinetic rate equation may be assumed for substrate consumption (r=k). This expression means that no diffusional limitations are taking place (see below for discussion of these aspects). Integrating equation (1) yields:

$$S - S_1 = k\Theta_1 \tag{3}$$

that is, the differences between substrate inlet and outlet concentrations remain constant during all of the considered period. The value of them being fixed by the enzyme activity and the recirculation flow rate q.

Substituting S_1 in equation 2 for its expression (equation 3) and integrating the resulting equation gives:

$$S = S_2^0 - Kt \tag{4}$$

where K is

$$K = k\Theta_1/\Theta_2 \tag{5}$$

If S is plotted against t, the slope will yield K. The estimation of the activity k may be carried out after the volume correction. It is interesting to note that equation 5 corresponds also to the integration of:

$$\frac{dS}{dt} = -K \tag{6}$$

that is, a balance set on the reservoir, assuming that the reaction takes place in that volume. Thus, as K is referred to the volume V_2, but actually the enzyme is confined to the volume V_1, it follows that , for activity estimation, it would be necessary to correct the volume (to multiply K by V_2 and divide it by V_1). Taking into account equation (5), it appears that in fact, these operations lead to an estimate of the activity k.

These results may lead to the false conclusion that the recirculation reactor operates in the same way as if all the enzyme were in a STR and only a volume correction is necessary to estimate the kinetic constants. This is not true as it will be shown in the following two sections, which consider the first order and the Michaelis-Menten kinetic models.

4. First Order Kinetic Model

In the case that the reaction rate, r, of substrate consumption follows a first order kinetic model and assuming no diffusional limitations, the kinetic equation will be:

$$r = kS \tag{7}$$

where k is the first order reaction rate constant.

When the reaction is carried out in a discontinuous stirred tank reactor, the kinetic parameter k can be estimated by fitting the experimental values of substrate concentration and the reaction time to the integrated equation 8:

$$S = S_0 e^{-kt} \tag{8}$$

where S_0 is the initial concentration of substrate.

Using a recycle reactor, the resolution of the model for the substrate consumption of the plug flow and stirred tank simultaneously will provide the equations to estimate the kinetic parameter k.

4.1. THE PLUG FLOW REACTOR EQUATION

Substituting equation 7 in the plug flow design equation (equation 1) and integrating the resulting expression, the following equation can be deduced for substrate consumption for the plug flow reactor:

$$S_1 = S e^{-k\Theta_1} \tag{9}$$

where Θ_1 is the residence time in the reactor.

This equation relates both time dependent variables, outlet S_1 and inlet S substrate concentrations in the plug flow reactor. As can be seen, the larger the recirculation flowrate, the closer the concentrations of S_1 and S are.

4.2. RESOLUTION OF THE MODEL

The outlet substrate concentration for the stirred tank, S, can be calculated by combining equations 2 and 9 and integrating the resulting expression:

$$S = S_0 e^{\left(1-e^{-k\Theta_1}\right)t/\Theta_2} \tag{10}$$

where So is the initial substrate concentration. Equation parameters are the residence time in both reactors (Θ_1 and Θ_2) and the kinetic parameter k. As can be seen, equation 10 relates the substrate tank concentration with time. This equation can be expressed in the following form:

$$S = S_0 e^{-Kt} \tag{11}$$

where K is a new apparent constant which is given by:

$$K = \frac{\left(1 - e^{-k\Theta_1}\right)}{\Theta_2} \tag{12}$$

As can be seen, K is not a kinetic constant, since it depends on either the recirculation rate q or the relative volumes of the plug flow reactor and the stirred tank. However, it may lead to confusion because, assuming that the system behaves as if all the enzyme were in the reservoir, and setting a batch balance in this tank:

$$\frac{dS}{dt} = -kS \tag{13}$$

Integration of this balance would yield equation 11, with k instead of K. Again a volume correction is necessary to really estimate k. This can be done, multiplying equation 12 by the factor Θ_2/Θ_1 . This would yield the following equation:

$$k = \frac{\left(1 - e^{-k\Theta_1}\right)}{\Theta_1} \tag{14}$$

Thus, an error in estimating k will be done, and this error will depend on the value of k and the value of Θ_1. Figure 2 shows the influence of the adimensional factor Θ_1k on the error percentage on estimating k. As can be seen, increasing the recirculation flow rate or having a low value for the true kinetic constant k, decreases the error. In other words, the recycle reactor is closer to the batch stirred tank reactor.

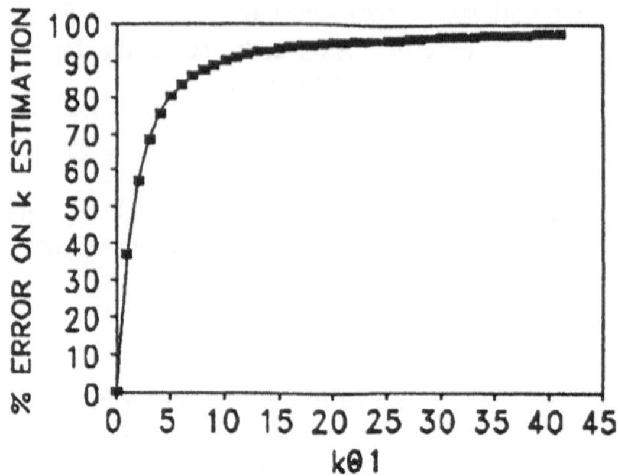

Figure 2. Error in k estimation assuming all the enzyme is in the reservoir.

5. Michaelis-Menten Kinetic Model

Consider the Michaelis-Menten mechanism expression for r, assuming again no diffusional limitations:

$$r = \frac{kS}{K_m + S} \tag{15}$$

Substitution of this rate equation model in a mass balance carried out in a batch, stirred tank reactor gives:

$$\frac{dS}{dt} = r = \frac{kS}{K_m + S} \tag{16}$$

Integration of equation 16 yields, after rearranging terms:

$$\frac{(S_0 - S)}{\ln(S_0/S)} = k\frac{t}{\ln(S_0/S)} - K_m \tag{17}$$

From this equation a simple linearization method can be applied to estimate the constants. The usual procedure would be to take samples from the stirred tank, make the appropriate calculations and plot the first member of the equation in front of $t/\ln(S_0/S)$. This would yield constants k and K_m. Of course a non-linear estimation method would yield better estimates, but for the scope of this discussion, this approach is good enough.

If the system behaved as if all the enzyme were in the reservoir, the supposedly true constants k and K_m would then be estimated after the volume correction. However, the equations which currently apply to this system are the following:

$$\frac{\Theta_1 k}{K_m} = \ln\frac{S_1}{S} + \frac{1}{K_m}(S_1 - S) \tag{18}$$

coming from the integration of the plug flow reactor substrate (equation 1), using the Michaelis-Menten expression for r. The second equation is the substrate balance in the stirred tank reactor:

$$\frac{dS}{dt} = \frac{1}{\Theta_2}(S - S_1) \tag{19}$$

Contour conditions for this equation are t=0, $S=S_0$. Its integration must be performed in combination with Equation 18 using numerical methods and for a set of given parameters.

5.1. SPREADSHEET RESOLUTION OF THE MATHEMATICAL MODEL

Resolution of equations 18 and 19 simultaneously requires numerical procedures for integrating and solving equations respectively. There are several programs available to solve mathematical models. However, not always are they easy to use. If they are, normally their capabilities are limited. However, spreadsheets are an available software, every time more powerful, easy to use and flexible enough to handle quite different kinds of problems (9,10). Since they appeared in the market, their applications have grown exponentially and today engineers and scientists, among other users, employ them for a variety of operations.

Five variables have been identified to study their influence on the error of Michaelis-Menten constant estimates: their own value (k and K_m), their initial concentration (S_0), the reactor retention time (Θ_1.) and the ratio between the reactor and reservoir volumes (V_{rat}). Table 1 shows this initial setting as they appear in the spreadsheet, together with the results observed with them.

TABLE 1. Initial setting of the spreadsheet, together with the results observed with them. Batch heading means the correction made on the assumed constants taking into account the reservoir and reactor volumes. The iteration control is the sum of exit reactor concentrations, which is controlled to be constant at the end of the manual iterative procedure.

DIFFERENTIAL REACTOR SIMULATION

SETTINGS		FITTING RESULTS			
			Batch	Estim.	Error (%)
Step	1				
Ks =	30	k	4	3.3	18.5
k =	40	Ks	30	33.4	11.5
S0 =	30				
Θ_1 =	0.2	ITERATION CONTROL	356.3153		
Volr	0.1				

Resolution of equations 18 and 19 is simple enough to be implemented in a spreadsheet. Table 2 shows how the calculations are performed. S is computed using a Runge-Kutta procedure of 4th order (each one of the four constants is calculated in columns G, H, I and J, respectively). Column K calculates the next S after the step is set initially. Column D reproduces column K, for presentation purposes. S_1 is calculated solving equation 18 by a Newton procedure. Thus, column M contains equation 18, column N its derivative and column O recalculates S_1. Values of S_1 to calculate the functions of columns M and N are taken from column E which at the same time gets its value from column O. A circle is formed and an iterative procedure has to be done until convergence. The iterations are performed manually, by pressing F9 until concentrations become constant.

TABLE 2. Spreadsheet integration of equation 19 with the Runge-Kutta method and simultaneous resolution of equation 18 with the Newton method. Layout of columns

COLUMNS

C	D	E	F	G	H	I	J	K	L	M	N	O
TIME	Reserv.	Reactor		RUNGE KUTTA 4TH ORDER				S2		NEWTON RAPHSON		S1
	S2	S1		rk1	rk2	rk3	rk4	funrk		f	f'	1-f/f'
0	30	26.13623						30		-2.6E-15	-2.14783	26.13623
1	28.48065	24.72407		-1.93188	-1.44891	-1.56965	-1.14705	28.48065		-1.2E-15	-2.21339	24.72407
3	25.56930	22.03378		-1.82355	-1.36766	-1.48163	-1.08273	25.56930		2.2E-15	-2.36154	22.03378
26	5.025614	3.985108		-0.56008	-0.42006	-0.45507	-0.33255	5.025614		2.2E-12	-8.52802	3.985108
28	4.236992	3.343355		-0.48249	-0.36187	-0.39202	-0.28648	4.236992		0.000000	-9.97302	3.343355
30	3.560629	2.797454		-0.41319	-0.30989	-0.33571	-0.24533	3.560629		0.000000	-11.7240	2.797454

Finally in another section of the spreadsheet not shown, two columns are set in order to perform the linear regression to estimate constants k and K_m using the wrong approach (equation 17). The values of this regression are used later to make the comparison between the real estimates and the estimates using the wrong assumption.

5.2 COMPARISON OF RESULTS OBTAINED USING THE MATHEMATICAL MODEL FOR THE STR AND THE RECIRCULATION REACTOR

Figure 3 presents the results of the simulation of the true model together with the simulation using the wrong model, that is using the batch approach, using the initial settings of the variables. The error on the constants estimated with the wrong assumption are in this case 18.5% for k and 11.5% for K_m. A more serious discrepancy is presented in Figure 4 which represents the same profiles shown in Figure 3, but instead of a reactor residence time Θ_1.of 0.1, a value of 0.04 has been tested. The respective error on the estimation of k and K_m are 85% and 1.8%, respectively. This variable has a large influence on estimation errors. Figure 5 shows this influence as a function of the values of. Θ_1. As can be seen, errors on k are increasing in both sides, that is at high and low values. On the other hand, errors in K_m are continuously increasing with Θ_1. It is important to notice that differences on k estimates become important, especially at high recirculation rates. As a consequence, if the system is operated at large pump flow rate values in order to avoid mass transfer problems, serious errors may occur in the kinetic constant k, if the true mathematical model is not considered.

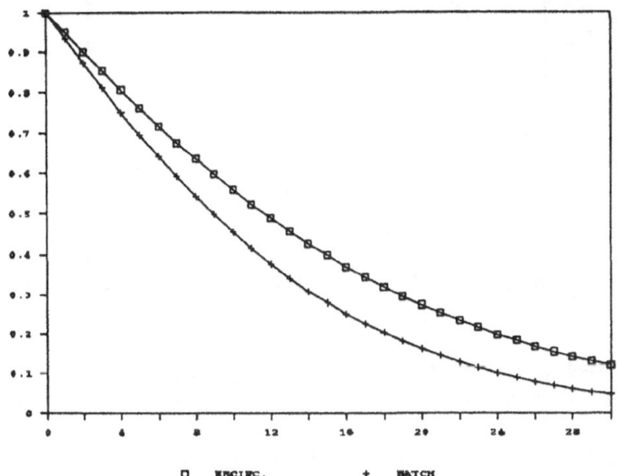

Figure 3. Dimensionless substrate concentration evolution with time, using a batch reactor and using a recirculation reactor. Conditions are those of Table 1.

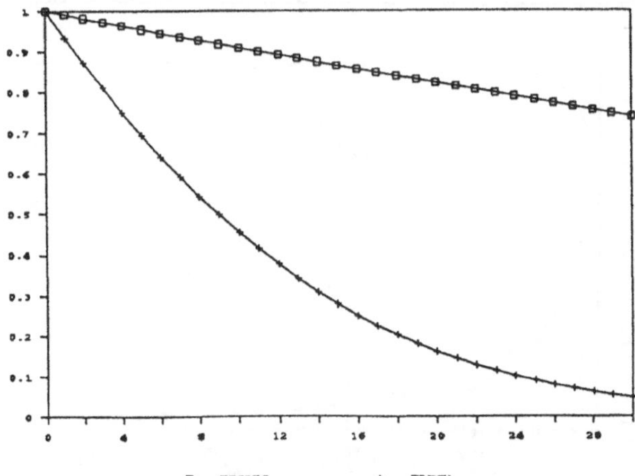

Figure 4. Dimensionless substrate concentration evolution with time, using a batch reactor and using a recirculation reactor. Conditions are those of Table 1 except Θ_1 =0.04.

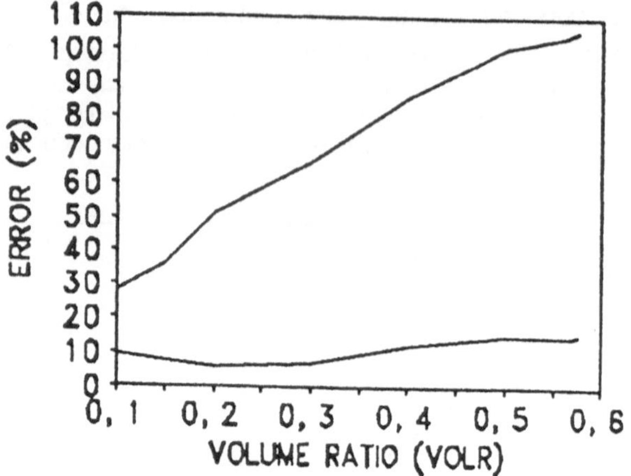

Figure 5. Error on estimating k and K_m as a function of the volume ratio.

Figures 6-9 present similar plots, studying the influence on the constant estimation errors volumes (V_{rat}), the initial substrate concentration (S_o) and the values of k and K_m, respectively. As can be seen, there are always differences which can lead to erroneous constant estimates. In fact, in all the cases analyzed, the conversion is lower when the batch system model is used for the constant determination. This leads to general underestimation of constant k if the system behavior is assimilated as that of a stirred tank batch reactor. However, the relative differences vary with the different values of the parameters.

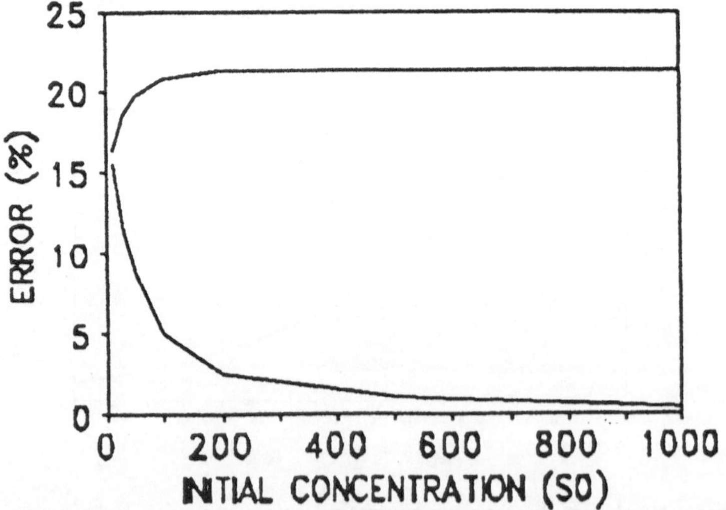

Figure 6. Error in estimating k and K_m as a function of the initial concentration.

The effect of the ratio between the reactor volume V_1 and the reservoir volume V_2 named V_{rat}, is considered in Figure 6. As can be seen, errors are decreasing with the decrease in the reactor volume. Important errors are produced in the estimation of k at high V_{rat}, whereas errors in K_m are kept around 10-15%.

Figure 7 examines the initial concentration S_0. Profiles obtained either using one or the other system are different. When S_0 increases errors in K_m become insignificant, whereas the contrary behavior is observed in the value of k. Nevertheless, in both cases, the concentration of S_0 introduces smaller errors in comparison to the other parameters.

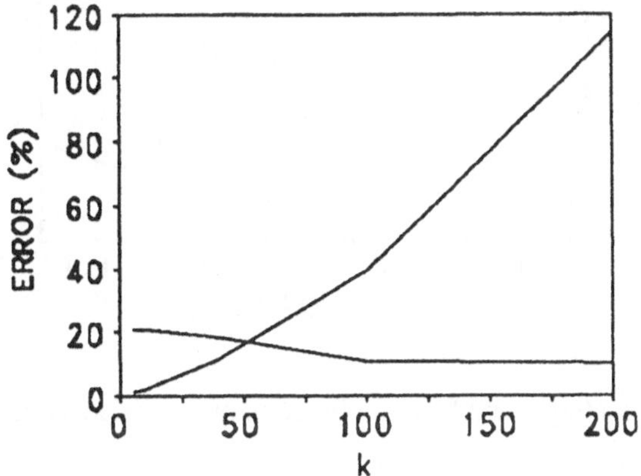

Figure 7. Error in estimating k and K_m as a function of the absolute value of k.

. The absolute value of the kinetic constant k does not introduce significant errors in its own estimate: errors from 20 to 10% at increasing values of k. However, serious errors are observed when estimating K_m, when the wrong assumption is adopted, at high values of k oscillating from 5 to more than 100% (see Figure 8).

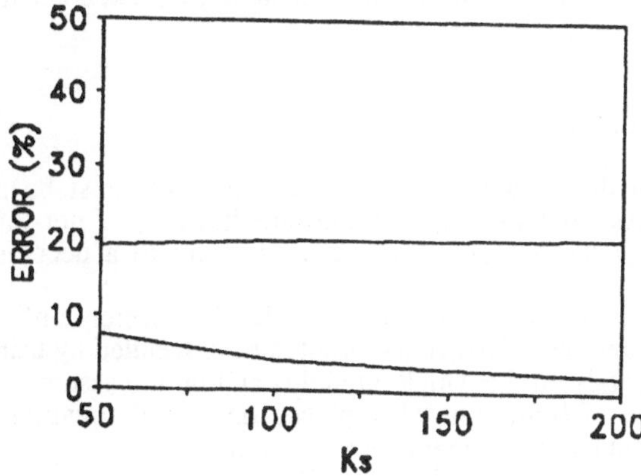

Figure 8. Error in estimating k and K_m as a function of the absolute value of K_m.

Finally, the effect of K_m is examined in Figure 9. The error in K_m increases as K_m decreases leading to error around 100%, whereas the error on k increases with Km in the range 10-20%.

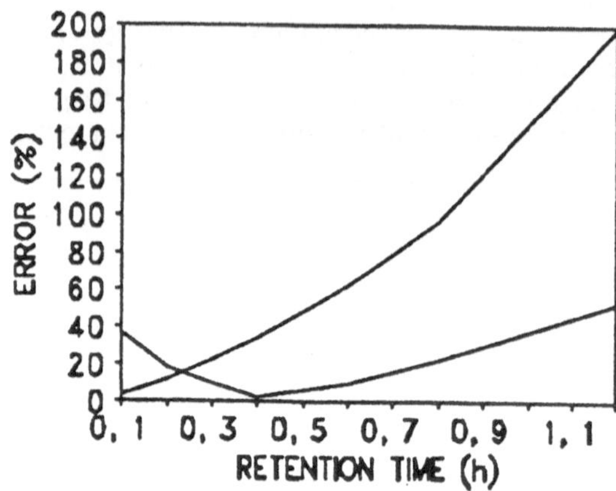

Figure 9. Error in estimating k and K_m as a function of the residence time.

5.3. CONSTANT FITTING WITH THE SPREADSHEET

Once the model has been set in the spreadsheet, it can be easily implemented a small section to fit the constant of the model. It requires only one column with the experimental results of S and a second column which computes the squared difference between the calculated S profile (through the model) and the experimental S. The sum of the squares can be manually minimized, by changing the values of k and Km following a simple optimization method. As the problem has only two unconstrained variables, the solution should be easy, although precautions should be taken because of the possible presence of local minima.

6. Diffusional Limitations

To really approach the model problem, considerations must be made whether the diffusion or the external mass transfer rate are limiting or not the conversion rate. Additionally, some deactivation could occur, leading to a decrease in the intrinsic reaction rate.

External mass transfer can be avoided with this reactor configuration by using a high recirculation flow rate. Diffusional problems are handled by using an effectiveness factor, e, which depends on the Thiele modulus and on the assumed particle geometry. Thus when this type of limitation takes place, the design equation (equation 1) must include the right formulation of the reaction rate term r:

$$r = (\text{rate model equation})*(\text{effectiveness factor}) \quad (20)$$

Thus, for the integration of equation 1, it is necessary to take into account the particular expression for the effectiveness factor, which in general will have a concentration term. For instance, for a first order reaction, the problem is very easy because the effectiveness factor is constant. Considering a spherical geometry, it is given by the following expression (11):

$$e = \frac{1}{(Th)}\left(\frac{1}{tgh(3(Th))} - \frac{1}{3(Th)}\right) \tag{21}$$

where (Th) is the Thiele modulus, a dimensionless expression given by:

$$(Th) = L(k/D)^{1/2} \tag{22}$$

where L is the characteristic dimension ($=R_p/3$) in the case of a spherical particle of radius R_p; k is the first order kinetic rate constant and D is the diffusion coefficient. As can be seen (Th) is constant and consequently, the effectiveness factor. Thus, integration of equation 1 can easily be carried out. Moreover, when (Th) < 0.04, that is, when the reaction kinetics are faster than the diffusion kinetics, e is practically 1 and no care is necessary for the diffusion effects. Values of (Th) over 4 gives place to serious limitations.

7. A Study Case: Enzymatic Peptide Synthesis

The enzymatic coupling reaction between Z-Tyr-OMe and Arg-NH_2 yielding the dipeptide Z-Tyr-Arg-NH_2, was considered as an example of the use of this device to gain kinetic information. α-Chymotrypsin, covalently immobilized to controlled pore glass (CPG) was used as a catalyst. A buffered aqueous monophasic reaction medium containing the water miscible organic solvent DMF was chosen as a reaction medium.

In this case, and in the range of substrate and nucleophile concentrations, the rate of acyl-donor (Z-Tyr-OMe) consumption was able to be approximated to a first order kinetic model with respect to substrate concentration (12). Conversion rate was also independent of the nucleophile concentration.

$$\frac{d[Z-Tyr-OMe]}{dt} = -k[Z-Tyr-OMe] \tag{23}$$

Several experiments were carried out using the recycle reactor. The experimental device used was equipped with a plug flow reactor of 0.314 ml of total volume and with a stirred tank reactor which contained 10 ml of reaction volume. A list of the generated constants k and K is presented in Table 3 as a function of the flow rate employed. As Table 3 shows, there are large differences between K and k, and the recirculation flow rate did not have any influence on the values of these constants. This is in agreement with the theoretical predictions since the volume ratio in this case was very small.

TABLE 3. Synthesis of Z-Tyr-Arg-NH2 by means of immobilized α-chymotrypsin. Kinetic constants obtained by fitting the experimental data to the model represented by the equations 11 and 12 at different recirculation flow rates q, using the experimental device shown in Figure 1. DWC means dry weight of catalyst (data from ref. 8)

q ($ml.min^{-1}$)	Dry weight of PFR bed (g)(DWC)	K (min^{-1})	k (min^{-1})	K' k/g DWC ($min^{-1}.g^{-1}$)
26.67	0.0835	0.0864	2.64	10.0
40.00	0.0835	0.1039	3.31	11.8
23.30	0.0835	0.0983	3.20	12.0
4.30	0.1075	0.1035	3.57	10.7
4.30	0.1083	0.1083	3.72	10.8

8. Conclusions

Spreadsheets can be easily used to solve a system of equations involving a numerical resolution procedure (Newton method) and a fourth order Runge Kutta method for integration.

If a recirculation system as that shown in Figure 1 is used, the obtained profiles differ from those of a batch system containing the same amount of enzyme. These differences are observed in a large range of conditions, that is varying the value of the kinetic constants or the controlled variables of the system (volumes, recirculation rates, etc.). Specifically, the recirculation rate directly influences the value of Θ_1., introduces the larger differences. In general, k is underestimated and K_m is overestimated if no precautions are taken to consider the true mathematical model of the system.

Although known, diffusional limitations are also a factor that must be taken into account if the true model is to be used and thus, representative results are to be obtained. Special care should be taken in assuring a large value for the Thiele modulus.

9. Notation

D Characteristic catalyst pore diameter
K_m Michaelis-Menten kinetic constant
k Michaelis-Menten kinetic constant
K Kinetic apparent constant
e Effectiveness factor of the catalyst
q Recirculation flow-rate
L Catalyst characteristic pore length
PFR Plug-flow reactor
r Intrinsic reaction rate

S_0 Initial substrate concentration
S_1 Substrate concentration at the reactor exit
S Substrate concentration
STR Stirred tank reactor
Th Thiele modulus
Θ_2 Reservoir residence time
Θ_1 Plug-flow reactor residence time
V_{rat} Reactor/reservoir volume ratio
V_1 Plug-flow reactor volume
V_2 Reservoir volume

10. References

(1) Whitesides, G.M., Wong, C-H (1985) Angew. Chem. Int. Ed. Engl. 24, 617-638 (and references cited therein).

(2) Jakubke, H.-D., Kulh, P. and Könnecke, A. (1985) Angew. Chem. Int. Ed. Engl. 24, 85-91.

(3) Akiyama, A. Bednarski, M., Kim, M.-J., Simon, E.S., Waldman, H., Whitesides, G.M. (1988) CHEMTECH (october), 627-634 (and references cited therein).

(4) Ford. J.R., Lambert, A.H., Cohen, W. and Chambers, R.P. (1972) Biotechnol. & Bioeng. Symp. No 3, 267-284.

(5) Vallat, I., Monsan, P., and Riba, J.P. (1986) Biotechnol. Bioeng. 28, 151-159.

(6) Clapés, P., Mata-Alvarez, J., Valencia, G., Reig, F., Haro, I., García-Antón, J.M. (1989) J. Chem. Technol. Biotechnol. 45, 271-278.

(7) Clapés, P., Mata-Alvarez, J., Valencia, G., Reig, F., Torres, J.L., García-Antón, J.M. (1989) J. Chem. Technol. Biotechnol. 45, 191-202.

(8) Clapés, P., Mata-Alvarez, J., Valencia, G., Reig, F., García-Antón, J.M. (1989) Biotechnol. Appl. Biochem. 11, 483-491.

(9) Mata-Alvarez, J., Labrés-Luengo, P. Clapés, P. (1987) Trends Biotechnol. 5(11), 294-296.

(10) Mata-Alvarez amd Llabrés, P. (1989). Ingenieria Química, 244, 213-215.

(11) Levenspiel, O. (1985). The chemical reactor omnibook. Ocatave Levespiel Publ. Corvallis, Oregon.

(12) Clapés, P. Valencia, G.,Garcia-Antón, J.M., Reig, F. Torres, J.L. and Mata-Alvarez, J. (1988) Biochim. Biophys. Acta. 953, 157-163.

PRODUCTION OF FUEL ALCOHOLS BY FERMENTATION: ENGINEERING AND PHYSIOLOGICAL BASIS FOR IMPROVING PROCESSES

G. GOMA, P. SOICAILLE, C. LAFORGUE, A. PAREILLEUX AND J. URIBELARREA
Institut National des Sciences Appliquees
Department de Genie Biochimique et Alimentaire, URA-CNRS 544
Complexe Scientifique de Rangueil
31077 Toulouse Cedex France

ABSTRACT. It is generally accepted that ethanol fermentation is limited by the inhibitory effect of the terminal product (ethanol). Similarly, acetone-butanol fermentations are limited by inhibition due to butanol and acetone. In both cases, the fermentation strategies of fermentation-extraction resulting from these observations when carried out do not give the expected gain in productivity since other inhibition factors are involved. These factors due to the presence of co-metabolites are even more inhibitory than the final products are not sufficiently understood and need to be studied and identified in a better way to have a rational approach in terms of microbial engineering. It is the aim of this lecture to discuss how it is possible to generate new processes.

Introduction: What Are The Problems?

THE INHIBITORY PHENOMENA IN ALCOHOLIC FERMENTATION

The general behavior of cell growth and production under inhibition by substrate and product is described in Figure 1. There is an inhibition by substrate and product. Obviously, the optimal trajectory to use, as best as possible for the cells, must be found by a strong work of process control. However, for optimization, a better knowledge of the inhibitions is needed. We studied the ethanol inhibition and showed that ethanol is not the only inhibitor. We found three kinds of inhibitors.

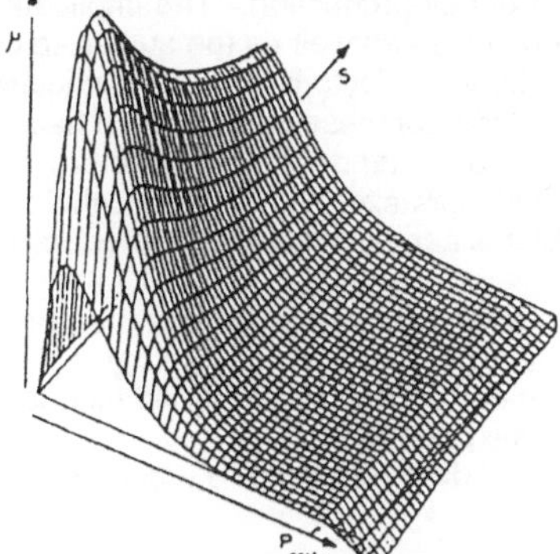

Figure 1. Product and substrate influence on the growth rate of *Saccharomyces cerevisiae*

A.R. Moreira and K.K. Wallace (eds.), Computer and Information Science Applications in Bioprocess Engineering, 435-470.

The global reaction of solvent production (ethanol alcoholic fermentation), acetone-butanol (acetonobutylic fermentation) can be represented by the following reactions.

$$\sum \alpha i S_i + \xrightarrow{Microorganisms} \sum \beta j Bj + \sum \gamma 1 P1$$

S, B, P are substrate (S), intermediary (B) and final product (P). The biomass is inhibited by S, B, or P. Synergistic inhibitions are observed which are now under identification. The mechanism of their effect must be understood.

THE MICROBIAL ENGINEERING APPROACH

Microbial utilization needs two complementary approaches. One is based on the identification of the limiting step and the optimization of the operating conditions. The other is based on the research of the higher concentration of active cells in the fermentor.

The first strategy needs studies on the limitations due to transport phenomena and to inhibitions by the substrate, end products and toxic by-products accumulated during cell growth. The second strategy uses cell recycle systems (cf. centrifuges, membranes, flocculation,...) or immobilized cells by adsorption on a packing or inclusion in a matrix. Some observations indicate that high cell concentration bioreactors are efficient but not proportionally to the increase in cell density. It is due to the inhibition of cells and decrease of their specific activity at high biomass concentrations. In another point of view, cell concentrations obtained in fermentations without cell recycle (3 g/l in anaerobiosis, 30 g/l in aerobiosis) are always lower than the theoretical maximum possible to reach: 140 g/l ± 20; so, by multiplying by 5 to 50, biological rates could be improved.

Cell density and specific rate. The specific rate γ (h^{-1}) of metabolism is defined as the instantaneous chemical reaction rate (g/l/h) per unit of cell dry weight (X/g/l). The biological activity can be expressed in terms of the rate of reaction r_C (g/l/h) for the component C (substrate S, end product P, cells X).

$$\gamma = \frac{r_c}{X} \quad \text{with } \mu = \frac{r_x}{X} \quad \text{(specific growth rate)}$$

$$\upsilon p = \frac{r_p}{X} \quad \text{(specific rate of product formation)}$$

For microbial growth limited by a single nutrient, Monod's equation is often sufficient to describe cell growth; however, a more accurate and general equation must be found. It is observed with recycled cells, a behavior described by Figure 2 which indicates beyond a critical cell concentration a strong inhibition of cell production. The implications of these observations are the cell concentration must be optimized as well as the substrate concentration in fed-batch. The deactivation or removal of inhibitory products is a priority objective for operating high rate biological processes. Of course knowledge of these inhibitors are necessary.

Our experience concerns on one hand, the lytic factors involved in acetonobutylic fermentation and on the other hand inhibitors excreted during alcoholic fermentations.

Mass and heat transfer problems can also occur due to velocity changes.

Cell Recycling and High Cell Concentration. In order to increase productivity significantly, it is necessary to remove both end inhibitor product P and inhibitory metabolite, and to increase the cell density. The achievement of high cell density is performed technologically by immobilizing or by recycling, which proceed of the same approach.

To accomplish this cell recycling devices generally used are described in Table 1.

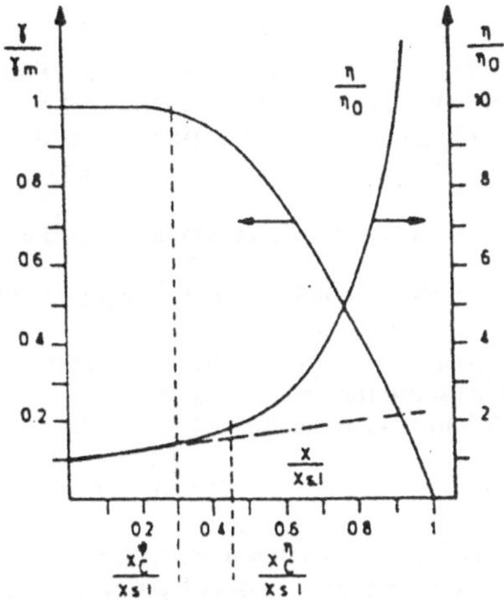

Figure 2. Relative specific rate of reaction γ/γ_m, and relative viscosity η/η_0 versus relative cell concentration X/X_0.

Table 1. Fermentation with cell recycling : cell recycling devices

	Cell density X g/l (dw)	Remarks
Immobilization	30 ± 10	adsorption inclusion clear medium are needed, limitaiton by mass transfer.
Flocculation	30 - 80	internal or external to the bioreactor simple technology.
Centrifugation	100 - 130	efficient maintenance of centrifuge
Membrane processes		important energetical cost of operating
a. microfiltration	> 100	attractive ; but severe membrane plugging.
b. ultrafiltration	> 100	attractive, but membrane plugging remains a bottle neck (though less than a)
c. dialysis		same remark that a limit of utilization with organism sensible to shearing
d. hollow fiber		Same remark than a)
Foaming		foams are enriched in cells of product.

CONCLUSIONS

The main progress in fermentation in the future will be obtained by introducing the concept of fermentation detoxification where unconventional inhibitors will be destroyed on line during the fermentation. This is possible by use of specific physiological properties or/and some physics (heat, adsorption, ...) treatment of fermentation broth separated during fermentation of the cells by the cell recycling operator.

Obviously, the technological concepts of fermentation developed in this program will be:

a) Fermentation cell recycling (membrane process, centrifuge flocculation)
b) Fermentation - cell recycling - detoxification
c) Extractive fermentation- cell recycling (the extraction of product is possible by treatment only on the fermentation broth).
d) Extractive fermentation - cell recycling-detoxification

Bioreactors of new generation are b,c,d types. Of course an optimal control is needed: more seriously than in classical fermentation. Loop reactors have a great future.

Connected with strong physiological studies these approaches seem to us promising.

We shall now explain what was our strategy to develop new processes and new ways of fermentation by searching mechanism of some bottlenecks and technological solutions which eliminate them.

PART 1: Ethanolic Fermentation

Introduction

The scientific literature contains an abundance of kinetic studies of the alcoholic fermentation of yeast. Both growth and fermentation end-product yields have been analyzed in relation to physico-chemical modifications to the culture medium brought about by accumulation of either end-product i.e. ethanol [12,24], or co-metabolites i.e. short-chain fatty acids [8,19], organic acids [18] and heavy esters [13]. The fermentation models resulting from such studies emphasize the role of carbon substrate limitation, viewed principally from the viewpoint of inhibitory phenomena.

The toxic effect of alcohol supplements on growth has never been sufficiently extreme to account for the reported inhibition encountered in the course of a typical fermentation [16]. It has likewise been demonstrated that the rate at which fatty acids, especially octanoic acid, accumulated in the culture broth cannot account for the decreased specific rates observed during fermentations [7]. The dynamic approach based upon the idea that the intracellular accumulation of endogenous ethanol takes place until a critical level is reached after which denaturing occurs does not take into account the reversible nature of the inhibition. Furthermore comparative kinetic studies of ethanol production and diffusion have made conclusions based upon intracellular ethanol levels controversial [2,9,11].

Identification of Inhibitions

ETHANOL INHIBITION

The medium described (enabled fermentations to occur in which total substrate consumption was possible after 10 hours with the production of 6.6 g/l biomass and an ethanol yield of 0.45 g/g. The viability of the cells remained 99% and the pH fell from 3.8 to 2.9. Maximum productivity was observed at the end of the culture and attained 6.9 g/l/h.

The specific production rate, v_p, remained fairly constant at 1.3 h^{-1} during the first 8 hours and then decreased, but as soon as 4 hours into the culture the growth rate (μ) began to decline progressively (Figure 3). The inhibitory effect of ethanol (P) on growth, quantified by supplementing fresh medium followed the linear relationship $\mu = \mu_m\left\{1-{}^l k_p(P)\right\}{}^l k_p$ = 0.0096 l/g.

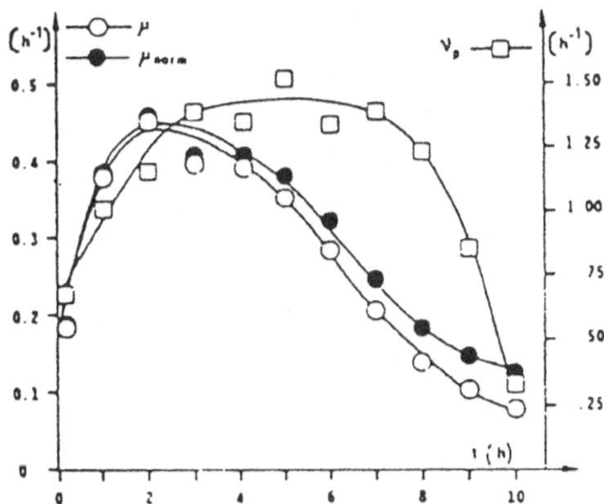

Figure 3. Evolution of the specific rates of production (vP) and growth (μ) during an ethanolic fermentation of *Saccharomyces cerevisiae* in batch culture. The specific growth rate corrected for the toxic effect of ethanol (μ_{norm}) is also shown.

Correction of the dynamic growth rates observed during fermentation to take into account this effect demonstrated that the diminished rates of growth observed after 4 hours of growth could not be explained by the toxicity of the ethanol produced (Figure 3)

FATTY ACIDS INHIBITION

After concentration of supernatants of culture, an extraction by ethylic ether was made for analysis of hydrophobic metabolites. After separation by TLC (chloroform/methanol/water) an inoculated growth medium agar (7%) was sprayed onto a plate. After incubation, the no-growth zones were observed and corresponding metabolites identified. Inhibitory effects were for short chain fatty acids (C_6, C_8, C_{10}) and phenyl-2-ethanol [4]. The other identified compounds (principally the n-butyl phenyl sulfonamide) were as a result of impurities within the growth medium constituents (the dynamic verification of their evolution during fermentation has been performed (See Appendix B).

The concentration of phenyl-2-ethanol reached at the end of the fermentation (on average 50 mg/l) did not agree with a significant inhibition. For the fatty acids, the kinetics of accumulation followed a linear relationship between production rate and acceleration or deceleration of growth (Figure 4). The excretion of fatty acids appeared when a net decrease of growth occurred by limitations (oxygen, carbon substrate).

The potentiality of inhibition of accumulated fatty acids agreed with the following relation: $\mu={}^*\mu_m\left\{1-{}^l k_{FA}(FA)\right\}$ and a reduction of the biomass yield. The evolution of the parameter ${}^l k_{FA}$ versus the ethanol concentration was in accordance with a synergy of inhibition between ethanol and fatty acids (Figure 5) [26]. However, a linear consumption of these added fatty acids was observed during aerated or non-aerated fermentations [28].

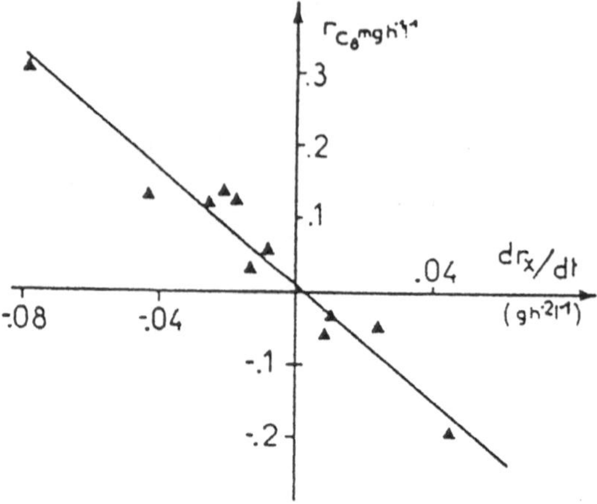

Figure 4. Linear relationship between the accumulation rate of octanoic acid (rC_8) and the acceleration of the growth dr_x/dt.

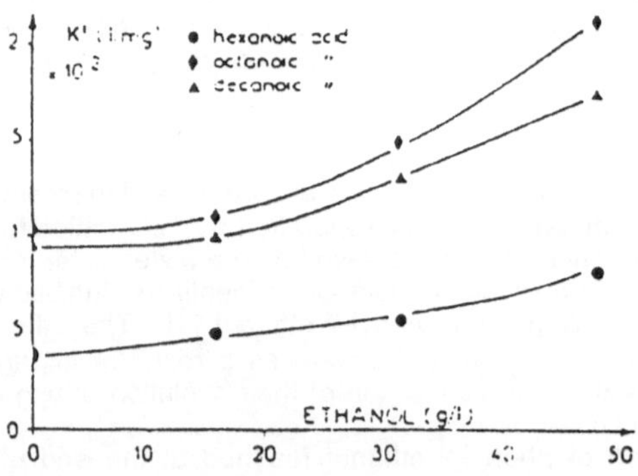

Figure 5. Inhibition by short-chain fatty acids: $\mu = {}^{*}\mu_m\left\{1 - {}^{l}k_{FA}(FA)\right\}$. Evolution of the inhibition constant ${}^{l}K_{FA}$ versus the ethanol concentration (P) for hexanoic, octanoic and decanoic acid.

EXTRACTIVE FERMENTATION

Batch cultures were performed with dynamic extraction (liquid-liquid extraction) of hydrophobic compounds so as to maintain their concentration below an inhibitory level and to shift the intra-exo cellular equilibrium (Figure 6) The residence time of extraction did not permit a decrease of ethanol concentration able to modify the kinetic parameters.

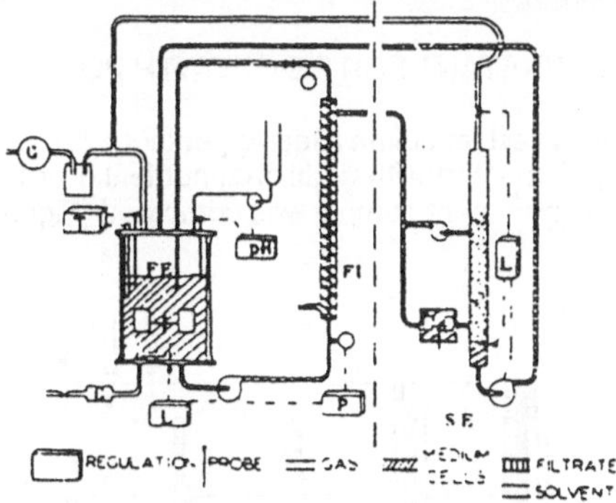

Figure 6. Experimental set-up for discontinuous cultivation with cell recycle by tangential microfiltration and liquid-liquid extraction.

No significant increase of productivities, yields or specific production rates (biomass, ethanol γ_p) were found. The fatty acid concentrations for the extractive fermentation were, on average, a hundred times smaller than those observed during a control culture (C_6: 0.6 mg/l, C_8: 1.96 mg/l, C_{10}: 0.2 mg/l) (Table 2). It was also confirmed that these fatty acids could not explain this discrepancy. Based on their kinetics of excretion resulting from a deceleration of growth, the nutritional potential of the medium was evaluated to determine whether or not nutritional limitations might contribute to the otherwise unexplainable low dynamic growth rates.

TABLE 2. Ethanol and Biomass Productivities for Batch Cultures with Fatty Acid Extraction

SOLVENT	ETHANOL PRODUCTIVITY g.l-1.h-1	BIOMASS PRODUCTIVITY g.l-1.h-1	um h-1	vpm g.g-1.h-1
HEXANE	1.8	0.19	0.38	0.95
DODECANOL	1.85	0.19	0.38	1.01
CONTROL	1.84	0.2	0.37	0.98

Characterization of Dynamic Limitations

One of the remarkable findings associated with dense cell cultures (achieved by coupling fermentation to membrane filtration techniques) has been the ability of this process to obtain extremely high cell concentrations (>200 g l^{-1}); significantly superior to biomass levels reported for batch fermentations. In addition, the application of various high dilution rates at comparable levels of both ethanol and carbon substrates has enabled specific parameters to be increased, and improvements have been shown to be proportional to the dilution rate [15].

This work has been interpreted as a removal of inhibition attributed to toxic co-metabolites due to the rapid replenishment of the fermentation medium, but by lifting such restrictions the problem of eventual dynamic limitations by nutrients other than the carbon substrate deserves consideration.

ANALYSIS OF CULTURE SUPERNATANT FOR RE-GROWTH POTENTIAL

The ability of a standard yeast inoculum (See Appendices A and C for details) to grow on supernatants (as taken, or supplemented with additional nutrients) or fresh medium adjusted to pH or ethanol concentration of supernatant sample was compared (Figure 7).

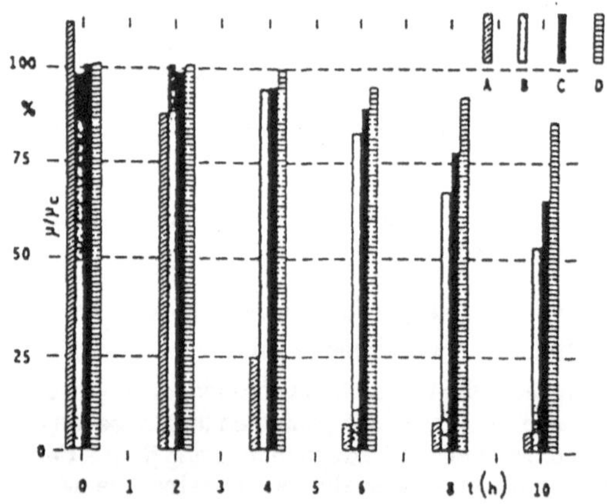

Figure 7. The average growth rates of a standard yeast inoculum in media taken at (or representative of) various times throughout a batch fermentation: A, "raw" supernatants; B, supernatants supplemented with vitamins, salts and casein hydrolysates; C, fresh media plus fermentation levels of ethanol; D, pH adjusted medium.

After supplementing the supernatant with amino acids, vitamins and salts (7B), growth rates comparable with those obtained for fresh medium adjusted to the same pH (7D) or ethanol (7C) were obtained throughout the fermentation time course. This progressive decrease can be attributed to the physico-chemical modification of the growth environment due entirely to the change in medium acidity and ethanol concentration.

However, comparison of the data obtained for "raw" supernatants (7A) and the supplemented supernatants (7B) clearly show that over the time course of the fermentation the medium became depleted in essential nutrients to such a level that *de novo* growth was greatly diminished. Moreover, this limitation became apparent in samples taken as soon as 4 hours of fermentation. By varying the additions made to supernatant samples taken over the course of a fermentation, the limiting nutrients were identified (Figure 8). Supplementing the samples with vitamins (8B) enabled growth of the standard inoculum to be restored to the same level as for addition of all nutrients (8B) indicating that vitamins alone were responsible for the nutrient limitations. Addition of biotin (8C) lifted the limitation apparent after 4 hours of fermentation and revealed a depletion after 6 hours of growth which could be overcome after adding either meso-inositol or nicotinic acid. Whilst it is true that these nutrient limitations could be described as regards the ability of supernatants to support the growth of a standard inoculum, it is no less true that within the fermentor growth continued despite the depleted vitamin level (Figure 9)

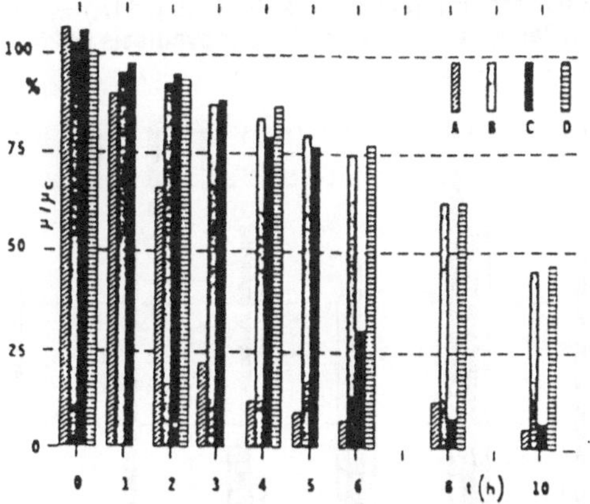

Figure 8. Average growth rates of standard yeast inoculum in fermentation supernatant samples taken throughout a batch fermentation: A, "raw supernatant; B, vitamin supplemented supernatant; C, Biotin supplemented supernatant; D, biotin and meso-inositol or nicotinic acid supplemented supernatant.

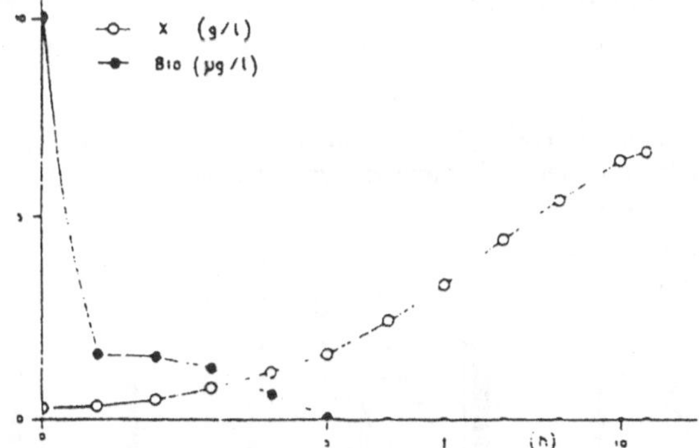

Figure 9. Evolution of residual biotin (Bio) and biomass (X) concentrations during batch fermentation.

The supernatant taken after 4 hours growth was able to support an initial growth rate of only 24% of that obtained with fresh medium, yet the original culture sustained a growth rate of 92% after 4 hours growth. It was therefore evident that the nutritional status of the medium cannot explain the observed kinetic behavior of the cells during alcoholic fermentation, and that the metabolic potential of the cells themselves required study.

CELL ANALYSIS

The kinetic behavior of yeast cells over the time period of the fermentation was analyzed by determining their growth rate (Figure 10) and their multiplication factor (Figure 11) after transfer to fresh medium diversly lacking in nutrients relative to these same parameters after transfer into complete medium. No matter when the cells were taken, an absence of casein hydrolysates reduced the growth rate by approximately 10% (10A). This was explained in view of further results with vitamins to the presence of 0.055 ug biotin per 10 g casein. The more extreme fall in the multiplication factor (on average 25%; Figure 11A) was most probably a result of the diminished direct assimilation of amino acids for cell synthesis [3,26].

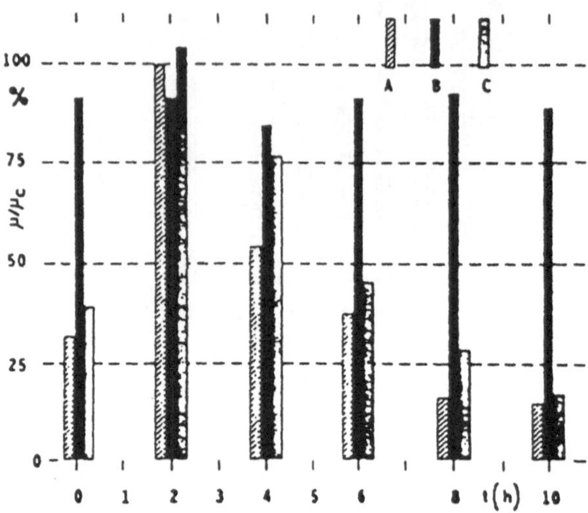

Figure 10. Kinetic behavior of cells taken at periods from a batch fermentation and transferred into fresh medium lacking nutrients: A, medium minus casein hydrolysates; B, medium minus all vitamins; C, medium minus biotin

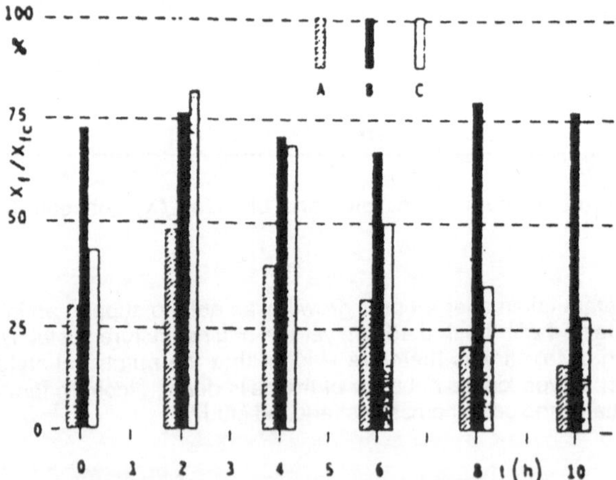

Figure 11. Evolution of the cell multiplication factor of cells taken throughout a batch- mode alcoholic fermentation and transferred into fresh medium lacking certain nutrients as described in Figure 9.

Despite the absence of biotin in the growth medium, the cells retained a growth potential which in terms of growth rate relative to the maximum growth rate (10C) was identical to that observed within the fermentor at the time of sampling. In light of these results, it must be supposed that an accumulation of intracellular biotin took place during the early stages of the fermentation leading to the deplete levels in the supernatants described above. The assimilation kinetics can thus be portrayed as an acceleration phase up to a point which a sufficient intracellular biotin concentration was reached to enable growth to occur at the maximum rate, followed by a progressive decrease in the growth rate related to the fall in intracellular biotin levels due probably to a dilution effect brought about by cell division. The kinetic characteristic obtained in the absence of all vitamins (10B) enables this hypothesis to be extended to cover other vitamins essential for growth, and in particular meso-inositol and nicotinic acid.

The cellular multiplication factor portrayed in Figure 11C shows that as early as 2 hours into the culture accumulated biotin levels within the cells was sufficient to assure the formation of 82% of the total biomass formed i.e. a multiplication factor close to 8. The lower vales obtained in the absence of all vitamins (11B) in which average multiplication factors representing 60% of those obtained in medium lacking only biotin were found, imply that vitamins other than biotin are not as essential for biomass synthesis.

It would appear that the kinetics of cell growth exaggerate the effect of ethanol and are principally a result of the intracellular vitamin concentrations and moreover such levels are reached in the first two to three hours of fermentation. Biotin has an important role in cell development and the estimated value of 1.8 ug biotin per gram biomass was compatible with the biotin levels of yeast.

BIOTIN ASSIMILATION AND GROWTH

Since biotin dependence described above resulted in depletion of the medium, the effect of initial biotin concentration on growth parameters was investigated under conditions whereby other vitamins were present in excess (See Figure 12).

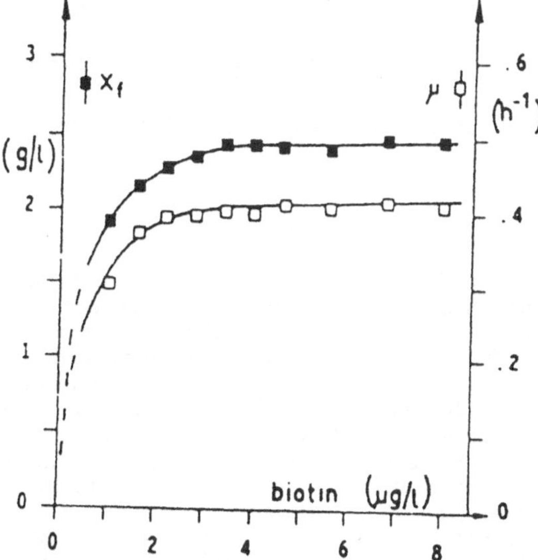

Figure 12. Biomass production (X) and specific growth rate (μ) versus the initial concentration of biotin (Bio).

The results obtained showed that the relationship between the maximum growth rate and the final biomass concentration persisted no matter what the initial vitamin concentration, confirming the dependence of these two parameters upon the accumulation capacity. Initial biotin concentrations in excess of 3-4 ug/l were sufficient to ensure maximum rates of growth and biomass levels, indicating that these levels were adequate to saturate the yeast. No further accumulation took place at higher biotin concentrations. Since at biotin levels in excess of 4 ug/l, the medium was presumably not depleted it must be asked why the saturating vitamin concentrations ensuring optimal growth could not be maintained throughout a larger period of the fermentation. It must be assumed that accumulation of vitamins was restricted to a specific phase of the culture determined by the physiological condition of the cells at the time of inoculation and that this state was altered when the biomass was biotin saturated.

In the first investigation, the effect of ethanol accumulation upon the capacity of biotin assimilation was quantified (Figure 13). The fraction of assimilated biotin lowered with the increase of ethanol concentration and the uptake of vitamin was stopped for a level of alcohol near 80 g/l.

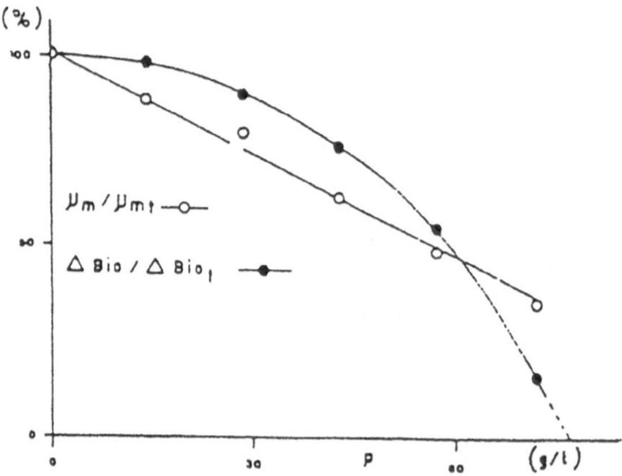

Figure 13. Specific growth rate (μ) and fraction of assimilated biotin)Bio) versus the ethanol concentration (P)

DISCUSSION

The growth kinetics of the yeast *S. cerevisiae* during alcoholic fermentation seem to be controlled by two distinct phenomena. The first is the well-documented modification of the physico-chemical properties of the fermentation medium by ethanol (and other inhibitory co-metabolites) and acidification. However, these factors alone cannot explain the phenomena at the poor ethanol concentrations obtained (50 g/l). It remains true that a critical threshold limit exists for ethanol above which growth is impossible, but this has been estimated to be in the order of 85 g/l for the strain used in this study. The second factor which exerts an influence on growth might be less predictable since it relates to the degree of intracellular accumulation of vitamins (especially biotin) associated with the early stage of growth and most probably the physiological state of the inoculum. Presence or absence of vitamins within the medium following the brief period of vitamin uptake cannot alone be used to stimulate the extent of growth. However, the relationship between growth rate and final biomass concentration is relatively predictable. The biochemical mechanism with the cell which controls vitamin levels and effects a halt to uptake following saturation of the biomass would appear to be a physiological limitation to the alcoholic fermentation.

Chemostat cultures maintained with high biotin levels (100ug/l) and under high aeration rates so as to ensure an oxidative metabolism instead of ethanolic fermentation attain high biomass levels [17]. Biochemical analysis has revealed that biotin-fixing capacity of yeast is linked exclusively to the concentration of the apoenzyme form of pyruvate carboxylase. The amount of the enzyme (either as apoenzyme or haloenzyme) remains constant for various concentrations of biotin varying between 0.5 and 15 ug/l [5]. According to these authors, the biotin-fixing capacity of vitamin deficient cells was in the order of 4.7 ug/g cells if a protein composition of 50% is assumed. This would correspond to total depletion of the medium under the inoculum conditions used here if biotin concentrations of 2.4 ug/l or less were used. It appears that the physiological state of the cells during oxidative respiration might allow uptake of the biotin, but under batch conditions this phase is short-lived due to a shift to fermentative metabolism in which uptake is subject to substrate repression by the high residual substrate levels. Only when the residual substrate has fallen below a critical level will biotin uptake be derepressed.

These speculative explanations of the mechanisms interacting to regulate vitamin accumulation provide a satisfactory model of cell behavior during fermentations using recycled cells. The continuous nature of the fermentation maintains a residual sugar concentration sufficiently low to lift the metabolic repression. The aeration conditions found necessary to maintain ethanol production are also those suitable for vitamin accumulation. The retention of the biomass enables dilution rates superior to the growth rate to be used and under such conditions low biotin levels in the medium are sufficient to meet the demand. Such systems establish conditions in which growth rates and as consequence specific growth rates can be improved since limitations are removed from the accumulated biomass [15].

Modelization of Ethanolic Fermentation

DYNAMIC NUTRITIONAL LIMITATION

For the batch alcoholic fermentation, the accumulation of end product (ethanol) and co-metabolites (short-chain fatty acids) alone cannot explain the growth kinetics of the yeast Saccharomyces cerevisiae at low ethanol concentrations. As demonstrated during the lag phase, a rapid depletion in biotin occurs resulting in the intracellular accumulation of the vitamin that in the coenzyme form controls the specific growth rate [27].

A mechanistic model based on inhibition and dynamic limitation phenomena is proposed to predict the behavior of the growth related to the medium status (Figure 14)

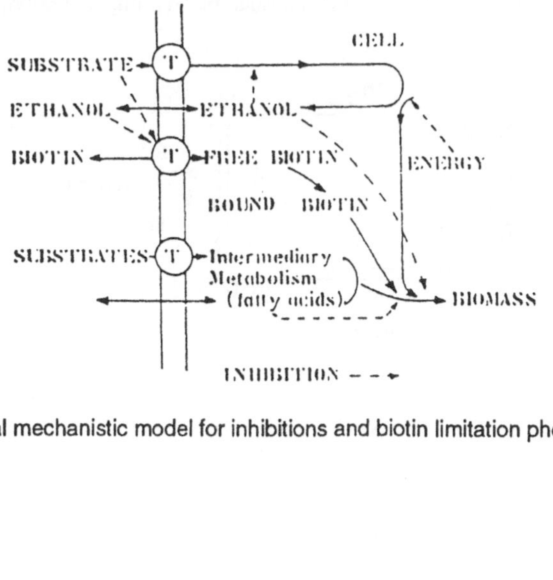

Figure 14. Verbal mechanistic model for inhibitions and biotin limitation phenomena.

Ethanol inhibition. The inhibition of the growth by the end-product is quantified by kinetic studies with added ethanol. The following linear models are retained to describe the ethanol (P) inhibition and the carbon source (S) limitation for the biomass and alcohol production:

$$\frac{r_X^*}{X} = \mu_{\max} S \frac{\left(1 - P/{}^{x}P_c\right)}{(k_s + S)} \tag{1}$$

and

$$\frac{r_P}{X} = \upsilon_{\max} S \frac{\left(P/(P_{k_1} + P) + P_{k_2} - P/P_{P_c}\right)}{(k_s + S)} \tag{2}$$

The equation 2 expresses a weak inhibitory effect below a concentration of 20 g/l followed by a linear inhibition. The productions (biomass and ethanol) stop when the alcohol concentration reaches a critical value (${}^{x}P_c$ and ${}^{P}P_c$) about 80 g/l.

Biotin limitation. The mathematical model that describes the dynamic limitation by the biotin includes a new physiological variable (a bonded biotin) and supposes the control of the specific growth rate by the biotin dependent holoenzymes [17].

A simple kinetic model of active transport (Figure 15) accounts for the several observations related to the biotin assimilation: kinetics, intra and extracellular biotin reation (Eq) to the equilibrium, ethanol effect on the partition coefficient. So the variation of the outer (B_o) and inner (B_i) biotin concentrations may be written:

$$dB_o/dt = X[B_i\text{-}EqB_o(1\text{-}P/{}^{b}P_c)]/[a+bB_i+cBo(1\text{-}{}^{b}P_c)+dB_iB_o(1\text{-}P/{}^{b}P_c)] \tag{3}$$

where a, b, c, d are linear functions of the kinetic constants K^1, K^{-1}, K^2, K^{-2}, D^1, D^{-1}, $D2$, D^{-2} Figure 15 and

$$dB_i/dt = (dB_o/dt)(1/V_c)\text{-}uB_i\text{-}r_{Bfix} \tag{4}$$

where V_c is the cellular volume, r_{Bfix} is the conversion rate of the free biotin into bonded biotin (B_{fix}) and the term uBi refers to the dilution of intracellular biotin brought about by cell division.

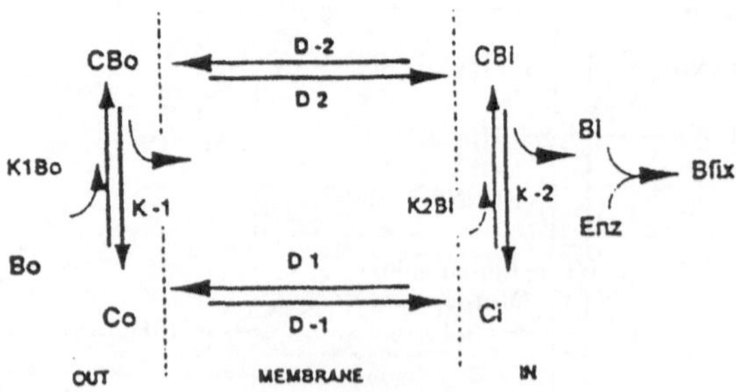

Figure 15. Model for biotin transport and bonded biotin conversion

The level of bonded biotin results of a two substrate enzymatic reaction (intracellular biotin and apoenzyme (Enz))

$$r_{Bfix} = AB_iEnz/(B+CB_i+dEnz+EB_iEnz) \quad (5)$$

with Enz = $Enz_{max}B_{fix}$ where Enz_{max} is the total fixation biotin capacity per cell. Model's parameters for the kinetics of biotin assimilation are estimated from experimental results to fit with the evolution of exocellular biotin and the Monod growth model.

As shown in Figure 16, the corrected ethanol inhibition specific growth rate (μ^*) versus intracellular biotin fits with a Monod model and it appears a minimal concentration (B_{min} = 0.66 ug/g) for that a growth stops.

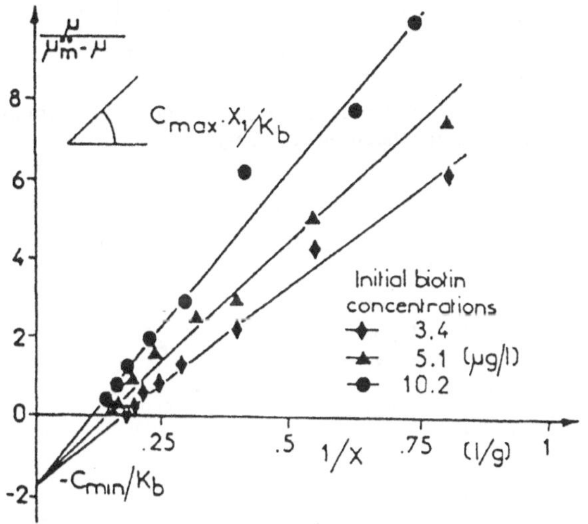

Figure 16. Experimental verification of linear proposed model for batch fermentation at several initial biotin concentrations. $A = \mu / \{\mu_{max}(1 - P/{}^{x}P_c)\}$, X: biomass.

$$r_s/X = -(r^{**}/X)(1/Y_x) - (r_P/X)(1/Y_P) - m \quad (6)$$

where Y_x and Y_p are the theoretical yields and m is the maintenance coefficient [21].

The numerical integration of the kinetic equations system permits a good prediction of the growth during batch fermentations (Figure 17). This mechanistic model can portray the acceleration phase connected with a biotin assimilation, the irreversibility of growth potential of cells ensures the cell proliferation in a depleted biotin medium (Figure 18).

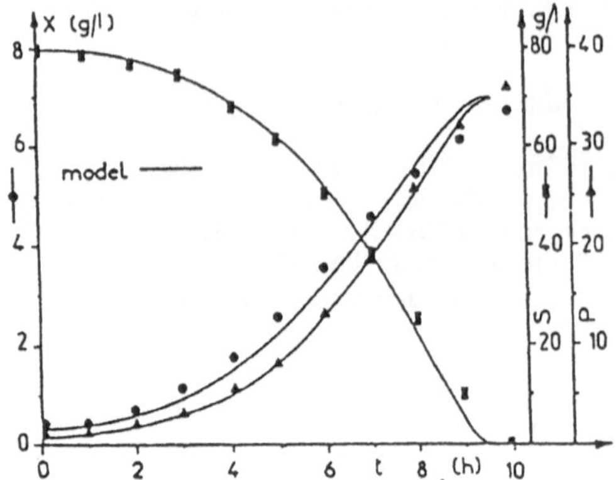

Figure 17. Experimental data and model simulation for batch fermentation (X, biomass; P, ethanol; S, glucose; initial biotin concentration 3.5 ug/l)

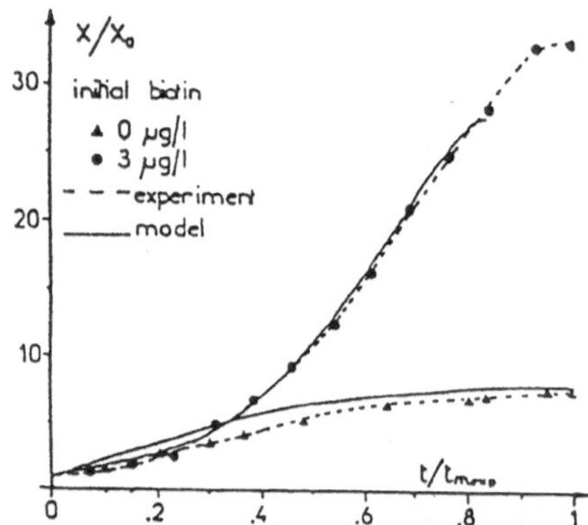

Figure 18. Experimental data and model simulation for two biotin concentrations. The growth was expressed as the ratio biomass/initial biomass and the reduced time was as the reaction time/final experiment time.

Cell recycle by tangential microfiltration. Continuous fermentation with recycle biomass (Figure 19) permits to reach important cell concentrations more than 300 g/l (dry matter) with an alcoholic volumic productivity higher than 80 g/l/h [15]. In such systems, the above mentioned model may be applied introducing the concept of available space, cell viability and steric limit for growth as a function of total cell concentration.

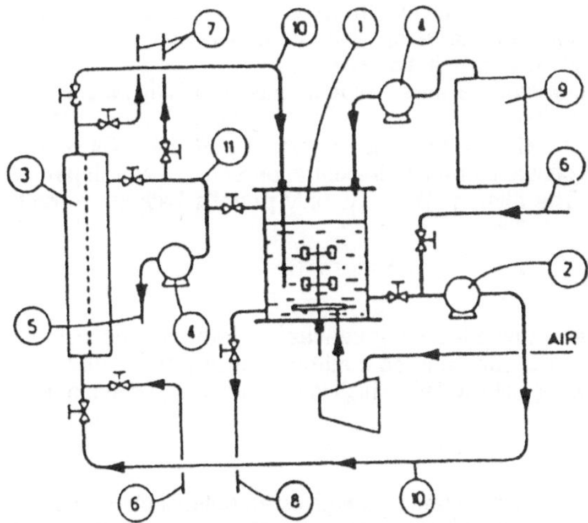

Figure 19. Cell recycle apparatus for continuous fermentation: 1, fermentor; 2, circulating pump; 3, microfiltration membrane; 4, peristaltic pumps; 5, filtrate outlet; 6, steam inlet; 7, retentate and filtrate bleeds; 8, sampling port; 9, feed tank; 10, retentate stream; 11, filtrate stream.

FERMENTATION BROTH VISCOSITY AND PERMEATE FLOW

Evolution of fermentation broth viscosity can be divided in three parts [14, 20]. For the first one, where biomass concentration is below 80±10 g/l and 280± 10 g/l, a second degree relationship between cell concentration and viscosity is suitable. Beyond a cell density of 280±10 g/l, lysis of a great proportion of yeast population increases apparent viscosity. Loose of filter specific permeate flow depends on viscosity (Figure 20), cell lysis, CO_2 production with membrane gas poisoning.

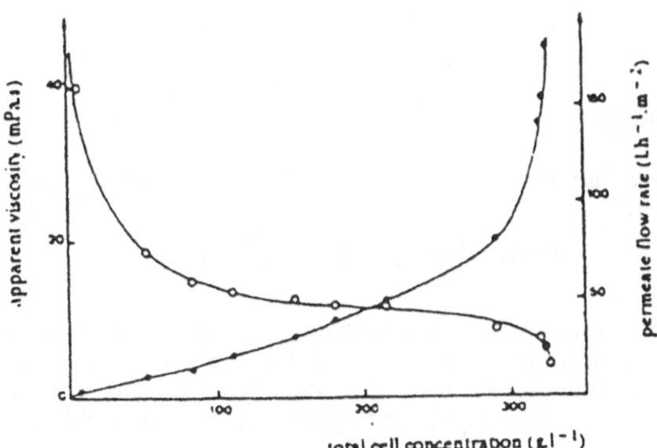

Figure 20. Apparent viscosity of fermentation broth (•) and permeate flow (°) versus total cell concentration.

In high cell concentration cultures, there are three classes of phenomena which contribute to the fall of specific parameters of substrate conversion, cell growth and metabolite: mechanic phenomena, inhibitions and limitation effects.

The synthetic medium used for these experiments is quickly insufficient when inoculum nutritive reserves are consumed that correspond to a total cell concentration of 30±10 g/l. Beyond this critical value, growth rate becomes constant and depends on dilution rate. Cell viability falls after a second critical biomass concentration value which depends on dilution rate and medium composition. The alimentation flow only permits keeping a fraction of the yeast population.

INFLUENCE OF DILUTION RATE

Dilution rate has two principal consequences on fermentation evolution. It fixes the alimentation flow of limiting substrates and so controls growth rate. It also fixes the viable cell concentration X_L obtained when viability falls which is proportional to dilution rate D:

$$X_L = D/K_{XL}$$

K_{XL} represents the value of the dilution rate which gives an optimal state to one gram of biomass per liter of fermentor. It is characteristic of medium nutritive capacity. But, in all cases, dilution rate depends on membrane plugging and technologic limits of the system.

KINETICS OF ALCOHOL PRODUCTION

For biomass concentration higher than 35±5 g/l, an exponential law is suitable to describe the influence of dilution rate and cell concentration on specific production rate:

$$\upsilon_p = \upsilon_{pm} \exp\left(-{}^{x}K_p X_v / D\right)$$

Before this value, biomass effect is no longer a preponderant parameter in the control of alcohol production and ethanol inhibition has a great influence. Ethanol specific production rate can be expresses by:

$$\upsilon_p = \upsilon_{pm}\left(1 - P/{}^{P}P_C\right)$$

KINETICS OF YEAST GROWTH

A first step in growth kinetic modeling is traduced by the following:

$$\mu = \mu_{\max}\left[S/(K_s + S)\right]\left(1 - P/{}^{x}P_c\right)$$

Inhibition effects linked to biomass are numerous. They include substrate sharing, toxic co-metabolite production and steric stress. One of the more critical problems is substrate limitation which is dependent on dilution rate and medium composition. Experimental results show a linear relation between the inverse specific growth rate and the viable cell concentration.

$$\mu = \mu_{max}(S/(K_S+S))(1-P/{}^{X}P_C)(K_X/(K_X+X_V))$$

To take into account inhibitory effect of steric stress on yeast growth at very high cell concentrations, another term is introduced. It shows the difference between viable and total cell concentrations and traduces the available space for yeast division.

$$\mu = \mu_{max}(S/(K_s+S))(1-P/P_{max})(K_x/(K_x+Xv)((X_M-X_T)/X_M)$$

EVALUATION OF CELL DEATH KINETICS

The death of a part of the cell population is observed when yeast concentration become higher than X_L (critical cell concentration which indicates apparition of dead yeast). Experimental results show that specific death rate of the yeast is proportional to the total cell concentration and seems to be dependent on the dilution rate.

$$D_r = (1/X_T)(dX_m/dt)$$

$$Dr = k_d\,(X_T-X_L)D$$

Model's parameters are deduced from experimental results that are available for the synthetic medium and for the yeast strain used. For a fixed dilution rate and a chosen bleed rate, the substrate concentration at the equilibrium is searched with the mathematical expressions of specific growth rate and ethanol production. The equation is solved by the Newton-Raphson method.

Simulations of the evolutions of ethanol, glucose, viable and total cell concentrations, for a partial cell recycle fermentation, in confrontation with experimental results show a difference average less than 20 percent which indicates that this kind of model can be considered accurate to perform the simulations and dimensioning of the system (Figures 21,22)

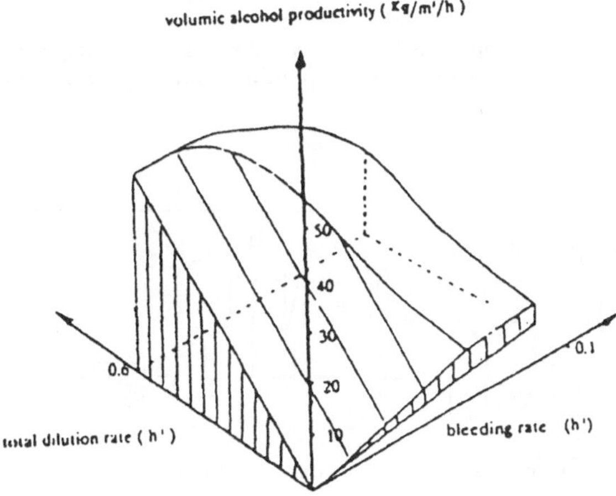

Figure 21. Continuous fermentation with cell recycle by tangential microfiltration. Ethanol productivity versus a total hydraulic dilution rate and the bleeding rate of biomass (Residual glucose concentration #0 for the cross hatched area)

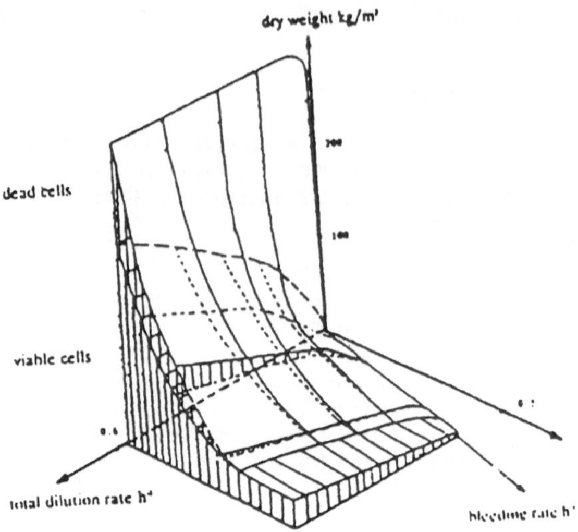

Figure 22. Continuous fermentation with cell recycle by tangential microfiltration. Viable and dead accumulated biomass versus a total hydraulic dilution rate and the bleeding rate of biomass.

Industrial Process of Ethanol Production by Cell Recycling

The development of the fermentation at high cell concentration by cell recycling with membrane process was performed on a real substrate with Beghin-Say (Ferruzi group) and Speichim (Figure 23)

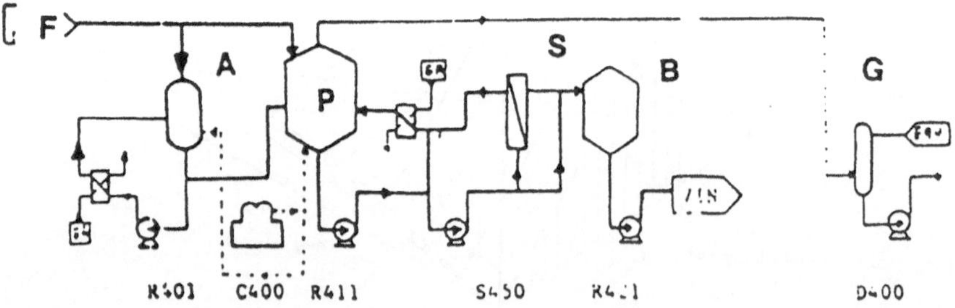

Figure 23. Improved process for rapid ethanolic fermentation (Beghin-Say, Speichim, INSA).
F, feed; A, aerated tank (ethanol dilution-cofactors assimilation); P, production tank; M, tangential microfiltration membrane; B, buffer tank; S, biomass recycle loop; G, gas treatment.

Evaluation of the processes are given in Table 3 for different raw materials.

TABLE 3. Results obtained with the Beghin-Say Speichim INSA process

SUBSTRATE	BEET MOLASSE	LOW GREEN SYRUP	CORN MALTODEXTRINE
Residence Time (h)	4.0	2.5	2.0
Ethanol Concentration (%V/V)	7.5	8.5	10.0
Yeast Concentration (Kg DW / m3)	30	60	85
Ethanol Productivity (Kg/m3.h)	15	27	40

Comparison between the behavior of the high cell bioreactor and the classical Speichim process is given in Table 4.

TABLE 4. Performance of Beghin-Say Speichim INSA process

	Standard Process (50% of cell recycling)	High cell concentration culture process
Ethanol Concentration (%V/V)	8	7.5
Residence Time of liquid (h)	20	4
Dilution rate (h-1)	0.05	0.25
Average residence time of yeast (h)	40	40
Yeast Concentration (Kg DW / m3)	6.6	30
Ethanol Productivity (Kg /m3.h)	3.2	15
Specific growth rate (h-1)	0.025	0.025
Specific rate of Ethanol production (h-1)	0.49	0.5
Yeast productivity (kg DW /m3.h)	0.15	0.75
Yeast produced (Kg/m3 of Ethanol)	4	4

PART II: Acetonobutylic Fermentation

Introduction

Clostridium acetobutylicum has been used for the production of acetone, butanol and ethanol from various carbohydrate substrates. Current research on this fermentation is moving in three directions: development of continuous fermentation technology (2,6), improvement of strain tolerance toward butanol (7,9(and understanding of the physiological relationships between organic acid metabolism and solvent production (8,11,14,20). Another important aspect of this fermentation is the production of bacteriocins and specifically autobacteriocins which are substances produced by microorganisms and which have antibiotic-like activity against the producing strain. This production has been previously described for an industrial strain of *Clostridium acetobutylicum* (3). Both biomass and solvent yields are affected by this clostridial autobacteriocin and therefore its physiological role and associated economic implications need to be more fully understood.

Autobacteriocin Production and Properties

AUTOBACTERIOCIN PRODUCTION USING BATCH CULTURE.

The production of extracellular autobacteriocins by C. *acetobutylicum* was shown to begin toward the end of the exponential growth phase and continued until cell lysis began (Figure 24).

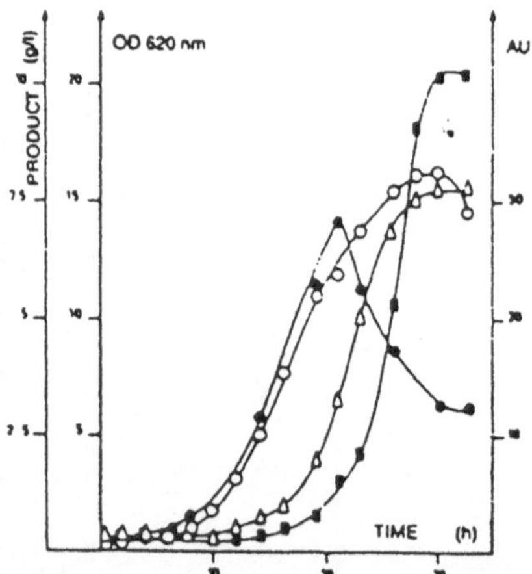

Figure 24. Production of butanol, butyric acid and autobacteriocin (AU) by C. *acetobutylicum* ATCC 824 during growth in a batch culture with 50 g/l of glucose as the source of carbohydrate and the pH controlled at 5.4. (∘) OD; ( ), butanol; (•) butyric acid; and (Δ) extracellular autobacteriocin.

EFFECT OF AUTOBACTERIOCINS SUPPLEMENT ON GROWTH OF *C. ACETOBUTYLICUM*

The effect of various concentrations of autobacteriocins (0-64 AU/ml) on the growth of C. acetobutylicum in the presence of various concentrations of butanol was studied. No significant effect linked to the autobacteriocin was observed for the exponential growth phase, but a more rapid lysis of the cells occurred at the end of the stationary growth phase. The extent of growth reached at stationary phase can be seen to be lower in the presence of additional autobacteriocin. The maximum specific rate of lysis during late stationary phase can be seen to be related to a synergistic effect involving both the concentration of butanol and the autobacteriocin level (Figures 25 and 26).

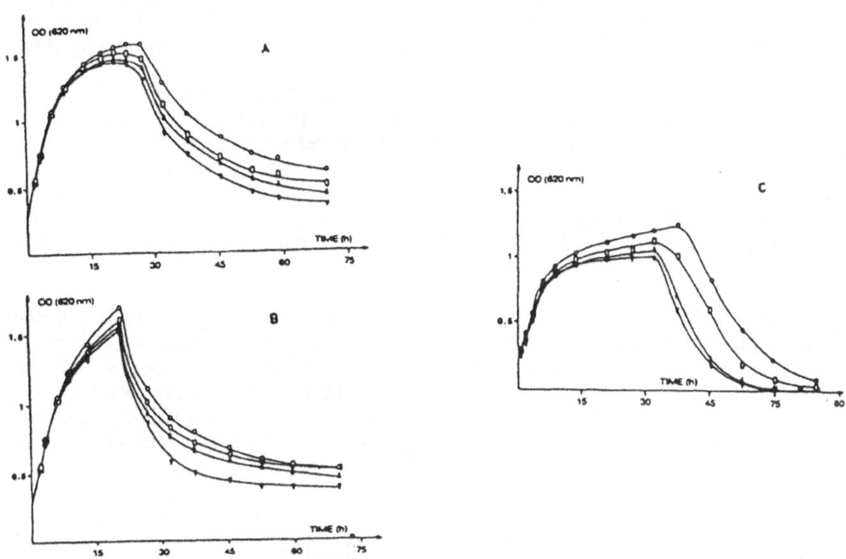

Figure 25. Effect of autobacteriocin on the growth of C. acetobutylicum during batch cultures in the presence of various concentration of butanol. (A) 0 g/l; (B) 5 g/l; (C) 12 g/l. (-○-) control; (-□-) 16 AU/ml; (-△-) 32 AU/ml and (-▽-) 64 AU/ml.

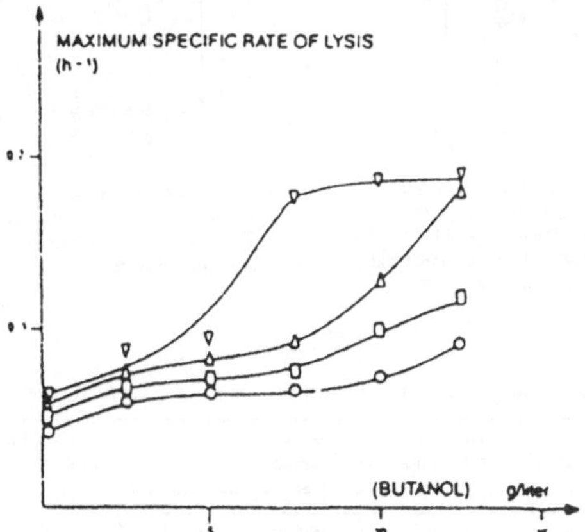

Figure 26. Effect of various concentrations of butanol and autobacteriocin on the maximum specific rate of lysis during the lysis phase of batch culture. Symbols as in Figure 25.

BUTANOL TOLERANCE AND AUTOBACTERIOCIN PRODUCTION

A strain was selected for its higher tolerance to butanol compared to the wild type ATCC 824. In consideration of the higher butanol tolerance a disappointedly low titer of butanol was produced by the new strain. Levels of autobacteriocin have been investigated for both strains.

Butanol resistant strain selection: A butanol resistant variant was obtained. This variant, G1, was tolerant to butanol concentrations of 18 g/l i.e. was able to maintain some growth and demonstrated Gram stain reaction and microscopic morphology similar to the parent strain, although this strain had acquired raffinose, sorbitol, rhamnose and trehalose fermenting ability.

Response of C. acetobutylicum to butanol. We characterized the behavior of *C. acetobutylicum* (wild type and G1) in response to a butanol challenge. The growth rate inhibition was an exponential function of butanol concentration between 5 g/l and 12 g/l for the ATCC 824 and 8 g/l and 16 g/l for G1. The specific growth rate of the parent strain was inhibited by 50% at 11.8 g/l butanol whereas 13.5 g/l butanol was necessary for this degree of inhibition for G1. At a butanol concentration of 14 g/l, the wild type had an apparent negative growth rate caused by cell lysis, whereas the resistant strain retained positive growth rates at butanol concentrations as high as 18 g/l.

Solvent and autobacteriocin formation during batch culture. The progression of cell density, glucose consumption and production of autobacteriocin and butanol were monitored during batch culture in 8% glucose synthetic medium for both wild type and G1 strains (Figures 27 and 28, Table 5). The maximum cell density, the final butanol and total solvents concentrations and the maximum autobacteriocin activity were higher for the variant than for the wild-type strain. The total yield of solvents was also better for G1 owing to an improved consumption of nutrients and acetic acids.

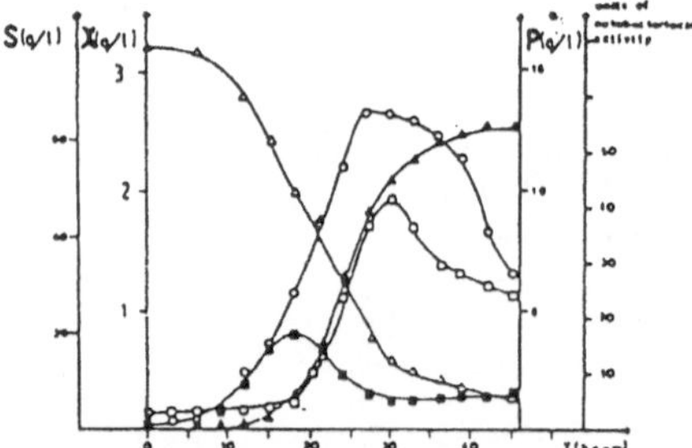

Fig. 27 Growth, products, and autobacteriocin formation of strain ATCC 824 on synthetic medium containing 8% (wt/vol) glucose. The initial pH was 5.5 and was regulated at 4.8: ○, cell concentration (scale X); △, glucose (scale S); ■, butyrate (scale P); ▲, butanol (scale P); and □, autobacteriocin activity.

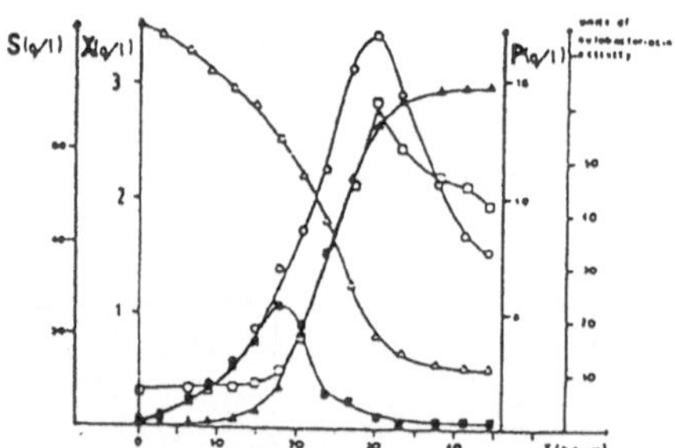

Fig. 28. Growth, products, and autobacteriocin formation of strain G1 on synthetic medium containing 8% (wt/vol) glucose. The initial pH was 5.5 and was regulated at 4.8. Symbols are the same as in Fig. 9

TABLE 5.

Comparison of *Clostridium acetobutylicum* ATCC 824 and G1 during batch cultures on synthetic medium

Strain	Glucose utilization (g · liter^{-1})	Butanol concentration (g · liter^{-1})	Acetone concentration (g · liter^{-1})	Ethanol concentration (g · liter^{-1})	Solvent yield[a]	Final butyric acid concentration (g · liter^{-1})	Final acetic acid concentration (g · liter^{-1})
ATCC 824	72 ± 1	12.6 ± 0.2	6.5 ± 0.3	0.8 ± 0.01	0.276	1.8 ± 0.1	1.7 ± 0.2
G1	75 ± 1	14.7 ± 0.2	7.4 ± 0.3	1.15 ± 0.01	0.31	0.3 ± 0.02	1.2 ± 0.2

[a] Calculated by dividing the grams of solvent produced by the grams of glucose utilized.

Correlation between produced butanol and the specific rates of growth and solvent production. **The plot of u versus the butanol produced (Figure 29) shows an apparent higher tolerance for G1 than ATCC 824. The growth rate of the ATCC strain remained positive at 13 g/l for butanol added prior to inoculation, but for butanol or butanol plus butyric acid (Table 6) produced during growth lower maximum values were reported. The differences were more important for G1 with 18 g/l for added butanol instead of 13 g/l for produced butanol and 13.6 g/l for produced butanol + butyric acid.**

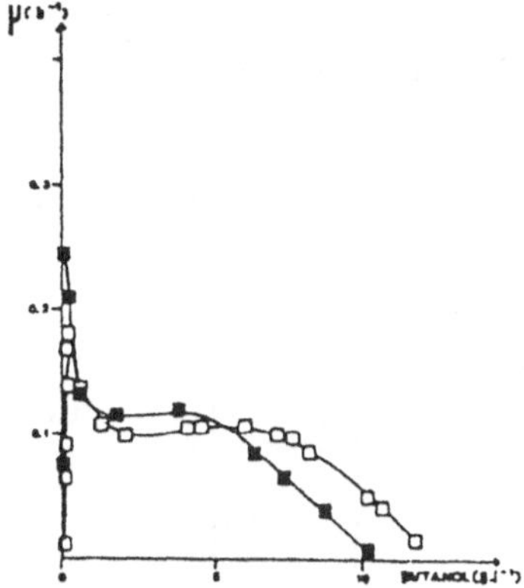

Fig. 29. Correlation between butanol produced during batch culture and the specific rate of growth of *Clostridium acetobutylicum* ATCC 824 (■) and G1 (□).

TABLE 6.

Concentration of principle parameters of the acetonobutylic fermentation at the point at which cell lysis begins

Strain	Biomass ($g \cdot liter^{-1}$)	Autobacteriocin activity[a]	Butanol ($g \cdot liter^{-1}$)	Butyric acid ($g \cdot liter^{-1}$)	Butanol plus butyric acid ($g \cdot liter^{-1}$)	Butanol tolerance[b] ($g \cdot liter^{-1}$)
ATCC 824	2.66	42	10.5	1.5	12	13
G1	3.4	62	13	0.6	13.6	18

[a] Expressed as the reciprocal of the highest dilution that did not give inhibition zone in the petri plate assay.
[b] Maximum concentration of butanol that permits growth during butanol supplement challenges.

Inactivation of the Autobacteriocin by Heat Treatment in Ultrafiltration Coupled Acetonobutylic Fermentation

EFFECT ON BIOMASS CONCENTRATION AND GROWTH

The evolution of biomass (dry weight) and residual glucose concentrations are represented against time for the two series of experiments, in Figures 30a and 30b, respectively. In the absence of glucose limitation for both pH 5.5 and 4.8 cultures, the cell concentration in the thermo-treated assays exhibited a maximal value about twofold higher than in the corresponding reference (7.1 g/l against 3.3 at pH 5.5 and 5.8 g/ against 2.8 at pH 4.8) for nearly the same duration of growth periods. If we consider the total liquid volume of the installation for thermo-treated experiments i.e. the broth plus the circulating ultrafiltrate, the calculated biomass concentration leads, after correction (multiplicative factor Ff/VT), to apparent maxima of 5.1 g/l for Tr.Th1 experiment and 3.9 g/l for Tr.Th2 corresponding respectively to 55% and 39% improvement compared with references.

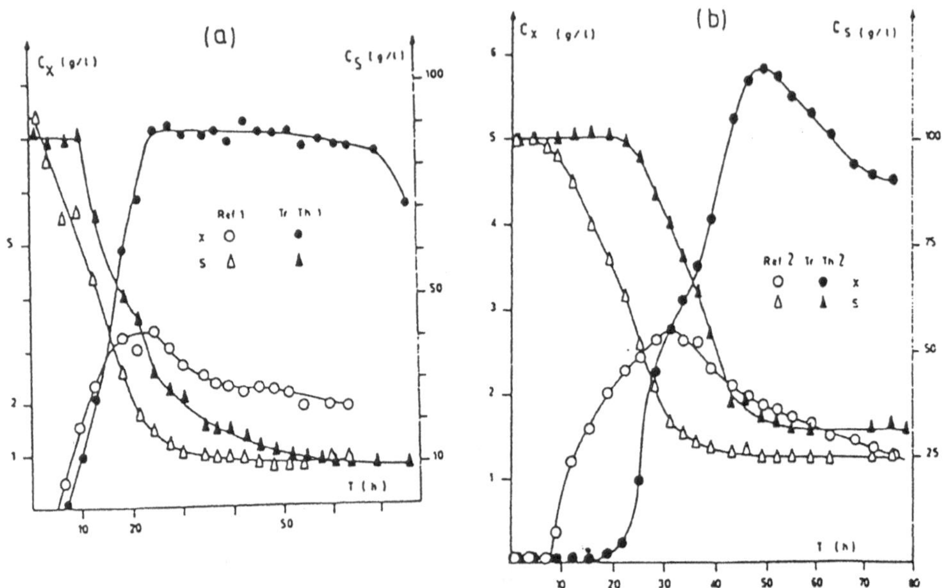

Figure 30. Variation in cell and glucose concentration versus time for control and heat-treatment experiments.(a) pH 5.5 and (b) pH 4.8.

These results are confirmed by the optical density evolutions shown in Figures 31a and 31b where the relationship with extracellular lytic activity is underlined: in the thermo-treated cultures, the increase of biomass production corresponded to a strong decrease of lytic activity.

The first conclusion of technological interest is the demonstration of the permeation of lytic enzymes through these UF membranes and their possible elimination from the fermentor, even by simple ultrafiltrate withdrawal in continuous mode.

In the present application of ultrafiltration, the separation of microorganisms allowed an effective detoxification of the culture by thermo-inactivating lytic enzymes in the permeate before recycling it. The reduction of the extracellular lytic activity permitted to maintain the growth rate and even to increase it during the fermentation, while it slowed down continuously in references. This prolongation with reactivation is pointed out in Figure 32 where evolution of u versus biomass concentration is represented . The growth rate was generally higher in treated cultures and a reactivation (secondary maximum) was observed for X= 2.8 g/l (39% of the X maximum value) and 23 UAA/ml. It can however, be noted that the rapid diminution of the growth rate during the first fermentation period was observed in all experiments, even when no lytic activity was detected. This early decrease could be caused by vitamin limitation (15) rather than inhibition.

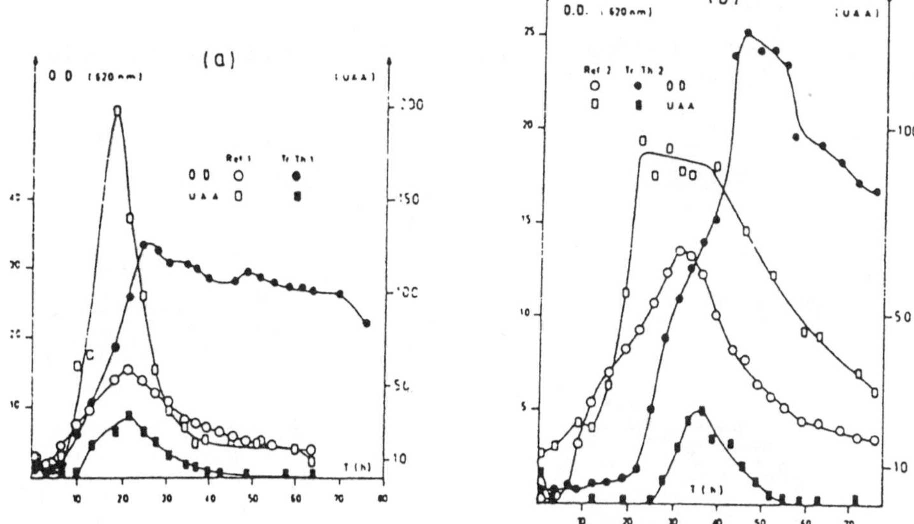

Figure 31. Variation in optical density and extracellular autolytic activity versus time for control and heat-treatment experiments. (a) pH 5.5 and (b) pH 4.8.

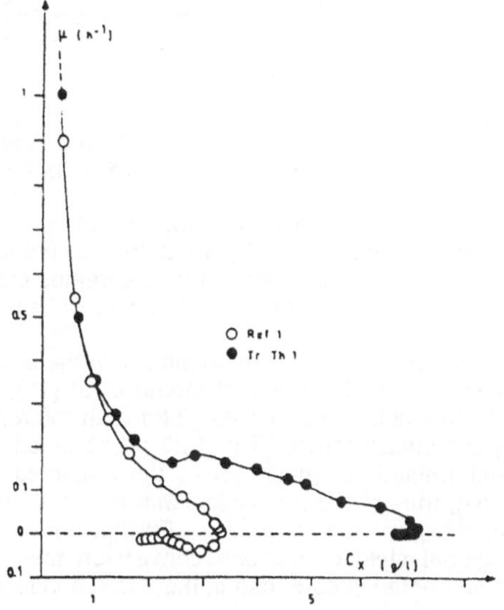

Figure 32. Variation in growth rate versus biomass concentration for control and heat-treatment at pH 5.5.

ACID AND SOLVENT PRODUCTION

The variations of butyric acid and solvent concentration in the broth are represented in Figures 33a and 33b, respectively.

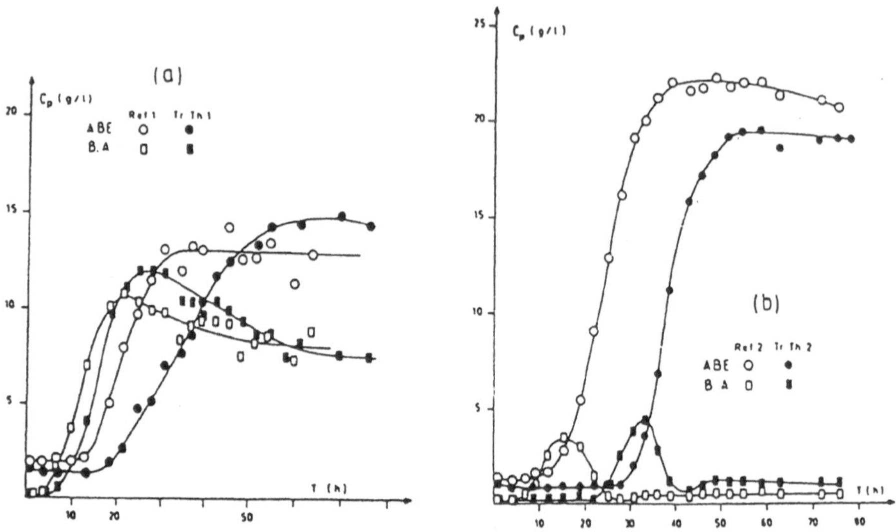

Figure 33. Variation in solvent (acetone, butanol and ethanol) and butyric acid concentration versus time for control and heat-treatment fermentations. (a) pH 5.5, (b) pH 4.8.

These results are commonly mentioned in the literature, with a relatively high production of acids at pH 5.5 and their quasi reconsumption at pH 4.8, the solvent formation being greatly enhanced at this lower pH (1,11). For pH 4.8 experiments, the treated culture presented a 35% enhancement of ABE production rate compared to the reference, while no significant difference was observed at pH 5.5.

The induction of solventogenesis appears more difficult in the case of thermo-treatment experiments (Figure 33); following the hypothesis of Monet et al (11), the evaluation of the concentration of non-ionized butyric acid leads to 1.25 g/l for both reference fermentations, but 1.35 and 1.55 g/ for the experimental assays. The 1.25 g/l obtained in the reference is in agreement with the 1.2 g/l determined previously (6) as the threshold induction value in UF continuous cultures. In contrast, the solvent production with thermo-treatment begins only for higher concentrations of uncharged butyric acids. Furthermore, the maximal butanol concentration reached the global yield of glucose conversion towards solvent were not necessarily higher in the thermo-treated culture than in the corresponding reference, in spite of the reduction of extracellular lytic activity.

Effect of Divalent Cations on Autolysin Production, Culture stability and Solvent Production in Chemostat Culture

EFFECT OF Mg^{++} ON METABOLIC ACTIVITY

It was mentioned that Mg^{++} played an important role in the tolerance of the culture to butanol (18). It was observed that increasing the concentration of magnesium led to an increase in both biomass and solvent productivities. However, beyond 2.5 mM Mg^{++}, there was no significant improvement in these parameters (Table 7).

TABLE 7. Influence of varying concentration of Mg^{++} on solvent productivity, biomass, glucose consumption in chemostat cultures of *Clostridium acetobutylicum* growing at a fixed rate of 0.26 h^{-1} (pH = 4.4, 35°C).

Concentration of Mg^{++} (mM)	Solvent productivity ($gl^{-1}h^{-1}$)	Biomass (gl^{-1})	Glucose consumption (gl^{-1})
0.8	1.5	1.84	20.38
2.5	2.02	2.28	25.8
4.0	2.06	2.35	26.3
6.0	2.07	2.40	26.7
8.0	2.09	2.46	27.0

EFFECT OF VARIOUS DIVALENT CATIONS ON METABOLIC ACTIVITY

Different divalent cations showed different effects on solvent productivity. Incorporation of Fe^{++} resulted in the maximum enhancement of solvent productivity. However, addition of Mg^{++}, Zn^{++} and Ca^{++} in the medium also showed significant improvements in solvent productivity. On the contrary, Mn^{++} at this concentration showed inhibitory effects as the biomass and solvent productivity were reduced remarkably compared to those of the control experiments.

INFLUENCE OF AMMONIUM IONS IN THE PRESENCE OF AN EXCESS OF DIVALENT CATIONS

It was demonstrated earlier (13) that increased concentrations of NH_4Cl improved the solvent productivity at high dilution rates. It can be noted (Tables 8 and 9) that at the increased concentration of NH_4Cl, the solvent productivities of 2.38 and 2.32 g/l, respectively were obtained with Fe^{++} and Mg^{++} respectively.

EFFECT OF PHOSPHATE AND POTASSIUM LIMITATION ON CONTINUOUS CULTURE STABILITY

Stability of cultures of *C. acetobutylicum* is one of the major points limiting the feasibility of continuous processes. Previous research (2,14) has demonstrated that long term stability of chemostat cultures of *C. acetobutylicum* could be sustained on medium supplemented with low concentrations of potassium together with low concentrations of phosphate. It is still questionable which of the two nutrients is limiting the culture. Experiments using low concentrations of

TABLE 8. Additional influence of NH_4Cl on the performance of chemostat cultures of *C. acetobutylicum* growing at fixed a growth rate of 0.23 hr^{-1} and high concentrations of Mg^{++} (pH 4.4, 35°C)

Concentration of NH_4Cl in the media (gl^{-1})	Concentration of Mg^{++} (mM)	Solvent productivity ($gl^{-1}h^{-1}$)	Glucose consumption (gl^{-1})	Biomass (gl^{-1})
1.5	0.8	1.66	24.71	2.02
1.5	2.5	2.08	30.74	2.22
2.0	2.5	2.17	32.1	2.35
2.5	2.5	2.32	33.97	2.46

TABLE 9. Additional influence of NH_4Cl on the performance of chemostat cultures of *C. acetobutylicum* growing at fixed a growth rate of 0.23 hr^{-1} and high concentrations of Fe^{++} (pH 4.4, 35°C)

Concentration of NH_4Cl in the media (gl^{-1})	Concentration of divalent ions		Solvent productivity	Biomass	Glucose consumptior
	Fe++ (mM)	Mg++ (mM)	($gl^{-1}h^{-1}$)	(gl^{-1})	(gl^{-1})
1.5	1.7	0.8	2.16	2.32	31.6
2.0	1.7	0.8	2.27	2.42	33.0
2.5	1.7	0.8	2.38	2.54	34.27

KH_2PO_4 (0.074 mM) indicated clearly that for all dilution rate used phosphate was not the limiting nutrient (Table 10)

TABLE 10. Influence of dilution rate on residual phosphate and potassium in the medium (pH 4.4, 35^oC, concentration of phosphate and potassium in the medium were 0.074 mM)

DILUTION RATE(H^{-1})	RESIDUAL PHOSPHATE(mM)	RESIDUAL POTASSIUM(mM)
0.025	0.065	ND
0.045	0.073	ND
0.108	0.116	ND
0.151	0.14	ND
0.174	0.17	ND

ND: not detectable

It appears, as shown in Table 11, that only the addition of potassium to chemostat cultures working at low concentrations of phosphate and potassium lead to an increase of sugar consumption and solvent production. It confirms that under low concentration of potassium dihydrogen phosphate, potassium is the limiting nutrient. Phosphate limited and potassium limited chemostat culture were then operated in order to study the stabilizing effect of these limitations. It appears that high stability can be obtained under phosphate or potassium limitation and that this high stability could be associated with the absence of extracellular autolytic activity

TABLE 11. Effect of different limiting conditions on the metabolic activity of *C. acetobutylicum* (dilution rate = 0.139 hr^{-1}, pH 4.4, 35 ^oC)

EXPERIMENTAL CONDITIONS G/L	TOTAL SOLVENTS G/L	TOTAL SUGAR CONSUMED G/G.H	SPECIFIC RATE OF SOLVENT PRODUC
POTASSIUM 9.42mM PHOSPHATE6.54mM	12.6	41.5	0.55
POTASSIUM 0.74mM PHOSPHATE6.54mM	1.6	11.2	0.37
POTASSIUM 0.74mM PHOSPHATE0.74mM	1.52	9.7	0.36
POTASSIUM 9.42mM PHOSPHATE0.74mM	4.1	18.4	0.53

Nomenclature

μ : Specific growth rate (h^{-1}).
μm : Maximum specific growth rate (h^{-1}).
ν_p : Specific ethanol production rate ($g.g^{-1}.h^{-1}$).
ν_{pm} : Maximum specific ethanol production rate ($g.g^{-1}.h^{-1}$).
B_I : Intracellular free biotin concentration ($ug.l^{-1}$).
B_O : Exocellular biotin concentration ($ug.l^{-1}$).
B_{min} : Minimal intracellular biotin concentration ($ug.l^{-1}$).
D :Hydraulic dilution rate (h^{-1}).
D_r : Death rate (h^{-1}).
K_d : Death constant (h^{-1}).
K_S : Monod's constant (substrate limitation) ($g.l^{-1}$).
K_B : Monod's constant (biotin limitation) ($ug.l^{-1}$).
K_X : Biomass inhibition constant ($g.l^{-1}$).
${}^{I}K_{FA}$: Short chain fatty acids inhibition constant ($l.mg^{-1}$).
${}^{I}K_p$: Ethanol inhibition constant ($l.g^{-1}$).
${}^{X}K_p$: Biomass inhibition constant for ethanol ($l.g^{-1}$).
m : Maintenance coefficient ($g.g^{-1}.h^{-1}$).
${}^{P}k_1$, ${}^{P}k_2$: Kinetic constants ($l.g^{-1}$).
C_6,C_8,C_{10} : repectively hexanoic, octanoic and decanoic acid.
C_6,C_8,C_{10} : repectively hexanoic, octanoic and decanoic acid concentrations ($mg.l^{-1}$).
FA : Short chain fatty acid concentration ($mg.l^{-1}$)
P : Ethanol concentration ($g.l^{-1}$)
${}^{P}P_b$:Critical ethanol concentration for the biotin assimilation ($g.l^{-1}$).
${}^{X}P_c = 1 / {}^{I}K_p$:Critical ethanol concentration for the growth ($g.l^{-1}$).
${}^{P}P_c$:Critical ethanol concentration for the alcool production ($g.l^{-1}$).
r_{Bfix} : Conversion rate of the free biotin into bonded biotin ($ug.l^{-1}.h^{-1}$).
r_P : Ethanol production rate ($g.l^{-1}.h^{-1}$).
r_X : Growth rate ($g.l^{-1}.h^{-1}$).
S : Substrate (glucose) concentration ($g.l^{-1}$)
Vc : Cellular volume (l)
X : biomass concentration ($g.l^{-1}$)
X_l : Maximum viable cell concentration ($g.l^{-1}$)
X_M : Maximum cell concentration ($g.l^{-1}$)
X_T : Total cell concentration ($g.l^{-1}$)
X_V : Viable cell concentration ($g.l^{-1}$)
Y_X, Y_P: Theoritical biomass and ethanol yields (gg^{-1})

superscript

* : corrected by ethanol inhibition.
** : corrected by biotin limitation

References for Part A: Ethanolic Fermentation

1 - BERTRAND A. Thesis, University of Bordeaux II.(1975) .

2 - DASARI G., E. KESHAVARZ, M.A. CONNOR, N.B. PAMMENT A reliable method for detecting the intracellular accumulation of fermentation products : application to intracellular ethanol analysis.Biotechnol.Letters, 7, 8, 541-546.(1985).

3 - DOMBEK K.M., L.O. INGRAM Nutrient limitation as a basis for the apparent toxicity of low levels of ethanol during fermentation.J.Indus.Microbiol., 1, 219-225.(1986).

4 - GENEIX C., S. LAFON-LAFOURCADE, P. RIBEREAU-GAYON Les causes, la prévention et le traitement des arrêts de la fermentation alcoolique.Connaissance Vigne Vin, 17, 3, 205-217.(1983)

5 - HAARASILTA S., E. OURA, H. SUOMALAINEN Pyruvate holo-and apocarboxylase content of biotin-deficient baker's yeast and the characteristics of the holoenzyme formation in permeabilized cells. Arch.Microbiol., 122, 121-127.(1979).

6 - HAWTHORNE B.H., R.D. JONES, P.A., BARRETT, T.E. KAVANAGH, B.J. CLARKEMethods for the analysis of C_4 to C_{10} fatty acids in beer wort and carbohydrates syrups. J.Inst.Brew., 92, 181-184. (1986) .

7 - HUNKOVA Z., A. FENCL Toxic effects of fatty acids on yeast cells: dependence of fatty acid concentration. Biotechnol.Bioeng., 19, 1623-1641.(1977).

8 - HUNKOVA Z., A. FENCL (1978) Toxic effects of fatty acids on yeast cells: possible mechanisms of action. Biotechnol.Bioeng., 20, 1235-1247.(1978)

9 - KOTYK A., A. ALONSO Transport of ethanol in baker's yeast.Fol.Microbiol., 30, 90-91.(1985)

10 - LAFFORGUE C. Fermentation alcoolique en bioréacteur a membrane,INSA Thesis, (1988).

11 - LOUREIRO V., H.G. FERREIRA On the intracellular accumulation of ethanol in yeast. Biotechnol.Bioeng., 25, 2263-2269.(1983).

12 - LUONG J.H.T. Kinetics of ethanol inhibition in alcohol fermentation. Biotechnol.Bioeng., 27, 280-285.(1984).

13 - MAIORELLA B., H.W. BLANCH, C.R. WILKE By-product inhibition effects on ethanolic fermentation by *Saccharomyces cerevisiae*. Biotechnol.Bioeng., 25, 103-121.(1983) .

14 - MALINOWSKI J., LAFFORGUE C., GOMA G. , J. Ferment. Technol.(1987).

15 - MOTA M., C. LAFFORGUE, P. STREHAIANO, G. GOMA Fermentation coupled with microfiltration : kinetics of ethanol fermentation with cell recycle. Bioproc.Eng., 2, 65-68.(1987).

16 - NOVAK M., P. STREHAIANO, M. MORENO, G. GOMA Alcoholic fermentation : on the inhibitory effect of ethanol. Biotechnol.Bioeng., 23, 201-211.(1981) .

17- OURA E. Effect of aeration intensity on biochemical composition of baker's yeast. Factors affecting the type of metabolism.Biotechnol.Bioeng., 16, 1197-1212.(1974).

18 - PONS M.N., A. RAJAB, J.M. ENGASSER Influence of acetate on growth and production control of *Saccharomyces cerevisiae* on glucose and ethanol.Appl.Microbiol.Biotechnol., 24, 193-198.(1986) .

19 - RIBEREAU-GAYON P. New developments in wine microbiology.Am.J.Enol.Vitic., 36, 1-10.(1984)

20 - SHIMMONS B.W., SVRECK W.Y., ZAJIC J.E., Cell concentration control by viscosity. Biotechnol. Bioeng., 18, 1793-1805.(1976).

21 - STREHAIANO P. Thesis, University of Toulouse III.(1973).

22 - TAYLOR G.T., B.H. KIRSOP Rapid methods for the analysis of the fatty acids of fermenting wort and beer (C_6 - C_{10}) and yeast (C_{16} - C_{18}).J.Inst.Brew., 83, 97-99.(1977).

23- URIBELARREA J.L., J. WINTER , G. GOMA and A. PAREILLEUX, Determination of maintenance coefficients of *Saccharomyces cerevisiae* Cultures with cell recycle by cross-flow membrane filtration, Biotechnol. Bioeng.,35,201-206, (1990).

24- VAN UDEN N. "Ethanol toxicity and ethanol tolerance in yeasts", Vol. 8 pp. 11-58 in G.T. TSAO (ed) On Fermentation Processes, Academic Press, INC.(1985).

25 - VIEGAS C.A., I. SA-CORREIRA, J.M. NOVAIS Synergistic inhibition of the growth of *Saccharomyces bayanus* by ethanol and octanoic or decanoic acids.Biotechnol.Letters, 7, 8, 611-614.(1985).

26 - VIEGAS C.A., I. SA-CORREIRA, J.M. NOVAIS Rapid production of high concentrations of ethanol by *Saccharomyces bayanus* : mechanisms of action of soy flour supplementation.Biotechnol.Letters, 7, 7, 515-520.(1985).

27 - WINTER J., LORET M.O. and J.L. URIBELARREA, Inhibition and growth factor deficiencies in alcoholic fermentation by *Saccharomyces cerevisiae*, Current Microbiol., 18, 247-252,(1989).

28 - WINTER J., Fermentation alcoolique par *Saccharomyces cerevisiae* : contribution à l'étude du contrôle de la dynamique fermentaire par l'inhibition et les facteurs nutritionnels, Thesis, Toulouse, (1988).

References for Part B: Acetonobutylic Fermentation

1)Andersch W, Bahl H, Gottschalk G (1982) Acetone-butanol production by *Clostridium. acetobutylicum* in an ammonium-limited chemostat at low pH values. Biotechnol Lett 4: 29-32

2)Bahl H, Andersch W, Braun K Gottschalk G(1982) Continuous production of acetone and butanol by *C. acetobutylicum* grown in continuous culture. Eur J Appl Microbiol Biotechnol 14: 17-20

3)Barber JM, Robb FT, Webster JR, Woods DR (1979) Bacteriocin production by Clostridium. acetobutylicum in an industrial fermentation process. Appl Environ Microbiol 37: 433-437

4)Bowles LK, Ellefson WL. Effects of butanol on *Clostridium acetobutylicum,* (1985) Appl. Environ. Microbiol., 50, 1165-1170

5)Costa JM Moreira AR (1983) Growth inhibition kinetics for the acetone-butanol fermentation, in :Foundations of Biochemical Engineering, (Blanch HW, Papoutsakis E.T, Stephanopoulos G), A.C.S. Symposium Series 207, Chap.24, 501-512, Washington D.C.

6)Ferras E, Minier M, Goma G (1986) Acetonobutylic fermentation : Improvement of performances by coupling continuous fermentation and ultrafiltration. Biotechnol Bioeng 28: 523-533

7)Hermann M, Fayolle F, Marchal R, Podvin L, Sebald M, Vandecasteele JP (1985) Isolation and characterization of butanol-resistant mutants of *Clostridium. acetobutylicum* .Appl Environ Microbiol 50: 1238-1243

8)Huang L, Gibbins LN, Forsberg CW (1985) Transmembrane pH gradient and membrane potential in *Clostridium acetobutylicum* during growth under acetogenic and solventogenic conditions. Appl Environ Microbiol 50: 1043-1047

9)Lin YL, Blaschek HP (1983) Butanol production by a butanol-tolerant strain of *Clostridium. acetobutylicum* in extruded corn broth. Appl Environ Microbiol 45: 966-973

10)Minier M, Grateloup R, Blanc-Ferras E, Goma G (1989) Extractive acetonobutylic fermentation by coupling ultrafiltration and distillation. Biotechnol Bioeng (in press).

11)Monot F, Engasser JM, Petitdemange H (1983) Regulation of acetone butanol production batch and continuous cultures of *Clostridium. acetobutylicum* . Biotechnol Bioeng Symp 13: 207-216

12)Monot F, Martin JR, Petitdemange H, Gay R (1982) Acetone and butanol production by *C. acetobutylicum* in a synthetic medium. Appl Environ Microbiol 44: 1318-1324

13)Soni BK, Soucaille P, Goma G (1987a) Continuous acetone butanol fermentation: a global approach for the improvement in the solvent productivity in synthetic medium. Appl Microbiol Biotechnol 25: 317-321

14)Soni BK, Soucaille P, Goma G (1987b) Effect of phosphate cycling on production of acetone-butanol by *Clostridium. acetobutylicum* in chemostat culture. Biotechnol Bioeng Symp 17: 239-252

15)Soni BK, Soucaille P, Goma G (1987c) Continuous acetone-butanol fermentation : influence of vitamins on the metabolic activity of *Clostridium. acetobutylicum* . Appl Microbiol Biotechnol 27: 1-5

16)Soucaille P, Goma G (1986) Acetonobutylic fermentation by *Clostridium. acetobutylicum* ATCC 824 : Autobacteriocin production, properties, and effects. Current Microbiol 13: 163-169

17)Soucaille P, Joliff G, Izard A, Goma G (1987) Butanol tolerance and autobacteriocin production by *Clostridium. acetobutylicum* . Current Microbiol 14: 295-299

18)Soucaille P, Minier M, Ferras E, Goma G(1984). Studies of the autolytic enzymes of *Clostridium acetobutylicum* and role in the butanol tolerance .3rd Europ. Congr. Biotechnol. Munich, RFA, 2, 85

19)Spivey MJ (1978) The acetone/butanol/ethanol fermentation. Process Biochem 13: 2-25

20)Terracciano JS, Kashket ER (1986) Intracellular conditions required for initiation of solvent production by *Clostridium. acetobutylicum* . Appl Environ Microbiol 52: 86-91

21)Van der Westhuizen A, Jones DT, Woods DR (1982) Autolytic activity and butanol tolerance of *Clostridium. acetobutylicum* . Appl Environ Microbiol 44: 1277-1281

22)Webster JR, Reid SJ, Jones DT, Woods DR (1981) Purification and characterization of an autolysin from *Clostridium. acetobutylicum* . Appl Environ Microbiol 41: 371-374

INDEX

A

B

C

I

K

L

M

N

O

P

Q

R

S

T

U

V

W

X

Y